D0152181

Lignin

ACS SYMPOSIUM SERIES **397**

Lignin

Properties and Materials

Wolfgang G. Glasser, EDITOR
Virginia Polytechnic Institute and State University

Simo Sarkanen, EDITOR
University of Minnesota

Developed from a symposium sponsored
by the Cellulose, Paper, and Textile Division
of the American Chemical Society
at the Third Chemical Congress of North America
(195th National Meeting of the American Chemical Society),
Toronto, Ontario, Canada,
June 5-11, 1988

American Chemical Society, Washington, DC 1989

Library of Congress Cataloging-in-Publication Data

Lignin: properties and materials.

(ACS symposium series, ISSN 0097-6156; 397)

"Developed from a symposium sponsored by the Cellulose, Paper, and Textile Division of the American Chemical Society at the Third Chemical Congress of North America (195th National Meeting of the American Chemical Society), Toronto, Ontario, Canada, June 5-11, 1988."

Includes bibliographies and indexes.

1. Lignin—Congresses. 2. Polymers—Congresses.

I. Glasser, Wolfgang, G., 1941– . II. Sarkanen, Simo, 1946– . III. American Chemical Society. Cellulose, Paper, and Textile Division. IV. Chemical Congress of North America (3rd: 1988: Toronto, Ont.) V. American Chemical Society. Meeting (195th: 1988: Toronto, Ont.) VI. Series.

TS933.L5L54 1989 661'.802 89-15069
ISBN 0-8412-1631-2

PRINTED IN THE UNITED STATES OF AMERICA

ACS Symposium Series

M. Joan Comstock, *Series Editor*

Foreword

The ACS SYMPOSIUM SERIES was founded in 1974 to provide a medium for publishing symposia quickly in book form. The format of the Series parallels that of the continuing ADVANCES IN CHEMISTRY SERIES except that, in order to save time, the papers are not typeset but are reproduced as they are submitted by the authors in camera-ready form. Papers are reviewed under the supervision of the Editors with the assistance of the Series Advisory Board and are selected to maintain the integrity of the symposia; however, verbatim reproductions of previously published papers are not accepted. Both reviews and reports of research are acceptable, because symposia may embrace both types of presentation.

Contents

WATER-SOLUBLE POLYMERS

PHENOLIC COMPOUNDS

POLYOLS, POLYURETHANES, POLYBLENDS, AND GRAFTS

APPENDIX AND INDEXES

Preface

THE INTRIGUING MACROMOLECULAR BEHAVIOR of lignin and its derivatives has yielded to ever greater measures of explicability; the properties of lignin-based polymeric materials have become amenable to modification over a considerable range, even in a predictable fashion. Significant developments in both macromolecular behavior and property modification owe a great deal to the application of physico-chemical techniques that have been made accessible through profound improvements in commercially available instrumentation and related supplies.

The flexibility achieved in both open-column and high-performance size-exclusion chromatography during the past 20 years has been invaluable to progress in documenting the macromolecular properties of lignin preparations. No less important have been the contributions from methods for absolute molecular weight determinations, such as low-angle laser light scattering, vapor pressure osmometry, and differential viscometry. It is curious that analytical ultracentrifugation occupies a more pivotal position in lignin chemistry now than before production of the classical Beckman model E instrument was discontinued more than half a decade ago.

Twenty-three years have elapsed since the American Chemical Society published a book devoted to papers from a symposium exclusively about lignin. In the preface to *Lignin Structure and Reactions* (Advances in Chemistry Series 59, 1966), the symposium chairperson remarked that "the current presentation of lignin structure is oversimplified. . . . more ingenious work is needed to establish the sequence of units—the most prominent singular task of future lignin research." Presumably, the author of these words had in mind the sequence of intermonomer *linkages* rather than the units themselves (which differ only in their aromatic methoxyl substitution pattern), and from this perspective his observation could be as true in 1989 as it was in 1966.

Lignin: Properties and Materials is illuminated by two keynote chapters: "The Lignin Paradigm" reminds readers that inadequacies still punctuate our understanding of lignin in both its native and its derivative configurations; on the other hand, "Specialty Polymers from

Lignin" draws attention to the spectacular vistas of new polymeric materials derived from this important biopolymer.

The advent of lignin-based polymeric materials as putative engineering plastics represents one of several technological advancements in the field during the past 10 years. The characterization of polymeric materials has relied heavily upon differential scanning calorimetry and dynamic mechanical thermal analysis, and scanning electron microscopy has provided invaluable confirmatory evidence about blend morphology. Successful incorporation as integral components, rather than as mere fillers or extenders, in useful polymeric materials clearly represents a potential for adding high value to industrial byproduct lignins. What "the most prominent singular task of future lignin research" would be in this context asks for speculation beyond the scope of a preface, but the answer may unfold in the productive developments that are destined to appear during the coming decade.

The book itself is divided into six sections. Collectively, the chapters provide a perspective that anticipates the future. The book's usefulness should extend to academic research groups and industrial organizations alike.

Acknowledgments

We acknowledge with gratitude the individual authors; Kathryn Hollandsworth, who has so skillfully and patiently prepared the final copy for this book; and Cheryl Shanks, ACS Books Acquisitions Editor. Particular thanks are due to Daishowa Chemicals, the Department of Energy, Empresa Nacional de Celulosas SA, Georgia–Pacific Corporation, ITT Rayonier, Westvaco Corporation, and Weyerhaeuser Corporation, whose generous support contributed so much in very real terms to ensuring the success of the symposium. It has been for us a distinct privilege to have been associated with such an interesting and absorbing enterprise.

WOLFGANG G. GLASSER
Department of Wood Science and Forest Products
Virginia Polytechnic Institute and State University
Blacksburg, VA 24061

SIMO SARKANEN
Department of Forest Products
University of Minnesota
St. Paul, MN 55108

February 21, 1989

MACROMOLECULAR STRUCTURE AND PROPERTIES

Chapter 1

The Lignin Paradigm

D. A. I. Goring

Department of Chemical Engineering and Applied Chemistry, University of Toronto, Toronto, Ontario M5S 1A4, Canada

Evidence supporting the original paradigm of lignin in wood as a random, three-dimensional network polymer is reviewed. More recent results which do not fit this simple picture are discussed. A modified paradigm is proposed in which lignin in wood is comprised of several types of network which differ from each other both ultrastructurally and chemically. When wood is delignified, the properties of the macromolecules made soluble reflect the properties of the network from which they are derived.

Lignin in wood may be considered to be a random three-dimensional network polymer comprised of phenylpropane units linked together in different ways. When wood is delignified the properties of the macromolecules made soluble reflect the properties of the network from which they are derived.

For several decades, the above has been the generally accepted paradigm for the structure of lignin in wood (1,2). In the present paper, some of the evidence in favor of this paradigm is reviewed. However, certain recently discovered aspects of the behavior of lignin do not fit easily into the simple concept of a random three-dimensional network polymer. These are discussed and an attempt is made to modify the original paradigm so that it agrees more closely with experimental observation.

Behavior of Lignin Compatible with the Paradigm of a Random, Three-Dimensional Network Polymer

Non-Crystallinity. Lignin seems to be amorphous (3). No one has ever reported any evidence of crystalline order in lignin. And the molecule does not appear to be optically active (3), unusual for a biopolymer. Such behavior is what might be expected from a random, three-dimensional network. However, with the more sensitive methods of analysis available today, it

0097–6156/89/0397–0002$06.00/0

might be a good idea to check whether this lack of crystallinity and optical activity is, indeed, absolute.

Insolubility. Then we have the notorious insolubility of lignin in virtually all simple solvents. As shown by Brauns (4), a small proportion (2-3%) of the lignin in wood can be dissolved in ethanol. If the wood is extensively milled, Björkman found that 50% or more of the lignin from spruce can be extracted with aqueous dioxane (5). However, the milling is so severe that it is likely that chemical bonds are broken (6). Thus, some type of covalent bond rupture seems to be necessary to make the lignin soluble. This would be expected of a three-dimensional network. Note that, once the network has been degraded, there is nothing unusual about the solubility of the fragments, as shown by Schuerch (7) and by Lindberg (8).

Molecular Weight. By reversing Flory's theory of trifunctional polymerization, it can be shown that, when a random crosslinked three-dimensional gel is degraded, there is a trend in the molecular weight of the fragments produced (9). Small molecules become soluble early in the degradation, while larger molecules are released later in the process. This usually happens when lignin in wood is made soluble by chemical treatment (10-12). The theory also predicts that the fractions made soluble will contain a substantial proportion of low molecular weight material with a polydisperse, high molecular weight tail (9). Fractionation of kraft lignins made soluble in a continuous flow process has confirmed this behavior (12). The degradation of a gel has been analyzed in more detail by Bolker and his coworkers (13-15) and by Yan and his coworkers (16-18). The results, in general, support the concept of lignin in wood as a random, three-dimensional network polymer.

Conformation of the Macromolecule. In solution, macromolecules can have a wide variety of shapes or conformations. The simplest is the solid sphere or Einstein sphere. It is a round ball, impermeable to solvent. The ball may be stretched into a prolate ellipsoid like a football or flattened into an oblate ellipsoid like a flying saucer. Many soluble proteins have conformations that approximate ellipsoids. If a prolate ellipsoid is stretched enough, it becomes a rod. Certain virus macromolecules are rodlike.

Then there are flexible linear polymers which curl up in solution to give a random cell. If the chain is stiff, such as in cellulose or in DNA, the coil becomes highly expanded.

Conformation in solution is indicated by the way in which the hydrodynamic properties of the macromolecules change with change in molecular weight. From trends in the intrinsic viscosity, the sedimentation coefficient, or the diffusion constant with molecular weight we can learn something about the conformation of the molecule in solution (19,20).

Where do soluble lignins fit with respect to conformation? They seem to be rather compact molecules in solution—the opposite of the highly expanded cellulose molecule. They are not as compact as a simple solid sphere. Yet, the chains of the lignin macromolecules in solution are more densely packed than those of a linear flexible polymer such as polystyrene

(19,20). Such behavior is what would be expected for soluble fragments of a random three-dimensional network polymer.

Before leaving the topic of conformation, it should be noted that there do not appear to be much interest in this property today. In earlier times, several schools included conformational studies in their characterization of soluble lignins (19). Recently, only Pla and his group (20) seem to be taking this property seriously. Yet it is important, both to the understanding of delignification and to the use of lignins as colloids and dispersants.

Conformation is also important in the use of size-exclusion chromatography. In the many papers which have been graciously dedicated to the author (21-23), there are ten or more in which GPC or HPLC has been used to determine the molecular weight and its distribution. It should be remembered, however, that the exclusion phenomenon depends not so much on molecular weight but rather on hydrodynamic volume. Thus, to give reliable molecular weights, the column should be calibrated, not with polystyrene, but with a more compact lignin-like molecule, preferably with narrow fractions of the soluble lignin derivative under study. The fractions used for calibration should be characterized independently by an absolute method such as ultracentrifuge sedimentation equilibrium or low angle light scattering. However, these absolute methods are time-consuming and expensive. Perhaps intrinsic viscosity should be looked at again. This parameter gives the effective hydrodynamic specific volume of the molecule. Benoit *et al.* (24) have shown that the product $[\eta]$ M is uniquely related to the elution volume. Thus, by measurement of intrinsic viscosity on selected fractions, a polystyrene calibration can be converted to a lignin calibration with little effort and simple equipment. It is interesting to note that two reports of the use of an on-line differential viscometer with GPC are included in the present symposium volume (25-26).

Behavior of Lignin Incompatible with the Paradigm of a Random, Three-Dimensional Network Polymer

From the discussion in the previous sections we see that the behavior of the protolignin in wood and the pattern of its subsequent degradation and solution are compatible, in a broad sense, with the paradigm of a random three-dimensional network polymer. There emerge, however, several aspects of the problem which do not fit easily with the simple paradigm given above. In the following sections some of these anomalies are discussed.

Molecular Weight Distribution. The gel degradation theory applied to a single network predicts that the sol fraction will be polydisperse but unimodal (9). Thus, there should be only one peak in the molecular weight distribution curve—except at low molecular weights when a substantial proportion of oligomers are present. Yet many authors have reported bimodal distributions of molecular weight in soluble lignins. Such bimodal behavior turns up not only in organosolv lignins (27-29) but also in lignins made soluble in the chemical pulping of wood (12,30). As the resolution of the chromatographic techniques are improved, the broad bimodal patterns are being resolved into distributions with several maxima (31-38). Some of these components are well-defined low molecular weight oligomers of the lignin macromolecule. Others seem to be of higher molecular weight and more indeterminate. Such paucidisperse behavior is not really compatible with a random network. It suggests a repetitive pattern of weak links which allows the polymer to be degraded into separate families of molecules, each with a characteristic molecular weight and, perhaps, chemical composition.

Association. Several authors have postulated that some lignins in solution are not covalent polymers in the normal sense, but rather an association of smaller molecules held together by various types of non-covalent linkages (30,34,36,39,40,41). S. Sarkanen and his coworkers have presented strong evidence that low molecular weight lignin fractions show considerably higher molecular weights in certain solvents (34,41). And the effect appears to be reversible.

The association phenomenon raises the possibility that lignin in wood is composed of low molecular weight molecules. When these are made soluble they associate into complexes of higher molecular weight. This picture is not really in accord with the idea of a random, three-dimensional network polymer. It should be noted, however, that a high molecular weight lignosulfonate gave constant molecular weight in solvents as different as 0.1 M aqueous sodium chloride and dimethyl sulfoxide, which indicates a covalent structure for the macromolecule (42).

Non-covalent association may not be the only way in which high molecular weight lignin is produced from smaller molecules. The conjugated, phenolic nature of the lignin monomers makes them prone to "condensation" reactions both in acid and alkaline media (44-46). Thus, delignification may be invariably accompanied by polymerization. Again, such behavior goes beyond the simple paradigm of the degradation of a three-dimensional network polymer. However, it should be remembered that very few examples have been reported in which the molecular weights of soluble lignins have been increased by further treatment in the medium used for delignification (47). If condensation does occur, perhaps it takes place rapidly, thus producing a network polymer which must then be degraded by the delignification reactions.

Topochemistry. Many authors have sought to characterize the chemical nature of lignin. This work has been extended into a study of the chemical structure of lignin in the various morphological regions of the cell wall. It now seems certain that in both hardwoods and softwoods there are differences between the chemical structure of lignin in the middle lamella and the secondary wall (48-56), although there are still many questions to be resolved (57-59). There are also chemical differences between the lignins in the fiber and vessel walls of birch wood (48,49,52,56,60). Terashima *et al.* (61) and Beatson (62) have suggested that these topochemical differences are related to the biosynthetic sequence of lignification in plant tissue.

The paradigm of an infinite random three-dimensional network polymer implies a uniform chemistry throughout. The topochemical behavior of lignin indicates that, from a structural chemical point of view, more than one type of network exists in wood. Such topochemical differences in chemical structure will have an important bearing on the course of delignification reactions and the properties of the resulting pulps (62-70).

Ultrastructure. Perhaps the most compelling reason for seeking an alternative paradigm for the nature of lignin is the ultrastructure of the polysaccharides in the cell wall. Their arrangement is anything but random. Microfibrils are probably single crystals of cellulose chains (71,72). Although the arrangement of the hemicelluloses is not known for certain, it seems likely that the molecules line up in the direction of the cellulose microfibrils (73-75). The microfibrils themselves are laid down in complex patterns (76,77). In the S2 layer, there is evidence of a lamellar ultrastructure with the interlamellar distance being equal to about twice the width of the microfibril, i.e., 7-10 nm (78-82). In view of the fact that most of it is in the secondary wall (83-85), how can lignin be regarded as a random network when it is the mirror image *on a molecular scale* of the highly organized polysaccharide ultrastructure? The lignin lamellae are probably about 2 nm thick, enough room for 2 to 4 phenylpropane monomers. It seems almost certain that constraints of space will impose non-randomness on the crosslinked network. Here it is interesting to note that Atalla has found evidence indicating that the aromatic rings in lignin are aligned preferentially in the direction tangential to the secondary wall (86).

We should note, also, that the lignin in the S2 layers is chemically bonded to the polysaccharide moiety (87-92). Such bonds occur not only in wood but may be formed during chemical pulping (93,94). Even if the lignin-carbohydrate bonds are restricted to the hemicelluloses (95), the regularity of these chain molecules will probably impose some non-randomness on the lignin structure.

The lamellar structure of the cell wall will probably affect the conformation of the lignin macromolecules dissolved when wood is delignified. This is particularly true when high molecular weight lignins diffuse out of the fibers during the washing of kraft pulp (96,97). The molecular weights of the fractions of lignin leached out of the cell wall are of the order of hundreds of thousands (98). Spherical lignin macromolecules of this mass would be too large to pass through the porous structure of the cell wall. It is possible that in the wall they are flatish covalently free fragments of the lignin lamellae, which diffuse through the interlamellar spaces into solution. Because the chains are flexible, these lamellar fragments fold in solution to an approximately spherical conformation. Such a conformation would be expected to be more compact than that of a random coil but more expanded than that of an Einstein sphere, in accord with the hydrodynamic behavior noted earlier in the paper. In support of this concept, we find that lignin macromolecules, if spread on the surface of a non-solvent (99,100) or deposited on a carbon-coated grid for electron microscopy (101), assume a flat conformation with a thickness of about 2 nm. It seems that the conformation of the macromolecules made soluble, reflects the ultrastructure of the cell wall from which they have been extracted.

Of course, the picture given in the preceding paragraph cannot apply to the almost pure lignin in the true middle lamella. Here the lamellar thickness is usually more than 100 nm (83,84). The concept of a random three-dimensional network polymer would seem to be appropriate for such a thick layer. However, the true middle lamella probably contains less than 20% of the total lignin (83,84,102). Thus, lignin from the secondary wall will be the dominant fraction in most preparations from whole wood.

A Modified Paradigm

From the foregoing, it seems that we need a better paradigm for lignin. The old, comfortable concept of an infinite random, three-dimensional network polymer is too simple to encompass everything we know about lignin and its soluble derivatives. To give a better fit with the experimentally observed behavior of lignin, the original paradigm may be modified as follows:

> Lignin in the true middle lamella of wood is a random three-dimensional network polymer comprised of phenylpropane monomers linked together in different ways. Lignin in the secondary wall is a nonrandom two-dimensional network polymer. The chemical structure of the monomers and linkages which constitute these networks differ in different morphological regions (middle lamella vs. secondary wall), different types of cell (vessels vs. fibers), and different types of wood (softwoods vs. hardwoods). When wood is delignified, the properties of the macromolecules made soluble reflect the properties of the network from which they are derived.

Of course, the modified paradigm given above is not entirely satisfactory. It is silent on the question of the association of lignin molecules. Also, it does not take into account the possibility of condensation reactions. The term "nonrandom" begs for a clearer definition which the paradigm does not provide. Perhaps the clearest concept in it is the two-dimensional network. After all, most nets are two-dimensional! In spite of its failings, the modified paradigm is probably the best we can do at the present time. We should not be surprised, however, that, as the behavior of the protolignin in wood is further elucidated, an entirely new paradigm will emerge.

Literature Cited

1. Nokihara, E.; Tuttle, M. J.; Felicetta, V. F.; McCarthy, J. L. *J. Am. Chem. Soc.* 1957, **79**, 4495.
2. Goring, D. A. I. In *Lignins*; Sarkanen, K. V.; Ludwig, C. H., Eds.; Wiley-Interscience: New York, 1971; p. 698.
3. Sarkanen, K. V. In *The Chemistry of Wood*; Browning, B. L., Ed.; Interscience: New York, 1963; p. 278.
4. Brauns, F. E. *J. Am. Chem. Soc.* 1939, **61**, 2120.
5. Björkman, A. *Svensk Papperstidn.* 1956, **59**, 477.
6. Chang, H.-M.; Cowling, E. B.; Brown, W.; Adler, E.; Miksche, G. *Holzforschung* 1975, **29**, 153.

7. Schuerch, C. *J. Am. Chem. Soc.* 1952, **74**, 5061.
8. Lindberg, J. J. *Suomen Kemistilehti* 1967, **B40**, 225.
9. Szabo, A.; Goring, D. A. I. *Tappi* 1968, **51**, 440.
10. Rezanowich, A.; Yean, W. Q.; Goring, D. A. I. *Svensk Papperstidn.* 1963, **66**, 141.
11. Yean, W. Q.; Goring, D. A. I. *Pulp Paper Mag. Can.* 1964, **65**, T-127.
12. McNaughton, J. G.; Yean, W. Q.; Goring, D. A. I. *Tappi* 1967, **50**, 548.
13. Bolker, H. I.; Brenner, H. S. *Science* 1970, **170**, 173.
14. Berry, R. M.; Bolker, H. I. *J. Pulp Paper Sci.* 1986, **12**, J16.
15. Argyropoulos, D. S.; Bolker, H. I. *J. Wood Chem. Technol.* 1987, **7**, 499.
16. Yan, J. F. *Macromolecules* 1981, **14**, 1438.
17. Yan, J. F.; Johnson, D. C. *J. Appl. Polym. Sci.* 1981, **26**, 1623.
18. Yan, J. F.; Pla, F.; Kondo, R.; Dolk, M.; McCarthy, J. L. *Macromolecules* 1984, **17**, 2137.
19. Goring, D. A. I. In *Lignins*; Sarkanen, K. V.; Ludwig, C. H., Eds.; Wiley-Interscience: New York, 1971; p. 705.
20. Pla, F.; Robert, A. *Holzforschung* 1984, **38**, 37.
21. *Holzforschung* 1986, Suppl., **40**, 5-157; 1986, **40**, 325-338.
22. *J. Wood Chem. Technol.* 1987, **7**, 1-131.
23. *J. Cell. Chem. Technol.* 1987, **21**, 215-327, 371-395.
24. Grubisic, Z.; Rempp, P.; Benoit, H. *Polym. Lett.* 1967, **5**, 753.
25. Siochi, E. J.; Haney, M. A.; Mahn, W.; Ward, T. C. In *Lignin: Properties and Materials*; Glasser, W. G.; Sarkanen, S., Eds.; American Chemical Society: Washington, DC, 1989.
26. Himmel, M. E.; Tatsumoto, K.; Oh, K. K.; Grohmann, K.; Johnson, D. K.; Chum, H. L. In *Lignin: Properties and Materials*; Glasser, W. G.; Sarkanen, S., Eds.; American Chemical Society: Washington, DC, 1989.
27. Brown, W.; Cowling, E. B.; Falkehag, S. I. *Svensk Papperstidn.* 1968, **71**, 811.
28. Wayman, M.; Obiaga, T. I. *Tappi* 1974, **57**, 123.
29. Wegener, G.; Fengel, D. *Wood Sci. Technol.* 1977, **11**, 133.
30. Lindström, T. *Colloid Polym. Sci.* 1979, **257**, 277.
31. Shaw, A. C.; Dignam, M. *Can. J. Chem.* 1957, **35**, 322.
32. Forss, K.; Fremer, K. E. *Pap. Puu* 1965, **47**, 443.
33. Gupta, P. R.; McCarthy, J. L. *Macromolecules* 1968, **1**, 495.
34. Connors, W. J.; Sarkanen, S.; McCarthy, J. L. *Holzforschung* 1980, **34**, 80.
35. Lewis, N. G.; Goring, D. A. I.; Wong, A. *Can. J. Chem.* 1983, **61**, 416.
36. Chum, H. L.; Johnson, D. K.; Tucker, M. P.; Himmel, M. E. *Holzforschung* 1987, **41**, 97.
37. Woerner, D.; McCarthy, J. L. *Tappi* 1987, **70**, 126.
38. Lewis, N. G.; Eberhardt, T. L.; Luthe, C. E. *Tappi* 1988, **71**, 141.
39. Gross, S. K.; Sarkanen, K. V.; Schuerch, C. *Anal. Chem.* 1958, **30**, 518.
40. Benko, J. *Tappi* 1964, **47**, 508.

41. Sarkanen, S.; Teller, D. C.; Hall, J.; McCarthy, J. L. *Macromolecules* 1981, **14**, 426.
42. Yean, W. Q.; Goring, D. A. I. *J. Appl. Polym. Sci.* 1970, **14**, 1115.
43. Lai, Y. Z.; Sarkanen, K. V. In *Lignins*; Sarkanen, K. V.; Ludwig, C. H., Eds.; Wiley-Interscience: New York, 1971; p. 182.
44. Marton, J. In *Lignins*; Sarkanen, K. V.; Ludwig, C. H., Eds.; Wiley-Interscience: New York, 1971; p. 658.
45. Gierer, J. *Holzforschung* 1982, **36**, 43.
46. Gellerstedt, G.; Lindfors, E.-L. *Nordic Pulp Paper J.* 1987, **2**, 71.
47. Argyropoulos, D. S.; Bolker, H. I. *J. Wood Chem. Technol.* 1987, **7**, 1.
48. Fergus, B. J.; Goring, D. A. I. *Holzforschung* 1970, **24**, 113.
49. Musha, Y.; Goring, D. A. I. *Wood Sci. Technol.* 1975, **9**, 45.
50. Yang, J.-M.; Goring, D. A. I. *Can. J. Chem.* 1980, **58**, 2411.
51. Hardell, H.-L.; Leary, G. J.; Stoll, M.; Westermark, U. *Svensk Papperstidn.* 1980, **83**, 44.
52. Hardell, H.-L.; Leary, G. J.; Stoll, M.; Westermark, U. *Svensk Papperstidn.* 1980, **83**, 71.
53. Cho, N. S.; Lee, J. Y.; Meshitsuka, G.; Nakano, J. *Mokuzai Gakkaishi* 1980, **26**, 527.
54. Whiting, P.; Goring, D. A. I. *Pap. Puu* 1982, **64**, 592.
55. Whiting, P.; Goring, D. A. I. *Wood Sci. Technol.* 1982, **16**, 261.
56. Saka, S.; Goring, D. A. I. In *Biosynthesis and Biodegradation of Wood Components*; Higuchi, T., Ed.; Academic: New York, 1985; p. 51.
57. Kolar, J. J.; Lindgren, B. O.; Roy, T. K. *Cell. Chem. Technol.* 1979, **13**, 491.
58. Obst, J. R.; Ralph, J. *Holzforschung* 1983, **37**, 297.
59. Westermark, U. *Wood Sci. Technol.* 1985, **19**, 223.
60. Wolter, K. E.; Harkin, J. M.; Kirk, T. K. *Physiol. Plant* 1974, **31**, 140.
61. Terashima, N.; Fukushima, K.; Takabe, K. *Holzforschung* 1986, **40** Suppl., 101.
62. Beatson, R. P. *Holzforschung* 1986, **40** Suppl., 11.
63. Procter, A. R.; Yean, W. Q.; Goring, D. A. I. *Pulp Paper Mag. Can.* 1967, **68**, T445.
64. Fergus, B. J.; Goring, D. A. I. *Pulp Paper Mag. Can.* 1969, **70**, T314.
65. Wood, J. R.; Ahlgren, P. A.; Goring, D. A. I. *Svensk Papperstidn.* 1972, **75**, 1.
66. Saka, S.; Thomas, R. J.; Gratzl, J. S.; Abson, D. *Wood Sci. Technol.* 1982, **16**, 139.
67. Beatson, R. P.; Gancet, C.; Heitner, C. *Tappi J.* 1984, **67**(3), 82.
68. Berry, R. M.; Bolker, H. I. *J. Wood Sci. Technol.* 1987, **7**, 25.
69. Westermark, U.; Samuelsson, B. *Holzforschung* 1986, **40** Suppl., 139.
70. Heazel, T. E.; McDonough, T. J. *Tappi J.* 1988, **86**(3), 129.
71. Manley, R. St.J. *J. Polym. Sci.* A-2 1971, **9**, 1025.
72. Revol, J.-F.; Goring, D. A. I. *Polymer* 1983, **24**, 1547.
73. Preston, R. D. In *The Formation of Wood in Forest Trees*; Zimmerman, M. H., Ed.; Academic: New York, 1964; p. 169.

74. Marchessault, R. H. In *Chimie et Biochimie de la Lignine, de la Cellulose et des Hémicellulose*; Les Imprimeries Réunies de Chambéry: Ed. Grenoble, 1964; p. 287.
75. Fengel, D. *Tappi* 1970, **53**, 497.
76. Dunning, C. E. *Tappi* 1969, **52**, 1326.
77. Kishi, K.; Harada, H.; Saiki, H. *Mokuzai Gakkaishi* 1979, **25**, 521.
78. Scallan, A. M. *Wood Sci.* 1974, **6**, 266.
79. Kerr, A. J.; Goring, D. A. I. *Cell. Chem. Technol.* 1975, **9**, 563.
80. Ruel, K.; Barnoud, F.; Goring, D. A. I. *Wood Sci. Technol.* 1978, **12**, 287.
81. Ruel, K.; Barnoud, F.; Goring, D. A. I. *Cell. Chem. Technol.* 1979, **13**, 429.
82. Fengel, D.; Shao, X. *Wood Sci. Technol.* 1984, **18**, 103.
83. Fergus, B. J.; Procter, A. R.; Scott, J. A. N.; Goring, D. A. I. *Wood Sci. Technol.* 1969, **3**, 117.
84. Wood, J. R.; Goring, D. A. I. *Pulp Paper Mag. Can.* 1971, **72**, T95.
85. Saka, S.; Goring, D. A. I. *Holzforschung* 1988, **42**, 149.
86. Atalla, R. H. *J. Wood Chem. Technol.* 1987, **7**, 115.
87. Fengel, D.; Wegner, G. *Wood: Chemistry, Structure, Reactions*; de Gruyter: Berlin, 1984; p. 167.
88. Meshitsuka, G.; Lee, Z. Z.; Nakano, J.; Eda, S. *J. Wood Chem. Technol.* 1982, **2**, 251.
89. Iversen, T. *Wood Sci. Technol.* 1985, **19**, 243.
90. Papadopoulos, J.; Defaye, J. *J. Wood Chem. Technol.* 1986, **6**, 203.
91. Joseleau, J.-P.; Kesraoui, R. *Holzforschung* 1986, **40**, 163.
92. Minor, J. L. *J. Wood Chem. Technol.* 1986, **6**, 185.
93. Glasser, W. G.; Barnett, C. A. *Tappi* 1979, **62**(8), 101.
94. Gierer, J.; Wännström, S. *Holzforschung* 1986, **40**, 347.
95. Eriksson, Ö.; Goring, D. A. I.; Lindgren, B. O. *Wood Sci. Technol.* 1980, **14**, 267.
96. Favis, B. D.; Choi, P. M. K.; Adler, P. M.; Goring, D. A. I. *Trans. Tech. Sect. Can. Pulp Pap. Assoc.* 1981, **7**, TR35.
97. Favis, B. D.; Goring, D. A. I. *J. Pulp Paper Sci.* 1984, **10**, J139.
98. Favis, B. D.; Yean, W. Q.; Goring, D. A. I. *J. Wood Chem. Technol.* 1984, **4**, 313.
99. Luner, P.; Kempf, U. *Tappi* 1970, **53**, 2069.
100. Goring, D. A. I. In *Cellulose Chemistry and Technology*; Arthur, J. C., Jr., Ed.; ACS Symp. Ser. No. 48; American Chemical Society: Washington, DC, 1977; p. 273.
101. Goring, D. A. I.; Vuong, R.; Gancet, C.; Chanzy, H. *J. Appl. Polym. Sci.* 1979, **24**, 931.
102. Fergus, B. J.; Goring, D. A. I. *Holzforschung* 1970, **24**, 118.

RECEIVED February 27, 1989

Chapter 2

Structure and Properties of the Lignin–Carbohydrate Complex Polymer as an Amphipathic Substance

T. Koshijima[1], T. Watanabe[1], and F. Yaku[2]

[1]Wood Research Institute, Kyoto University, Gokasho, Uji, Kyoto 611, Japan

[2]Government Industrial Research Institute, Osaka 563, Japan

It has been found that lignin-carbohydrate complexes (LCC's) consist of sugar chains and relatively small lignin fragments attached as pendant side chains; they have number-average molecular weights of ca. 6000-8000. The linkage between sugar and lignin was determined to be mainly of the benzyl ether type by a newly developed method using DDQ oxidation. Some of the LCC's exhibit a strong tendency to form micelles or aggregates in aqueous solution due to hydrophobic and also electrostatic interactions.

Since Björkman first proposed the existence of lignin-carbohydrate complexes (LCC's) as species incapable of separation into the respective components by selective chemical treatments or special purification techniques, experimental results supporting the presence of LCC's in wood have been reported by Merewether (1), Koshijima (2), Wegener (3), and Yaku (4). Thereafter, a lot of experiments designed to isolate fractions containing LCC's from partly degraded wood preserving the native configuration to different extents have been conducted. Among them, Björkman (5), Brownell (6), Koshijima (7), Yaku (8,9), Eriksson (10) and Watanabe (11) have extracted LCC's from finely divided wood powder from which milled wood lignin had been extracted previously. By contrast, Lundquist (12), Azuma (13), Mukoyoshi (14), Takahashi (15) and Kato (16) have separated the LCC's contained in a milled wood lignin fraction extracted with 80% aqueous dioxane from very fine wood powder.

Meanwhile, Freudenberg (17) was the first person who demonstrated the formation of an addition compound from a quinonemethide and sucrose during enzymatic dehydrogenation of coniferyl alcohol in a concentrated sucrose solution. Thereafter, Tanaka (18) observed the formation of a benzyl ester between the quinonemethide of a dilignol and a uronic

0097–6156/89/0397–0011$06.00/0

acid. Katayama (19) isolated a compound in which glucose is linked to coniferyl alcohol through a benzyl ether bond. Analogous compounds were also obtained by Feckel (20). Since those experiments were carried out in the presence of 20 to 500 times the amount of coniferyl alcohol in relation to sugar, those results indicated only the possibility of chemical linkage formation between lignin precursors and sugars.

Noteworthy results have been obtained from electron microscopic observations of LCC's by Fengel (21) and Kosíková (22). When comparing these two schematic illustrations, coiled and wound fibrils of polysaccharides are buried in the lignin matrix in both cases, but the polysaccharide fibril in Fengel's scheme is multiply connected to a big lignin particle. In Kosíková's scheme, a fibril is attached to a relatively small lignin particle at only one position. Despite so many investigations of LCC's, no chemical proof has been advanced so far about the kind of linkage connecting the sugar to the lignin moiety. Also, no distinct macromolecular confirmation has yet been provided for the schematic molecular forms that have been proposed on the basis of electron microscopic observations.

In this paper, the molecular shape and micelle or aggregate formation of LCC molecules, the nature of the sugar-lignin linkage, and the molecular weights of LCC's will be documented on chemical or physicochemical grounds.

New Methods for Extraction of LCC's and Isolation of LCC Oligomers

The use of high-boiling point solvents, such as dimethylformamide or dimethylsulfoxide used by Björkman for extraction of LCC's from spruce wood, may lead to denaturation or partial degradation of LCC's in the course of solvent removal or concentration. We have developed the following two methods for easily obtaining LCC's. One involves isolation of LCC's from an aqueous 80% dioxane extract obtained from finely divided wood powder. After separating the precipitate of milled wood lignin from the aqueous solution remaining after dioxane removal, the LCC's in the supernatant are taken out and purified by using a pyridine-acetic acid-water-chloroform solvent system to remove sugar-free lignin. The LCC-W thus obtained is analogous to Björkman LCC in respect to chemical composition and molecular weight, but differs in yield which ranges from 0.75 to 1.1% of wood (13).

The second way consists of hot-water extraction of finely divided wood powder, previously extracted with aqueous dioxane for removing milled wood lignin. The resultant LCC-WE is obtained in 9.3-10.0% yield and does not differ from Björkman LCC in relation to chemical composition and molecular weight (11) (see Table I). This fact is very significant in demonstrating that LCC's containing up to 18% lignin are able to be extracted with hot water alone from wood powder.

In order to estimate the frequency of lignin-sugar linkages in pine LCC's, LCC oligomers were attempted to be prepared from LCC-WE of pine wood. The greatest problems to be solved were the separation of

Table I. Chemical Composition and Properties of Pine LCC-W

	Lignin-Carbohydrate Complexes			
Components	LCC-WE	C-1-M	C-1-A	C-1-R
Recovery (%)	9.3[a]	43.4[b]	48.7[b]	2.1[b]
Carbohydrate content (%)				
Neutral sugar	80.0	95.5	76.0	41.5
Uronic acid	4.2	N.D.[c]	6.4	1.9
Lignin content (%)	17.9	3.7	26.6	43.6
Acetyl content (%)	3.3	7.6	N.D.[c]	N.D.[c]
$[\alpha]_D^{20}$	-15.5°	-28.2°	-11.4°	-8.0°
S(S)	N.D.[c]	0.9	0.8	N.D.[c]
$\overline{M}_w$	1.2×10^4	1.2×10^4	1.1×10^4	N.D.[c]
$\overline{M}_n$	7.6×10^3	7.5×10^3	6.7×10^3	N.D.[c]

[a] Values are expressed as weight percentages of the wood meal extracted with 80% aqueous dioxane.
[b] Values are expressed as weight percentages of LCC-WE.
[c] Not determined.

LCC oligomers from oligosaccharides and the prevention of reaggregation of LCC oligomers through hydrophobic interactions. Chromatography of an enzyme-degraded C-1-A fraction was successful on a Toyopearl HW-40S column which had been already demonstrated to adsorb only lignin-containing oligosaccharides (23). The mixture of LCC oligomers and oligosaccharides was prepared by hydrolyzing the fraction C-1-A that had been fractionated from LCC-WE with purified Cellulosin AC (protein 93.7%) and then with a mixture of purified Meicelase (protein 92.2%) and Cellulosin AC at 40°C for 72 hours. The enzyme-degraded products were introduced onto the Toyopearl HW-40S column, which was eluted with water to remove the oligosaccharide components (A-ESW) and subsequently with 50% aqueous dioxane to recover the adsorbed LCC oligomers (A-ESD), which contained no contaminating lignin or oligosaccharide (see ref. 23 and Table II).

Estimation of Lignin-Sugar Linkages in LCC's

In 1982, Oikawa *et al.* reported that 2,3-dichloro-5,6-dicyanobenzoquinone (DDQ) reacts with 4-methoxybenzyl ether to give the alcohol quantitatively (24). Provided that DDQ attacks specifically at the *p*-alkoxybenzyl ether linkage, direct evidence can be obtained for the occurrence of benzylic lignin-carbohydrate linkages which may be present in LCC molecules. Nine model compound analogs of LCC's were synthesized and subjected to reaction with DDQ at 50°C for 1 hr or 40°C for 24 hrs in 50% dioxane solution (25). Further, model compounds (I-IV) were also synthesized to check the effect of substituents at the *para* positions of the guaiacyl moieties upon the release of sugar residues by DDQ oxidation (Fig. 1). A two-hour reaction of the model compounds with DDQ at the boiling temperature

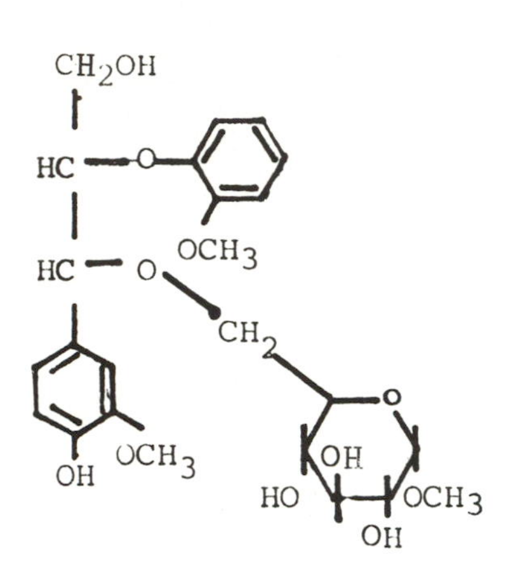

(I)

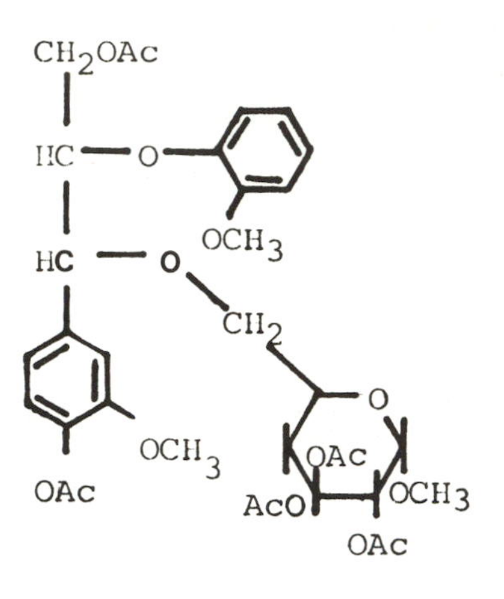

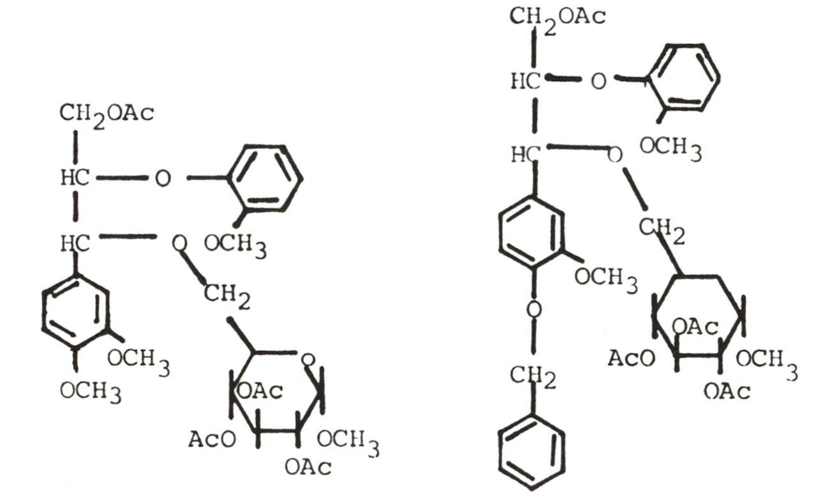

(III)

Figure 1. Synthesized model compounds.

Table II. Neutral Sugar Composition of LCC Subfractions

Components	Lignin-Carbohydrate Complexes	
	C-1-A	A-ESD
Carbohydrate content, %	76.0	8.3
Carbohydrate composition[a]:		
L-Arabinose	7.3	7.5
D-Xylose	49.5	17.6
D-Mannose	27.0	48.7
D-Galactose	7.2	7.7
D-Glucose	9.1	18.6

[a] Values are expressed as mole percentages of the total neutral sugars.

of the dichloromethane-water (18:1) mixture employed was optimum (26). The LCC models 3-methoxy-4-hydroxy (I) and 3-methoxy-4-benzyloxy benzyl ether (IV) were oxidatively decomposed by DDQ, while the *p*-acetoxy LCC model compound II was inert to the oxidation process because of the electron withdrawing inductive effect of the acetoxy group, as reported by Becker. Furthermore, all of the β-ether linkages in the LCC models I-IV were stable during the DDQ oxidation. However, the LCC model compound III having the 3,4-dimethoxyphenyl moiety was stable during DDQ oxidation under these conditions. Probably, this is because the oxidation potential of the LCC model III is not low enough to form a charge transfer complex with DDQ. In fact, when the *para* position was substituted with the more electron donating benzyloxy group, quantitative oxidative cleavage was observed in the LCC model IV. Most of the *para*-substituents in lignins are far more electron donating than the methoxyl group and, in many cases, than the benzyloxy group in their effects. Thus, it may be concluded that DDQ effectively decomposes nonphenolic benzyl ether linkages in lignins. Glycosidic bonds between sugar residues, however, were inert to DDQ oxidation (24,25).

To identify which sugar hydroxyls participate in benzyl ether linkages to lignin, Watanabe and Koshijima (23) developed a new method which involves acetylation of LCC's or LCC oligomers followed by DDQ oxidation and Prehm's methylation. Before application of this method to LCC's, it was confirmed by using 1,2,3,4-tetraacetyl- and 1,2,3,6-tetraacetyl-D-glucose that DDQ does not cause acetyl group removal from or migration in the sugar and that Prehm's methylation does not induce acetyl migration. According to this method (Scheme 1), the LCC oligomers A-ESD were subjected to analysis of the type of linkage present and the corresponding monomethylated sugars are summarized in Table III. The sites of benzyl ether linkages to the sugar moieties of the LCC's are predominantly at C-6 in hexoses and mostly at C-3 and somewhat less so at C-2 in pentoses. The sugars liberated by this reaction are the ones located at the α-carbons of lignin phenylpropane units etherified at the *p*-hydroxyl position. Most of the non-phenolic benzyl ether linkages are easily broken by the action of DDQ due to the enhanced electron-donating properties of *para* substituted phenylpropane units. However, phenolic hydroxyls are acetylated during the first step of this method in phenolic benzyl ethers and the enhanced electron withdrawing properties of the acetoxy groups make cleavage of the benzyl ether linked sugar residues (as shown in Scheme 1) difficult.

Table III. Methyl Ethers of Monosaccharides from the Hydrolyzate of Methylated LCC Oligomers (A-ESD)

Methylated Sugars	Mol %[a]
2-*O*-methyl-D-xylose	9.2
3-*O*-methyl-D-xylose	24.2
Total xylose	33.4
2-*O*-methyl-D-mannose	2.8
6-*O*-methyl-D-mannose	41.6
Total mannose	44.4
2-*O*-methyl-D-glucose	2.0
6-*O*-methyl-D-glucose	20.3
Total glucose	22.3

[a] Based on total methylated sugar components identified.

Molecular Shape of LCC's

The molecular sizes of LCC's extracted from finely divided wood powder of 10 to 30 nanometer diameter particle size may be somewhat smaller than originally the case owing to vigorous mechanical action during the course of milling. However, this action affects both lignin and polysaccharide molecules evenly (2), and so it does not seem to be the case that only specific linkages are cleaved by milling.

First, fractionation of the LCC's extracted from pine (*Pinus densiflora*) wood was attempted according to the method of Björkman (5) into three fractions by means of adsorption chromatography with a DEAE Sephadex column. In the case of pine LCC's, the fraction C-1-M eluting first with water was composed of neutral LCC's (fraction C-1-M-1, 5% yield) consisting of mannose, glucose, arabinose, xylose and lignin, as well as lignin-free acetyl glucomannan (45% yield). Subsequent elution with one molar ammonium carbonate solution liberated acidic LCC's (fraction C-1-A, 25-30% yield) which constituted the major portion of the pine LCC's consisting of 12-13% lignin, about 6% uronic acid and 76% neutral sugars. The last elution (fraction C-1-R, 2-5% yield) was effected by using 10 molar acetic acid, which released a fraction containing around 44% lignin and 43% sugars.

The following experiments were conducted using fraction C-1-A, which was thought to be representative of the LCC's (17). The acidic LCC fraction, C-1-A, was partially hydrolyzed with a cellulase preparation, Cellulosin AC, and changes in the lignin and sugar distributions were analyzed

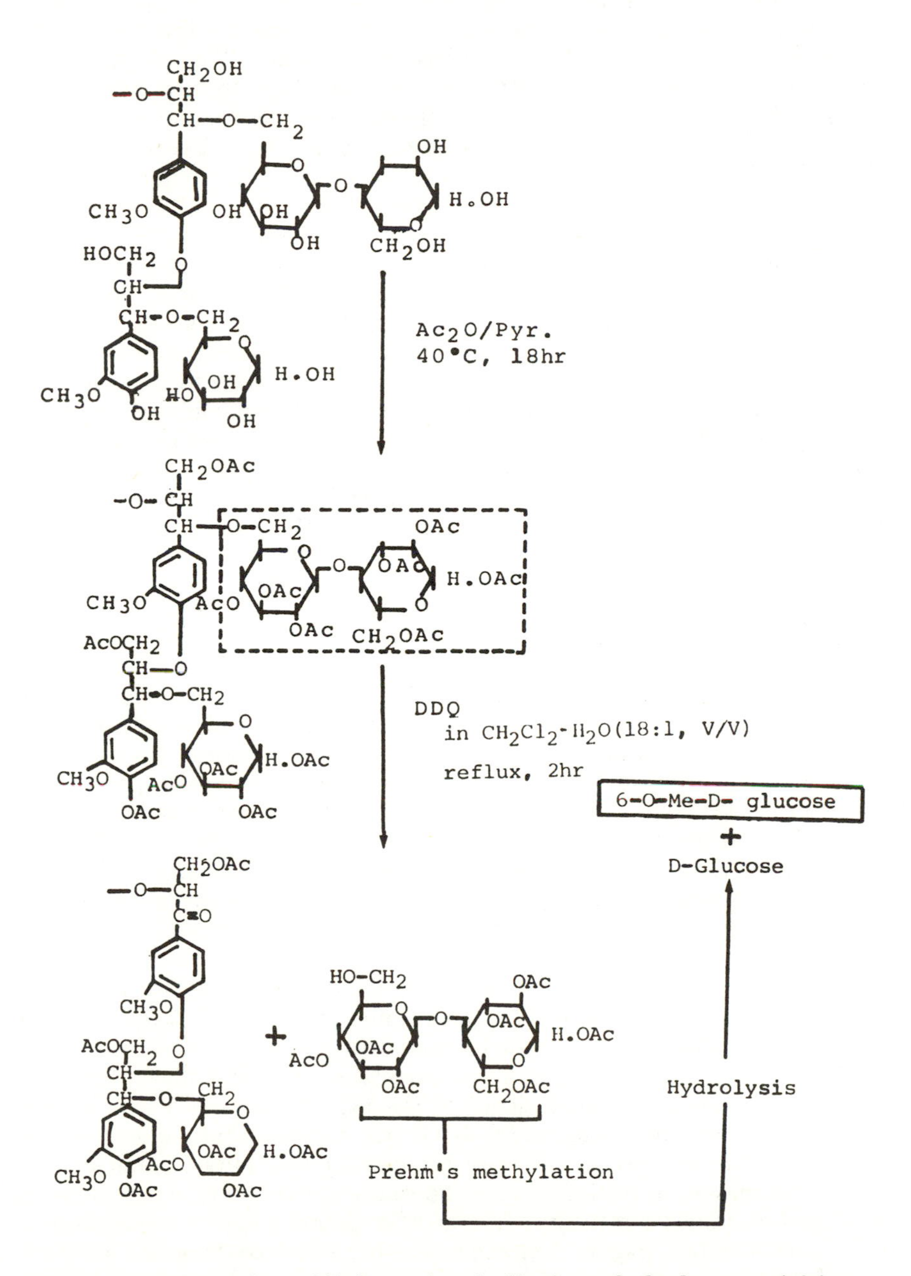

Scheme 1. Estimation of linkage sites in lignin-carbohydrate moieities.

by means of Sephadex column chromatography. Figure 1, where the lignin component is marked by a dotted line, shows that not only the polysaccharide but also even the lignin component was broken down to smaller molecules by the action of the cellulase preparation. This was quite unexpected, since the Cellulosin AC has been confirmed to have intense carboxymethylcellulase, hemicellulase and β-glucosidase as well as avicelase activities but no ligninase or peroxidase activities. Figure 2a shows the profile of fraction C-1-A from a Sephadex G-15 column. The major fraction in Figure 2A was secured, partially hydrolyzed with Cellulosin AC and rechromatographed (Fig. 2b). When peaks I and II in Figure 2b were isolated and again hydrolyzed with the same cellulase preparation, the lignin components again separated into four fractions in the lower molecular weight region of the profile as shown in Figure 2c (4). This phenomenon is easily understandable by assuming the lignin component of the LCC's to consist of low molecular weight pendant-like fragments attached to a sugar chain (Fig. 3). In other words, Cellulosin AC could break down only the interconnecting sugar chain between lignin moieties, with the result that the lignin components appeared to be cleaved by the cellulases and separated into several peaks corresponding to different molecular sizes on the gel-filtration chromatograms (4). This view was supported by the viscosimetric behavior of the neutral LCC (C-1-M-1) solution in water (9).

Micelle or Aggregate Formation and Hydrophobic Interactions of LCC Molecules in Aqueous Solution

When hydrophilic and hydrophobic groups are present in a single molecule and their ratio and distribution over the molecule are suitable, the molecules associate with each other and form micelles or aggregates in aqueous solution. As LCC molecules have both hydrophobic lignin and hydrophilic sugar moieties, it is reasonable to consider that they should associate to form aggregates or micelles. Figure 4 displays the relationship between electrical conductivity of the LCC solution and concentration, which changes sharply at the point where the micelles or aggregates are formed, confirming indeed that such entities do occur. The critical micelle concentration (c.m.c.) was 0.035% in this case (8).

Meanwhile, pinacyanol chloride, a kind of pigment, has two visible absorption maxima, the α-band (λ_{max}605 nm) and β-band (λ_{max}550 nm). These λ_{max}'s shift to longer wavelengths with increasing molar extinction coefficient as the LCC concentration increases. Figure 5 illustrates plots of the molar extinction coefficients of the LCC-pinacyanol chloride solution against the concentration of C-1-A-1, the purified acidic LCC fraction. The discontinuity (corresponding to the c.m.c.) in Figure 5 confirms that micelle formation occurs. Figure 5 shows the same c.m.c. value, 0.035%, as Figure 4, indicating that micelle or aggregate formation takes place in LCC solutions (8). Figure 6 provides another instance demonstrating micelle formation or aggregation of LCC's in solution as a result of cationic detergent-LCC interactions. The LCC's used here were isolated together with milled wood lignin from finely divided pine wood with an aqueous diox-

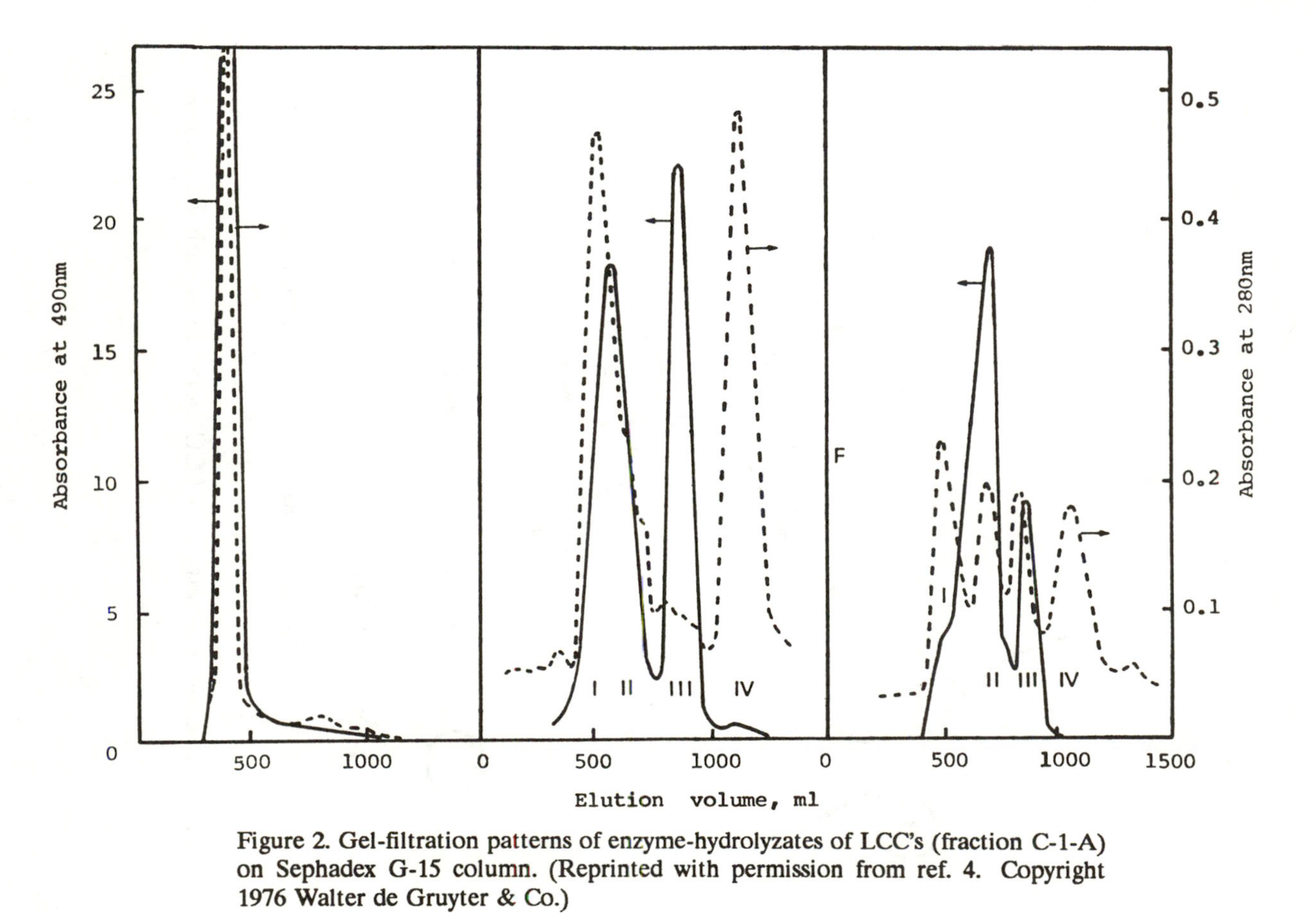

Figure 2. Gel-filtration patterns of enzyme-hydrolyzates of LCC's (fraction C-1-A) on Sephadex G-15 column. (Reprinted with permission from ref. 4. Copyright 1976 Walter de Gruyter & Co.)

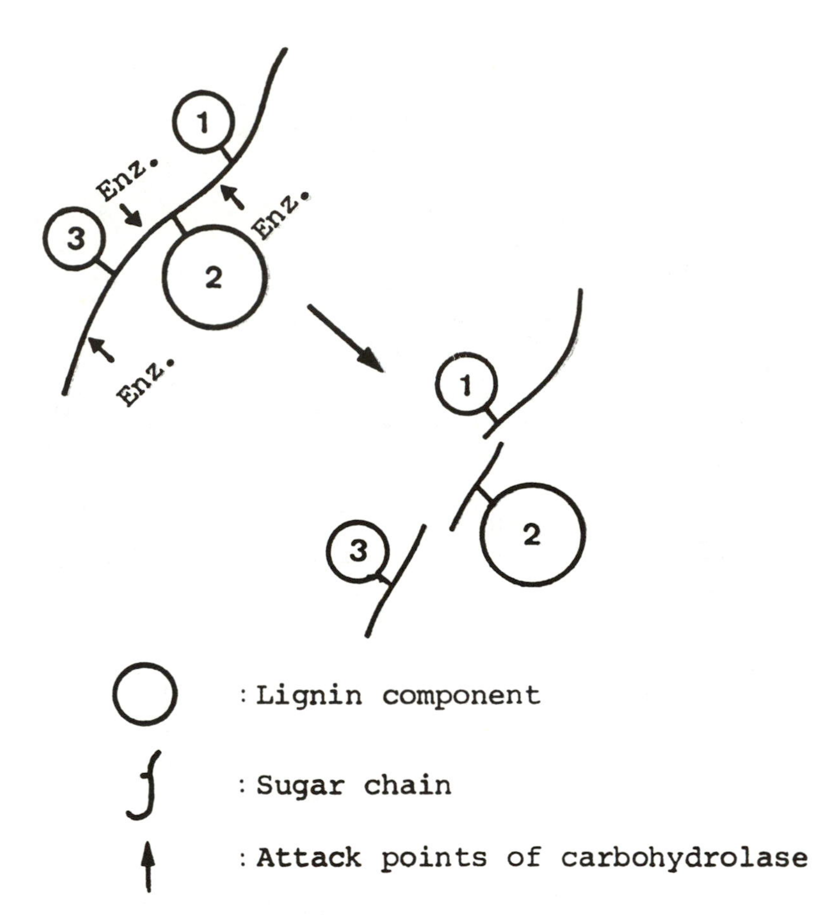

Figure 3. Degradation pattern of LCC molecule by carbohydrolases.

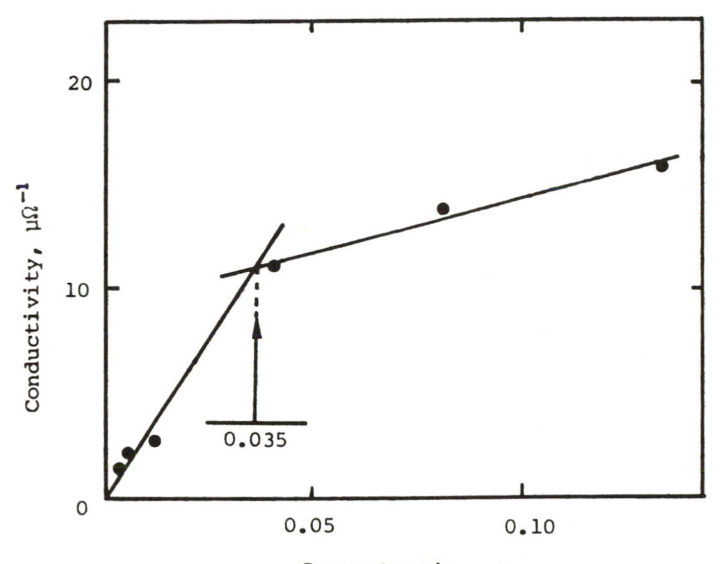

Figure 4. Plot of conductivity against LCC (fraction C-1-A-1) concentration. (Reprinted with permission from ref. 8. Copyright 1979 Walter de Gruyter & Co.)

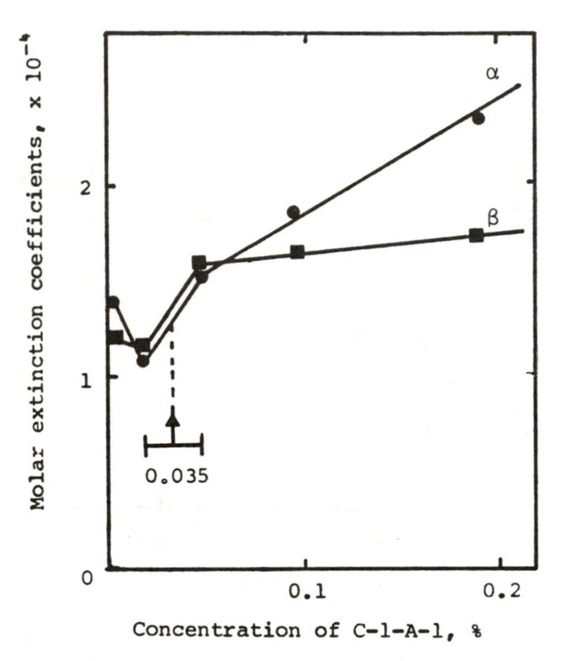

Figure 5. Plot of molar extinction coefficient against LCC (fraction C-1-A-1) concentration. (Reprinted with permission from ref. 8. Copyright 1979 Walter de Gruyter & Co.)

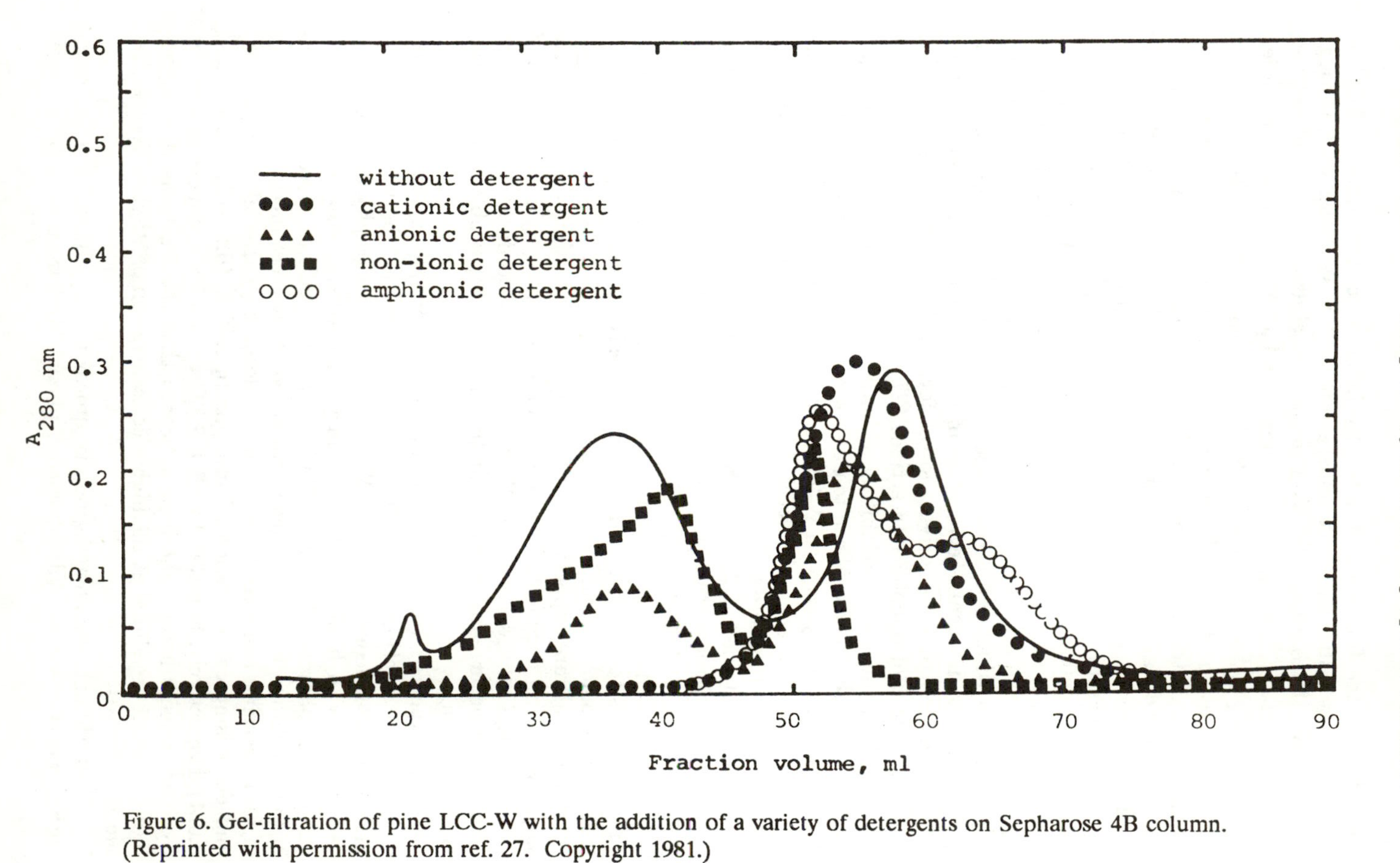

Figure 6. Gel-filtration of pine LCC-W with the addition of a variety of detergents on Sepharose 4B column. (Reprinted with permission from ref. 27. Copyright 1981.)

ane solution; the fraction is denoted by LCC-W in this chapter. LCC-W was fractionated into three fractions, namely W-1, W-2 and W-3, through a Sepharose 4B column (13). W-1 is the one eluted at the void volume, W-2 ($\overline{M}_n = 2 \times 10^5$) contained 51% lignin and 38% neutral sugar moieties, and W-3 ($\overline{M}_n = 3 \times 10^3$) included 21% lignin and 73% sugars.

The addition of 0.1% of a variety of detergents to the eluents in gel-filtration chromatography using a Sepharose 4B column resulted in the decrease or disappearance of the higher molecular weight fraction W-2, while the lower molecular weight fraction W-3 remained at its initial position in the chromatogram. W-3 appeared in the chromatogram alone when cationic or amphionic detergents were used as additives (27). As the fraction W-2 had a number-average molecular weight in the region of 10^5 and the remaining W-3 is of the order of 10^3, W-2 corresponds to micelles or aggregates of pine LCC's which are decomposed by interactions with cationic detergents. Further, it became evident from this experiment that electrostatic interactions contribute as driving forces for micelle or aggregate formation in addition to hydrophobic interactions (27).

The LCC molecules exhibiting a tendency to form micelles or aggregates are thought to be confined to those with lignin:sugar ratios around 1:1. This is exemplified by fraction W-2 here. On the other hand, LCC molecules containing a larger amount of carbohydrate and less lignin would lack the ability to form micelles as in the case of those LCC molecules in fraction W-3 which contained 70% sugar and 20% lignin on average (Fig. 7b). Figure 7 depicts the structural determinants of the LCC molecules and their micelles or aggregates proposed on the basis of the summarized results.

The LCC molecules proposed here furnish a chemical and physicochemical basis to the images obtained by Kosíková and Fengel from electron microscopic observations. As shown in Figure 7, some of the sugar chain terminals are expected to link glycosidically to lignin moieties since sugar-free lignin fragments have been isolated by enzymatic degradation of C-1-A using Cellulosin AC (4). However, there is no direct evidence concerning the occurrence of glycosidic linkages, so this problem remains an open question.

It is expected from the molecular shapes of LCC's that strong hydrophobic interactions due to the lignin moieties act between the LCC molecules in aqueous solution. When the chromatography of LCC's was carried out with hydrophobic stationary phases using columns of phenyl and octyl Sepharose, 80.4% and 77.6% of the charged LCC's were adsorbed onto the column, which were being eluted with a solvent mixture composed of an increasing concentration of ethyl cellosolve (0, 15, 30, 45 and 50%) and decreasing concentration of ammonium sulfate (0.8, 0.6, 0.4, 0.2 and 0.0M), respectively, adjusted to pH 6.8. The chemical compositions of the five components hereby obtained indicated that the fractions with higher lignin content are eluted at the higher concentration of ethyl cellosolve (Table IV; ref. 15).

Even though one phenyl group is approximately equivalent to a three-carbon alkyl chain in hydrophobicity, phenyl Sepharose adsorbs a larger

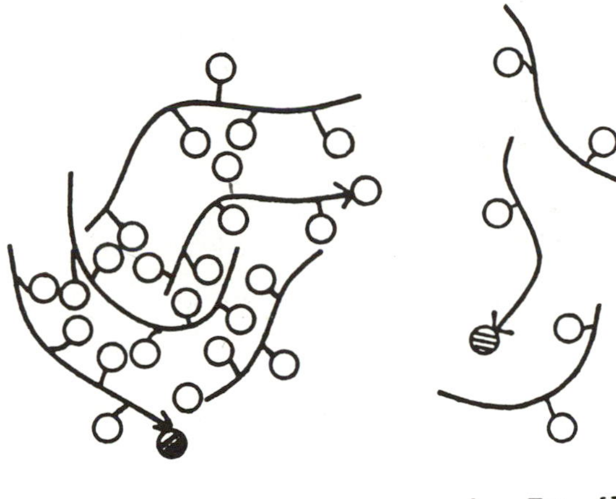

: Lignin fragment

: Carbohydrate moiety linking non-glycosidically to lignin fragment

: Carbohydrate moiety linking glycosidically to lignin fragment

Figure 7. Structural determinants of LCC molecules and their micelle (or aggregate) formation. (Reprinted with permission from ref. 27. Copyright 1981.)

Table IV. Properties of Fractions Obtained from Hydrophobic Chromatography of Pine LCC's

	Phenyl-Sepharose						Octyl-Sepharose				
Component	LCC-W	P-I	P-II	P-III	P-IV	P-V	O-I	O-II	O-III	O-IV	O-V
Lignin content[b] (%)	40.5	8.0	22.6	34.0	43.5	50.5	8.8	28.4	38.8	44.9	60.5
Neutral sugar[c] (%)											
L-Arabinose	22.5	23.4	19.0	13.9	17.0	13.4	30.6	14.7	15.0	15.4	24.0
D-Xylose	18.2	24.5	22.6	37.4	22.3	13.7	17.8	16.0	12.4	22.7	15.6
D-Mannose	29.2	25.8	28.8	13.9	28.6	32.5	18.5	31.5	27.7	18.0	32.2
D-Galactose	19.1	15.0	19.3	20.5	23.0	18.0	10.2	25.3	32.1	23.1	21.3
D-Glucose	11.0	11.3	10.4	14.3	9.2	22.4	23.4	12.5	12.6	20.7	7.9
Recovery[d] (%)		17.5	7.8	12.7	29.7	32.3	12.7	25.3	36.5	14.6	10.9

[a]Details of the procedure are given in the Experimental Section.
[b]Percentage of the dry weight of each lignin-carbohydrate complex.
[c]Percentage of the total neutral sugar.
[d]Percentage of the eluted lignin-carbohydrate complex.
Source: Reprinted with permission from ref. 15. Copyright 1982 Elsevier.

amount of LCC's than octyl Sepharose, indicating that π-π interactions due to the lignin aromatic ring moieties of the LCC's act in addition to hydrophobic interactions involving the carbon atoms of lignin side chains.

The number-average molecular weights were 6×10^3 for pine LCC's, and 6×10^3 and 2×10^5 for the species (present in a 1:1 ratio) comprising beech LCC's (28). The beech LCC molecules seem to consist of very few but bigger lignin moieties compared with those of pine LCC's.

The progress of lignin biosynthesis in the presence of oligo- or polysaccharides in the primary cell wall would necessarily engender the formation of sugar-linked lignins or lignin-interconnecting sugar chains, provided that lignin biosynthesis involves quinonemethide intermediates. The resulting products are all lignin-carbohydrate complexes in a broad sense, but the LCC's which can be extracted by the usual methods will be limited to those with less than a 50% lignin content. A part of the chemical structures of the interunit linkages in pine LCC's is presented in Figure 8.

The role of LCC's in living plant tissues is presumed to be related to the prevention of water-soluble hemicelluloses from dissolving out of the cell wall by the formation of micelles or aggregates that immobilize sugar chains, and the solubilization of water-insoluble material such as lignin, thereby enabling it to move to any place in the cell.

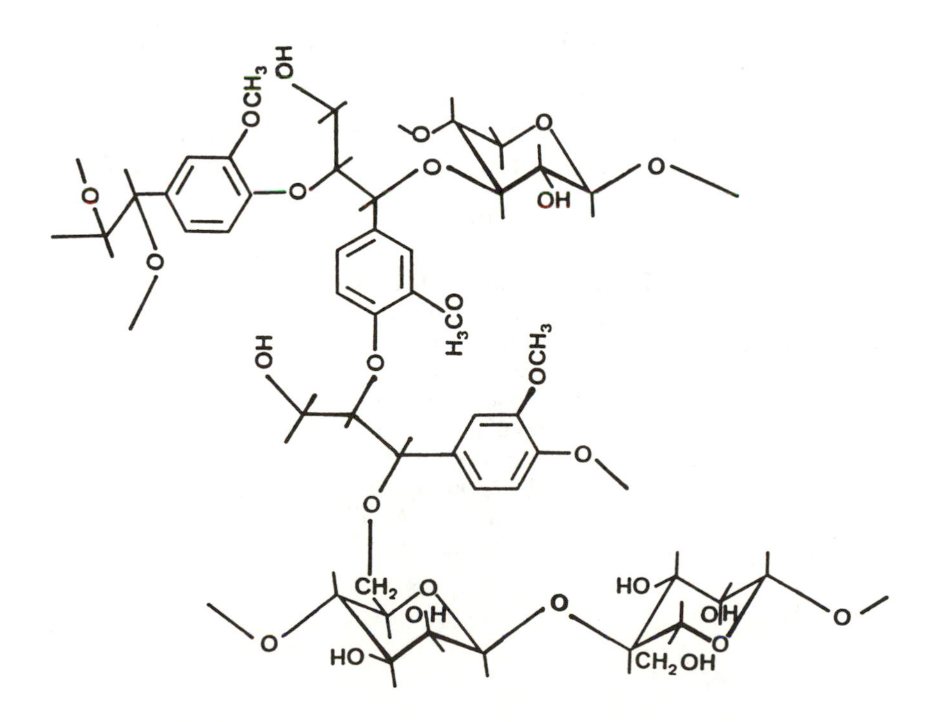

Figure 8. Structural features of the interunit linkages in the lignin-carbohydrate moieties in pine LCC's.

Literature Cited

1. Merewether, J. W. T. *Holzforschung* 1957, **11**, 65-80.
2. Koshijima, T.; Taniguchi, T.; Tanaka, R. *Holzforschung* 1972, **26**, 211-17.
3. Wegener, G. Ph.D. Thesis, Universitat München, 1975.
4. Yaku, F.; Yamada, Y.; Koshijima, T. *Holzforschung* 1976, **30**, 148-56.
5. Björkman, A. *Svensk Papperstidn.* 1957, **60**, 243-51.
6. Brownell, H. H. *Tappi* 1971, **54**, 66-71.
7. Koshijima, T.; Yaku, F.; Tanaka, R. *Appl. Polym. Symp.* 1976, **28**, 1025-39.
8. Yaku, F.; Tsuji, S.; Koshijima, T. *Holzforschung* 1979, **33**, 54-9.
9. Yaku, F.; Tanaka, R.; Koshijima, T. *Holzforschung* 1981, **35**, 177-81.
10. Eriksson, O.; Lindgren, B. O. *Svensk Papperstidn.* 1977, **80**, 59-63.
11. Watanabe, T.; Azuma, J.; Koshijima, T. *Mokuzai Gakkaishi* 1987, **33**, 798-803.
12. Lundquist, K.; Simonson, R.; Tingsvik, K. *Svensk Papperstidn.* 1980, **83**, 452-4.
13. Azuma, J.; Takahashi, N.; Koshijima, T. *Carbohydr. Res.* 1981, **93**, 91-104.
14. Mukoyoshi, S.; Azuma, J.; Koshijima, T. *Holzforschung* 1981, **35**, 233-40.
15. Takahashi, N.; Azuma, J.; Koshijima, T. *Carbohydr. Res.* 1982, **107**, 161-8.
16. Kato, A.; Azuma, J.; Koshijima, T. *Chem. Lett.* 1983, 137-40.
17. Freudenberg, K.; Harkin, J. M. *Chem. Ber.* 1960, **93**, 2814-9.
18. Tanaka, K.; Nakatsubo, F.; Higuchi, T. *Mokuzai Gakkaishi* 1979, **25**, 653-9.
19. Katayama, Y.; Morohoshi, N.; Haraguchi, T. *Mokuzai Gakkaishi* 1980, **26**, 358-62.
20. Feckel, J. Ph.D. Thesis, Universitat München, 1982.
21. Fengel, D. *Svensk Papperstidn.* 1976, **79**, 24-8.
22. Kosíková, B.; Zakutna, L.; Joniak, D. *Holzforschung* 1978, **32**, 15-8.
23. Watanabe, T.; Kaizu, S.; Koshijima, T. *Chem. Lett.* 1986, 1871-4.
24. Oikawa, Y.; Yoshida, T.; Yonemitsu, O. *Tet. Lett.* 1982, **23**, 885-8; Yonemitsu, O. *Yuki-Goseikagaku* 1985, **43**, 691-6.
25. Koshijima, T.; Watanabe, T.; Azuma, J. *Chem. Lett.* 1984, 1737-40.
26. Watanabe, T.; Koshijima, T. *Mokuzai Gakkaishi* 1989, **35**, in press.
27. Koshijima, T.; Azuma, J.; Takahashi, N. *The Ekman Days* 1981, **I**, 67.
28. Takahashi, N.; Koshijima, T. *Wood Sci. Technol.* 1988, **22**, 177-89.

RECEIVED May 29, 1989

Chapter 3

Heterogeneity of Lignin

Dissolution and Properties of Low-Molar-Mass Components

Kaj Forss, Raimo Kokkonen, and Pehr-Erik Sågfors

The Finnish Pulp and Paper Research Institute, P.O. Box 136, 00101 Helsinki, Finland

On heating pre-extracted spruce wood meal for 48 h with a 60:40 v/v mixture of dioxane and 0.5 *M* phosphate buffer pH 6.8, low molar mass lignins (i.e., hemilignins and breakdown products of high molar mass glycolignin) amounting to 24 percent of total lignin are dissolved. By extraction of the solution with *n*-hexane the dioxane is transferred to the hexane phase together with hexane-dioxane soluble lignins. Removal of dioxane results in precipitation of dark brown hydrophobic lignin. Water soluble lignins remain in solution. When wood is heated first with the buffer solution alone and then with the mixture of dioxane and buffer solution, only small amounts of the hydrophobic lignin dissolve. It seems that during heating with the aqueous buffer solution the hydrophobic lignin, which may form the residual lignin in kraft pulp, becomes irreversibly bound to the fibers.

It has earlier been shown that spruce wood (*Picea abies*) lignin is a group of compounds, consisting of 80-85% polymeric carbohydrate-bound lignin, which we are designating glycolignin and 15-20% of a group of low molar mass lignins, monomers, dimers and oligomers, which we are collectively designating hemilignins (1).

The purpose of this work is to elucidate the role of hemilignins and glycolignin in the formation of color in pulp. The aim of the present part of the work was to develop a selective method for dissolution of low molar mass lignins without dissolving the glycolignin.

Heating wood with an acid bisulfite solution causes sulfonation of both the hemilignins and the glycolignin. The hemilignins are sulfonated and dissolved before the glycolignin. In all probability, due to its bonding with carbohydrates, the dissolution of glycolignin is strongly affected by the pH

0097–6156/89/0397–0029$06.00/0

of the cooking liquor. For instance, on heating wood with a bisulfite solution of pH 5.5 at 130°C, both hemilignins and glycolignin are sulfonated but dissolution is limited almost entirely to the hemilignin sulfonates. However, by subjecting the wood material to a subsequent acid-catalyzed hydrolysis, the glycolignin-carbohydrate bonds are broken and the glycolignin sulfonic acids dissolve (1).

The conclusion that all low molar mass lignins are hemilignins is, however, an oversimplification since it has been shown that, on heating an aqueous solution of glycolignin sulfonic acid acidified with hydrochloric acid or sulfur dioxide and bisulfite, a monomeric sulfonated hydrolysis product is formed (2).

A comparison of the rates of dissolution shows that glycolignin is dissolved much faster in the kraft process than in the acid bisulfite process (3). In spite of this, kraft pulp contains more residual lignin than sulfite pulp and is more difficult to bleach.

This observation leads to the view that the so-called residual lignin may not originate from undissolved glycolignin but from low molar mass lignins, i.e., hemilignins or breakdown products from glycolignin deposited on the fibers at an early stage of the cook.

In order to investigate this possibility, a series of kraft cooks was performed. Cotton wool was placed in each digester. The chlorine number, (ISO 3260), of the cotton wool in Figure 1 shows that the deposition of lignins takes place at an early stage of the cook. These colored lignins could not be removed by washing the cotton wool with sodium hydroxide solution.

This result supports the hypothesis that the residual lignin in kraft pulp is formed from some hemilignins or the degradation products of glycolignin. In sulfite pulping, sulfonation of the hemilignins and the hydrolysis product of glycolignin may prevent these compounds from reacting further and from becoming deposited on the fibers.

In mechanical wood pulping, the glycolignin remains undissolved, whereas some of the hemilignins are dissolved, their solubilities in water being the main limiting factor. It is thus possible that some of the hemilignins are also responsible for the yellowing of mechanical pulp.

Delignification of Spruce Wood Meal on Heating with Water

In order to show the amount and molar masses of lignins that can be dissolved on heating wood with water, 5 g portions of wood meal extracted with cyclohexane-ethanol (4) were heated with 100 mL water at 150°C.

Figure 2 shows that delignification reaches a maximum of 8% after 4 h of heating. Prolonged heating caused a redeposition of lignins on the wood material.

During heating the pH dropped below 4 due to the formation of acetic acid. The acid caused hydrolysis of hemicelluloses and formation of furfural. It is thus possible that the delignification was caused by hydrolytic cleavage of interunit linkages. Figure 3 shows that the dissolved lignins

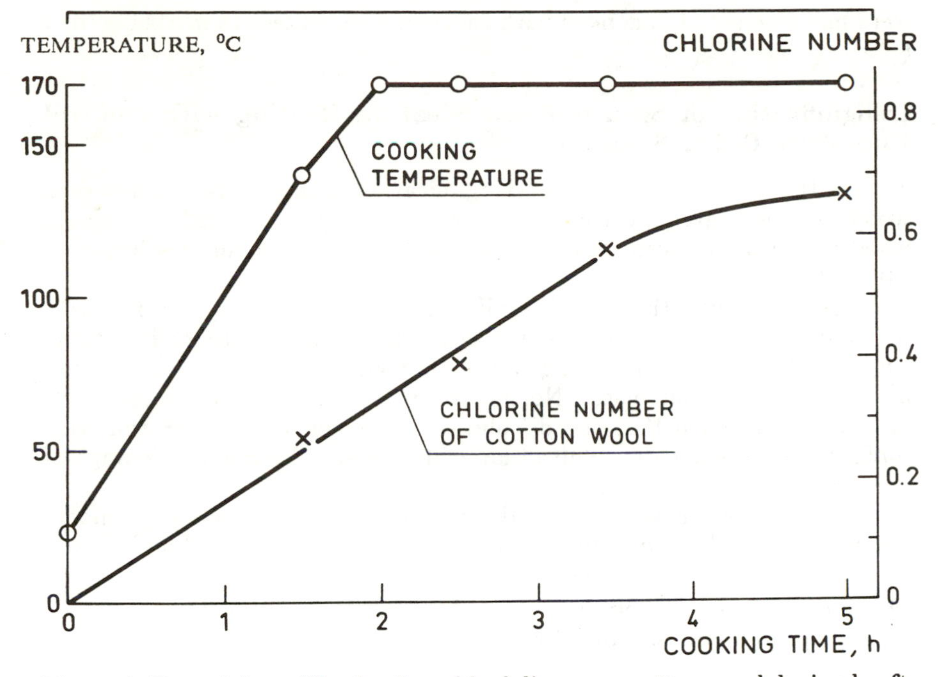

Figure 1. Deposition of lignins from black liquor on cotton wool during kraft cooking.

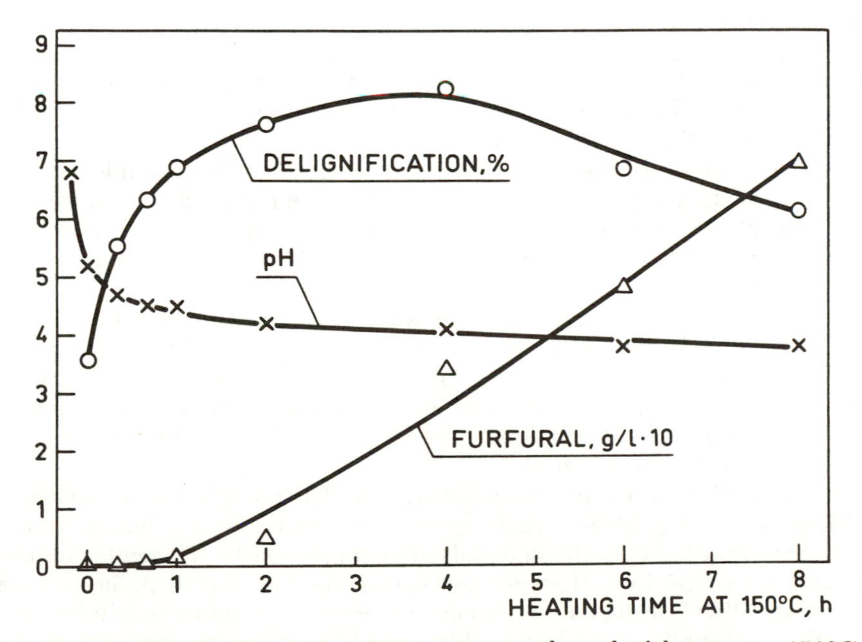

Figure 2. Delignification on heating spruce wood meal with water at 150°C.

were monomers and oligomers with molar masses of less than 1000 g/mol (7).

Delignification of Spruce Wood Meal on Heating with Neutral Phosphate Buffer Solution

To test the assumption that the delignification taking place when spruce wood meal is heated with water is a result of acid catalyzed hydrolysis, wood meal samples were heated with 0.5 *M* phosphate buffer solution at pH 6.8 (Fig. 4).

Figure 4 shows that 8% delignification was reached by heating with buffer solution but that the rate was slower than when heating with water. Figure 5 shows that, on heating with buffer solution, the lignins dissolved are also of low molar mass. No furfural was formed in this experiment. It can be concluded that the dissolution of lignins with water and with buffer solution is not the result of an acid-catalyzed hydrolytic cleavage of chemical bonds.

However, as shown by Sakakibara *et al.* (5), α-*O*-4 bonds in model compounds dissolved in water-dioxane (1:1) are readily cleaved on heating.

Delignification of Spruce Wood Meal on Heating with Mixtures of Phosphate Buffer and Dioxane

Since the reason for the limited degree of delignification reached on heating wood with water or with buffer solution may be the low solubility of lignins in water, wood meal was heated with mixtures of phosphate buffer and dioxane (Fig. 6).

Figure 6 shows that, on heating wood meal with buffer solution for 6 h at 150°C, 6% of the total lignin was dissolved, but on heating with dioxane only 4% dissolved. However, on heating with a mixture of buffer and dioxane 14% of the lignin dissolved.

Using reversed-phase chromatography (Fig. 7), it was possible to show that, on heating with buffer solution, most of the lignins dissolved were of a hydrophilic nature, eluting between 0 and 30 minutes, whereas dioxane preferentially dissolved hydrophobic lignins, which eluted between 30 and 80 minutes.

As heating with buffer solution was found preferentially to dissolve hydrophilic molecules, an attempt was made to delignify by heating wood meal successively with buffer and with a buffer-dioxane mixture. Figure 8 shows that heating with buffer resulted in a maximum delignification of about 8% after 8 h, at which point some of the dissolved lignin was redeposited on the wood material.

After 48 h of heating with buffer, 5% of the lignin was in solution. When the wood residue was subsequently heated in dioxane-buffer, an additional 3% of the total lignin dissolved instantaneously. Thereafter almost no lignin was dissolved. However, by heating wood meal with buffer-dioxane solution, 24% of the lignin was dissolved after 48 h. It can be concluded that heating with buffer alone irreversibly bound lignin to the wood material.

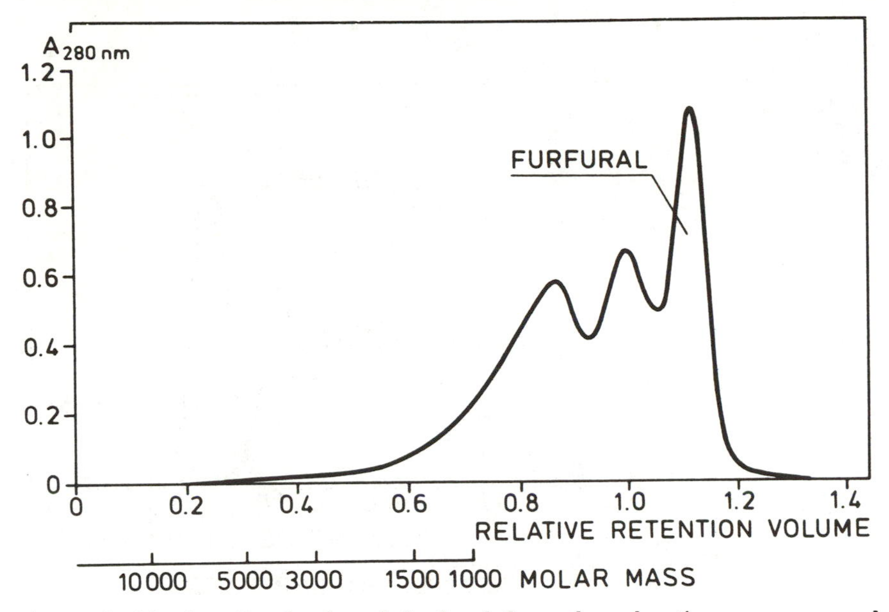

Figure 3. Lignins dissolved and furfural formed on heating spruce wood meal with water for 6 h at 150°C. Column: Sephadex G-50. Eluent: 0.5 *M* NaOH.

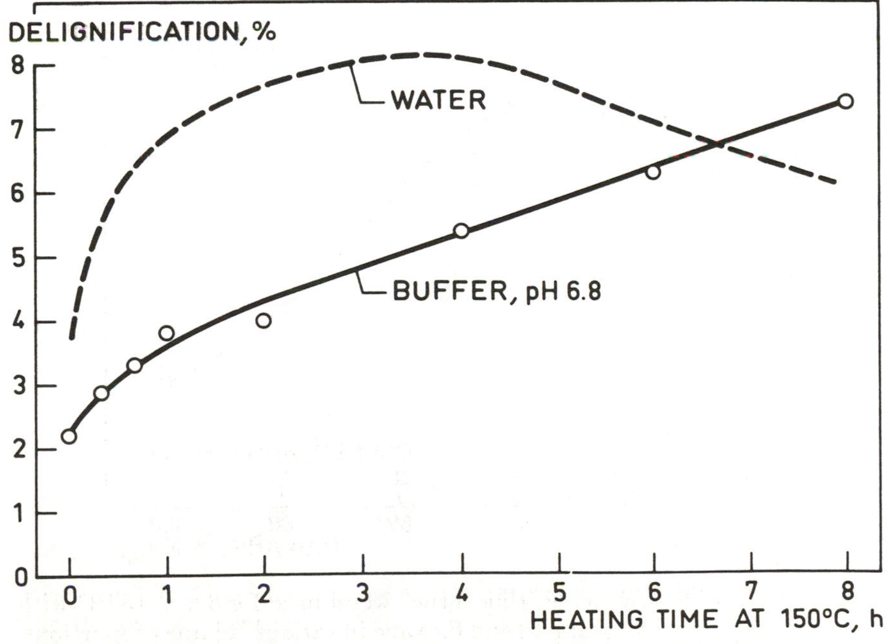

Figure 4. Delignification on heating spruce wood meal with phosphate buffer, pH 6.8, at 150°C.

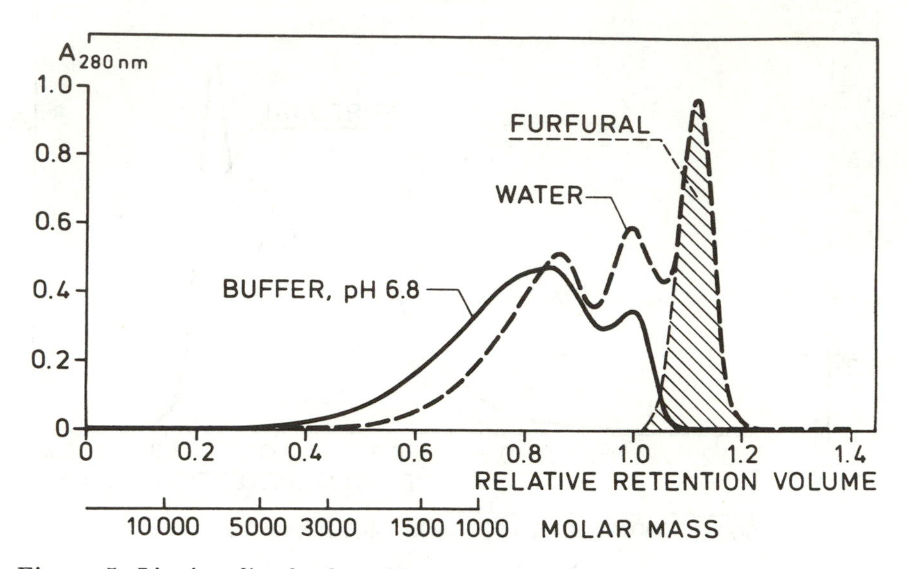

Figure 5. Lignins dissolved on heating spruce wood meal with phosphate buffer, pH 6.8, for 6 h at 150°C. Column: Sephadex G-50. Eluent: 0.5 *M* NaOH.

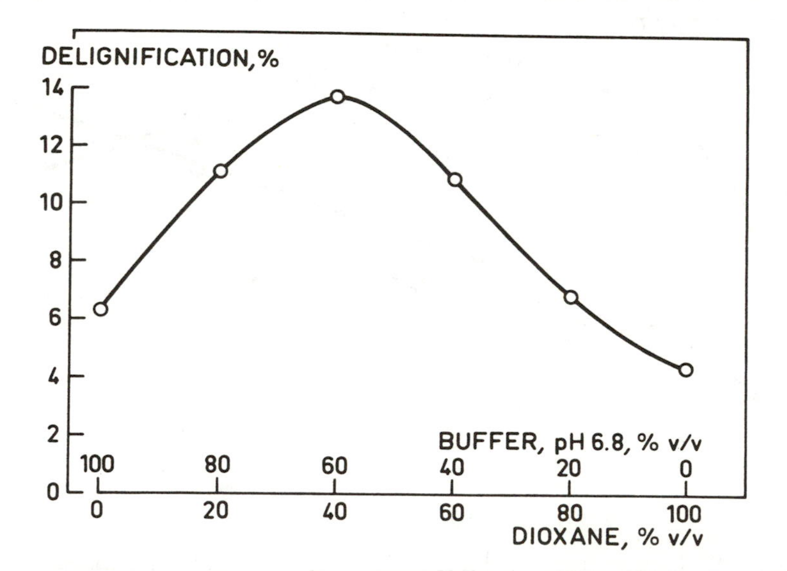

Figure 6. Delignification on heating spruce wood meal for 6 h at 150°C with 0.5 *M* phosphate buffer (pH 6.8) and dioxane in various volume proportions.

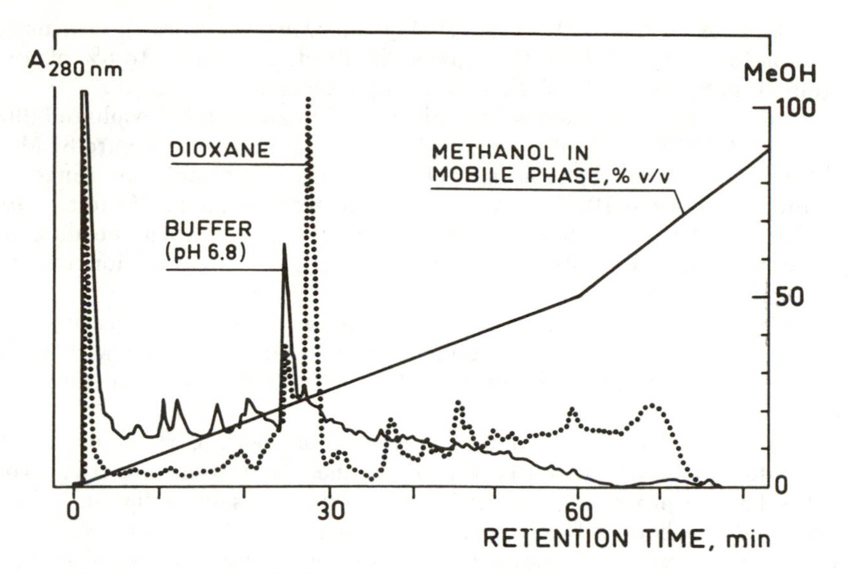

Figure 7. Lignins dissolved on heating spruce wood meal with phosphate buffer 0.5 *M* (pH 6.8) and with dioxane for 6 h at 150°C. Column: Spherisorb ODS. Elution: Gradient elution with phosphate buffer-methanol.

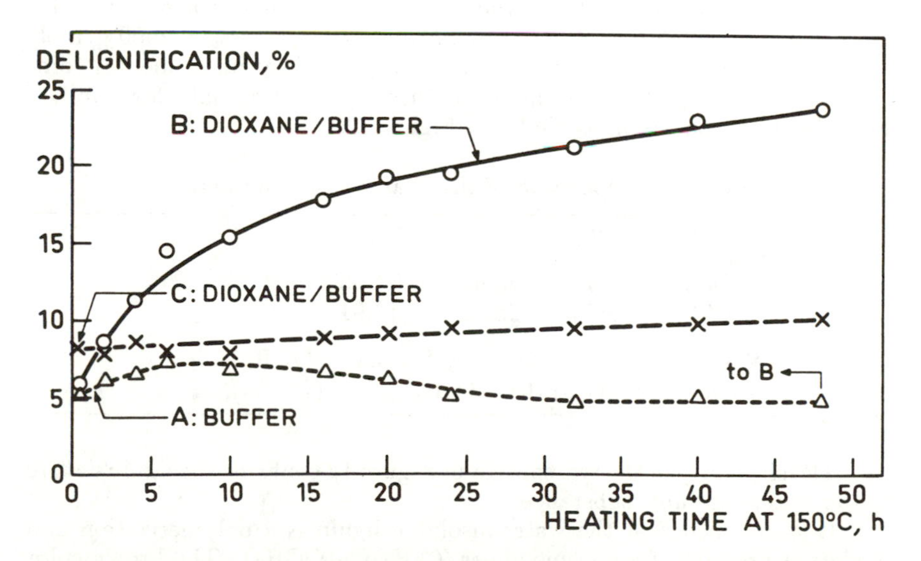

Figure 8. Delignification on heating (A) spruce wood meal with phosphate buffer (pH 6.8), 150°C; (B) spruce wood meal with dioxane-phosphate buffer (60:40% v/v); (C) wood residue from (A) (48 h) with dioxane-phosphate buffer (60:40% v/v).

Curve A in Figure 9 reveals that, on heating wood meal with buffer solution for 48 h at 150°C, the lignins dissolved, amounting to 5% of total lignin (Fig. 8), are hydrophilic to varying degrees.

On heating wood meal with a mixture of dioxane and phosphate buffer for 48 h at 150°C, 24% of the lignin dissolved, as shown in Figure 8. Most of this lignin is hydrophobic and eluted in the retention time range 60-90 minutes (Curve B). However, when the wood residue obtained after heating wood meal with phosphate buffer was heated with the dioxane-phosphate buffer mixture, only moderate further delignification was obtained (Curve C).

Figure 9 thus shows that most of the lignin dissolved on heating with the buffer-dioxane solution is bound to the wood on heating with buffer alone. This hydrophobic lignin is eluted in the retention time range 60-90 minutes.

By extraction of the solution obtained on heating wood meal with buffer-dioxane solution for 16 h with *n*-hexane, the dioxane was transferred to the hexane phase together with hexane-dioxane soluble lignins. These amounted to about 20% of the dissolved lignins and to about 3% of the total lignin in wood. Removal of the dioxane resulted in precipitation of hydrophobic lignin, as revealed by Figure 10. This lignin amounted to 30% of the dissolved lignins and about 6% of the total lignin in wood. This hydrophobic lignin is dark brown, insoluble in water, incompletely soluble in dioxane and in 0.1 *M* sodium hydroxide, but soluble in a dioxane-water mixture (60:40 v/v).

Figure 11 reveals that the water-insoluble lignin has a broad molar mass distribution with an average molar mass close to $\overline{M}_w = 2000$ g/mol.

Its nature as a lignin component is confirmed by its elemental composition and IR-spectrum compared with corresponding data for a milled wood lignin preparation, Table I and Figure 12.

Table I. Elemental Composition of the Water-Insoluble Lignin

C	H	O	CH_3O
64.37	5.48	28.67	14.03
64.22	5.45	28.79	13.91

Water-insoluble lignin: $C_9H_{7.46}O_{2.47}(OCH_3)_{0.83}$
Milled wood lignin (*P. abies*) (6): $C_9H_{8.83}O_{2.37}(OCH_3)_{0.96}$

The similarity of the two spectra in Figure 12 confirms the lignin nature of the water-insoluble substance.

It is assumed that the water-insoluble lignin is a polymerization and oxidation product of a C_{10} monomer, $C_9H_{11}O_3(OCH_3)$. The brown color may be the result of loss of 0.17 methyl groups and hydrogen atoms during the formation of a quinonoid structure and loss of 0.7 moles of water.

On heating wood meal for 16 h with the dioxane-phosphate buffer, 17% of the total lignin in wood dissolved. As revealed by Figure 13, about

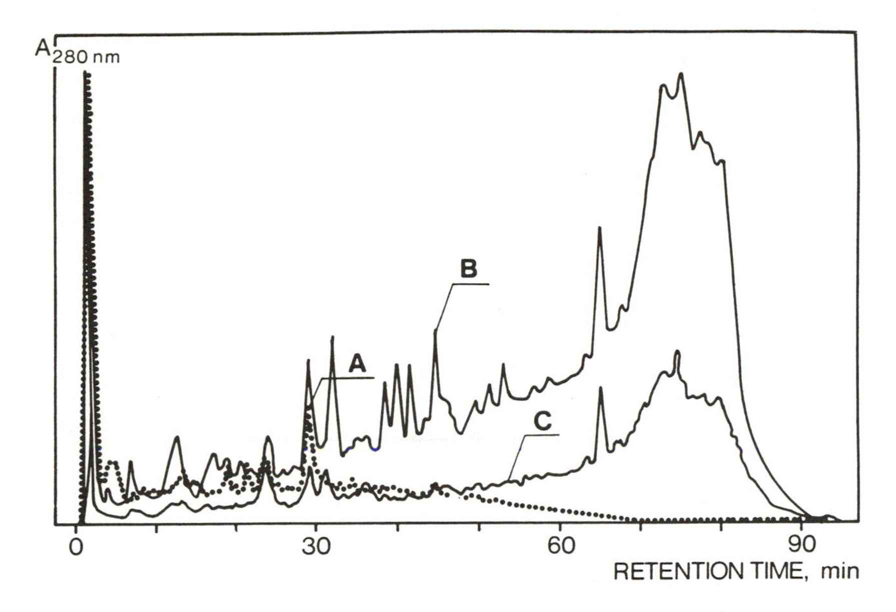

Figure 9. Lignins dissolved on heating for 48 h at 150°C: (A) spruce wood meal with 0.5 *M* phosphate buffer (pH 6.8); (B) spruce wood meal with dioxane-phosphate buffer (60:40% v/v); (C) wood residue from (A) (48 h) with dioxane-phosphate buffer (60:40% v/v). Column: Spherisorb. Elution: Gradient elution with phosphate buffer-methanol.

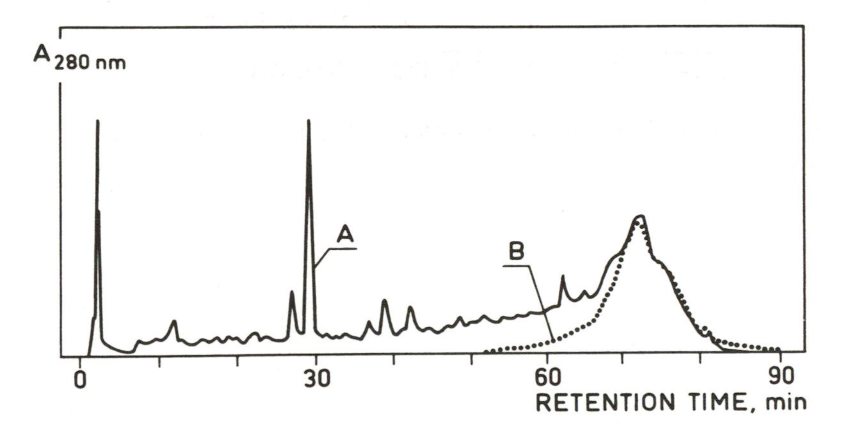

Figure 10. Lignins dissolved on heating spruce wood meal with dioxane-phosphate buffer (60:40% v/v) for 16 h at 150°C: (A) Lignins dissolved; (B) Water-insoluble lignin. Column: Spherisorb ODS. Elution: Gradient elution with phosphate buffer-methanol.

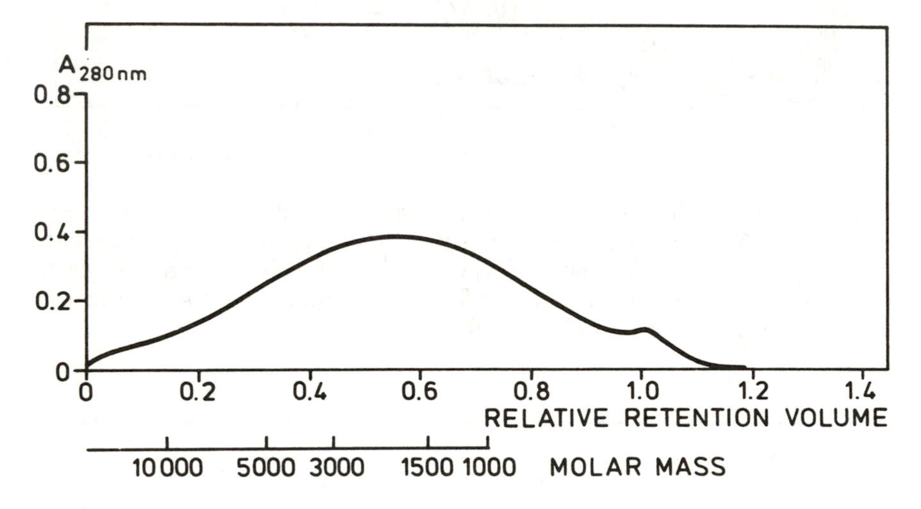

Figure 11. Water-insoluble lignin. Column: Sephadex G-50. Eluent: 0.5 *M* NaOH.

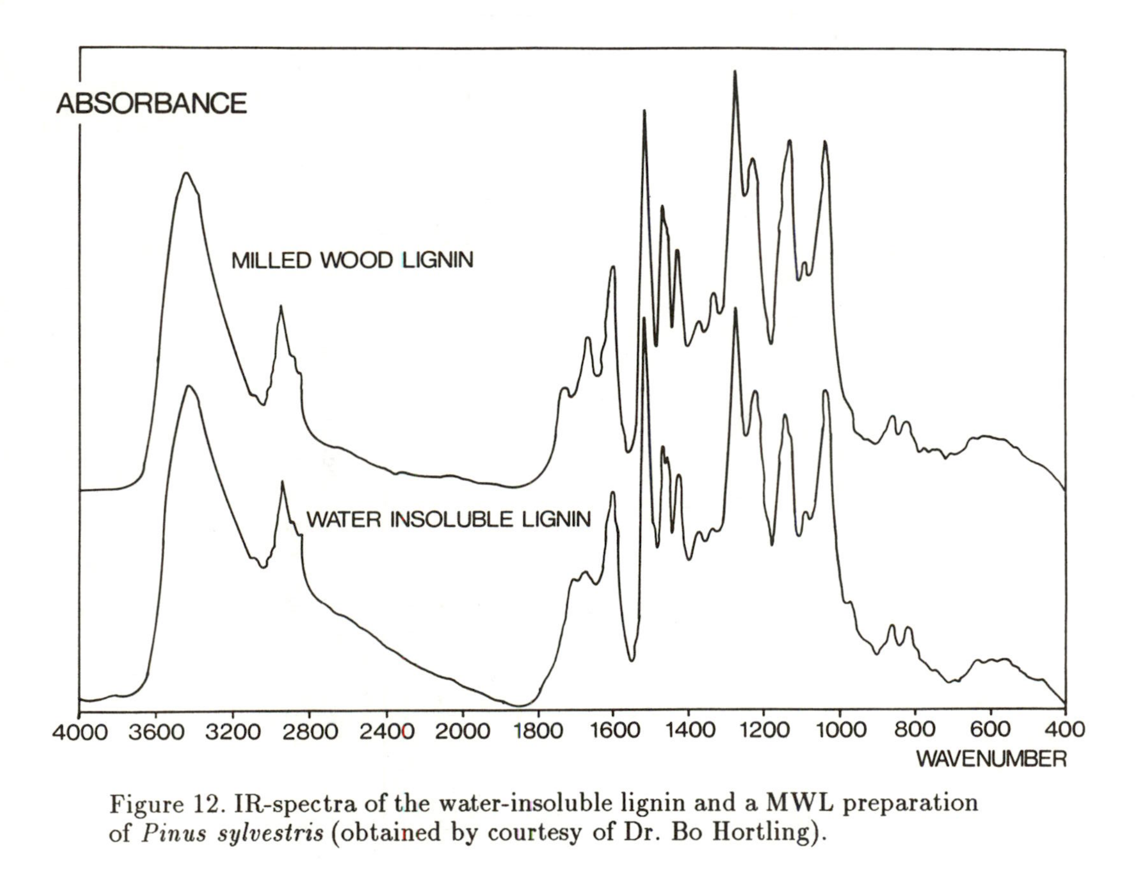

Figure 12. IR-spectra of the water-insoluble lignin and a MWL preparation of *Pinus sylvestris* (obtained by courtesy of Dr. Bo Hortling).

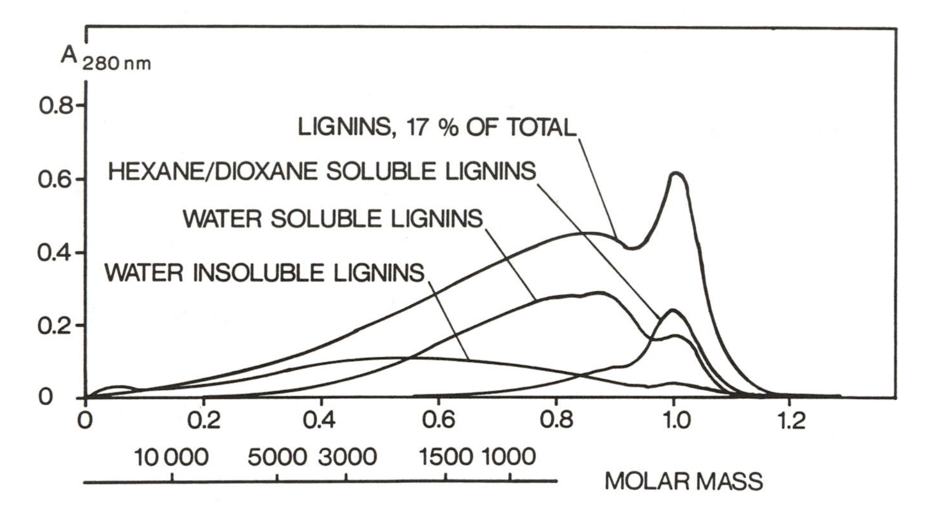

Figure 13. Lignins dissolved on heating spruce wood meal with dioxane-phosphate buffer (60:40% v/v) for 16 h at 150°C. Column: Sephadex G-50. Eluent: 0.5 *M* NaOH.

50% of the dissolved lignin consisted of water-soluble lignins. Most of these were of rather low molar mass.

The water-insoluble lignin amounted to about one-third, and the hexane-dioxane soluble lignins (which were mostly monomers) to about 20%, of the lignins dissolved.

Conclusion

These results support earlier findings that spruce wood lignin is a heterogeneous group of molecular species. It has been shown that the low molar mass hemilignins, possibly together with breakdown products from high molar mass glycolignin, are dissolved on heating wood meal with a mixture of neutral buffer solution and dioxane. Part of the lignin dissolved is a water-insoluble dark brown substance. If heating with dioxane-buffer is preceded by heating with buffer in the absence of an organic solvent, this substance becomes bound to the wood material. It is possible that this lignin forms the so-called residual lignin in kraft pulp. The high molar mass glycolignin is not dissolved on heating wood meal with dioxane-buffer solution.

Literature Cited

1. Forss, K.; Fremer, K.-E. *Tappi* 1964, **47**, 485-93.
2. Fogelberg, B. C.; Forss, K.; Oila, U. *Pap. Puu* 1969, **51**, 731-3; 735-6.
3. Forss, K.; Stenlund, B. *Kemian Teollisuus* 1971, **28**, 757-61.
4. Fengel, D.; Pryzklenk, M. *Holz als Roh-und Werkstoff* 1983, **41**, 193-4.
5. Sakakibara, A.; Takeyama, H.; Morohoshi, N. *Holzforschung* 1966, **20**, 45-7.
6. Björkman, A.; Person, B. *Svensk Papperstidn.* 1957, **60**, 158-69.
7. Forss, K.; Kokkonen, R.; Sågfors, P.-E. "Determination of the Molar Mass Distribution of Lignins by Gel Permeation Chromatography"; *ACS Symp. Ser.* 1989, *this volume*.

RECEIVED March 17, 1989

Chapter 4

Supercritical Fluid Extraction of Lignin from Wood

Lixiong Li and Erdogan Kiran[1]

Department of Chemical Engineering, University of Maine, Orono, ME 04469

Reactive extraction of lignin from red spruce has been studied using supercritical methylamine and methylamine-nitrous oxide binary mixtures. The wood residues and precipitated fractions after extractions have been characterized by chemical and spectroscopic procedures. The molecular weights and molecular weight distributions of the extracted lignins have been determined by gel permeation chromatography. The effect of extraction time, temperature, pressure, and composition of extraction fluid on molecular weights of the extracted lignins has been studied. The molecular weight distribution of the lignins extracted by methylamine-nitrous oxide binary mixture is observed to be similar to that of kraft lignin (Indulin AT). In contrast, lignins extracted by pure methylamine display much broader molecular weight distributions. Molecular weight distributions become broader with an increase in extraction time, temperature, or pressure.

Supercritical fluid extraction is a new separation technique that finds a number of applications in the natural products, biochemicals, food, pharmaceuticals, petroleum, fuel, and polymer industries (1-8). There is now an interest in applying this technology in the pulp and paper industry (9,10). In a recent comprehensive study on the interaction of supercritical fluids with lignocellulosic materials, it has been shown that lignin can be not only extracted from wood by reactive supercritical fluids but also separated as solid products in solvent-free form by reducing the extraction fluid pressure from a supercritical to subcritical level (11,12).

[1]Address correspondence to this author.

0097–6156/89/0397–0042$06.00/0

Among the various extraction fluids, supercritical methylamine and methylamine-nitrous oxide binary mixtures have been found to be especially effective in the selective removal of lignin from wood. This has been verified by chemical (Klason lignin determination), thermal (thermogravimetric analysis), and spectroscopic (infrared) analyses of the wood residues and the extracted fractions. It has been shown that lignin removal increases with extraction time, temperature, pressure, and the methylamine content (in the binary mixture of methylamine-nitrous oxide). A motivation in the use of binary mixtures involving one of the components with selective reaction capability toward lignin (such as methylamine) is the possibility of inducing a limited fragmentation of lignin upon which its dissolution is achieved in the supercritical fluid media. If this is in fact the case, the molecular weight of the lignins may be expected to be larger than that of lignins obtained by conventional pulping processes. To test this hypothesis, information on the molecular weight and molecular weight distribution (MWD) of lignin from supercritical fluid extraction is needed. The present paper is focused on this particular aspect.

The literature on molecular weight and MWD of lignins resulting from various conventional chemical treatments is quite extensive (13-15). In the selection of proper solvents for GPC analysis of lignin, information on solubility of lignins in various solvents documented by many researchers (17-19) is very useful. The commonly used solvents are either aqueous solutions (20-23) or organic compounds such as tetrahydrofuran (THF), dimethylformamide (DMF), dimethylsulfoxide (DMSO), and dioxane (24-28). More recently, the use of high performance gel permeation chromatography with styrene-divinylbenzene copolymer gel columns to characterize lignin molecular weight has been reported (29-31).

Among the various solvents, DMF is reported to be the most effective organic solvent to dissolve kraft lignins (19). It has also been suggested that association effects of nonacetylated lignin molecules in DMF can be substantially reduced by adding lithium bromide to DMF solvent (31-33). In the present study, GPC analyses were carried out in DMF +0.1M LiBr using Waters Ultrastyragel columns. GPC results of the extracted lignins have been interpreted with respect to elution volumes, and compared with the results of kraft pine lignin (Indulin AT) to distinguish different molecular weight features.

Experimental

Extraction System. The flow-through extraction system used in this study is shown in Figure 1. The system is operable up to 400 bar at 200°C. It consists of solvent delivery systems (Fluid 1, Fluid 2, Fluid 3), a flow-through reactor (FR), a set of separator traps (TP1, TP2), and the temperature and pressure control units. The reactor, traps, micrometering valves, and tubing connections are housed in a heated oven.

In a typical extraction experiment, a known amount of wood (about 3 g) in sawdust form is first loaded into the reactor. Both the reactor and the separators are heated under a gentle nitrogen flow (10 cc/min) to the

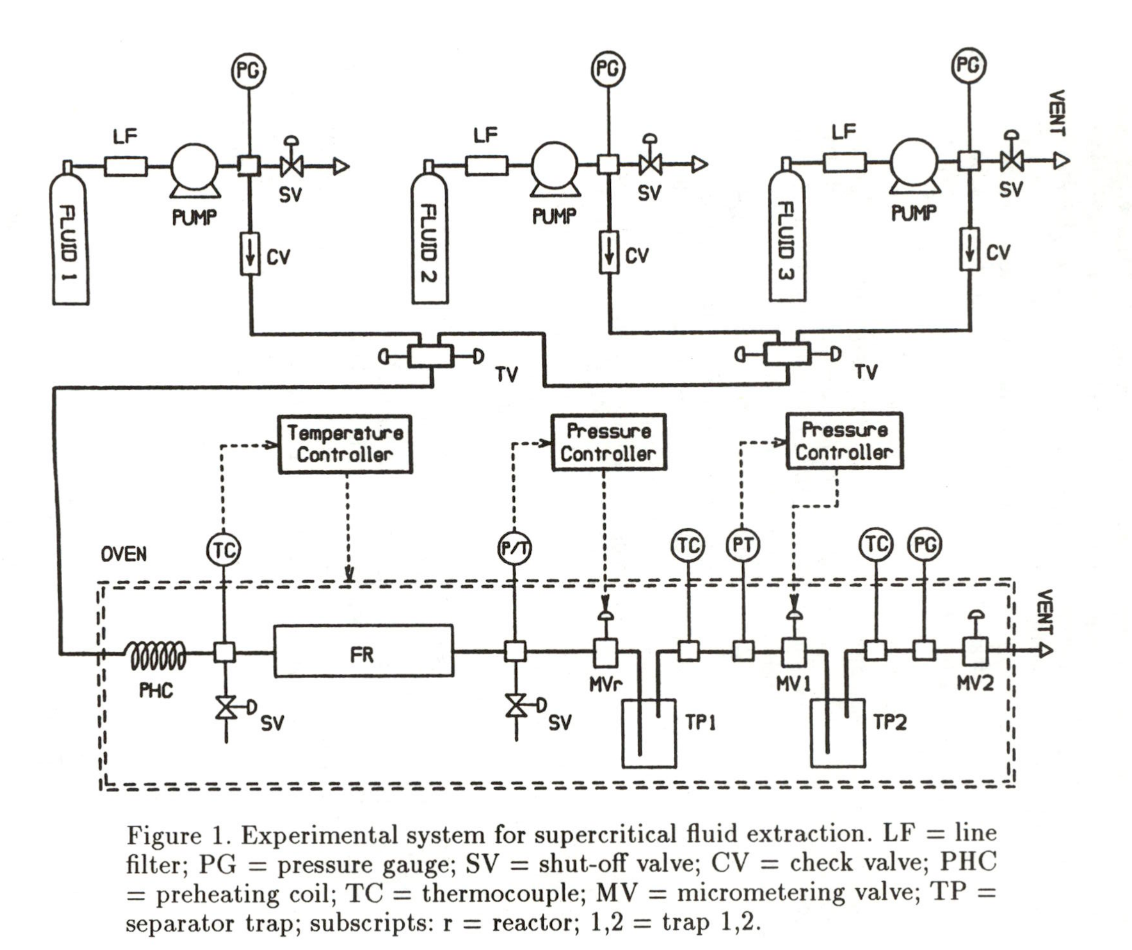

Figure 1. Experimental system for supercritical fluid extraction. LF = line filter; PG = pressure gauge; SV = shut-off valve; CV = check valve; PHC = preheating coil; TC = thermocouple; MV = micrometering valve; TP = separator trap; subscripts: r = reactor; 1,2 = trap 1,2.

desired extraction temperature. Then the extraction fluid is introduced into the reactor. By adjusting the micrometering valve (MV_r), the pressure in the reactor is maintained at a desired level. After the reaction, the system pressure is reduced to atmospheric level either directly or in a stagewise fashion with the aid of the micrometering valves MV_1 and MV_2. The extraction is continued for a specified time. At the end of each run, the wood residue in the reactor (FR) and the precipitate in the traps (TP1 and TP2) are collected and analyzed.

In the present study reactor temperature, pressure, and extraction time are varied in the range from 175 to 185°C, 170 to 275 bar, and 0.5 to 5 hr, respectively. The pressure in the first trap is maintained at 1 bar.

Materials. The wood sample, red spruce (*Picea rubens*), was obtained locally and used in sawdust form collected from a 1 mm sieve. Kraft pine lignin (Indulin AT) was obtained from Westvaco.

The extraction fluids, nitrous oxide (Airco, 99.9 wt.% purity) and methylamine (Linde Specialty Gases, 98.0 wt.% purity) were used without further purification.

Eluent solution (DMF +0.1M LiBr) for GPC analysis was prepared with HPLC grade dimethylformamide (Burdick and Jackson) and lithium bromide (Fisher Scientific Co.), which were used without further purification.

GPC performance was tested using polystyrene standards dissolved in the same solvent. The polystyrene standards (average molecular weights: 300,000, 100,000, 50,000, 35,000, 17,500, 4,000, 2,000 with polydispersity < 1.06) were obtained from the Pressure Chemical Company.

Critical Properties. The critical temperature, pressure and volume for methylamine, nitrous oxide and their binary mixtures were experimentally determined and have been previously reported (34). The critical temperatures of the mixtures are intermediate between those of the pure components (T_c methylamine = 156.9°C; T_c nitrous oxide = 36.5°C). The critical pressure goes through a maximum between the pure component values (P_c methylamine = 7.43 bar; P_c nitrous oxide = 72.4 bar). The maximum (92.5 bar) is observed at about 46 wt.% methylamine content. The extraction conditions reported in the present study are all above the critical T and P of the fluids used.

Chemical and Spectroscopic Analyses. Acid insoluble lignin (Klason lignin) contents of wood species before and after extractions were determined using a modified procedure suitable for small samples (35).

A Digilab FTIR spectrophotometer (Model FTS-60) was used to obtain IR spectra of samples before and after extractions. Standard KBr pellets containing 1% by weight sample were used.

GPC Analysis. Molecular weight characterizations were carried out using a Waters 840 Gel Permeation Chromatograph equipped with both an ultraviolet (UV) (Model 481) and a refractive index (RI) detector (Model 410). Two Ultrastyragel columns, in the running order of 1,000 and 10,000Å pore size, were used. To prevent the column from clogging with impurities that may exist in the mobile phase, a 2-micron guard column was connected in series before the 1000Å column.

The instrument was operated at 1.0 cc/min solvent flow with 30 min run time. The columns and the sample and reference cell in the RI detector were maintained at 40°C. The sample concentration was 0.1% (w/v) and the injection volume was 25 μl. The wavelength of the UV detector was set at 268 nm.

Because lignins contain a large number of phenolic, methoxyl, and aryl ether functional groups (36,37), interactions between lignin molecules themselves, and between lignin molecules and gel material, may not be completely avoided regardless of the nature of a given solvent. Such molecular interactions, especially in the case of underivatized lignins, result in a back-pressure rise across the columns and tailing in the chromatogram which may be observed when the system is continuously used over an extended time period. A consequence of column back-pressure rise is a reduction in the actual solvent flow rate and an increase in the observed elution volumes. In the present study, to test the reliability of elution times, standard polystyrene samples were run in between several runs with lignin samples.

Results and Discussion

Figure 2 shows the extent of dissolution of red spruce in methylamine, the amount of precipitate collected in the first trap upon complete depressurization to 1 bar, and the Klason lignin content in the wood residue after extraction, as functions of extraction time. The total dissolution and precipitation are normalized with respect to oven dry weight of initial wood. The extraction conditions were 185°C, 275 bar, and 1 g/min solvent flow rate. As shown in the figure, dissolution initially increases with time and levels off at about 28% by weight. The precipitates which were collected as solids follow a similar trend. The Klason lignin content of the wood residue decreases with extraction time, from an initial value of 26.5% down to 10.1% after 5 h of extraction.

The residues and precipitates from the above time dependent extraction study were further characterized by FTIR. The most distinctive IR absorbance band for lignin is observed at 1510 cm^{-1} due to aromatic ring vibrations (37). As shown in Figure 3, the absorption intensity of the residues at 1510 cm^{-1} decreases with increasing extraction time. When compared to the spectrum of red spruce at this same absorbance region, it is easy to see that the relative amount of lignin in the wood samples decreases to a lower level after methylamine extraction. Figure 4 shows the IR spectra of Indulin AT and the precipitates from methylamine extraction of red spruce and Indulin AT, respectively. We observe from this comparative plot that the IR spectra of the precipitates from methylamine extraction of red spruce and methylamine extraction of Indulin AT are nearly identical, suggesting that the precipitate from methylamine extraction of red spruce is indeed lignin-like material. The figure also shows the IR spectrum of the precipitate from ammonia extraction of red spruce. The spectrum of the

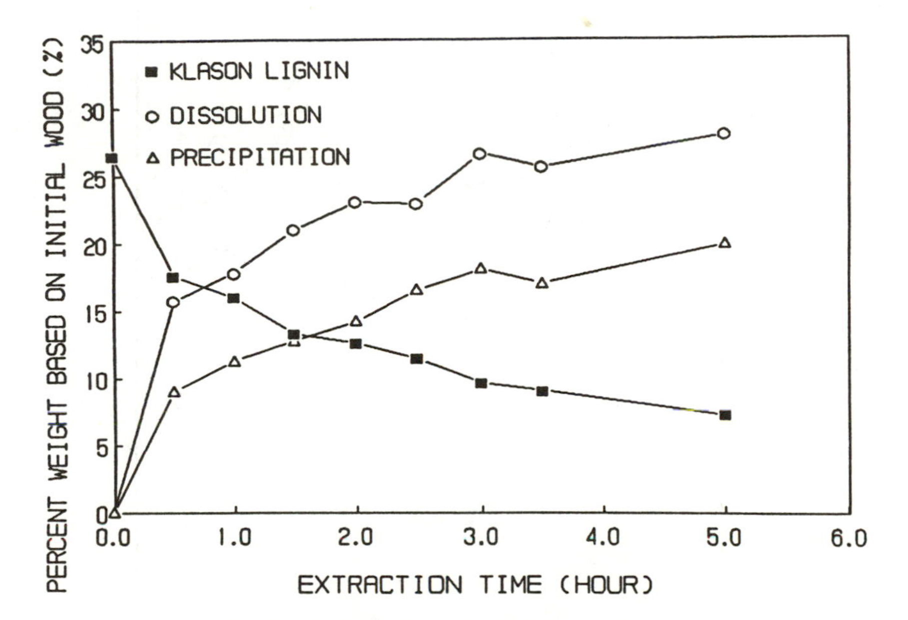

Figure 2. Dissolution (%) of red spruce in methylamine (O), Klason lignin content (%) in the residues after extraction (■), and the amount (%) of precipitates in the first trap (at 1 bar) (Δ), as functions of extraction time. Extraction conditions: 185°C, 275 bar, and 1 g/min solvent flow rate.

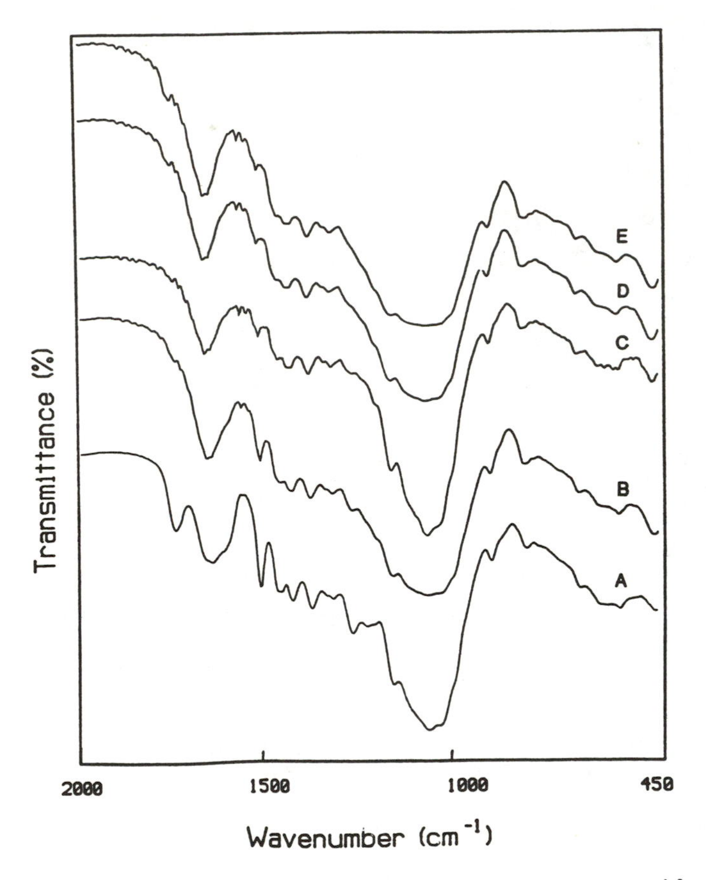

Figure 3. Infrared spectra of red spruce (A), and red spruce residues after 1 h (B), 2 h (C), 3 h (D), and 5 h (E) methylamine extraction at 185°C, 275 bar and 1 g/min solvent flow rate.

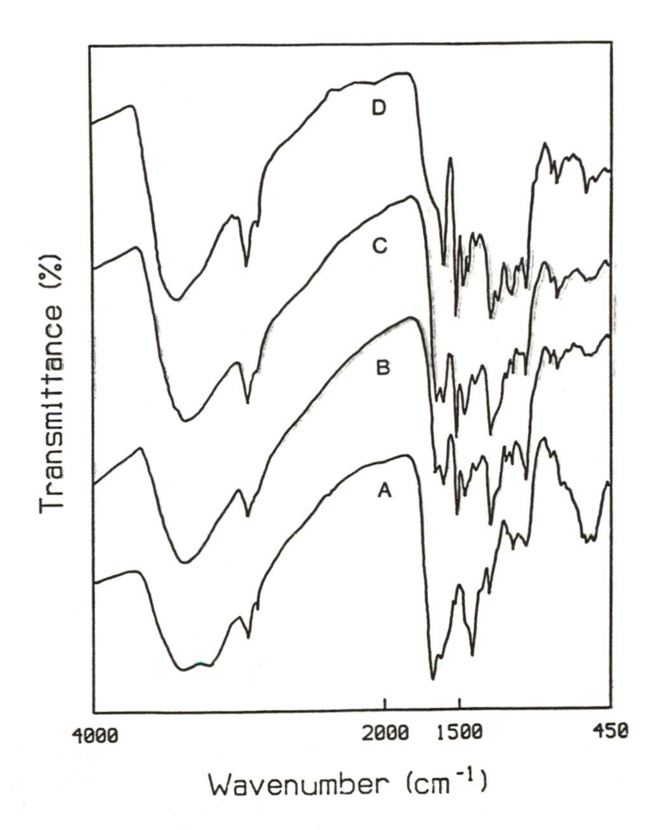

Figure 4. Infrared spectra of kraft lignin, Indulin AT (D), and the precipitates collected in the first trap (at 1 bar) after methylamine extraction of Indulin AT (C), methylamine extraction of red spruce (B), and ammonia extraction of red spruce (A). Extraction conditions: 2 h extraction at 185 °C, 275 bar, and 1 g/min solvent flow rate.

precipitate from ammonia extraction displays a different absorbance pattern, indicating that this precipitate may be a mixture of dissolved lignin as well as carbohydrates from red spruce. Further details of extraction results with other fluids and lignocellulosic materials and the results of chemical analyses are presented elsewhere (11,12).

Having identified the precipitate from methylamine extraction of red spruce as being primarily extracted lignins, we investigated their molecular weights and molecular weight distributions by GPC. All GPC results in this paper are presented on the basis of relative sample elution volume (indicative of molecular weight) and relative elution volume range and intensity (indicative of molecular weight distribution). The sample chromatograms presented in Figures 5 to 10 are traces of the UV absorbance detected at 268 nm. These chromatograms show the effects of extraction time, temperature, pressure, and composition on the molecular weight of the extracted lignin.

Figure 5 shows the comparative chromatograms of five lignin samples obtained from methylamine extraction of red spruce. The extraction conditions were maintained at 185°C, 275 bar, 1 g/min solvent flow rate, except that extraction time was varied from 0.5 h to 5 h. As the extraction time increases, the detector signal intensity corresponding to lower molecular weight range (at higher elution volume) becomes more distinct, and the base of the sample peaks extends more to the left (indicated by small arrows). These changes in the chromatograms are indicative of an increase in the relative fraction of both small and large sized molecules, which lead to a broader MWD. Thus, longer extraction time with methylamine appears to produce lignins with broader MWD.

The effect of extraction temperature on the MWD's of the extracted lignins is shown in Figure 6. The lignin samples were obtained from methylamine extraction of red spruce at temperatures ranging from 175 to 185°C, while maintaining other extraction conditions at 275 bar, 3 h, and 1 g/min solvent flow rate. The figure shows that the MWD's of the lignins become broader with increasing extraction temperature. In a similar way, an increase in extraction pressure leads to a broader MWD of the lignins (Fig. 7). These lignins were obtained from methylamine extraction of red spruce at 185°C. The extraction time was 3 h at 1 g/min solvent flow rate in the pressure range from 172 to 275 bar.

Figure 8 shows the comparative chromatograms of lignin samples obtained from extraction of red spruce and methylamine-nitrous oxide binary mixture at five different compositions (0.2, 0.4, 0.6, 0.8, 1.0 weight fraction of methylamine, at 185°C, 275 bar, 2 h, and 1 g/min solvent flow rate). The apparent MWD's of the lignins from the binary solvent extractions are narrower than those of the lignins obtained with pure methylamine extraction. On the other hand, lignins from pure methylamine extraction appear to have the largest average molecular weight among the lignins shown in Figure 8.

To compare lignins produced from supercritical fluid extraction of wood with lignins from conventional pulping processes, a kraft pine lignin

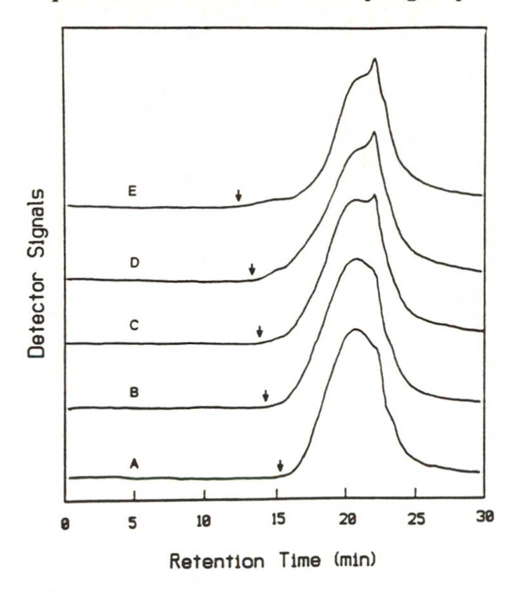

Figure 5. GPC analyses of the precipitates collected in the first trap (at 1 bar) after 0.5 h (A), 1 h (B), 2 h (C), 3 h (D), and 5 h (E) extraction of red spruce with methylamine at 185°C, 275 bar, and 1 g/min flow rate.

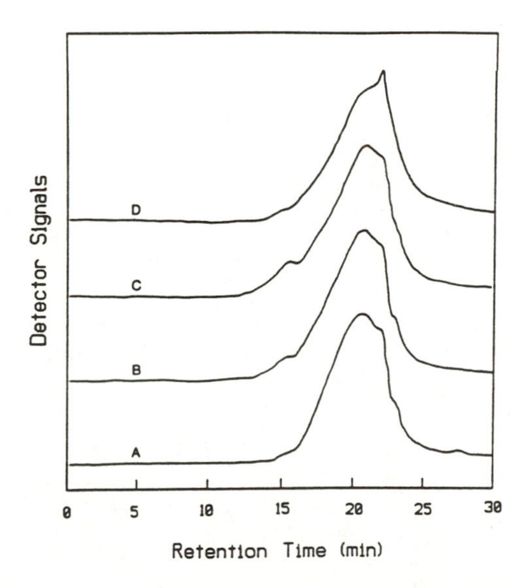

Figure 6. GPC analyses of the precipitates collected in the first trap (at 1 bar) after extraction of red spruce with methylamine at 170°C (A), 175°C (B), 180°C (C), 185°(D). Other conditions maintained at 3 h extraction time, 275 bar, and 1 g/min solvent flow rate.

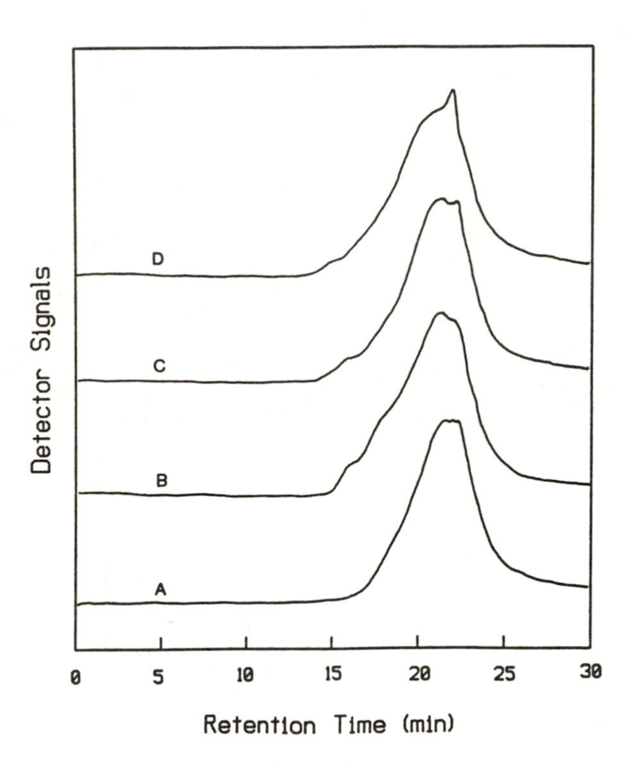

Figure 7. GPC analyses of the precipitates collected in the first trap (at 1 bar) after extraction of red spruce with methylamine at 172 bar (A), 207 bar (B), 241 bar (C), 275 bar (D). Other conditions maintained at 3 h extraction time, 185°C, and 1 g/min solvent flow rate.

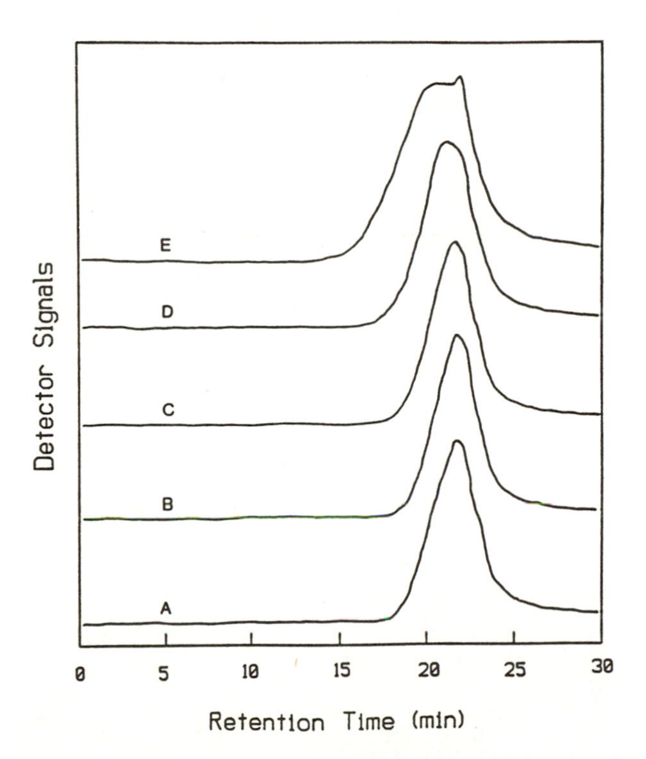

Figure 8. GPC analyses of the precipitates collected in the first trap (at 1 bar) after extractions of red spruce with binary mixtures of methylamine-nitrous oxide at methylamine weight fractions of 0.2 (A), 0.4 (B), 0.6 (C), 0.8 (D), 1.0 (E). Other conditions maintained at 2 h extraction time, 185°C, 275 bar, and 1 g/min solvent flow rate (The extraction conditions are above the critical T and P of the binary mixtures; see ref. 34.)

(Indulin AT from Westvaco) and the precipitate from methylamine extraction of Indulin AT were also analyzed with GPC. As shown in Figure 9, the precipitates from 1 h, 3 h, and 5 h methylamine extraction of Indulin AT at 185°C, 275 bar, and 1 g/min solvent flow rate do not show an appreciable difference in their apparent MWD's from that of Indulin AT. This indicates that no significant chemical transformations that may alter molecular weight distribution of Indulin AT occur during methylamine extractions.

In Figure 10, the chromatograms of the precipitates from methylamine and methylamine-nitrous oxide extraction of red spruce, and methylamine extraction of Indulin AT are compared along with the chromatogram of Indulin AT. The lignins obtained from methylamine extraction of red spruce has the broadest apparent MWD. Extraction with methylamine-nitrous

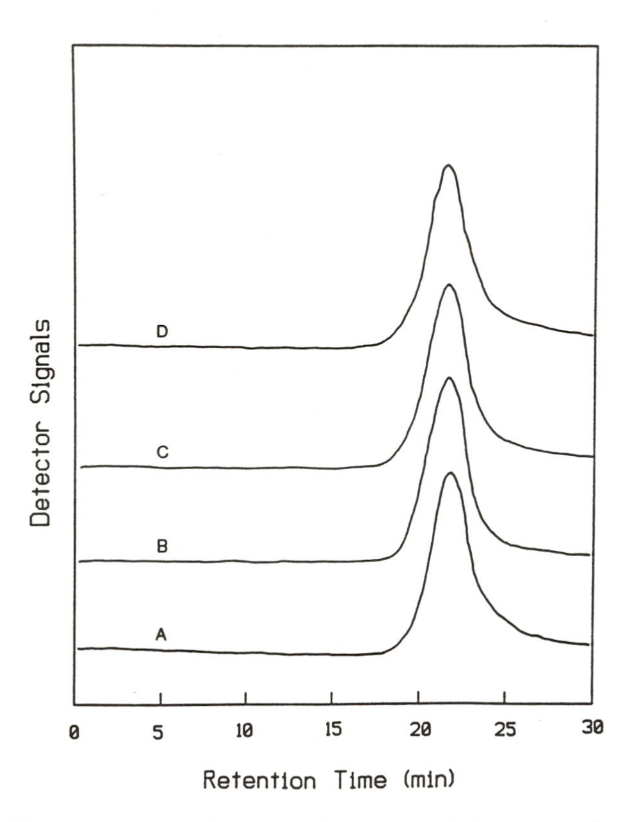

Figure 9. GPC analyses of kraft pine lignin, Indulin AT (A), and the precipitates collected in the first trap (at 1 bar) after 1 h (B), 3 h (C), and 5 h (D) extraction of Indulin AT with methylamine at 185 °C, 275 bar, and 1 g/min solvent flow rate.

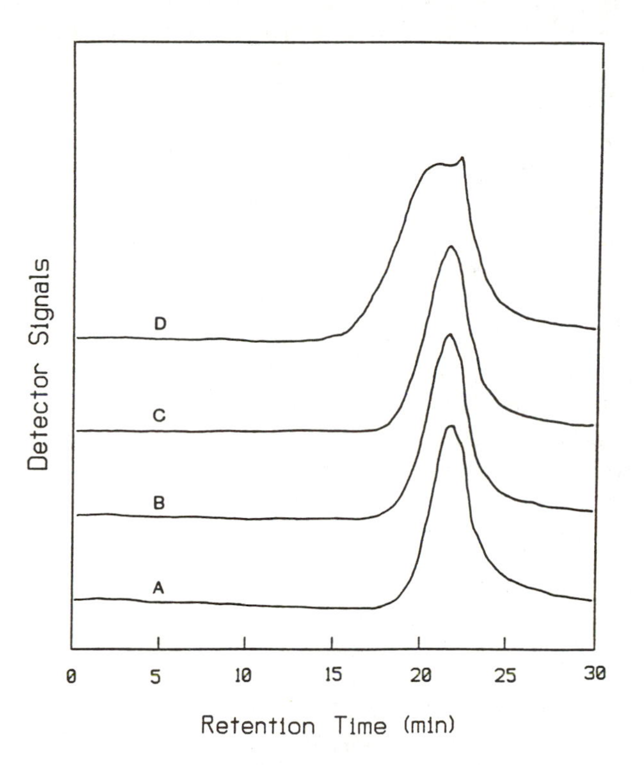

Figure 10. GPC analyses of Indulin AT (A), and the precipitates corresponding to chromatogram C in Figure 9 (B), chromatogram C in Figure 8 (C), and chromatogram D in Figure 5 (D).

oxide binary mixture produces lignins which display an apparent MWD similar to that of Indulin AT. These observations are important in that by manipulating the composition of the extraction fluid, the molecular weight distribution of the lignin can be altered and possibly regulated.

Conclusions

The results that have been presented indicate that supercritical fluid extraction can achieve not only separation of lignin from wood, but also may permit control of the lignin molecular weight and MWD by manipulation of extraction temperature, pressure, time, or the composition of the extraction solvents. Among these extraction variables, the influence of extraction time and solvent composition are greater. The lignins produced from methylamine extraction of red spruce generally show higher average molecular weight than that of kraft pine lignin (Indulin AT). In the binary (methylamine-nitrous oxide) solvent system, lignins with narrower MWD are obtained with an increase in nitrous oxide fraction. An increase in extraction time, pressure or temperature tends to broaden the MWD of the lignins obtained from either methylamine or methylamine-nitrous oxide extraction.

Acknowledgments

This research has in part been supported by the National Science Foundation (Grant No. CBT-8416875).

Literature Cited

1. Schneider, G. M.; Stahl, E.; Wilke, G., Eds. *Extraction with Supercritical Gases*; Verlag Chemie: Deerfield Beach, FL, 1980.
2. Williams, D. F. *Chem. Eng. Sci.* 1981, **36**, 1769.
3. Randall, L. G. *Sep. Sci. Technol.* 1982, **17**, 1.
4. Paulaitis, M. E.; Krukonis, V. J.; Kurnik, R. T.; Reid, R. C. *Rev. Chem. Eng.* 1983, **1**, 178.
5. Paulaitis, M. E.; Penninger, J. M. L.; Grey, R. D.; Davidson, P., Eds. *Chemical Engineering at Supercritical Conditions*; Ann Arbor Science: Ann Arbor, MI, 1983.
6. Penninger, J. M. L.; Radosz, M.; McHugh, M. A.; Krukonis, V. J., Eds. *Supercritical Fluid Technology*; Elsevier: New York, 1985.
7. McHugh, M.; Krukonis, V. *Supercritical Fluid Extraction, Principles and Practice*; Butterworths: Boston, 1986.
8. Stahl, E.; Quirin, K. W.; Gerard, D. *Dense Gases for Extraction and Refining*; Springer-Verlag: New York, 1988.
9. Kiran, E. *Tappi* 1987, **70**, 23-24.
10. Kiran, E. *Am. Papermak.* 1987, **11**, 26-28.
11. Li, L.; Kiran, E. Paper presented at AIChE Ann. Meet., New York, NY, Nov., 1987.
12. Li, L.; Kiran, E. *Ind. & Eng. Chem.-Res.* 1988, **27**, 1301-12.
13. Kringstad, K. P.; Månsson, P.; Mörck, R. *SPCI Intl. Symp. Wood % Pulp. Chem.* (Stockholm) Preprints 1981, **5**, 91-93.
14. Kolpak, F. J.; Cael, J. J. *Proc. TAPPI Ann. Meet.* (Atlanta) 1983, p. 55.
15. Kolpak, F. J.; Cietek, D. J.; Fookes, W.; Cael, J. J. *Proc. IPUAC Macromol. Symp.* (Amherst, MA) 1982, p. 449.
16. Wagner, B. A.; To, T.; Teller, D. E.; McCarthy, J. L. *Holzforschung* 1986, **40**, Suppl., 67-73.
17. Schuerch, C. *J. Amer. Chem. Soc.* 1952, **74**, 5061-67.
18. Brown, W. J. *Appl. Polym. Sci.* 1967, **11**, 2381-96.
19. Kim, H. Ph.D. Thesis, Univ. of Maine, Orono, ME, 1985.
20. Simonson, R. *Svensk Papperstidn.* 1967, **21**, 711-14.
21. McNaughton, J. G.; Yean, W. Q.; Goring, D. A. I. *Tappi* 1967, **50**, 548-52.
22. Gupta, P. R.; McCarthy, J. L. *Macromolecules* 1968, **1**, 236-44.
23. Pellinen, J.; Salkinoja-Salonen, M. *J. Chromat.* 1985, **322**, 129-38.
24. Hütterman, A. *Holzforschung* 1978, **32**, 108-11.

25. Connors, W. J. *Holzforschung* 1978, **32**, 145-47.
26. Connors, W. J.; Sarkanen, S.; McCarthy, J. L. *Holzforschung* 1980, **34**, 80-85.
27. Marchessault, R. H.; Coulombe, S.; Morikawa, H.; Robert, D. *Can. J. Chem.* 1982, **60**, 2372-82.
28. Lewis, N. G.; Goring, D. A. I.; Wong, A. *Can. J. Chem.* 1983, **61**, 416-420.
29. Faix, O.; Lang, W.; Salud, E. C. *Holzforschung* 1981, **35**, 3-9.
30. Himmel, M.; Oh, K.; Sopher, D. W.; Chum, H. L. *J. Chromat.* 1983, **267**, 249-65.]
31. Chum, H. L.; Johnson, D. K.; Tucker, M. P.; Himmel, M. E. *Holzforschung* 1987, **41**, 97-108.
32. Sarkanen, S.; Teller, D. C.; Hall, J.; McCarthy, J. L. *Macromolecules* 1981, **14**, 426-34.
33. Glasser, W. G.; Barnett, C. A.; Muller, P. C.; Sarkanen, K. V. *J. Agric. Food Chem.* 1983, **31**, 921-30.
34. Li, L.; Kiran, E. *J. Chem. & Eng. Data* 1988, **33**, 342-44.
35. Effland, M. J. *Tappi* 1977, **60**, 143.
36. Sarkanen, K. V.; Ludwig, C. H., Eds. *Lignins: Occurrence, Formation, Structure and Reactions*; Wiley: New York, 1971.
37. Fengel, D.; Wegener, G. *Wood: Chemistry, Ultrastructure, Reactions*; Walter de Gruyter: New York, 1984.

RECEIVED March 17, 1989

Chapter 5

Determination of a Polymer's Molecular Weight Distribution by Analytical Ultracentrifugation

John J. Meister[1] and E. Glen Richards[2]

[1]Department of Chemistry, University of Detroit, Detroit, MI 48221–9987
[2]Research Division, Veterans Administration Hospital, 4500 South Lancaster Road, Dallas, TX 75216

A method has been developed for the calculation of detailed and absolute molecular weight distributions for complex polymer samples. The method requires that a sedimentation velocity experiment be performed on at least three dilute concentrations of the polymer in solvent. Rayleigh interference pattern photos must be taken at equal times during the experiment. Three fringes of the interference pattern are digitized, averaged, and fitted with a polynomial. The derivative of the resulting curve, when combined with a function obtained from the Mark-Houwink equation, produces a differential molecular weight distribution. The distribution can be used to calculate any molecular weight average or moment of the distribution. Application of the method to a commercial poly(1-amidoethylene) standard gave a limiting viscosity number that matched the experimental value to within 2 percent and weight average molecular weight that bracketed the value claimed by the supplier.

Many physical or chemical properties, such as heat of reaction or index of refraction, are single valued. However, some properties are multiple valued and extend over a range of temperature, energy or mass. Examples of this phenomenon are glass transition temperature, bond energies, or molecular weight. These properties are called "distributed" and occur frequently in complex chemicals such as polymers. These chemicals have a distribution of values for melting point, reaction rate, or molecular weight and each such property must be represented by a "distribution." This paper describes how to determine such a distribution for the molecular weight of a homopolymer composed of a mixture of molecules having different numbers of repeat units in them.

0097–6156/89/0397–0058$06.75/0

In order to describe how this procedure works, distributions will first be discussed in more detail and then molecular weight distributions will be briefly discussed. Many properties of samples are actually a distribution of values. In a powdery, crystalline solid, individual crystals come in different sizes. Further, different measures of this "crystal size" control common behaviors of the crystalline sample. A single dimension of the crystals, a length or diameter, will be a good measure of filterability of the crystals. The square of this dimension, which is proportional to *surface area*, will correlate with and describe the rate of solution of the crystals. The cube of this dimension, a function of *volume* of the crystals, will specify how a crystal will sediment under a gravitational force since volume times density gives crystal mass. The *distribution of lengths cubed* will tell how the *total* sample will sediment. Thus, the rate at which a crystalline sample settles in a vessel or under a centrifugal force will be given by a sample-specific curve which shows, for all crystals, how many crystals of a given volume or, equivalently, of a given length cubed there are in this particular sample.

Macromolecules are very much like the crystalline powder just described. A few polymers, usually biologically-active natural products like enzymes or proteins, have very specific structure, mass, repeat-unit sequence, and conformational architecture. These biopolymers are the exceptions in polymer chemistry, however. Most synthetic polymers or storage biopolymers are collections of molecules with different numbers of repeat units in the molecule. The individual molecules of a polymer sample thus differ in chain length, mass, and size. The molecular weight of a polymer sample is thus a distributed quantity. This variation in molecular weight amongst molecules in a sample has important implications, since, just as in the crystal dimension example, physical and chemical properties of the polymer sample depend on different measures of the molecular weight distribution.

It is usually not convenient to use an entire curve of values to give a molecular weight of a sample, however. There are different measures of a molecular weight distribution that provide a formula to collapse the distribution into a single number. These formulas are usually called molecular weight averages and are useful in predicting a polymer's properties, such as viscosity, tensile strength, or viscoelasticity. The formulas which provide one number to represent a distribution, give *averages* of the distribution in the same units as the property. These are *not* methods to calculate moments of the distribution since a distribution moment has different units than the distributed property.

The first average in common use is number-average molecular weight. This number is obtained by dividing the total mass of the sample by the number of molecules in it. The formula for this average is

$$\overline{M}_n = \sum_{i=1}^{n^*} m_i N_i \Big/ \left(\sum_{i=1}^{n^*} N_i \right)$$

where $\overline{M}_n$ is the notation for number-average molecular weight, n^* is the largest number of repeat units in any molecule of the sample, N_i is the number of molecules with i repeat units in the molecule, and m_i is the mass of a molecule with i repeat units. Ranking samples of a polymer by number average molecular weight will also rank the samples by how much each would decrease the freezing point of a solvent, how much each would increase the boiling point of a solvent, or how much osmotic pressure a solution of the sample would generate when placed inside a semipermeable membrane. These properties are all functions of the free energy of a solution. The capacity of $\overline{M}_n$ to rank samples of a polymer in order of magnitude of change produced in these properties shows that the number of polymer molecules in a molecular weight distribution controls solution free energy. Number of molecules in the molecular weight distribution thus is a controlling variable for the thermodynamics of polymer solutions.

The second average is viscosity average molecular weight, $\overline{M}_v$. This expression is obtained by using the exponent from the limiting viscosity number-molar mass relationship, α, as a power for the molecular weight of each molecule in the distribution. The formula for this average is

$$\overline{M}_v = \left(\sum_{i=1}^{n^*} m_i^{(1+\alpha)} N_i \Big/ \left(\sum_{i=1}^{n^*} N_i m_i \right) \right)^{1/\alpha}$$

This average shows how the distribution of molecules would affect the volume fraction that is polymer in a polymer solution and control flow in a melt of the polymer sample.

The third average is weight-average molecular weight, $\overline{M}_w$. This measure is obtained by summing up the mass of each molecule times the weight fraction of that molecule in the sample.

$$\overline{M}_w = \sum_{i=1}^{n^*} m_i^2 N_i \Big/ \left(\sum_{i=1}^{n^*} N_i m_i \right)$$

By having each molecule's mass contribute in proportion to the weight fraction of the sample, the value of this average changes more when the number of high molecular weight molecules in the distribution changes, as compared to the preceding two averages. The weight average of a distribution will give a molecular weight which will rank how samples of a polymer sediment in a gravitational field or scatter light from solution.

A fourth average molecular weight is the zeta-average, $\overline{M}_z$. While this formula can be written in terms of weight fractions of molecules in a sample, this formulation of the average gives none of the physical significance produced with the weight-average. The formula for zeta average molecular weight is

$$\overline{M}_z = \sum_{i=1}^{n^*} m_i^3 N_i \Big/ \left(\sum_{i=1}^{n^*} N_i m_i^2 \right)$$

The value of this average is controlled by the high molecular weight end of the polymer distribution. $\overline{M}_z$ ranks the viscoelastic properties of a sample and gives the effect of molecular weight distribution on recoverable compliance (capacity to recover shape, (1)) through the polydispersity factor, $(\overline{M}_z\overline{M}_{z+1})/(\overline{M}_w)^2$. The magnitude of the different molecular weight averages given above will, for a given sample, increase in the order in which they appear above.

As this discussion of molecular weight distributions and their averages indicates, the properties of a polymer are direct functions of the way the individual molecular masses are distributed in a sample. The molecular weight distribution and its averages are critical to understanding the nature and behavior of polymers. Most important for the results of this paper, all of the above averages can be calculated from the molecular weight distribution that can be determined by analysis of the Rayleigh interference pattern produced from a sedimentation velocity experiment. The procedure for obtaining such a distribution will now be described.

Ultracentrifugal Sedimentation

The sedimentation velocity experiment is conducted with the macromolecular solution and solvent placed in separate sector-shaped chambers in the cell of an analytical ultracentrifuge rotor. The rotor is accelerated to the desired speed and sedimentation of the solute is allowed to occur. All of the molecules sediment to the centrifugal "bottom" of the cell. After a certain time the meniscus is cleared of solute, thereby creating a "boundary" in the solution with a flat plateau region of uniform concentration and a thin layer of sedimented solute against the cell bottom. The shape of the boundary is determined by diffusion of the solute, concentration-dependence of sedimentation, the Johnston-Ogston effect and heterogeneity of the solute (2,3). During the progression of the experiment, the plateau becomes more dilute because of diffusion and the sector-shaped cell width and centrifugal field increasing with radius.

A mathematical expression for the results of a typical experiment is shown in Figure 1. The simplest experiment is to obtain an average sedimentation coefficient by measuring the movement of the boundary, r_n, with time. From the definition of the sedimentation coefficient,

$$s = \frac{1}{\omega^2 r}\frac{\partial r}{\partial t} = \frac{1}{\omega^2}\frac{\partial ln\ r}{\partial t} \tag{1}$$

where s is the sedimentation coefficient, $\omega = 2\pi(RPM)/60$, r is the radius measured from the center of rotation, and t is the time. The sedimentation coefficient has units of time (sec) and is measured in units of svedbergs (s) where 1 s = 10^{-13} sec. The sedimentation coefficient is calculated from the slope of a plot of the logarithm of boundary position versus time. Usually the boundary position is assumed to be the position of the maximum ordinate of the peak that is obtained with an optical system which records the derivative pattern of the refractive index. Using a more sophisticated treatment of points along the boundary, one can obtain the weight-average sedimentation coefficient for the solute at the concentration corresponding to the plateau region.

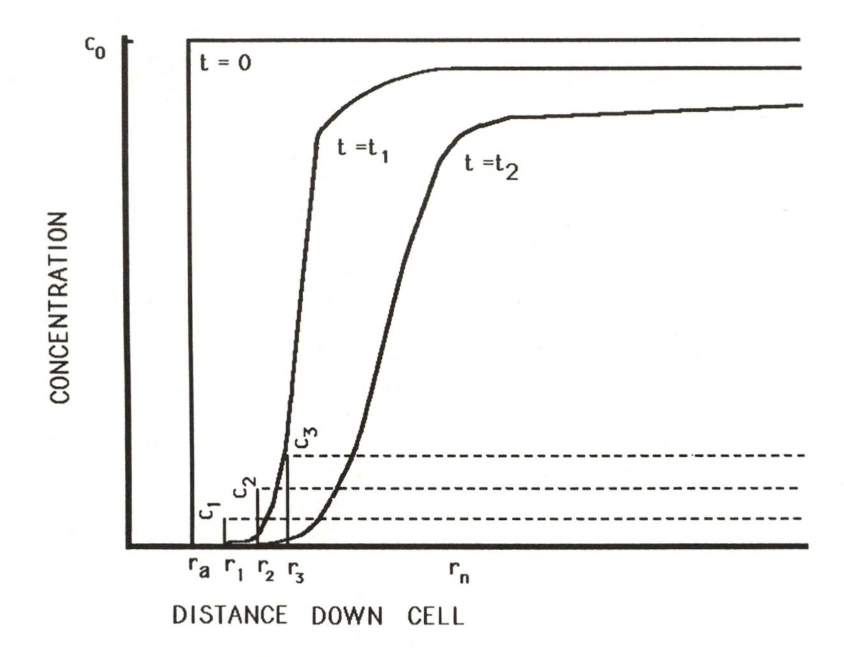

Figure 1. A diagram of several Rayleigh fringes placed on an image of the centrifugation cell. Distortion of the fringe from a straight line is caused by a concentration gradient produced by spinning the cell about the point $r = 0$.

Since the theories for the interpretation of sedimentation velocity experiments require the sedimentation coefficient at zero concentration, a series of experiments is performed at different concentrations. The value at zero concentration (or "limiting sedimentation coefficient"), s_0, is obtained from plots using either of these equations,

$$1/s = (1 + k_s c)/s_0 \tag{2}$$

$$s = s_0(1 - k_s c) \tag{3}$$

where k_s is a positive term that represents the given system. Generally, the former equation covers a wider concentration range.

The molecular weight M_r of a macromolecule can be calculated from the sedimentation coefficient using the Svedberg equation,

$$M_r = Nsf/(1 - \overline{v}\rho) \tag{4}$$

where N is Avogadro's number, f is the frictional coefficient, $\overline{v}$ is the partial specific volume of the solute, and ρ is the density of the solution. If the values of all the parameters are known, one can calculate the molecular weight for a homogeneous material. If the macromolecule is heterogeneous, an indeterminate average is obtained.

During the 1950's, a number of theoretical methods were put forth for the determination of sedimentation distributions of heterogeneous macromolecules, neglecting or including the effect of diffusion. For macromolecules of sufficient size that diffusion is negligible, it is sufficient to perform several velocity experiments at different concentrations and somehow extrapolate "corresponding" points on the pattern to zero concentration. Several approaches made use of parameters more easily obtainable with the Schlieren optical system generally used at that time, but they are of uncertain validity (2-4). With concentration data obtained from the Rayleigh optical system, the extrapolation procedure based upon relative concentration is reasonable and the necessary data manipulations are "easy" to perform. Relative concentrations can be determined because the Rayleigh optical system corrects for solvent component gradients or pressure gradient effects because it records index of refraction differences between sedimenting solvent and solution.

Before continuing, it is important to understand the complexity of the processes that warp the shape of the boundary, even if diffusion is negligible. The Johnston-Ogston effect will be described with the aid of Figure 1. In the figure, the pattern at $t = t_2$ has been divided into three components in the lower concentration zone. The first component (between r_a and r_1) starts at $C = 0$, rises to C_1 as a step and continues to the cell bottom. The second component starts at a concentration of zero and rises at the second step to C_2, with the total concentration beyond the step being $C_1 + C_2$. Since the material in the first step sedimenting beyond the second step moves at a slower rate (because of the higher concentration), more material enters the second step than leaves it. Thus, the magnitude of the first step increases; in other words, its apparent concentration is higher. The effect moves along the boundary, the net effect being an upward shift of the pattern. However, the total concentration remains unchanged (4).

The concentration dependence of sedimentation causes an additional reshaping of the boundary, called boundary sharpening. The faster moving material is at a higher concentration than the slower material, with the result that the boundary is progressively sharpened as the concentration increases; that is, it shifts to the left. The dependence of sedimentation constant on concentration and on the concentration of other molecules around the sedimenting molecule, which causes the above effects, are too difficult to measure. In some systems with spherical macromolecules, approximate corrections have been made (5). The extrapolation of s to zero polymer concentration should eliminate the major part of these effects.

Conversion of Sedimentation Patterns to Distribution Patterns

The most common symbolism for characterizing sedimentation distribution patterns is the use of G(s) (where G(s) represents the weight fraction of material having a sedimentation coefficient less than or equal to the given value) for the integral distribution, and g(s) for the differential distribution, where g(s) = dG(s)ds. The integral distribution is simple to obtain from Rayleigh-fringe data. We will assume that the concentration of polymer is proportional to refractive index, which the optical system records.

First the radial position is converted to units of sedimentation coefficient. This is accomplished by using the integrated form of Equation 1.

$$s = \frac{1}{\omega^2 t} ln(r_i/r_a) \tag{5}$$

where r_a is the position of the meniscus at the top (low r) of the solution. Equation 5 relates sedimentation coefficients to a ratio of radial distance along the cell, r/r_a, and time, t, at angular velocity, ω. This equation allows position along the cell to be plotted as s instead of r. The number of fringes, J, seen in the interference pattern is a function of solute concentration, c_2, specific refractive index increment, $\partial n/\partial c_2$, cell thickness along the optical path, a, and wavelength of the incident light, λ:

$$J = \frac{a(\partial n/\partial c_2)c_2}{\lambda} \tag{6}$$

Equation 6 allows the number of fringes to be converted to standard concentration units. The concentration values in fringes must be corrected for radial dilution. The number of data points n serves to divide the profile into zones. Each zone, $\Delta C_i = C_i - C_{i-1}$, is multiplied by x_i^2/x_a^2, where $x_i = (x_i + x_{i-1})/2$ and each x is the distance down the fringe pattern recorded on film that corresponds to a given r position in the spinning cell. The notation x_a thus denotes the position of the meniscus. The corrected relative concentration CR_i is

$$CR_i = \sum_{j=1}^{i} \Delta C_j (x_j^2/x_a^2)/C_n \tag{7}$$

where C_n is the corrected concentration in the plateau region, namely

$$C_n = \sum_{k=1}^{n} \Delta C_k (x_k^2/x_a^2) \tag{8}$$

At any i, CR_i is, of course, a weight fraction concentration of solute based on or referenced to the concentration of the plateau region. When considered as a weight fraction concentration, CR_i will be denoted as w_i. A plot of w_i versus s is the integral molecular weight distribution for a sample.

To obtain the distribution extrapolated to zero concentration, the distribution at each concentration is divided into a number of zones within the weight fraction zone 0 to 1. Then for each zone a plot of s or $1/s$ versus the sample concentration is made and extrapolated to obtained the sedimentation coefficient at zero concentration, s_0 . A plot of weight fraction versus s_0 is the corrected integral distribution at zero concentration. The differential distribution, $\partial c/\partial s$, can be obtained by fitting groups of points with a sliding least mean squares cubic fit.

The conversion of the sedimentation distribution to a molecular weight distribution using the Svedberg equation requires knowledge of the frictional coefficient and partial specific volume of each weight fraction. For a synthetic linear polymer, the partial specific volumes of all species should be the same, which leaves the frictional coefficient to be estimated from a random-coil model. One approach is to use the relation containing the sedimentation coefficient, limiting viscosity and molecular weight (2,6),

$$\frac{s[\eta]^{1/3}}{M^{2/3}} = \Phi^{1/3}P^{-1}\frac{(1-\overline{v}\rho)}{\eta N} \tag{9}$$

where $[\eta]$ is the limiting viscosity, η is the viscosity of the solvent and $\Phi^{1/3}P^{-1}$ is a constant of value 2.5×10^6.

Combining Equation 9 with the Mark-Houwink equation (7, 8),

$$[\eta] = K'\overline{M}^{\alpha} \tag{10}$$

gives

$$\overline{M}_x = (ks)^{3/(2-\alpha)} \tag{11}$$

where $\overline{M}_x$ is a molecular weight average determined by the standard molecular weights used to determine K' and α in Equation 10 and

$$k = \frac{K'^{1/3}\eta N}{\Phi^{1/3}P^{-1}(1-\overline{v}\rho)} \tag{12}$$

Note the difference between K', the Mark-Houwink constant, and k defined in Equation 12. The differential sedimentation distribution can be transformed to the differential molecular weight distribution by taking the derivative of Equation 11 with respect to s, to get

$$\frac{\partial M}{\partial s} = \frac{3k^{3/(2-\alpha)}}{(2-\alpha)} s^{(1+\alpha)/(2-\alpha)} \quad (13)$$

with the result that

$$\frac{\partial c}{\partial M} = \frac{(\partial c/\partial s)}{(\partial M/\partial s)} \quad (14)$$

Note the implications of Equation 14. If the change in amount of polymer with respect to the molecular weight of that polymer, $\partial c/\partial M$), is needed, it can be found in a two step process. First determine $(\partial c/\partial s)$ and then determine $(\partial M/\partial s)$. The function $(\partial c/\partial M)$ is often desired since it is the plot of the number of molecules of a given molecular weight versus molecular weight, or the *differential molecular weight distribution*. Further, $(\partial c/\partial s)$ is the slope of the Rayleigh interference pattern in a sedimentation velocity experiment. If this slope is quantified and the value of $(\partial M/\partial s)$ is obtained from Equation 13, then the differential molecular weight distribution can be found for a given sample. Details of this procedure are given here.

Experimental

Viscometry. Viscosities of aqueous polymer solutions were measured using a Cannon-Fenske #50 viscometer immersed in a 20°C water bath. The limiting viscosity number was determined from 5 viscosity measurements using the Huggins equation (9). The limiting viscosity number of aged poly(1-amidoethylene) in 0.01 M aqueous Na_2SO_4 was 2.45 dL/g.

Materials and Solution Preparation. The poly(1-amidoethylene) used in all experiments was a "molecular weight standard" supplied by Polysciences, Inc., as material 8249, batch 93-5. The polymer was dried for 3 hr. under a vacuum of < 10 Pa at a temperature of 25°C. The polymer was dispersed on a vortex of 0.01 M Na_2SO_4 solution and stirred for 1 day. The solution was then centrifuged to remove undissolved polymer particles and the preweighed centrifuge tube was dried and reweighed. The polymer concentration of the master sample was calculated from the weight of polymer retained in the solution. The concentration of the master sample was 2,115 ppm or 0.2115 g/dL.

All solutions were stored for 60 days before use. This delay was necessary for accurate results since Narkis (10) had already shown that poly(1-amidoethylene) solutions are *not* molecularly disperse until 54 days after preparation. These solutions age by losing intermolecular entanglements and becoming monomolecular solutions (10). Poly(1-amidoethylene) solutions are known to lose viscosity with time (11). Several authors have attributed this viscosity loss to oxygen or radical degradation of the polymer (11), but Francois (12) has shown that changes in viscosity only occur in solutions made from broad-molecular-weight-distribution poly(1-amidoethylene). Since very narrow-molecular-weight-distribution poly(1-amidoethylene) produces a stable solution viscosity and since Narkis (10) has shown that the original solution viscosity can be obtained by precipitating and redissolving the polymer, it would appear that solution viscosity loss is caused by slow disentangling of a broad-molecular-weight polymer mixture. All solutions used here were aged to insure complete dissolution and uniformity in sampling the actual distribution of the polymer.

Sedimentation Velocity Experiments. The sedimentation velocity experiments were carried out in a Beckman Model E ultracentrifuge equipped with Rayleigh interference optics and a digital electronic system that measures the rotor temperature to 0.01° with control to ±0.03°C (13). The optical system was aligned according to the procedure of Richards *et al.* (14). A Rayleigh mask with 0.4 mm slit width (instead of 0.8 mm) improved the accuracy of fringe measurement by increasing the number of fringes in the pattern. The two sectors of the charcoal-filled Epon centerpiece were filled with a microliter syringe with the solvent side filled sufficiently higher to permit resolution of the solution meniscus on the fringe pattern. The rotor with cell and counterbalance was controlled at 20.0°C in the centrifuge chamber (no vacuum) for at least 10 min. The rotor was accelerated to 5200 rpm, where a baseline photograph (Kodak Metallograph plates) was obtained during a 30 second pause. The rotor was then accelerated at constant amperage to the operating speed of 44,000 rpm, whereupon another baseline picture was taken. During acceleration, additional heating was used to maintain the temperature constant at 20.0°C and the clock measuring the time of sedimentation was started when the rotor reached 2/3 of the final speed.

Sedimentation pictures were taken starting at 20 to 40 min. after clock start and were taken at 20 min. intervals until 160 min. had elapsed. The experiments with the poly(1-amidoethylene) samples were carried out under identical conditions, insofar as possible.

Reading of Rayleigh Fringe Patterns. Fringe patterns were read with a Nikon Profile Projector model 6C equipped with digital micrometers (accurate to 1.25 μm) and a dual-photocell light-difference detector mounted on the screen that locates fringe centers (15). The reading of patterns is semi-automatic, being controlled by an Altair 8080 microprocessor computer with the data recorded on a floppy disc. After alignment of the pattern on the projector stage, the positions of the counterbalance reference edges and solutions menisci are located and recorded.

The Rayleigh fringe pattern that is obtained from the ultracentrifuge using a Rayleigh mask with 0.4 mm slit width has only about nine readable fringes in the y direction. The fringes are equally spaced, except for progressive warpage in the position of fringe centers away from the center of the diffraction envelope; hence, it is necessary to confine readings to the central region of the envelope. However, improved accuracy is achieved by averaging as many fringes as possible, in this case three.

For the reading of the fringe profile (usually the light fringes on the negative photograph) in the solution region of the pattern, the approximate center of the fringe envelope in the y direction is located. The reading of fringes in the y direction is then confined to the central distance corresponding to three fringe spacings (the distance between fringes being about 280 μm). The reading of fringes is begun near the meniscus (Figure 2). The x-axis of the pattern is the long axis in the interference pattern. The distance across the fringe image is the y-axis. At each x position, the position of the fringe centers is transferred to the computer by a signal from the light detector as the stage is moved in the y direction, across the light path. The x and average y values and the fringe spacings are stored by the computer. About 150 points across the boundary and into the plateau are read at equal x increments. As the lower of the three fringes being measured rises to within 1/2 fringe of the center of the envelope (middle of the fringe pattern in the y direction), one drops down one fringe in order to stay in the central region. This need to drop "down" a fringe in order to stay in the y-middle of the fringe pattern is caused by the sigmodal distortion of the Rayleigh fringe pattern caused by the sedimenting boundary. The profile that results from this procedure is discontinuous, but a simple statement in the subsequent computer program converts the data to the correct profile. By measuring the fringe pattern in this way, three high-accuracy fringes are converted to a set of (x, y) data points which are the numerical equivalent of the plot of three fringes from the low r point to the high r point in the spinning cell. This data set can then be smoothed and averaged to give a complete plot of the fringe profile (and the sedimenting boundary which represents c as a function of r) from the meniscus of the centrifuged solution to the bottom of the cell.

The low- and high-speed baseline patterns and the 80 and 100 min. sedimentation patterns for each solution were read in the same manner. The data were transferred to a Radio Shack Model III computer for further treatment.

Calculations

A computer program converted the fringe data into concentration in units of fringes by division of the y values by the average fringe separation. The sedimentation patterns were corrected for differences in optical path through the two cell compartments by subtracting the high speed baseline, using an interpolative procedure. The high or low speed baseline could be used in this correction step. Both baselines were tested and found to give the same results for the higher concentration samples but to differ slightly in the effect on the low concentration sample's patterns. The high speed baseline adjustment was chosen for correcting all data sets to avoid biasing the data. The x values of the corrected patterns were then transformed into distance r from the center of rotation according to the relation

$$r = r_c + (x - x_{ave})/MF \qquad (15)$$

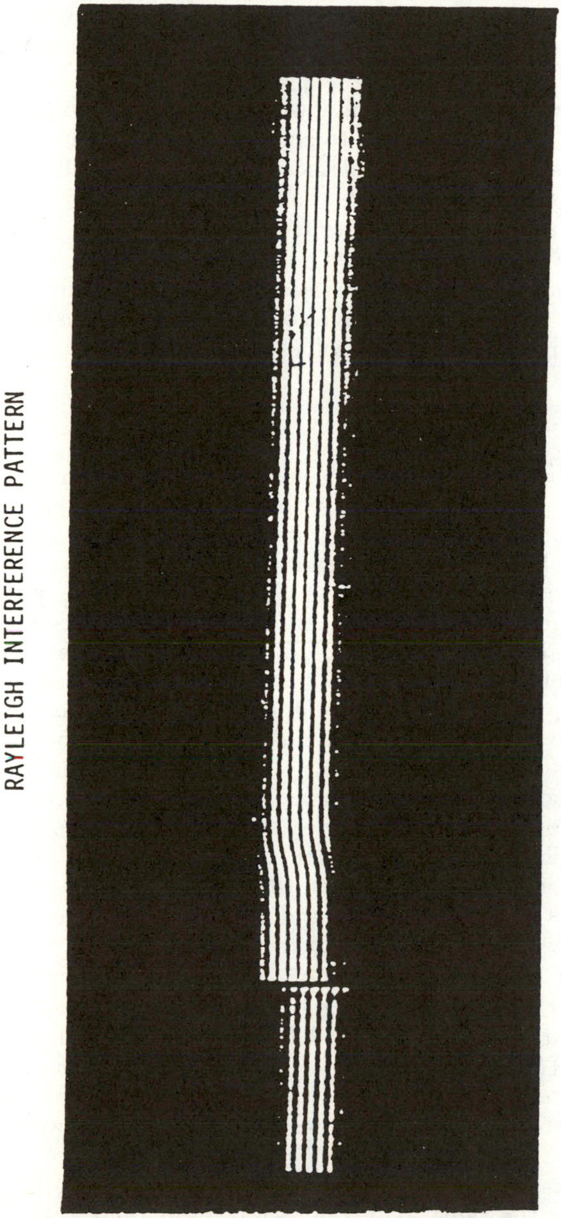

RAYLEIGH INTERFERENCE PATTERN

Figure 2. A print of a photograph of the Rayleigh fringe pattern produced by sedimentation of a graft copolymer of starch from aqueous brine. (For more information see ref. 17.)

where r_c is the distance of the center of the cell from the axis of rotation (6.5 cm), x is the fringe position in micrometer units, x_{ave} is the average of the inner and outer reference positions in micrometer units, and MF is the optical magnification factor. The fringe values are corrected for radial dilution and the radial positions are transformed into sedimentation units as described earlier.

The same program fits the data with a sliding 15 point least mean squares cubic fit, using two passes to smooth the data. The use of fourth or fifth power fits for a variation of the number of points fitted (13 to 21 points) did not significantly improve the results. The average deviation for the fringe value (after averaging three fringes) was about 0.005 fringes and 0.001 fringes for the smoothed values.

Other computer programs were used to plot the distribution profiles, interpolate these curves for extrapolation to zero concentration, generate the molecular weight distribution profiles, and to plot them.

Results

Five solutions of a poly(1-amidoethylene) sample at concentrations ranging from 0.0422 to 0.2111 g/dL were sedimented in the analytical ultracentrifuge. Examination of the patterns at different times revealed that the 80 min. patterns had a definite plateau region, but there was still a small amount of material near the meniscus. For the 100 min. patterns, the plateau was nearing the cell bottom, and the meniscus region was almost clear of solute. Later patterns exhibited a definite region near the meniscus with no solute, but no plateau near the bottom. It is possible to splice the distribution patterns obtained at different times, but errors can lead to a discontinuity in the overlap region. Because the distribution patterns obtained at 80 and 100 min. were nearly the same, it was decided to treat only those and ignore any small amount of material that may not have sedimented: its contribution to the average viscosity and molecular weight should be negligible. After integration of these patterns, the molecular weight distribution curve showed that the material at the boundary had molecular weights of less than 9,000 and the material already sedimented had a molecular weight of 5,000,000 or more.

The starting and ending regions of the integral distribution patterns were examined to determine the fringe values at these levels. The starting value for each pattern was subtracted from the fringe values for the rest of the pattern. The plateau values for the 80 and 100 min. patterns for each solution agreed to within 0.02 fringes. With the zero and plateau concentrations determined, the patterns were converted to weight fraction distributions.

Since the integral distribution patterns for the 80 and 100 min. patterns were nearly the same, only the results from the former are shown in Figure 3. The differential patterns (not shown) obtained from the sliding 15 point least mean squares cubic fit were also nearly the same, with considerable noise for the two solutions of lowest concentration.

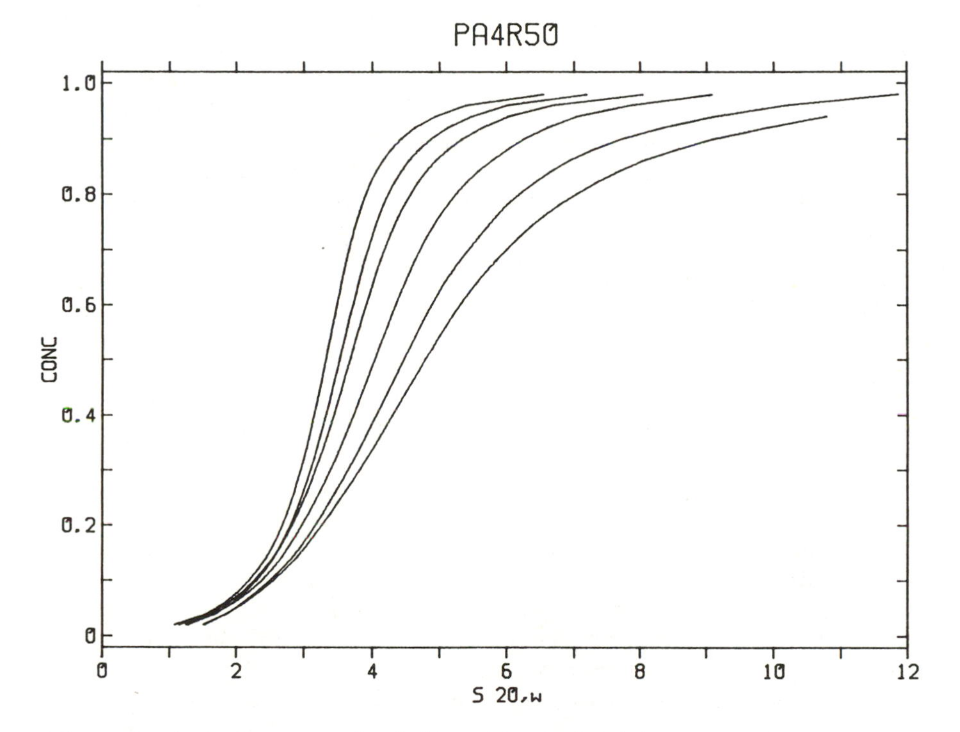

Figure 3. Concentration of poly(1-amidoethylene) in a centrifuged sample plotted as a function of sedimentation coefficient.

The five integral patterns were interpolated to obtain 49 values of the boundary ranging from 0.02 to 0.98 weight fraction of the polymer sample. The extrapolations of $1/s$ versus c to zero concentration for the nine weight-fraction concentration levels ranging from 0.1 to 0.9 are shown in Figure 4. The noise in the position of the data points relative to the line are random. For the corresponding plot for s versus c (not shown) there was definite curvature in the position of the points. The integral plot for the 80 min. pattern is the right-hand lowest line on Figure 3.

The results of the sliding 7 point least mean squares fit are shown in Figure 5. The accuracy of the data is indicated by the closeness of the actual data points (squares) to the smoothed points (continuous line) for the differential distribution curve given in Figure 6. One cannot tell whether the small convolutions which are revealed in the corresponding differential distribution pattern are real, or represent artifacts or errors in the data.

Finally, the integral molecular weight distribution is shown in Figure 7 and differential molecular weight distribution is shown in Figure 8. These transforms were made using Equations 11 and 14 with the Mark-Houwink constants $K' = 6.31 \times 10^{-5}$ and $\alpha = 0.8$ for weight-average, molecular weight samples as measured by Shawki and Hamielec (16). The shape of these curves are what one would expect for a fraction of poly(1-amidoethylene) obtained from a free-radical polymerized polymer. The high molecular weight tail of the distribution is a common and characteristic aspect of this polymer. It should be noted that the transformation of sedimentation coefficient as given in Figures 5 and 6 to molecular weight as given in Figures 7 and 8 generates a profile with shifted position and relative magnitude. The peak at 4s seen in Figure 6 shifts to a molecular weight peak of 220,000 in the molecular weight distribution given in Figure 8. The one s unit displacement to the shoulder at 5s seen in Figure 6, which implies a significant weight fraction of sample sedimenting with coefficients between 4s and 5s, truncates to a minor inflection in the differential molecular weight curve given in Figure 8, while the monotonic tail from 5.5s to 14s in Figure 6 expands to the long 1,000,000 to 4,500,000 molecular weight tail of Figure 8.

The average limiting viscosity number and weight-average molecular weight were calculated by using Equations 10 and 11. These values increase with an increase in the number of zones into which the sedimentation fringe curve is broken, because increasing the zones leads to greater inclusion of a region that is much higher in molecular weight. The values obtained for 10, 20, and 50 zones are shown in Table I for both the 80 and 100 min. patterns. The values for the 80 min. patterns rise more sharply with the number of zones. The sharper rise is due to the fact the earlier patterns reveal the existence of larger molecules that are sedimented past the plateau at a later time. From the data in the table one can estimate an average molecular weight of about 5.5 to 6.5 × 10^5 and a limiting viscosity of 2.2-2.5.

Since increasing the number of zones used to measure the above averages incorporates more of the high molecular weight tail of the polymer sample, the measurement with the highest number of zones should be the

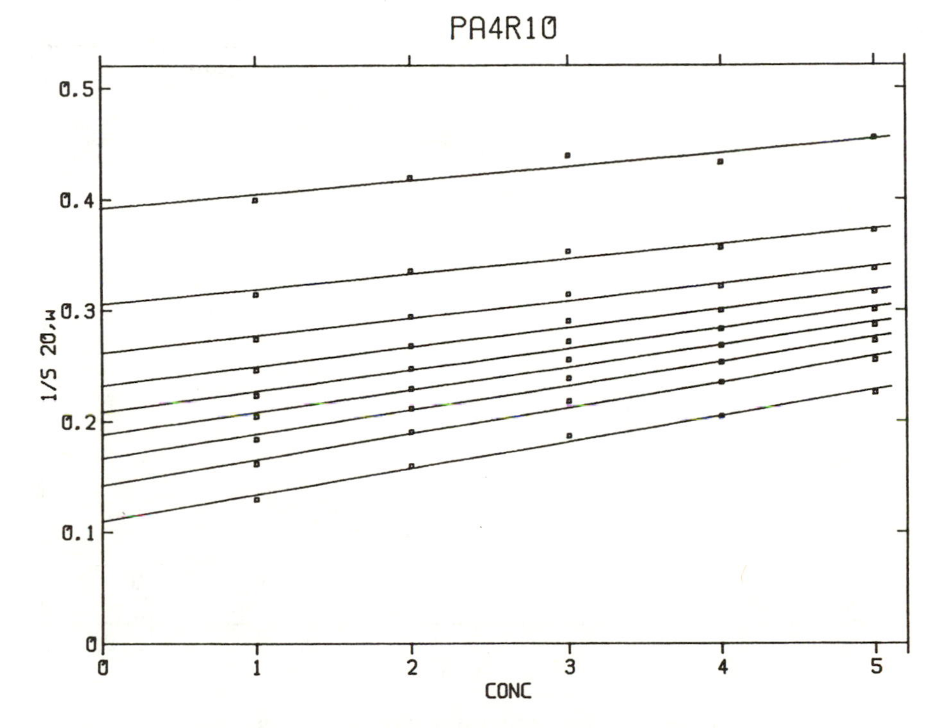

Figure 4. Determination of limiting sedimentation coefficient for 9 weight fractions of the poly(1-amidoethylene) standard based on equation 2.

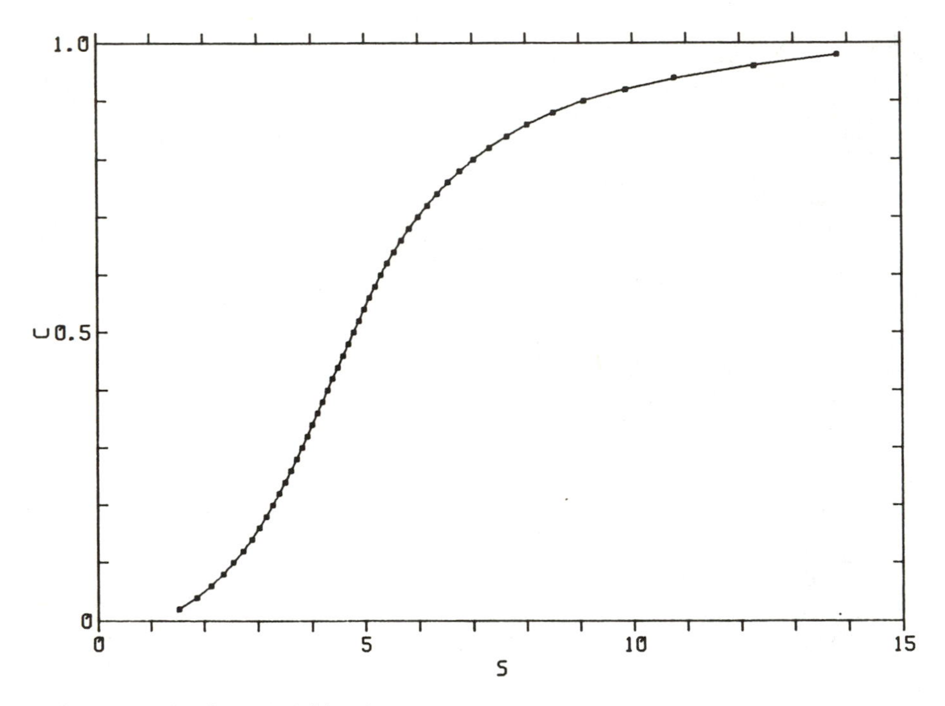

Figure 5. A plot of G(s), the integral sedimentation distribution pattern with *s* shown in svedbergs.

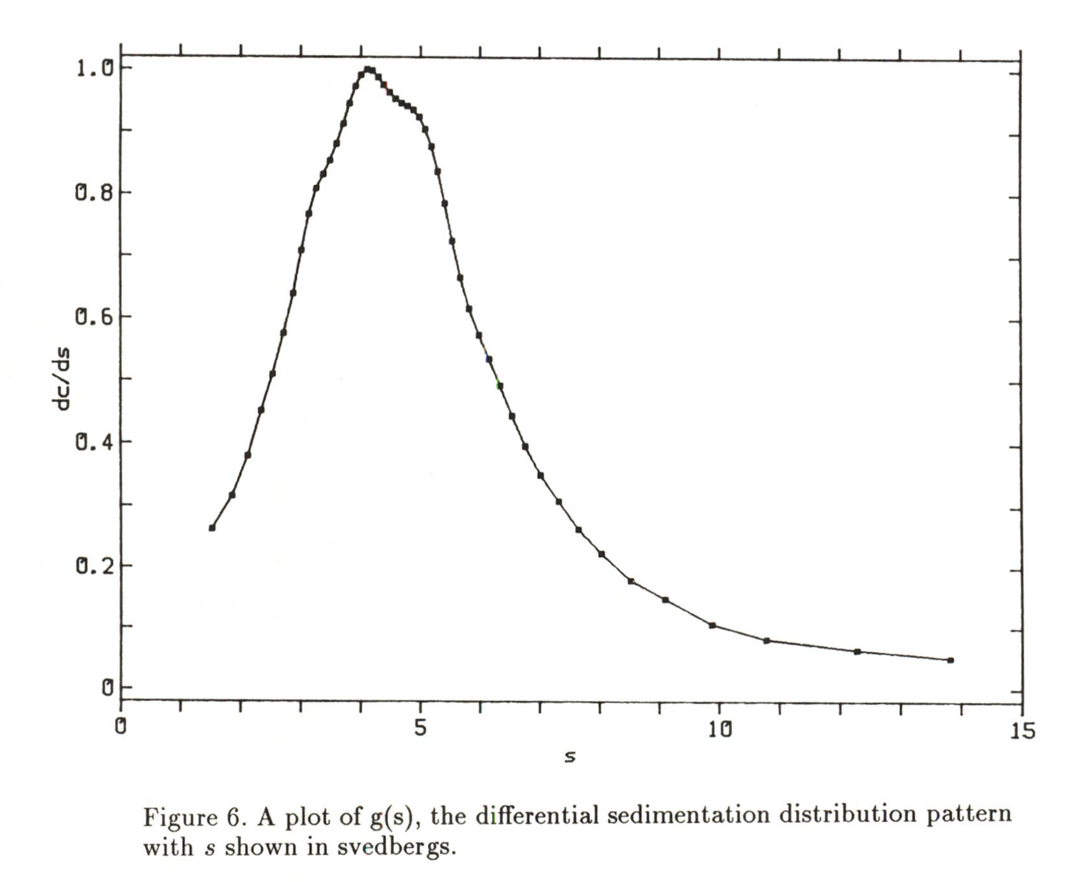

Figure 6. A plot of g(s), the differential sedimentation distribution pattern with *s* shown in svedbergs.

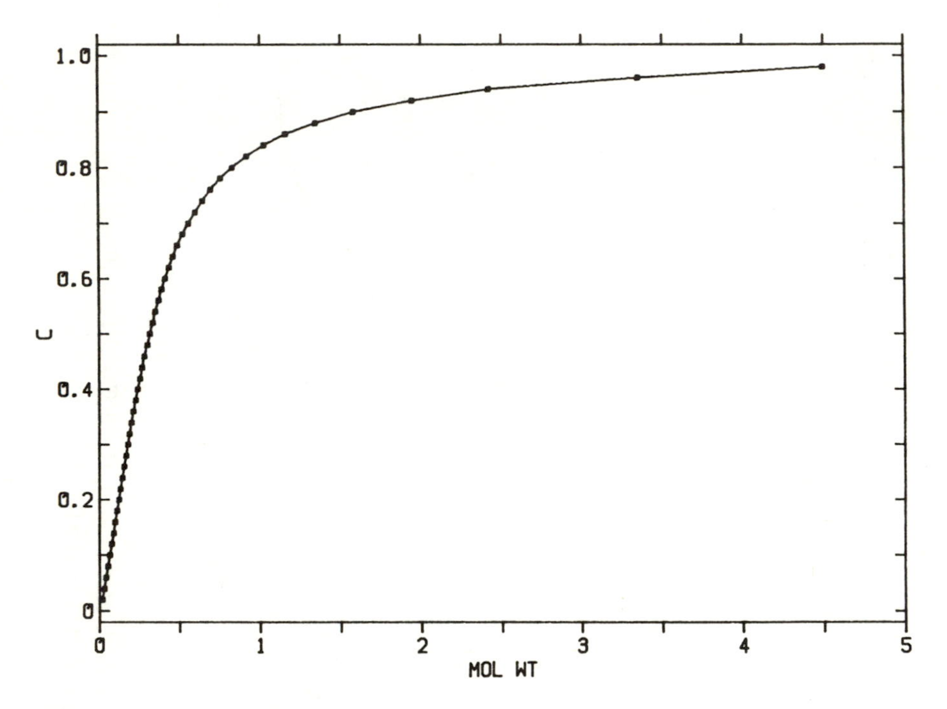

Figure 7. A plot of the integral molecular weight distribution for the poly(1-amidoethylene) standard.

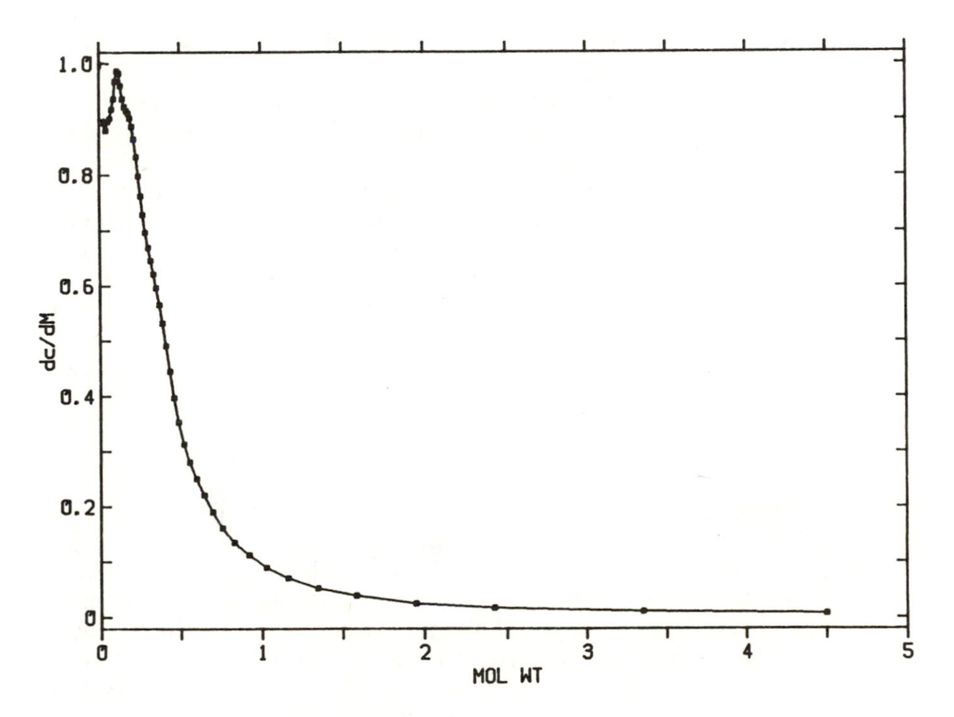

Figure 8. A plot of the differential molecular weight distribution for the poly(1-amidoethylene) standard.

Table I. Average Limiting Viscosity Number and Weight-average, Molecular Weight for the Poly(1-amidoethylene) Sample

	$\overline{M}_w \times 10^{-5}$		$[\eta]$	
Number of Zones	80 min.	100 min.	80 min.	100 min.
10	5.14	4.90	2.18	2.11
20	6.31	5.61	2.39	2.22
50	6.42	5.61	2.52	2.29

better measure of the actual properties of the polymer sample. Furthermore, the fringe pattern with the largest plateau region and a clear meniscus will again include the largest fraction of the sample's molecular weight distribution. The best data points should then come from the 80 min. pattern analyzed with 50 zones. The sample was supplied as a 500,000 molecular weight standard with $\overline{M}_w = 476,000$ as determined by light scattering. The value found from the distribution curve integration is 642,000, significantly higher than the label rating for the standard and 35% higher than the molecular weight found by light scattering. The several averages for the molecular weight distribution of the 80 min. pattern analyzed with 50 zones are $\overline{M}_n = 127,000$, $\overline{M}_w = 642,000$, and $\overline{M}_z = 2,011,000$. The loss of the high molecular weight end of the distribution with longer spinning times is verified by a detailed analysis of the data. The sedimentation coefficients of the two bottom zones of the 80 min. run were 49th = 12.69 and 50th = 14.50. The sedimentation coefficient of the bottom zone of the 100 min. run was 50th = 12.24. Thus, in 20 minutes of spinning, material with s between 12.24 and 14.5 was spun out of solution. The $\overline{M}_w$ at 80 min., based on only the first 49 points, was 551,000 or 98.3% of the $\overline{M}_w$ from all zones of the 100 min. run. The method allows all molecular weight fractions of a sample to be observed when the fringe pattern is photographed after spinning times of 40 min. or less. Further, this method will allow a clearer and more quantitative distinction to be made between contaminants and sample. In light scattering, if the sample scatters too much light, it is filtered or spun to remove "dust." If the sample were contaminated by macroscopic particulates, ultracentrifugation would show a discontinuous high molecular weight tail. This contaminant could then be excluded from calculation of molecular weight averages. If the "dust" to be removed for a light scattering experiment is the high molecular weight tail of the molecular weight distribution, this method would quantitatively demonstrate that. A light scattering experiment would not show that the results had been biased by removal of some of the sample.

The limiting viscosity number for the 80 min. pattern with 50 zones is 2.50 dL/g as compared to an experimental value of 2.45 dL/g. The experimental limiting viscosity number was determined from the data of Table II.

The ratio of calculated $[\eta]$ to experimental $[\eta]$ is 1.020. This result is in strong contrast to the molecular weight values determined from the

Table II. Viscosity of the Samples Spun in the Ultracentrifuge

Sample Number	Concentration (ppm)	Relative Viscosity (dimensionless)
1	2120	1.497
2	1690	1.400
3	1250	1.302
4	849	1.205
5	412	1.102
Solvent*	0	1.000

* Viscosity of solvent was 1.005×10^{-3} Pas.s at 20°C.

distribution since the $[\eta]$ values match to within 2.0%. Since the rheological properties of any polymer are controlled by the high molecular weight tail of the molecular weight distribution, the $[\eta]$ values provide very strong support for the validity of the method. Also, the $[\eta]$ value obtained from experiment was determined on the polymer solutions tested for $\overline{M}_w$.

Errors

There are a number of factors to be considered in the evaluation of the measurements. First, the average deviation of the fringe positions as calculated from the averaging of three fringes is about 0.004 fringes. Thus, correction of the sedimentation patterns with a baseline pattern doubles the deviation to about 0.01 fringes. After the sliding least mean squares cubic fit, the average deviation is reduced to 0.001 fringes.

However, there is a persistent variable problem in ultracentrifuge measurements, namely the instability of the cell at high centrifugal fields arising from distortion of plastic centerpieces and window gaskets. The usual test for cell stability is examination of baselines at low speed and just after attaining the operating speed. (Sometimes additional baselines are obtained after removing the rotor, shaking the rotor without removing the cell to redistribute the sedimented material, and obtaining new baseline patterns.) If the low and high speed patterns are the same, one is more confident that the baseline is correct. If they differ, one assumes that the high-speed baseline is the better of the two, but there is always the possibility that the pattern continued to change further with time. Moreover, after disassembly and reassembly of a cell, the baseline pattern usually changes; hence, it must be measured for every experiment. For these experiments, the baseline patterns deviate from one another by, at most, 0.05 fringes. Baseline error leads to warpage of the interference fringe pattern; the magnitude of this cannot be determined. An alternative method, sedimentation equilibrium ultracentrifugation, does not require the high rotor speeds that produce warpage.

In these experiments, the lack of a well-defined supernatant region leads to an uncertainty in the weight-fraction values for each solution, with the error being less in the steeper parts of the curves. The plateau region was sufficiently free of random noise that the fringes values are accurate to probably better than ±0.02 fringes. The magnitude of this fringe error in molecular weight units depends on the concentration of polymer in the sample and position with respect to the plateau region of the spun solution. As an example of the effect of this uncertainty, however, an error of .02 fringes would be a molecular weight error of 0.3% for measurements on sample 2 in the plateau region.

Conclusions

A method has been developed for the calculation of detailed and absolute molecular weight distributions for complex polymer samples. The method enjoys the advantages of speed and comparative convenience with respect to equilibrium ultracentrifugation and the much needed features of absolute measurement and condition independence with respect to gel permeation chromatography. The method requires that a sedimentation velocity experiment be performed on at least three different concentrations of the polymer in solvent. The solutions must be dilute and the Rayleigh interference pattern photos of the sedimenting samples must be taken at equal times during the experiment. Three fringes of the interference pattern are digitized, averaged, and fitted with a polynomial. The polynomial can be differentiated to give a curve which, when combined with a function obtained from the Mark-Houwink equation, produces a differential molecular weight distribution. The differential form of the distribution can be integrated to give an integral molecular weight distribution. Either distribution can be used to calculate molecular weight averages or moments of the distribution. Application of the method to a commercial poly(1-amidoethylene) standard gave a limiting viscosity number that matched the experimental value to within 3% and weight average molecular weight that bracketed the value claimed by the supplier.

Acknowledgments

This work was partially supported by the National Science Foundation under grants CPE-8260766 and CBT-8417876.

Literature Cited

1. Mark, J. E.; Eisenberg, A.; Graessley, W. W.; Mandelkern, L.; Koenig, J. L. *Physical Properties of Polymers*; Am. Chem. Soc.: Washington, DC, 1984; pp. 111-137.
2. Schachman, H. K. *Ultracentrifugation in Biochemistry*; Academic Press: New York, 1959.
3. Fujita, H. *Mathematical Theory of Sedimentation Analysis*; Academic Press: New York, 1962.
4. Fujita, H.*Foundations of Ultracentrifugal Analysis* (Chemical Analysis Series Monographs, Vol. 42); Wiley: New York, 1975.
5. Gralen, N.; Lagermalm, G. *J. Phys. Chem.* 1952, **56**, 514.

6. Mandelkern, L.; Krigbaum, W.R.; Scheraga, H.A.; Flory, P.J. *J. Phys. Chem.* 1952, **20**, 1392.
7. Mark, H. *Z. Elektrochem* 1934, **40**, 499.
8. Houwink, R. *J. Prakt, Chem.* 1940, **157**, 15.
9. Huggins, M. L. *J. Amer. Chem. Soc.* 1942, **64**, 2716-18.
10. Narkis, N.; Rebhun, M. *Polymer* 1966, **7**, 507-12.
11. MacWilliams, D. C.; Rogers, J. H.; West, T. J. In *Water Soluble Polymers*; Bikales, N. M., Ed.; Plenum: New York, 1973; 106-24.
12. Francois, J.; Sarazin, D.; Schwarts, T.; Weill, G. *Polymer* 1979, **20**, 969-75.
13. Richards, E.G.; Rockholt, D.L.; Richards, J.H. *Fed. Proc.* 1977, **36**, 854.
14. Richards, E.G.; Richards, J.H. *Anal. Biochem.* 1974, **62**, 523.
15. Richards, E.G.; Teller, D.C.; Hoagland, V.C. Jr.; Haschemeyer, R.H.; Schachman, H.K. *Anal. Biochem.* 1971, **41**, 214.
16. Shawki, S. M.; Hamielec, A. E. *J. Appl. Polym. Sci.* 1979, **23**, 3323.
17. Meister, J. J.; Sha, M.-L.; Richards, E. G. *J. Appl. Sci.* 1987, **33**, 1873-85.

RECEIVED March 17, 1989

Chapter 6

Molecular Weight Distribution of Aspen Lignins Estimated by Universal Calibration

M. E. Himmel[1], K. Tatsumoto[1], K. K. Oh[1], K. Grohmann[1], D. K. Johnson[2], and Helena Li Chum[2]

[1]Applied Biological Sciences Section, Biotechnology Research Branch, Solar Fuels Research Division, Solar Energy Research Institute, 1617 Cole Boulevard, Golden, CO 80401
[2]Chemical Conversion Branch, Solar Fuels Research Division, Solar Energy Research Institute, 1617 Cole Boulevard, Golden, CO 80401

This study describes the application of differential viscometry as a GPC detector to the problem of determining molecular weight distributions of acetylated hardwood lignins in tetrahydrofuran. Molecular weight distributions of ball-milled, organosolv, alkali-extracted/mild acid hydrolyzed, and alkali-extracted/steam exploded aspen lignins were estimated using universal calibration. Low molecular weight lignin model compounds (synthetic phenyl-tetramers and Igepals^{TM}) were found to fit universal calibration. Fractions from preparative GPC, when analyzed by universal calibration, yield molecular weight distributions which add to a similar value to that found for the unfractionated parent sample.

Lignins are irregular phenylpropane polymers that represent approximately 20-30% by weight of the available polymeric content of hardwood tree stems (1-3). This material offers, therefore, a valuable resource that must be utilized as fully as possible if the full value of harvested tree crops is to be attained.

The understanding of the macromolecular properties of lignins requires a reliable method for estimating the molecular weights (MW or M) and distribution of molecular weights (MWD) in a suitable solvent. Suitable solvents must be defined here as those that minimize interactions of solute-solute (aggregation), solute-solvent, and solute-column packing material. Although important contributions have been made to this field historically through packed-bed and high performance size exclusion chromatography (HPSEC) (4-17), the design of a chromatography system (solvent and stationary phase) that performs optimally has not been reported. Indeed, the more hydrophobic solvents such as dioxane and tetrahydrofuran (THF), which work well to minimize solute-column and solute-solute interactions,

0097–6156/89/0397–0082$06.00/0

perform poorly at the task of solubilizing lignins over a wide range of MW. However, these solvents work very well with polystyrene-divinylbenzene (e.g., μ-Styragel) column packing materials, causing no perceptible column deterioration (18). Apart from these important limitations presented by conventional "GPC" methodology to the study of lignin MWD is the issue of the complexity of these polymers even in ideal solvents. Conventional GPC is effective in estimating MW of unknown polymers of similar or identical chemical structures to those used to calibrate columns. The elution of polymers of unknown (or imprecisely known) chemistry, degree of branching, shape, degree of solvation, and degree of repeating units (as in chemical copolymers) can be treated only "phenomenologically" with conventional SEC. The result of the size exclusion method (if chemical interactions can be assumed to be negligible) is the separation of solutes on the basis of their respective hydrodynamic radii, R_h. This information, alone, is not of great utility.

Size exclusion chromatography has been greatly enriched recently by the advent of two commercial detectors, the real-time differential viscometer (DV) and the low angle laser light scattering (LALLS) photometer. The only DV detector currently available is offered by Viscotek in Porter, TX, as the model 100.

This study reports the first application of universal calibration via HPSEC-DV to four acetylated hardwood lignins obtained from aspen (*Populus tremuloides*) wood meal by ball milling and solvent extraction; steam explosion followed by alkaline extraction; organosolv pulping followed by water extraction of the associated sugars; and dilute sulfuric acid hydrolysis followed by sodium hydroxide extraction.

Materials and Methods

Chemicals and Standards. All chemicals and HPSEC eluants used in this study were obtained from major chemical suppliers (J. T. Baker, Fisher Scientific, and Aldrich). The THF used was Fisher HPLC grade. The MW standards used to calibrate the three column system were obtained from American Polymer Labs, Mentor, Ohio [polybutadienes (narrow MWD): PB900, PB1K, PB3K, PB5K, PB23K, PB43K; poly-α-methylstyrenes (narrow MWD): PAMS6K, PAMS23K, PAMS66K; polymethylmethacrylates (broad MWD): PMMA17K, PMMA35K, PMMA75K, PMMA100K] and from Polymer Labs, England [polystyrenes (narrow MWD): PS1250, PS1700, PS2450, PS3250, PS5050, PS7000, PS9200, PS11600, PS22K, PS34K, and PS68K; and polymethylmethacrylates (narrow MWD): PMMA3000, PMMA10K, PMMA27K, PMMA60K, and PMMA107K]. Three synthetic polystyrene star-polymers from Polysciences, Warrington, Pennsylvania, were also used [Mn = 7000, lot # 55687; Mn = 59,200, lot # 71520; Mn = 126,900, lot # 55690; and Mn = 116,700, lot # 55689]. Other MW standards examined included four phenyl-tetramers which were supplied as generous gifts from Dr. J. A. Hyatt at Eastman Kodak Labs. These model compounds were prepared by a modified enolate addition method (19) and include biphenyl tetramer hexaacetate ($C_{54}H_{66}O_{20}$,

MW = 1034), biphenyl tetramer hexaol (MW = 782), β-O-4 tetramer heptaacetate ($C_{55}H_{66}O_{22}$, MW = 1078) and β-O-4 tetramer heptaol (MW = 784). The Igepal™ (GAF Corp., sold through Aldrich) standards F.W. = 749 and 1982 were also examined. Two synthetic polymers prepared by anion-initiated polymerizations of a quinonemethide according to the procedure of Chum *et al.* (20) were treated as intermediate MW lignin model polymers.

Lignin Samples. Ball-milled (BM) aspen lignin was prepared following the procedure of Lundquist *et al.* (21). The yield of purified milled wood lignin obtained was usually about 10% w/w that of ethanol/benzene-extracted aspen wood.

Alkaline-extracted/steam-exploded (AESE) aspen lignin samples were prepared from steam exploded wood samples (55 s residence time at 240°C) obtained from Iotech Corp. Exploded wood pulp was treated with a series of carbon tetrachloride and alkaline extractions (12).

Alkaline-extracted/acid hydrolysis (AH/NaOH) lignin samples were prepared by subjecting aspen wood flour to a one hour cook at 120°C in 0.05N sulfuric acid (22), followed by mixing the clarified supernatant with 1% w/w NaOH at 25°C with a Waring blender. The insoluble lignins were precipitated by addition of acid and water washes (32% yield).

The organosolv (OS) lignin was prepared from the liquor obtained by treating aspen wood flour with a 70:30 MeOH:water (v/v) extraction at 165°C for 2.5 hours in a rocking autoclave as described in ref. 23.

Lignin samples were acetylated following a method developed by Gierer and Lindeberg (24) which allows quantitative recovery of lignins. Lignin samples were stored frozen during the course of the study. Throughout the study, freshly prepared solutions were investigated. However, no time-dependence of MW data was observed in any of the techniques employed when samples were occasionally reexamined after initial preparation.

Chromatography System. The HPSEC-DV system used in this study consisted of a Beckman Model 100A dual-piston HPLC pump fitted with external pulse dampening, a Beckman Model 210 injection valve fitted with a 250 μL loop, an SSI injection valve filter, a Hewlett-Packard Model 1037A high sensitivity RI detector, and a Viscotek Model 100LC differential viscometer. For studies of lignin concentration effects, a Knauer UV detector set at 280 nm was used as well. The chromatography column system was composed of three 7.8 × 30 mm columns (Beckman μ-Spherogel, 10,000, 1,000, and 500Å) connected in series in order of increasing pore size. Calculations were performed using the Viscotek Unical 2.71 software. All injections on the HPSEC-DV system were made by overfilling the 250μL loop, thereby providing a true 250 μL injection.

Narrow and broad MW standards were injected onto the HPSEC-DV system at concentrations near 1 mg/mL and 2 mg/mL, respectively. Initially, in order to obtain a usable differential pressure chromatogram, the lignin samples were injected at concentrations near 20 mg/mL, with an instrument (A-D amplifier) gain setting of 1 (0-1.0 volt Full Scale). As the work proceeded, a gain setting of 2 (0-0.1 volt FS) allowed examination of lignin samples at concentrations of 4 and 8 mg/mL. As a test for lignin association in THF at these concentrations, the organosolv lignin was chromatographed at initial injection concentrations of 0.5 and 1 mg/mL as well. Concentrations of all standard and sample solutions studied were precisely determined by weighing the dry materials to the nearest 0.0005 mg using a Sartorius Ultramicro balance Model 4504 MP8. Lignin samples examined by HPSEC-DV were made to concentration immediately before injection.

The organosolv lignin samples were subjected to preparative column chromatography in THF using a two column, μ-Styragel system from YMC, Japan (5 cm ×200 cm, 500 and 1000Å). A Beckman Model 110B pump, Model 210 injection valve with a 2 mL loop, and a Model 153 UV detector with semi-preparative flow cell were used with these columns. Sample loadings were usually 60 mg. Fractions were collected with an Isco Foxy fraction collector with preparative capability and stored in Kimax screw-topped glass tubes (25 x 150 mm) with Teflon-lined caps. Before chromatographic analysis these fractions were pooled into five master fractions and concentrated by roto-evaporation at 25°C. Values found for the extinction coefficient of organosolv and ball milled lignin by conventional gravimetric analysis, ϵ_{270nm} (g $mL^{-1}cm^{-1}$) = 15,300 and 10,100, respectively, were used to estimate the dried-weight equivalent of the concentrated fractions recovered from preparative chromatography. These concentration values are critical for meaningful estimation of MW using the Unical software.

Calculation of Results from Differential Viscometry and SEC. The familiar relationship first expressed by Mark (25) and Houwink (26) in the 1940's is central to the concept of universal calibration first suggested by Benoit *et al.* (27). In this relationship,

$$[\eta] = K'M^a \tag{1}$$

$[\eta]$ is the intrinsic viscosity, and K' and a are known as the Mark-Houwink constants and are specific to a polymer-solvent-temperature system. For flexible, linear polymers, values of a are limited to the range 0.50 to 0.80.

Considering the derivations of equation (1), it can be predicted that all molecules having the same value of $[\eta]M$ would have the same value of v_h, the hydrodynamic volume. Also, if v_h is the parameter that uniquely determines the elution volume, V_e, these molecules should have the same elution volume. The arguments presented by these authors do not predict that the relationship between these parameters should necessarily be linear. Most universal calibration curves shown in the literature that cover 4 to 6 decades of M show a definite upward curvature at high values of M (28).

Sources of error in this approach arise from both experimental and theoretical grounds. Modern theories of SEC retention mechanism are based on the assumption that the size exclusion process uniquely determines the elution volume, and yet the possibility of reversible adsorption is difficult to dismiss and, where it occurs, errors in the interpretation may easily result. As a warning for the application of universal calibration methodology, Cassassa (29) indicates in a later paper that the quantity $[\eta]M$ is not a truly universal elution parameter for SEC, but that both theory and experience indicate that good results can be obtained for eluting species of similar type (e.g., rod-like macromolecules of similar cross-sectional dimension in a restricted size range or linear flexible polymer chains). Cassassa predicts from theory, however, that over restricted ranges of M, a common $[\eta]M$ dependence between random coil polymers and rod-like structures should exist. Divergence often increases, however, when considering fit using data over three orders of magnitude in M (30).

Application of universal calibration to unknown polymers using the Viscotek Unical software, once the column system has been calibrated with narrow MWD standards, is quite straightforward. A master calibration file of narrow MWD standards was developed which incorporates the "peak parameter" values calculated from one (or averaged from several) well behaved narrow MWD standard. These values correct for chromatographic mismatch of the two detectors (RI and differential pressure) used in the system and lead to the calculation of values that approximate a correction for peak broadening (τ) and peak tailing (σ). These peak parameters represent effects specific to the chromatographic system used in each laboratory, and were determined for this study to be 0.256 mL, 0.280 and 0.256 for σ, τ (V), and τ (C), respectively. The concentration value for each sample processed by this procedure must be known accurately, as this term enters into calculations of reduced viscosity, measured here directly as specific viscosity, and the MW averages. An assumption central to the data processing is that under chromatographic conditions sample dilution is sufficient to assume that the reduced viscosity approximates the intrinsic viscosity (or limiting viscosity number). The software can be used to calculate the Mark-Houwink plots ($[\eta]$ versus M) for each standard polymer series. All polymer standards are then used to construct a universal calibration plot of $[\eta]M$ versus elution volume. The software can also be used to recalculate the values of $\overline{M}_n$, $\overline{M}_w$, $\overline{M}_z$, and $\overline{M}_{z+1}$ for the narrow MWD standards used to construct the curve (approximately ±10% deviation for $\overline{M}_w$ is observed in these recalculated values when compared with the values entered initially). Molecular weight averages are found for unknown polymers (accepting the limiting assumptions discussed above) in a similar way.

Application of Unical software also requires the selection of chromatographic baselines, thus selecting the specific data taken for further analysis. In the studies reported here, we choose to analyze only the polydisperse envelope from lignin elution, so that the distinct component which often elutes near V_t (the total column volume) was not included in the analysis.

Results and Discussion

Universal Calibration. The aspen wood lignin samples chosen for this study were prepared by organosolv, steam explosion, dilute acid hydrolysis, and ball-milling procedures.

Calibration curves were developed for HPSEC-DV using polymer standards including narrow MWD polystyrenes, polybutadienes, polymethylmethacrylates, and poly-α-methylstyrenes. A set of broad MWD polymethylmethacrylates was also examined. All five standard curves indicated very good fit to a linear function over the range of MW tested. The Mark-Houwink parameters for a and $-K'$ were found to be 0.73 and 4.35, 0.60 and 3.76, 0.72 and 3.87, 0.69 and 4.26, 0.73 and 4.55 for the PS, PAMS, PB, PMMA, and PMMA-b standards, respectively. All Mark-Houwink parameters, a and K', with the exception of a for the poly-α-methylstyrenes, compare closely with those published by Haney and Armonas (31) using identical conditions and instrumentation.

A universal calibration plot (log $[\eta]M$ vs. elution volume) using these five standards series was constructed (Fig. 1). Several other standard MW polymers appropriate to lignin model studies were also examined. These included two Igepals, four polystyrene star polymers, and four synthetic phenyltetramers (two biphenyls and two β-O-4 linked tetramers). Of all the standards examined, only the polystyrene star polymer preparation indicated paucidispersity. Here, the elution of the lower MW component (usually in preponderance) was considered in the calculations. With the exception of one high MW star polymer, all these compounds were found to fit universal calibration at least as well as the commercial polymer standards (Fig. 1). Indeed, the fit of the low MW phenyltetramers (MW $\approx$ 800-1000) was important. The calibration curve constructed for use in this study shows little or not curvature over the five decade range of $\log[\eta]M$. This observation is consistent with that of other workers (28,31), where the lowermost portion of such curves approaches linearity, while over a wider range of $\log[\eta]M$ some upward curvature is evident. Unical 2.71 software allowed the calculation of a universal calibration curve from a broad MWD standard, PMMA17K-b. This curve appears in Figure 1 as a dashed line. Although some deviation at data extremes is apparent, the fit near the lignin elution region is nearly identical to the curve from narrow calibration.

MWD of Acetylated Lignins. The first attempts at lignin chromatography employed 250 μL injections of samples made to 20 mg/mL. This procedure produced acceptable differential pressure signals; however, the very high concentration was undesirable as it is known to induce solute-solute interaction. After resetting the amplifier gain to 2 (a setting of 1 being the standard "default" value for the instrument), differential pressure chromatograms with lignins at loadings of 2 mg (250 μL injections from 8 mg/mL stock solutions) were very well behaved (Figs. 2-4). Inspection of Figures 2 and 3 reveals that the relative magnitude of the differential pressure signal at higher MW is greater than that from the differential refractive index detector. The four lignins were also injected on the chromatography system at a loading of 1 mg (250 μL from a 4 mg/mL stock solution). These dual chromatograms proved to indicate the limiting sensitivity for the broadly polydisperse lignin samples studied here. Although the "smoothing" routines in Unical were capable of rendering these noisy differential chromatograms usable, it was clear that a more dilute injection sample would not be meaningful. An example of the best dual chromatogram from a 1 mg loading (before smoothing) is shown in Figure 4U.

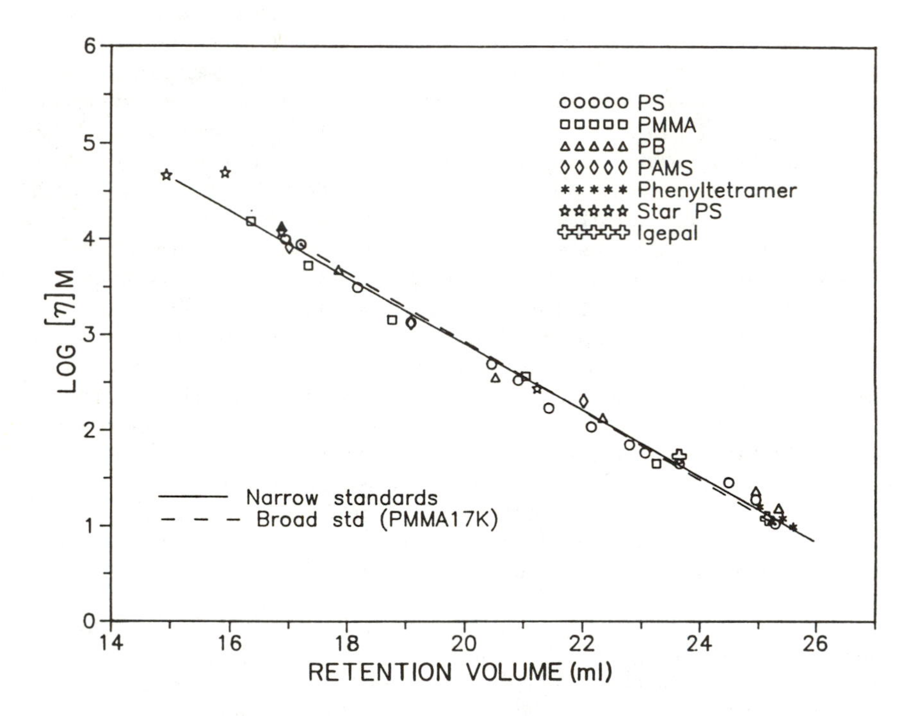

Figure 1. Master universal calibration curve obtained with a Beckman μ-Spherogel column system in THF. The fit of narrow MWD standards, polystyrene star polymers, and a single broad MWD standard (calculated with Unical 2.71 software) are shown.

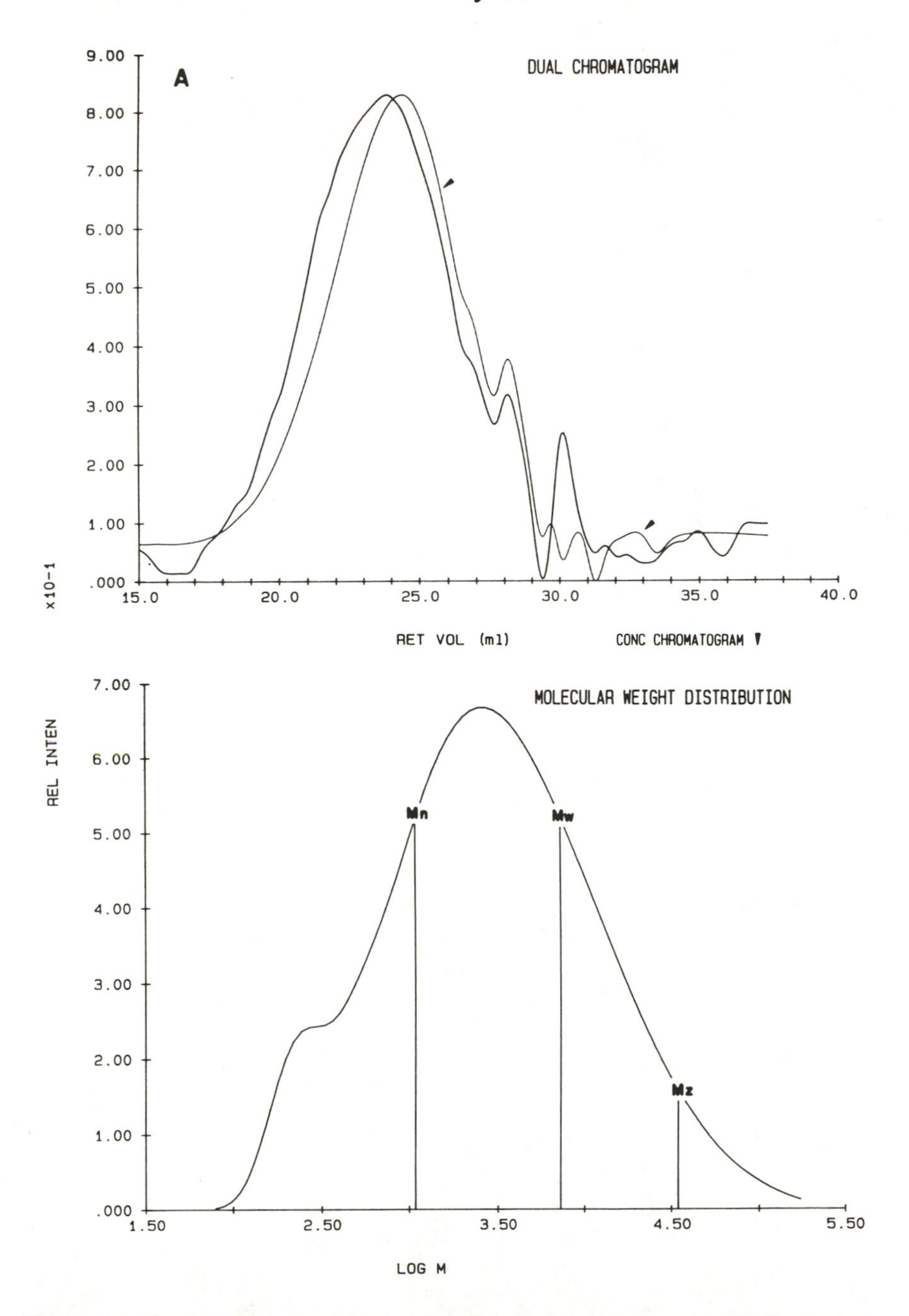

Figure 2A. Dual chromatograms showing the elution of aspen AESE lignin from the HPSEC-DV system. A 250 μL sample was injected from a freshly prepared 8 mg /mL stock solution. Viscotek A/D amplifier gain setting of 2 and RI detector setting of 1x. Calculated molecular weights are also shown.

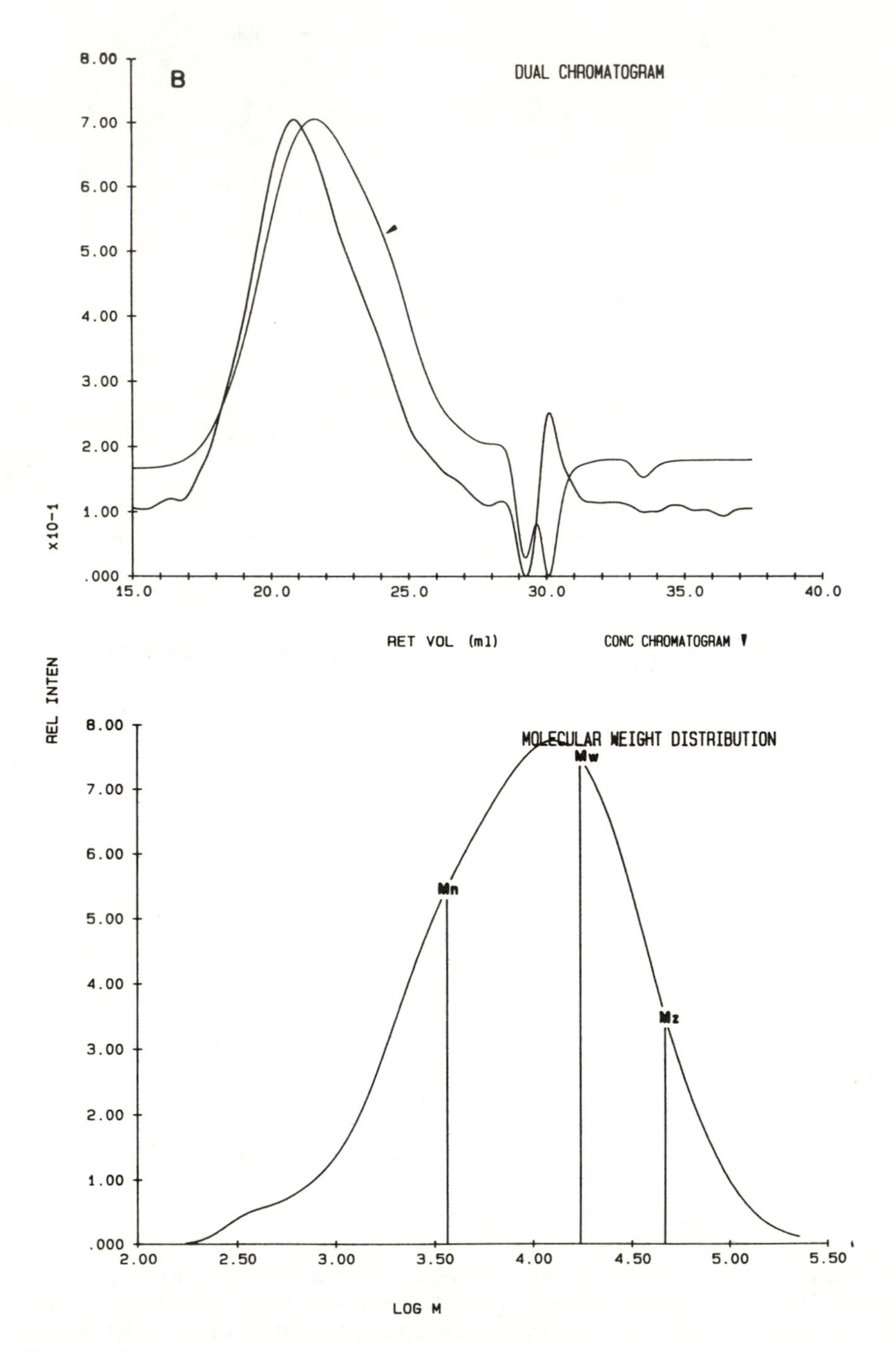

Figure 2B. Dual chromatograms showing the elution of ball-milled lignin from the HPSEC-DV system. Sample loading was the same as in Figure 2A.

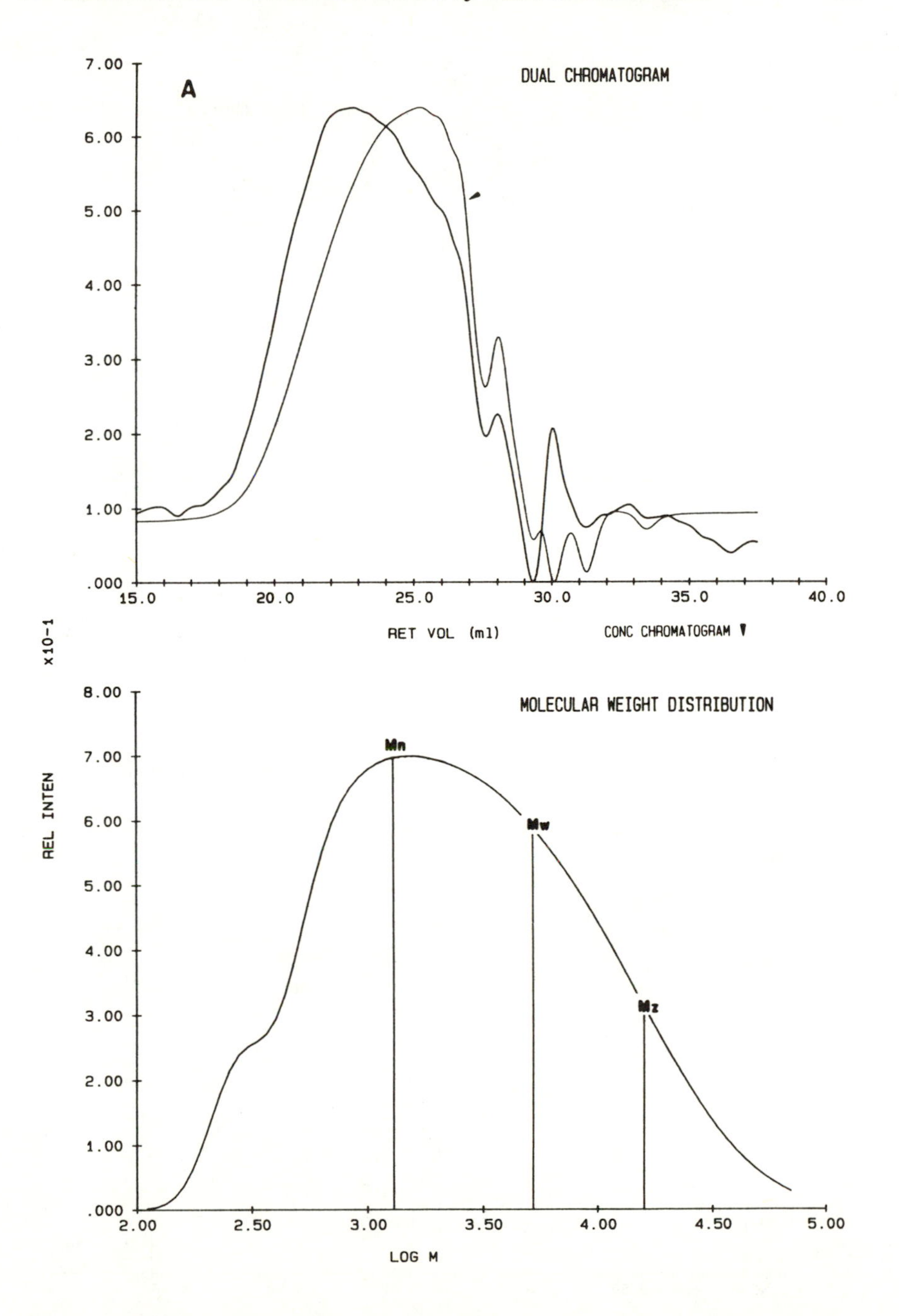

Figure 3A. Dual chromatograms showing the elution of aspen organosolv lignin from the HPSEC-DV system. Sample loading was the same as in Figure 2A. Calculated molecular weights are also shown.

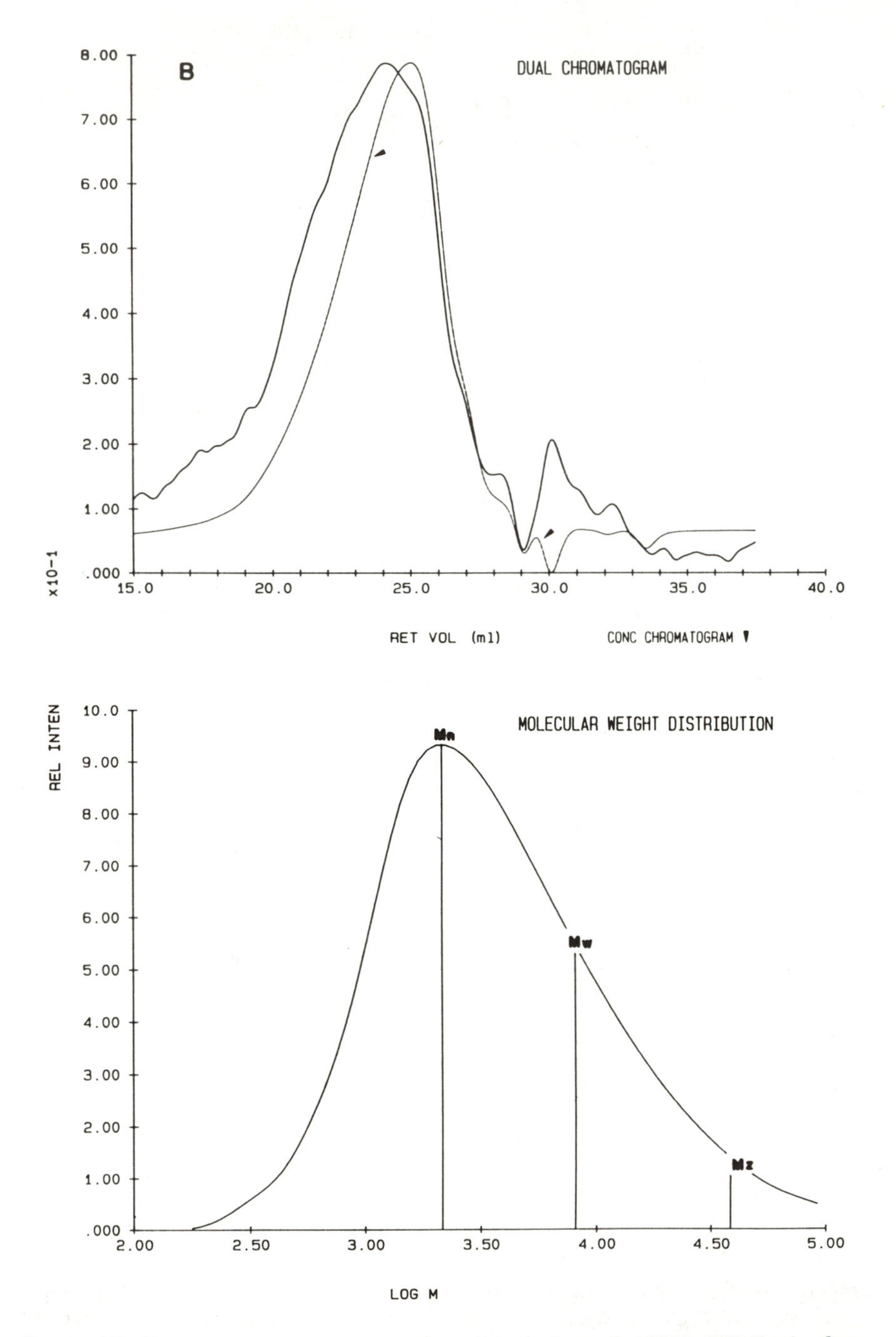

Figure 3B. Dual chromatograms showing the elution of AH/NaOH lignin from the HPSEC-DV system. Sample loading was the same as in Figure 2A.

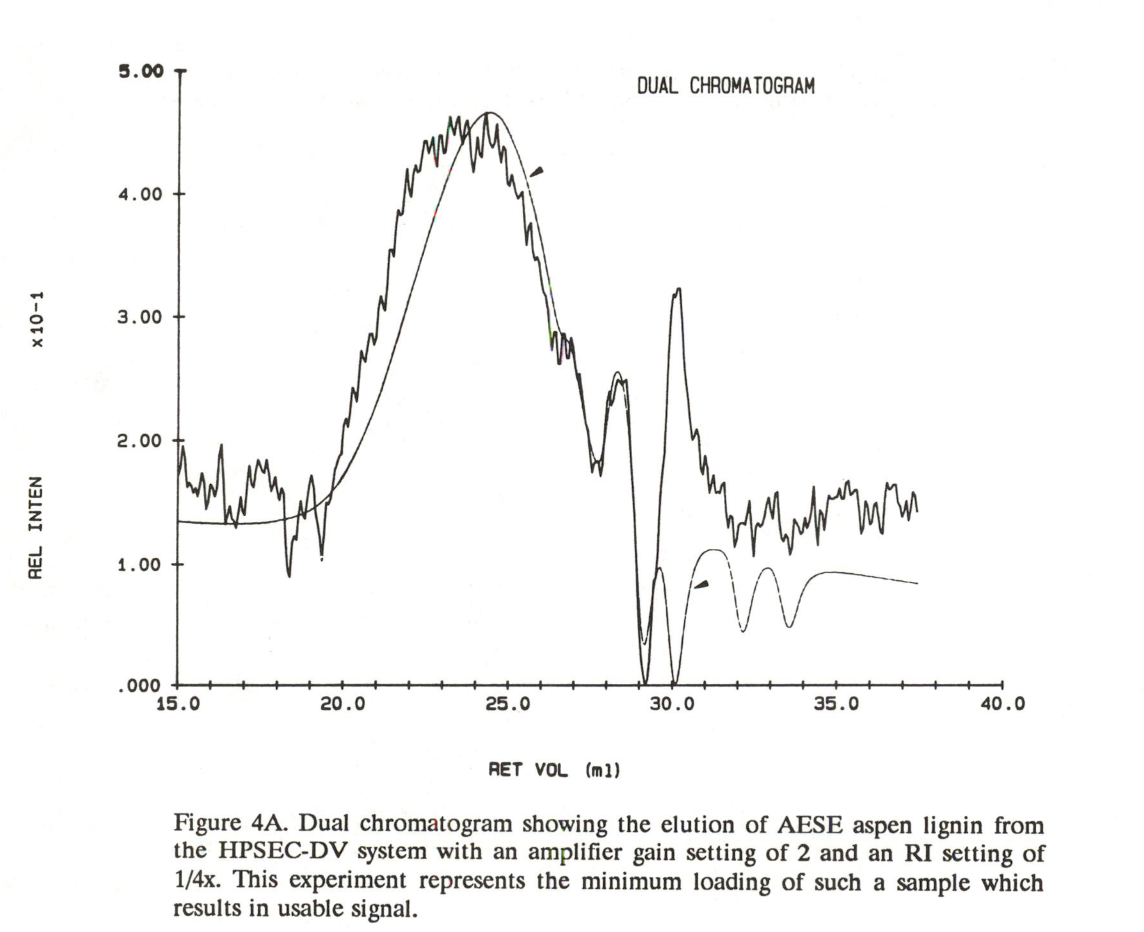

Figure 4A. Dual chromatogram showing the elution of AESE aspen lignin from the HPSEC-DV system with an amplifier gain setting of 2 and an RI setting of 1/4x. This experiment represents the minimum loading of such a sample which results in usable signal.

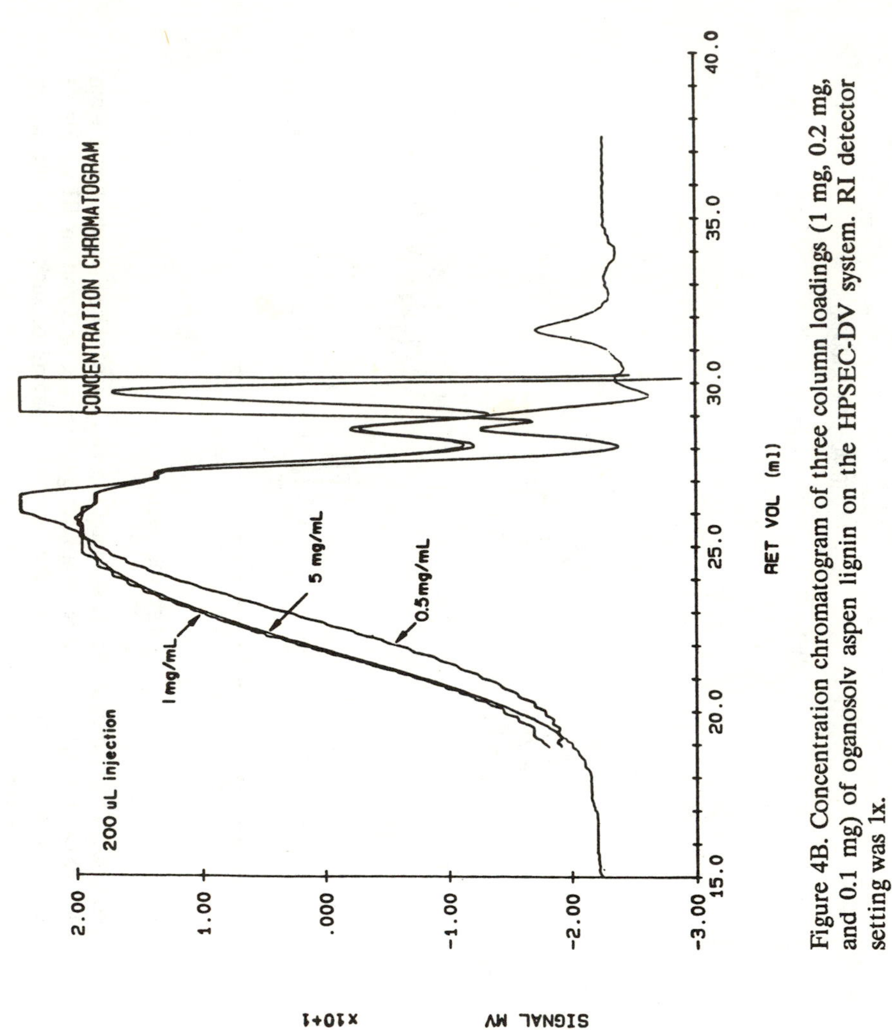

Figure 4B. Concentration chromatogram of three column loadings (1 mg, 0.2 mg, and 0.1 mg) of oganosolv aspen lignin on the HPSEC-DV system. RI detector setting was 1x.

Values for system "peak parameters" were found using a narrow distribution polystyrene standard (PS68K) before calculating MWD data for the lignin samples from universal calibration. To check software and instrument operation, several narrow MWD polystyrene and one broad MWD polymethylmethacrylate standards were treated as unknown samples and subjected to analysis with the universal calibration curve assembled from all polymer standards files. It was found that the MWD could be estimated for the "recalculated" polymer standards with errors between ±5 and 10% of the original value indicated by the supplier of the standard (e.g., $\overline{M}_w$ for PS11K and $\overline{M}_w$ and $\overline{M}_n$ for PMMA17K-*b*).

Table I illustrates the molecular weight averages found from universal calibration (narrow standards) for four aspen lignins and two quinonemethide-derived polymers using the Unical software. The polydispersities were the same within the experimental errors for all lignin samples. In contrast, the polydispersities found for the quinonemethide-derived polymers by universal calibration were near 1.1. The molecular weight averages found for the four acetylated lignins studied by universal calibration were substantially larger than those determined from previous work using conventional GPC (e.g., AESE and BM lignins from refs. 7 and 12 had approximately one-third those values found by HPSEC-DV in the present study).

Table I. MWD of Acetylated Aspen Lignins and Model Compounds from Universal Calibration with Narrow Standards [1]

Samples/loading	$\overline{M}_n$	$\overline{M}_w$	$\overline{M}_z$	$\overline{M}_v$	$\overline{M}_w/\overline{M}_n$
AESE/1 mg	1100	7300	34500	3300	6.7
AESE/2 mg	1900	7100	27000	—	3.7
OS/1 mg	1300	5200	16000	3200	4.0
OS/2 mg	1000	4400	18000	—	4.4
AH/NaOH/1 mg	2200	8100	38000	4600	3.7
AH/NaOH/2 mg	1300	6600	34000	—	5.0
BM/1 mg	3600	17300	46800	11500	4.7
BM/2 mg	9000	22000	47000	—	2.4
QM 34/0.25 mg	8700	10300	12400	9900	1.1
QM 33/0.25 mg	14700	16700	20000	16000	1.1

[1] Obtained in THF at 20°C with RI detection and Unical 2.71 software (Viscotek, Inc.). For high loadings the system was set at RI = $1x$ and 20 PAFS, gain = 2; injections (250 μL) made from 8 mg/mL stock solutions. For low loadings the detector settings were RI = $1/4x$ and 20 PAFS, gain = 2; injections (250 μL) made from 4 mg/mL stock solutions.

The issue of column loading was further investigated by injecting 200 μL samples from stock organosolv lignin solutions of 5, 1 and 0.5 mg/mL.

The composite chromatogram shown in Figure 4L clearly indicates that an apparent increase in higher MW content occurs when column loadings increase from 0.1 mg to 0.2 mg lignin. The 0.1 mg loading represents the sensitivity limit, however, of the high sensitivity refractive index detector used in this study. However, the 0.2 and 1 mg loadings (where most lignin data were collected in this study) were nearly identical in distribution (see Figure 4L). Comparison of the curves in Figure 4L with closely eluting pairs of standard polymers indicated that the apparent increase in $\overline{M}_w$ induced by this concentration effect would be $< 10\%$.

Chromatographic Fractionation of Organosolv Lignin. The preparative μ-Styragel column from YMC was loaded with 60 mg of organosolv lignin. The resulting elution profile is shown as an insert in Figure 5. This figure shows the relative distribution of the chromatographic fractions pooled to generate the five master fractions used for further study. These fractions were chosen so that the relative areas of all five zones were nearly identical. Figure 5 also shows the superposition of the resulting chromatographic analysis of four of these fractions. Fraction number five was omitted from this analysis because this peak was not included in the standard procedure used in establishing baselines for the four native lignins described earlier. The total weight average molecular weight for the entire distribution (32) was estimated using the relationship

$$\overline{M}_{w,tot} = \Sigma c_i M_i / \Sigma c_i \tag{2}$$

where c is the concentration (here in milligrams) of each master fraction, i, and M is the estimated $\overline{M}_w$ of each master fraction from universal calibration. The value of $\overline{M}_{w,tot}$ found using these four master fractions was 3800. When considering the $\overline{M}_w$ from the chromatography of the unfractionated organosolv lignin at a 1 mg loading was estimated to be 4400, the value obtained from the organosolv fractions is in good agreement.

This experiment was designed to examine the possible bias the broad polydispersities of the lignin samples may have on estimation of MWD by universal calibration. These data indicate that no such contribution exists, since the summation of the individual fractions of narrow(er) dispersity lead to values of MWD similar to those found using Unical software for the unfractionated lignin sample.

Conclusions

Although evidence exists that concentration effects may be important with even acetylated lignins in THF, the effect of increasing column loadings from 1 to 2 mg seems unlikely as the cause of the variance in MW shown in Table I. This observation illustrates the more general problem in current SEC-based "absolute" MW measurement: that of a limited concentration window for analysis. The limiting value for sample concentration appears to be near 1 mg per injection for HPSEC-DV, which is comparable to the 0.2-1 mg per injection range usable in HPSEC-LALLS (33). For studies

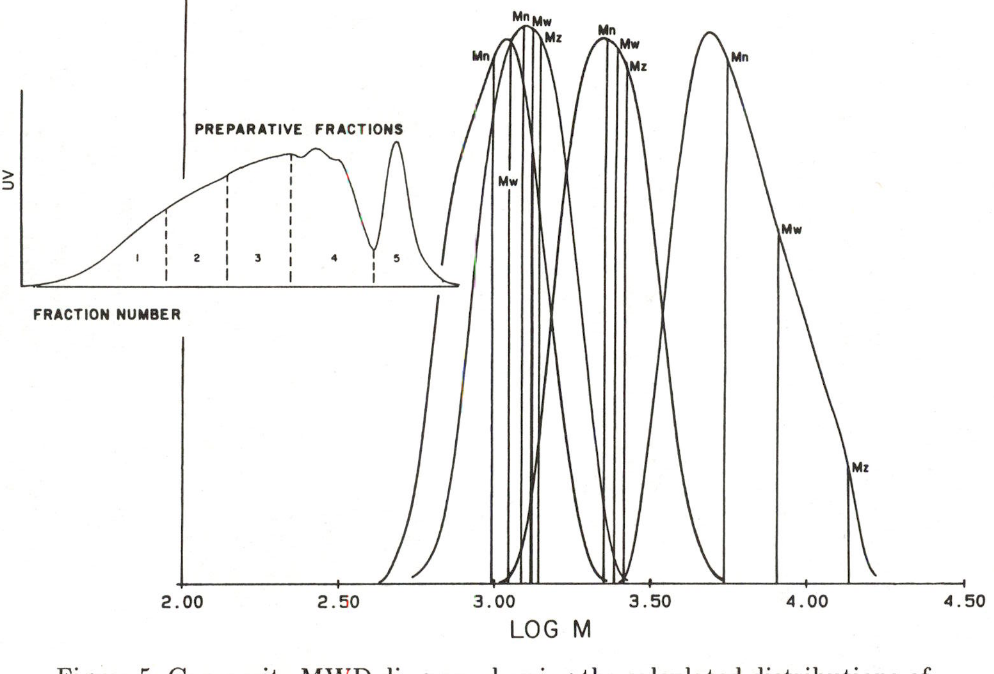

Figure 5. Composite MWD diagram showing the calculated distributions of four of the five master fractions obtained from preparative chromatography of organosolv aspen lignin (fraction number 4 to 1 from left to right). The $\overline{M}_w$ values for the master fractions from left to right are 1,110, 1,310, 2,420, and 8,050, respectively. The insert shows the elution profile of organosolv aspen lignin from the YMC preparative μ-Styragel column (5 × 200 cm). The 30 fractions collected were pooled into the five master fractions shown.

such as these, an increase in sensitivity of one order of magnitude would be highly valued and should be considered an area of focus for suppliers of SEC detection equipment.

The determination that the low MW, acetylated aspen lignins examined actually fit universal calibration, however, must be deferred to future studies comparing these data to results from LALLS and sedimentation equilibrium analysis (if possible).

As a result of the dependence of universal calibration on column elution behavior (i.e., anomalous behavior due to adsorption or exclusion), the contribution of the polymer "core" and "shell" components (33, 34) to hydrodynamic behavior must be fully understood if competent analysis of block copolymers and branched heteropolymers is to be made. It is hoped that with the advent of appropriate MW, composition, and branched polymer standards, the limits of fit of universal calibration to biopolymers such as lignin can be judged.

Acknowledgments

We thank Dr. J. A. Hyatt for the gift of synthetic phenyltetramers and Max Haney for helpful discussions. This work was funded by the Biochemical Conversion Program at the DOE Biofuels and Municipal Waste Technology Division through FTP No. 658.

Literature Cited

1. Brauns, F. E.; Brauns, D. A. *The Chemistry of Lignin*; Academic Press: New York, 1960.
2. Sarkanen, K. V.; Ludwig, C. H., Eds. *Lignins: Occurrence, Formation, Structure and Reactions*; Wiley-Interscience: New York, 1971.
3. Sarkanen, K. V. In *The Chemistry of Wood*; Browning, B. L., Ed.; R. E. Krieger Publ. Co.: Huntington, 1975.
4. Marchessault, R. H.; Coulombe, S.; Hanai, T.; Morikawa, H.; Robert, D. *Can. J. Chem.* 1982, **60**, 2372.
5. Kolpak, F. J.; Cietek, D. J.; Fookes, W.; Call, J. J. *J. Appl. Polym. Sci., Polym. Sci. Symp.* 1983, **31**, 491.
6. Himmel, M. E.; Oh, K. K.; Sopher, D. W.; Chum, H. L. *J. Chromatogr.* 1983, **267**, 249.
7. Faix, O.; Lange, W.; Beinhoff, O. *Holzforschung* 1980, **34**, 174.
8. Faix, O.; Lange, W.; Salud, E. C. *Holzforschung* 1981, **35**, 3.
9. Lange, W.; Schweers, W.; Beinhoff, O. *Holzforschung* 1981, **35**, 119.
10. Meier, D.; Faix, O.; Lange, W. *Holzforschung* 1981, **35**, 247.
11. Lange, W.; Faix, O.; Beinhoff, O. *Holzforschung* 1983, **37**, 63.
12. Chum, H. L.; Johnson, D. K.; Tucker, M. P.; Himmel, M. E. *Holzforschung* 1987, **41**, 97.
13. Pellinen, J.; Salkinoja-Salonen, M. *J. Chromatogr.* 1985, **328**, 299.
14. Connors, W. J.; Sarkanen, S.; McCarthy, J. L. *Holzforschung* 1980, **34**, 80.
15. Sarkanen, S.; Teller, D. C.; Hall, J.; McCarthy, J. L. *Macromolecules* 1981, **14**, 426.

16. Tirtowidjojo, S. M.S. Thesis, University of Washington, Seattle, WA, 1984, Parts 1 and 2, pp. 1-119.
17. Sarkanen, S.; Teller, D. C.; Stevens, C. R.; McCarthy, J. L. *Macromolecules* 1984, **17**, 2588.
18. Yau, W. W.; Kirkland, J. J.; Bly, D. D. *Modern Size Exclusion Chromatography: Practice of Gel Permeation and Gel Filtration Chromatography*; John Wiley & Sons: New York, 1979.
19. Hyatt, J. A. *Holzforschung* 1987, **41**, 363.
20. Chum, H. L.; Johnson, D. K.; Palasz, P. D.; Smith, C. Z.; Utley, H. P. *Macromolecules* 1987, **20**, 2698.
21. Lundquist, K.; Ohlsson, R.; Simonson, R. *Svensk Papperstidn.* 1975, **78**, 390.
22. Grohmann, K.; Torget, R.; Himmel, M. E. *Biotech. Bioeng. Symp.* 1986, **17**, 135.
23. Chum, H. L.; Johnson, D. K.; Ratcliff, M.; Black, S.; Schroeder, H. A.; Wallace, K. In *Proc. Intl. Symp. Wood and Pulp. Chem.*; Can. Pulp and Paper Assoc.: Vancouver, 1985; pp. 223-227.
24. Gierer, J.; Lindeberg, O. *Acta Chem. Scand.* 1980, **34**, 161.
25. Mark, H. *Der Feste Korper*; Hirzel: Leipzig, 1938; p. 103.
26. Houwink, R. *J. Prakt. Chem.* 1941, **157**, 15.
27. Benoit, H.; Grubisic, Z.; Rempp, P.; Decker, D.; Zilliox, J. G. *J. Chim. Phys.* 1966, **63**, 1507.
28. Gallot-Grubisic, Z.; Rempp, P.; Benoit, H. *J. Polym. Sci.* 1967, **5**, 753.
29. Cassassa, E. F. *Macromolecules* 1976, **9**, 182.
30. Frigon, R. P.; Leypoldt, J. K.; Uyeji, S.; Henderson, L. W. *Anal. Chem.* 1983, **55**, 1349.
31. Haney, M. A.; Armonas, J. E. In *Proc. GPC Symp. '87*; Waters Chromatography Corp.: Milford, MA, 1987; pp. 523-544.
32. Schachman, H. K. *Ultracentrifugation in Biochemistry*; Academic Press: New York, 1959.
33. Jordan, R. C.; Silver, S. F.; Sehon, R. D.; Rivard, R. J. *ACS Symp. Ser.* 1984, **245**, 295.
34. Bi, L. K.; Fetters, L. J. *Macromolecules* 1976, **9**, 732.

RECEIVED February 27, 1989

Chapter 7

Molecular Weight Determination of Hydroxypropylated Lignins

E. J. Siochi[1,2], M. A. Haney[3], W. Mahn[3], and Thomas C. Ward[1,2,4]

[1]Department of Chemistry, Virginia Polytechnic Institute and State University, Blacksburg, VA 24061
[2]Polymeric Materials and Interfaces Laboratory, Virginia Polytechnic Institute and State University, Blacksburg, VA 24061

The number average molecular weight of hydroxypropylated lignins (HPL) was determined using gel permeation chromatography with a differential viscosity detector (GPC/DV). Vapor phase osmometry was used to provide comparative number average molecular weight values to verify the results obtained by GPC. GPC/DV proved to be a reliable and convenient tool for the determination of absolute molecular weights of lignin derivatives. The results revealed that a time dependent association of HPL molecules was occurring in solution.

Lignin is an amorphous biopolymer second in natural abundance only to cellulose. It is composed of phenylpropane units linked primarily through ether bonds and constitutes 15-40% of the dry weight of wood. A combination of the abundance of lignins, its versatility due to the variety of sources available and the recent interest in renewable resources have opened up research into the potential applications of lignins in many fields (1-5). Due to the inherent strength of lignin, it has been favored by researchers studying structural materials. Lignins have been used as prepolymers for the modification of synthetic polymers and as substitutes for various polymeric materials (3, 6-8). Where the incorporation of lignins had adverse effects on the mechanical properties of the resulting products, hydroxyalkylation of the lignin employed has been found to improve the material characteristics (6).

In any structure/property studies on potential applications of lignins, an important parameter is the molecular weight and molecular weight distribution (MWD). In recognition of this fact, a substantial number of papers

---

[3]Current address: Viscotek Corporation, 1032 Russell Drive, Porter, TX 77365
[4]Address correspondence to this author.

0097–6156/89/0397–0100$06.00/0

have been published which report the use of vapor phase osmometry (VPO) (8-11) and ultracentrifugation (12-16) for this purpose. More recently, gel permeation chromatography (GPC) has gained advocates for lignin molecular weight determination due to its ease of use and the short analysis time (9,12,17-25). Although GPC is highly popular, the published data treatments typically yield only *relative* and not *absolute* molecular weights. This is the result of a calibration scheme using easily accessible polystyrene standards whose conformations differ from lignins (14,26,27). While low angle laser light scattering (LALLS) has been employed to avoid calibration problems, other difficulties subsequently emerge. These necessitate corrections for absorbance, fluorescence and polarization (9,28). In recent years, a number of viscosity detectors have been developed for use with GPC in conjunction with the necessary concentration sensitive detectors (29-40). Although such detectors are commercially available (39,40), their application to lignin MWD determination is rare. This lack of usage may be related to the consensus that lignins have a three-dimensional network structure which would not obey universal calibration, on which GPC/DV is based.

The objective of this present work was to investigate the feasibility of using GPC/DV for absolute molecular weight determination of hydroxypropylated lignins. In order to verify the validity of the universal calibration method, vapor phase osmometry (VPO) was used to provide reference number average molecular weight values. Comparisons with LALLS results have also been made and will be reported in another publication.

Experimental

Materials. Hydroxypropyl derivatives of red oak, aspen and a hardwood kraft lignin were studied. There were two samples of red oak. One, designated as "RO:PO", is identical to the red oak HPL, but was made in larger quantities to allow preparative fractionation. The aspen was an organosolv lignin obtained from Biological Energy Corporation of Valley Forge, Pennsylvania. The hardwood kraft lignin was supplied by Westvaco, Charleston, South Carolina. All of the above samples were derivatized according to the procedure of Wu and Glasser (6).

Vapor Phase Osmometry. A Wescan Model 233 vapor phase osmometer was used to obtain number average molecular weights. The lignin solutions were made up with HPLC grade tetrahydrofuran (THF) and shaken manually until the solutions were clear. The experiments were conducted at 30°C. Number average molecular weights were determined by multistandard calibration (41), a procedure found to greatly enhance reproducibility and accuracy of the results. Experiments were conducted immediately after sample preparation and three days later.

Gel Permeation Chromatography. Polymer Laboratories narrow distribution polystyrene standards with nominal molecular weights of 1250, 2150, 3250, 5000, 9000, 34500, 68000 and 170000 g/mole were dissolved in HPLC grade THF. These standards were used for the construction of the universal calibration curve.

Approximately 2.5 mg/ml solutions of the lignins were prepared in HPLC grade THF. The solutions were made up immediately before GPC was conducted and shaken manually for a couple of minutes until the solutions were a clear brown color. GPC analysis was carried out on a Waters 150C ALC/GPC equipped with an optical deflection type differential refractive index detector having a sensitivity of 1×10^{-6} ΔRI units and a Viscotek Model 100 differential viscosity detector. The chromatographic conditions were: flow rate 0.9 ml/min, injection volume 200 μl, with Column Resolution repacked microstyragel columns having pore sizes of 500 Å, 10^3 Å, 10^4 Å, 10^5 Å and Ultrastyragel™ columns with pore diameters of 100 Å and 10^6 Å connected in series. Temperature was set at 30°C for both the GPC and the differential viscosity detector. Experiments were conducted on the first day and the third day of sample preparation.

Results and Discussion

Vapor Phase Osmometry. The calibration curve used to determine number average molecular weights by VPO is shown in Figure 1. The standards used were polystyrene with nominal molecular weights of 1250, 2150 and 5000 g/mole and sucrose octaacetate whose molecular weight is 678.6 g/mole. To use the calibration curve, a ΔV (voltage change related to the lowering of solvent vapor pressure in solution) versus concentration plot for the HPL sample was generated. The slope of such a graph was determined and used to obtain the number average molecular weight by interpolation from the calibration curve. Details of this unconventional multistandard calibration will be published elsewhere (41).

The results of the molecular weight determination of the hydroxypropylated lignins by VPO are shown in Table I. It may be noted that for all samples, there was about a 20% increase in apparent molecular weights between freshly prepared solutions and those tested three days later. It is postulated that such an increase was due to a time dependent association of the hydroxypropylated lignin molecules in solution.

Gel Permeation Chromatography. An example of a GPC/DV dual chromatogram is shown in Figure 2. The bolder trace is the DRI signal while the finer trace is due to the viscosity signal. It may be noted that the viscosity signal is significantly noisier than the DRI trace. This is a result of the HPL sample molecular weight being near the lower detection limit of the differential viscosity detector.

A summary of the results obtained from GPC/DV on the first day and the third day after solution preparation is shown in Table II. For Red Oak and RO:PO, the number average molecular weight decreased by approximately 6% from the first day to the third day. However, the intrinsic viscosities increased. According to traditional polymer solution theory (42), the product of intrinsic viscosity and molecular weight yields the hydrodynamic volume; specifically, it has been shown that the molecular weight that

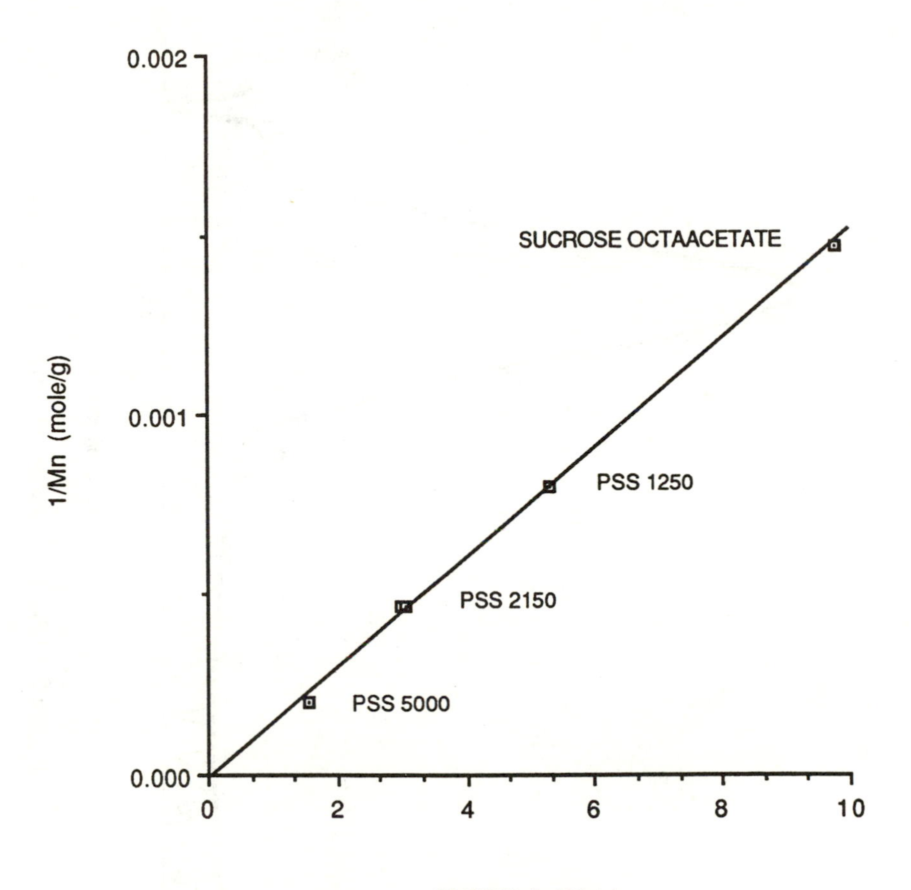

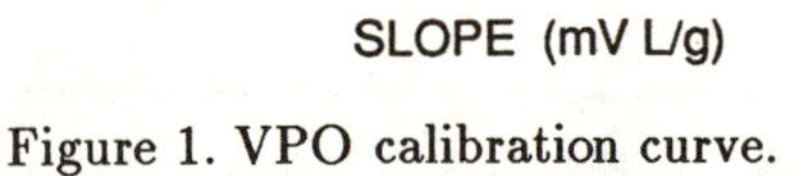

Figure 1. VPO calibration curve.

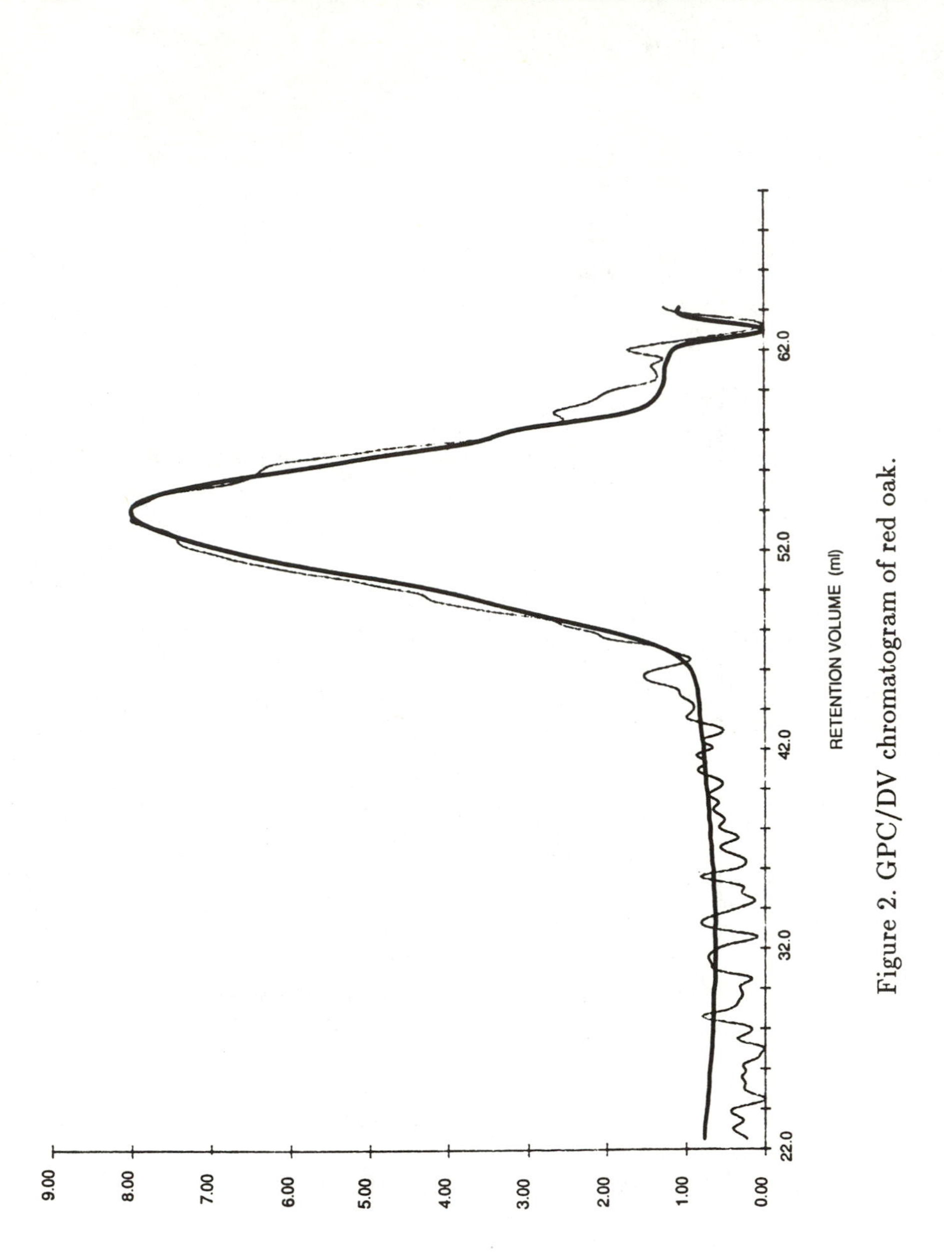

Figure 2. GPC/DV chromatogram of red oak.

Table I. $\overline{M}_n$ from Vapor Phase Osmometry

Sample	Day	$\overline{M}_n$ (g/mole)	% Diff.
Red Oak	1	1416	17.3
	3	1684	
RO:PO	1	1108	19.5
	3	1348	
Aspen	1	1393	19.3
	3	1691	
Westvaco	1	1499	21.0
	3	1850	

should be used in this expression is $\overline{M}_n$ (43). Therefore, a significant increase in the hydrodynamic volume for RO:PO over time is indicated. The number average molecular weights for aspen and Westvaco HPL's increased by about 20% from the first day to the third day, which is consistent with the results obtained by VPO. However, the intrinsic viscosity remained almost unchanged. Nevertheless, this brought about a dramatic increase in hydrodynamic volumes for both specimens. It is recognized that in the case of red oak the small change in hydrodynamic volume may be within experimental error; however, the overall trend in these results reveals that there was an increase in the hydrodynamic volumes of the lignins over time. This observation may be evidence of the presence of time dependent association of the HPL molecules in solution.

Table II. Results from GPC/DV

Sample	Day	$[\eta]$ (dl/g)	$\overline{M}_n$ (g/mole)	$[\eta]$M (dl/mole)
Red Oak	1	0.041	1535	62.9
	3	0.045	1433	64.5
RO:PO	1	0.032	1567	50.1
	3	0.041	1473	60.4
Aspen	1	0.044	1591	70.0
	3	0.045	1924	86.6
Westvaco	1	0.047	1597	75.1
	3	0.044	1951	85.8

Summary and Conclusions

A comparison of the molecular weights obtained from VPO and GPC/DV is shown in Table III. The results reveal that the values of number average molecular weights obtained by GPC/DV compare favorably with those obtained from the VPO. Thus, GPC/DV is a reliable technique of obtaining average molecular weights for lignins quite easily; more important, the absolute MWD is also revealed. In addition, because GPC/DV yields intrinsic viscosity as well, one has the benefit of obtaining hydrodynamic volume information. This advantage allowed for the acquisition of evidence in support of the presence of association in lignin solutions.

Table III. $\overline{M}_n$ from VPO and GPC/DV in g/mole

Sample	Day	VPO	GPC/DV
Red Oak	1	1416	1535
	3	1684	1433
RO:PO	1	1108	1567
	3	1348	1473
Aspen	1	1393	1591
	3	1691	1924
Westvaco	1	1499	1597
	3	1850	1951

Acknowledgments

We would like to thank Professor Wolfgang G. Glasser of the Department of Wood Science and Forest Products at Virginia Polytechnic Institute and State University for providing the lignin samples and the National Science Foundation for financial support of this project through grant #CBT-8512636.

Literature Cited

1. Soni, P. L.; Gaur, B. *J. Sci. Ind. Res.* 1984, **43**, 589.
2. Gray, R. L.; Parham, R. A. *CHEMTECH* 1982, **12**, 232.
3. Glasser, W. G.; Kelley, S. S. In *Encyclopedia of Polymer Science and Engineering*; Kroschwitz, J. I., Ed.; Wiley: New York, 1984; Vol. 8, p. 795.
4. Meister, J. J. In *Renewable Resource Materials, New Polymer Sources*; Carraher, C. E., Jr.; Sperling, L. H., Eds; Plenum: New York, 1986; p. 287.
5. Wilson, J. D.; Hamilton, J. K. *J. Chem. Ed.* 1986, **63**, 49.
6. Wu, C. F. L.; Glasser, W. G. *J. Appl. Polym. Sci.* 1984, **29**, 1111.
7. Rials, T. G.; Glasser, W. G. *Holzforschung* 1986, **40**, 353.
8. Glasser, W. G.; Barnett, C. A.; Sano, Y. *Appl. Polym. Symp,* 1983, **37**, 441.
9. Kolpak, F. J.; Cietek, D. J.; Fookes, W.; Cael, J. J. *Appl. Polym. Symp.* 1983, **37**, 491.
10. Brown, W. *J. Appl. Polym. Sci.* 1967, **11**, 2381.

11. Glasser, W. G.; Barnett, C. A.; Muller, P. C.; Sarkanen, K. V. *J. Agric. Food Chem.* 1983, **31**, 921.
12. Sarkanen, S.; Teller, D. C.; Hall, J.; McCarthy, J. L. *Macromolecules* 1981, **14**, 426.
13. Favis, B. D.; Yean, W. Q.; Goring, D. A. I. *J. Wood Chem. Technol.* 1984, **4**, 313.
14. Goring, D. A. I. In *Lignins: Occurrence, Formation, Structure and Reactions*; Sarkanen, K. V.; Ludwig, C. H., Eds.; Wiley: New York, 1971; Ch. 17.
15. Obiaga, T. I.; Wayman, M. *J. Appl. Polym. Sci.* 1974, **18**, 1943.
16. Connors, W. J.; Sarkanen, S.; McCarthy, J. L. *Holzforschung* 1980, **34**, 80.
17. Himmel, M. E.; Oh, K. K.; Sopher, D. W.; Chum, H. L. *J. Chromatogr.* 1983, **267**, 249.
18. Sarkanen, S.; Teller, D. C.; Stevens, C. R.; McCarthy, J. L. *Macromolecules* 1984, **17**, 2588.
19. Meister, J. J.; Nicholson, J. C.; Patil, D. R.; Field, L. R. *Polym. Mat. Sci. Eng.* 1986, **55**, 679.
20. Mörck, R.; Yoshida, H.; Kringstad, K. P.; Hatakeyama, H. *Holzforschung* 1986, **40** (Suppl.), 51.
21. Sarkanen, S.; Teller, D. C.; Abramowski, E.; McCarthy, J. L. *Macromolecules* 1982, **15**, 1098.
22. Huttermann, A. *Holzforschung* 1978, **32**, 108.
23. Forss, K. G.; Stenlund, B. G.; Sågfors, P. E. *Appl. Polym. Symp.* 1976, **28**, 1185.
24. Pellinen, J.; Salkinoja-Salonen, M. *J. Chromatogr.* 1985, **322**, 129.
25. Wagner, B. A.; To, T.; Teller, D. E.; McCarthy, J. L. *Holzforschung* 1986, **40** (Suppl.), 67.
26. Pla, F.; Robert, A. *Holzforschung,* 1984, **38**, 137.
27. Goring, D. A. I. In *Lignins: Properties and Materials*; Glasser, W. G.; Sarkanen, S., Eds.; American Chemical Society: Washington, DC, 1989.
28. Pla, F. Personal communication, 1988.
29. Trowbridge, P.; Brower, L.; Seeger, R.; McIntyre, D. *Polym. Mat. Sci. Eng.* 1986, **54**, 80.
30. Styring, M. *Polym. Mat. Sci. Eng.* 1986, **54**, 88.
31. Kuo, C.; Provder, T.; Koehler, M. E.; Kah, A. F. *Polym. Mat. Sci. Eng.* 1986, **54**, 80.
32. Haney, M. A.; Armonas, J. A.; Rosen, L. *Polym. Mat. Sci. Eng.* 1986, **54**, 75.
33. Yau, W. W. *Polym. Mat. Sci. Eng.* 1986, **54**, 74.
34. Tinland, B.; Mazet, J.; Rinaudo, M. *Makromol. Chem. Rapid Commun.* 1988, **9**, 69.
35. Letot, L.; Lesec, J.; Quivoron, C. *J. Liq. Chrom.* 1980, **3**, 427.
36. Lecacheux, P.; Lesec, J.; Quivoron, C. *J. Appl. Polym. Sci.* 1982, **27**, 4867.

37. Malihi, F. B.; Kuo, C.; Koehler, M. E.; Provder, T.; Kah, A. F. In *Size Exclusion Chromatography: Methodology and Characterization of Polymers and Related Materials*; Provder, T., Ed.; ACS Symp. Ser. No. 245; American Chemical Society: Washington, DC, 1984; p. 281.
38. Hamielec, A. E.; Meyer, H. In *Developments in Polymer Characterization–5*; Dawkins, J. V., Ed.; Elsevier: England, 1986, p. 95.
39. Haney, M. A. *J. Appl. Polym. Sci.* 1985, **30**, 3023.
40. Haney, M. A. *J. Appl. Polym. Sci.* 1985, **30**, 3037.
41. Gorce, J. N.; Siochi, E. J.; Ward, T. C., to be published.
42. Hiemenz, P. C. *Polymer Chemistry: The Basic Concepts*; Marcel Dekker: New York, 1984; Ch. 9.
43. Hamielec, A. E.; Ouano, A. C. *J. Liq. Chrom.* 1978, **1**, 111.

RECEIVED May 29, 1989

Chapter 8

Molecular Weight Distribution Studies Using Lignin Model Compounds

D. K. Johnson[1], Helena Li Chum[1], and John A. Hyatt[2]

[1]Solar Energy Research Institute, 1617 Cole Boulevard, Golden, CO 80401
[2]Research Laboratories, Eastman Chemicals Division, Eastman Kodak Company, Kingsport, TN 37662

High performance size exclusion chromatography (HPSEC) has been extensively employed in studies of the molecular weight distributions of lignins. Tetrahydrofuran (THF) is the most commonly used mobile phase with polystyrene-divinyl benzene HPSEC columns. Under these conditions the apparent molecular weight distributions of lignins may be determined relative to the standards employed, usually polystyrenes. There are, however, many lignins that are partially or totally insoluble in THF that require a solvent of greater solvating power. Many problems exist in using such eluants, e.g., dimethyl formamide alone or in the presence of added salts, including solute-solute association, and interactions between solute and solvent, column gel and solvent, and calibration standards and column gel. These interactions will be reported in light of their effect on the elution of lignin model compounds obtained from stepwise syntheses (molecular weights of 200-1100) and quinonemethide polymerizations (apparent weight-average molecular weights of 2000-7000).

The HPSEC of homogeneous linear polymers using polystyrene-divinyl benzene gels and relatively non-polar solvents such as tetrahydrofuran (THF) is well understood and can be used for the determination of molecular weight distributions (MWD) of similar polymers (1,2). The estimation of lignin MWD is, however, complicated by the interaction of several components in the HPSEC system (3). Lignins are irregular, multiply branched polymers containing various polar functionalities and a number of different interunit linkages which are a marked function of the method of lignin isolation. Thus, lignin solubility in THF varies from zero to 100%. To increase

0097-6156/89/0397-0109$06.00/0

solubility and prevent interactions arising from the presence of hydroxyl (alcoholic and phenolic) groups, lignins are often derivatized. Lignins are mostly acetylated (4,5), but have also been methylated (6) and silylated (7) prior to HPSEC. This study focuses on acetylated lignins and model compounds but also includes underivatized lignin model compounds and some other derivatives. Connors *et al.* (8) examined the use of solvents of higher solvating power for the SEC of lignins, e.g., dimethyl formamide (DMF) and dimethyl sulfoxide. However, the observation of multimodal elution profiles in these solvents, even when using derivatized lignins where hydrogen bonding between solute and solvent molecules should not occur, complicates the use of these solvents. Elution of this type has been ascribed to solute-solute association. To overcome these associative effects, salts such as LiCl and LiBr were added to give more unimodal elution profiles. The mechanism of association of lignin molecules and effects of adding electrolytes has been discussed by Sarkanen *et al.* (9).

Changes in solvent can also alter the size and shape of molecules which must strongly influence a technique relying on size exclusion for measurement of molecular weight. Molecular size may also be altered by hydrogen bonding of solvent molecules to solute molecules. This has been observed in the HPSEC of underivatized phenols with THF as solvent (10,11). Pellinen and Salkinoja-Salonen (7) have examined the HPSEC of a number of lignin model compounds and their acetylated and silylated derivatives using THF as eluant. They found no clear evidence of association or adsorption of these compounds to the column gel; however, the relationship between molecular weight and elution volume, while similar for both sets of derivatives, was different from that of the underivatized model compounds which eluted earlier by comparison. They postulated that this was due to strong solvation of the underivatized model compounds. It would appear, then, that lignins should be derivatized before HPSEC analysis; however, the elution of the derivatized model compounds did not coincide with that of the polystyrene standards.

HPSEC as a technique relies on transformation of elution time or volume into a scale of molecular weight, by observing the elution of polymer standards of known molecular weight or MWD. This implies that the relationship between size and molecular weight of the standards and polymers under investigation should be at least similar. With linear homopolymers, standards can usually be found so that good calibrations can be made. With more complicated polymers, e.g., branched polymers and copolymers, calibration becomes more inexact. There are only a few lignin model compounds of low molecular weight available for calibration and so other polymers such as polystyrenes have been used. More polar solvents such as DMF further complicate the use of chemically different standards as they may well behave differently than the lignins being studied, through changes in molecular size and from interactions with the column gel. The use of more polar solvents can also affect the column gel causing swelling of the gel beads and excessive back pressure. Any changes in pore size of the gel will obviously have a large effect on SEC although the current gel polymers are highly cross-linked to minimize this effect.

Recently, efforts have been made to produce calibration standards of higher molecular weight that are chemically similar to lignins, by step-wise syntheses (12), anion-initiated polymerization of quinonemethides (13), and preparative HPSEC of acetylated lignins (14). Knowledge of the molecular weights of these materials is either built into the method of preparation or determined by absolute methods such as sedimentation equilibrium measurements.

This study compares elution of lignin model compounds and model polymers in two solvents of similarly high solubility parameter (9.9 for THF and 12.1 for DMF), but with quite different fractional polarities (0.075 for THF and 0.77 for DMF) (15). Elution of model compounds is compared to that of acetylated lignins so that inferences may be made concerning molecular structure and the various interactions that occur between lignins and the HPSEC system. Any evidence of association of the model compounds and model polymers is of particular interest. The effects of modifying DMF by addition of LiBr and formic acid are also assessed particularly to overcome associative effects. Narrow MWD standards are included to study the difference in their elution compared to the lignin model compounds as solvent composition is changed.

Experimental

High performance size exclusion chromatography was performed using a Hewlett-Packard HP1090 liquid chromatograph containing an HP1040A ultraviolet diode array (UV) detector after which was connected an HP1037A refractive index (RI) detector.

Two columns (30×0.8 cm each) containing polystyrene-divinyl benzene gel beads (10μm diameter) with nominal pore diameters of 500Å (Altex, μ-Spherogel) and 100Å (Polymer Laboratories, PL-Gel) were used connected in series. These columns were chosen to give maximum resolution in the molecular weight range of the model compounds to be studied. The columns were maintained at 28°C and injections (5-10 μL) were made with an automatic injector of samples (1-2 mg/mL) that were dissolved in the eluant being used. The solvents used were chromatographic grade THF and DMF (Burdick and Jackson or similar quality). Formic acid (Aldrich, 96% ACS reagent grade) and lithium bromide (Aldrich, 99+%, anhydrous) were used without further purification for the higher ionic strength DMF eluants. The five eluants studied were pure THF and DMF, a mixture of THF and DMF (1:1 on a volume basis), DMF containing formic acid (5 wt%) and DMF containing lithium bromide (0.1 M). A flow rate of 1.0 mL/min was used throughout this study. The reproducibility of retention volume was measured by following the elution of toluene included in many samples as an internal standard.

The polystyrene (PS) and polymethylmethacrylate (PMMA) narrow molecular weight standards were obtained from three kits (Polymer Laboratories, S-L-10, S-M-10 and M-M-10) covering the range from 3,000,000 to 500. Five nonylphenyl-terminated polyethylene oxides (Aldrich, Igepals) with molecular weights of 4600 to 300 were also studied. The monodisperse lignin model compounds were prepared by a modified enolate addition method (12). Other monodisperse lignin model compounds were prepared as described by Himmel *et al.* (16). The synthetic lignin model polymers (α,β-bis(0-4-aryl) ether bonded) were prepared by anion-initiated polymerization of a quinonemethide according to the procedure of Chum *et al.* (13). The structures of the lignin model compounds and polymers are shown in Figure 1. The lignin samples were prepared by organosolv pulping of aspen wood in aqueous methanol (70 vol%) with sulfuric acid catalyst (0.05 M) as described by Chum *et al.* (17). Before analysis the lignins were quantitatively acetylated using a method developed by Gierer and Lindeberg (18).

Results and Discussion

With THF as the solvent the retention times of the PS, PMMA and Igepal standards were obtained. Figure 2 compares the elution of these three sets of standards on the column set used, and shows a curve obtained by a non-linear least squares fitting of this data. Considering the differences in chemical structure of these polymers, it is surprising that such a good fit was found. This suggests that the change in hydrodynamic volume with molecular weight of these polymers is similar.

Figure 3 compares the elution in THF of the lignin model compounds having free phenolic groups (squares) to those in which the phenolic groups were derivatized (stars). The solid line that is shown is the calibration curve obtained by non-linear least squares fitting of the retention times of the standard PS, PMMA and Igepal polymers. The model compounds with derivatized phenolic groups eluted later than predicted by this curve. Pellinen and Salkinoja-Salonen (7) observed a similar result using a different set of lignin model compounds which they acetylated and silylated. A best fit of the derivatized models is shown as the dashed line in Figure 3 and appears to parallel the calibration curve. The reason for the later elution of the derivatized models relative to the polymer standards is most likely that they have a relatively smaller hydrodynamic volume per unit molecular weight than the polymer standards. Alternatively, elution of the derivatized lignin models may have been delayed by adsorption to the column gel. When using THF as solvent, adsorption has been reported for highly condensed aromatic compounds (10,11). The dashed line could be used as a calibration for acetylated lignins but covers too small a range of molecular weights even for low-molecular-weight lignins.

Apart from one compound (II), the lignin model compounds that had free phenolic groups eluted at close to the retention times predicted by the calibration curve from the polymer standards and not from the derivatized model compounds. This could simply be a result of the underivatized models having a similar variation in hydrodynamic volume with molecular weight as the polymer standards. However, it is to be expected that solvation of the underivatized model compounds should occur with THF as solvent (10), with hydrogen bonding of one THF molecule to each under-

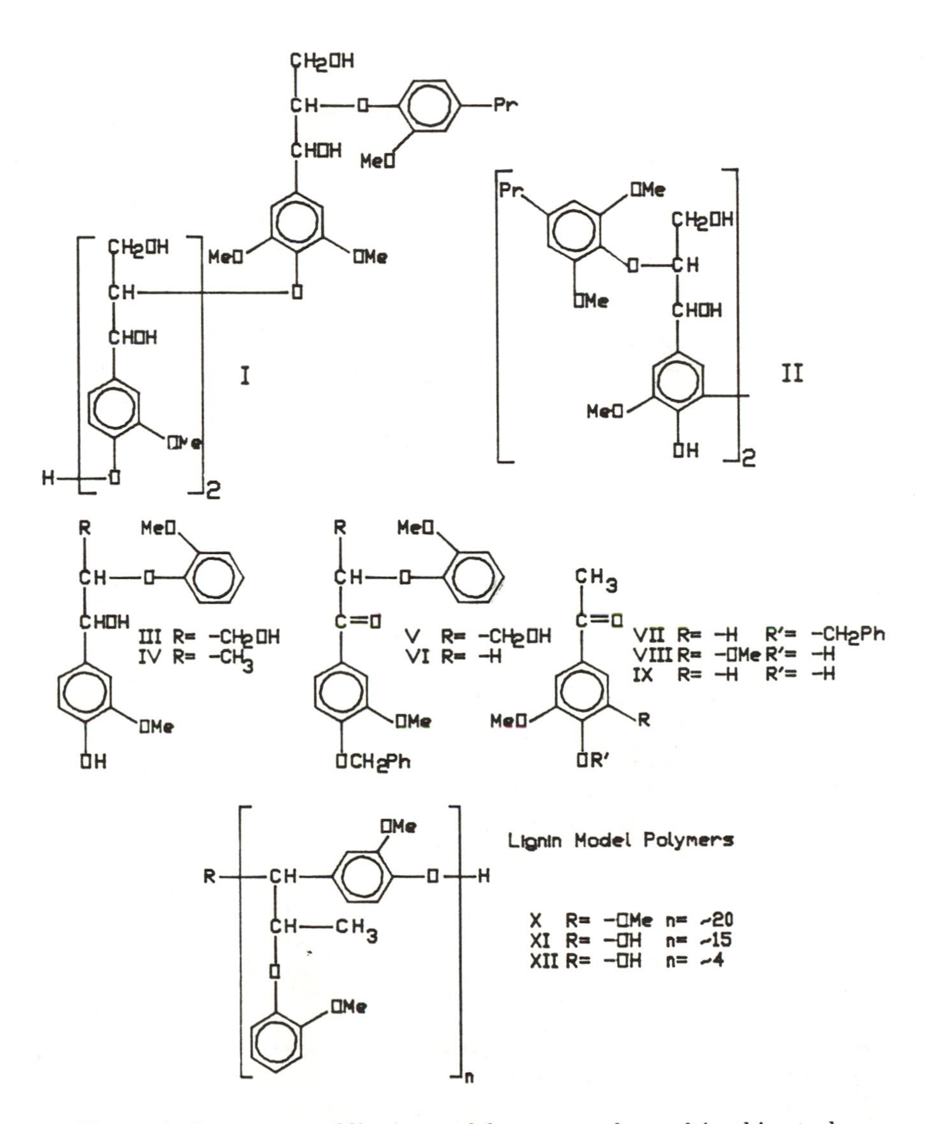

Figure 1. Structures of lignin model compounds used in this study.

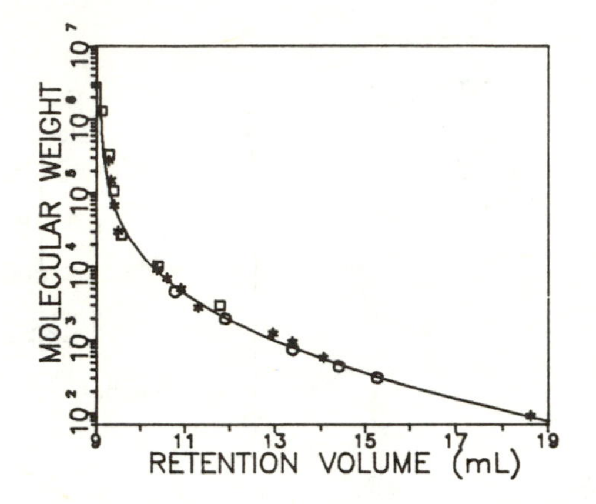

Figure 2. Calibration of column set with THF as eluant using polystyrenes (⋆), polymethylmethacrylates (□) and Igepals (○).

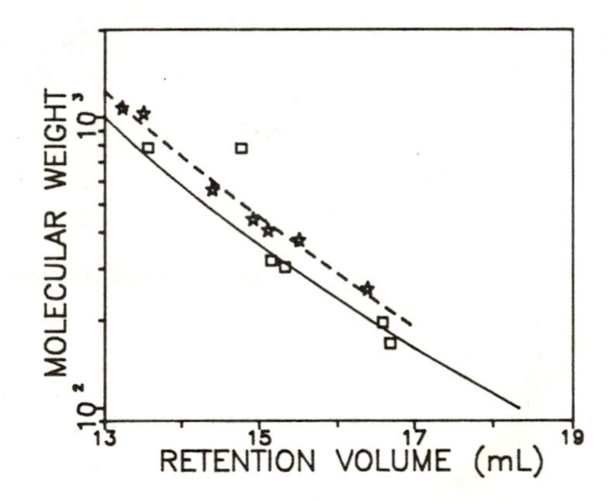

Figure 3. Elution of derivatized (⋆) and underivatized (□) lignin model compounds with THF as eluant.

ivatized OH group, the degree of hydrogen bonding being higher the more acidic the OH group. This would make the underivatized models large for their molecular weight relative to the derivatized model compounds and explain why they elute earlier than the derivatized models. The behavior of compound III, the biphenyl tetramer hexaol, appears anomalous.

The effects on the column parameters, void volume (V_o) and total permeation volume (V_t), of changing the eluant were also examined. The void volume was determined by following the retention volumes of the largest PS (3000K) and PMMA (1300K) polymers. As can be seen from Table I, the void volume appeared to increase slightly as the polarity of the solvent was increased, although the change in retention volumes could have other causes, especially as the PS seemed to increase more than the PMMA.

Table I. Investigation of Changes in V_t and V_o

	Retention Volume (mL)				
	THF	THF/DMF	DMF	DMF/LiBr	DMF/HCO_2H
Hexane	17.45		26.74	28.08	
Toluene	18.61	20.07	23.63	24.20	24.36
Dioxane	18.91	19.04	20.90	21.23	21.19
Acetone	18.25	18.22	19.48	19.65	19.65
Me THF	18.41	18.24	23.07	23.67	23.59
THF		18.19	22.91	23.35	23.32
Me Formate	18.61		19.34	19.50	19.52
DMA	18.74	18.25	19.30	19.26	19.19
DMF	18.88	18.36			
DEF		18.28		19.44	19.43
V_t	18.30	18.30	19.40	19.40	19.40
3000K PS	8.99	8.99	9.01	9.26	9.26
1300K PMMA	9.13	9.11	9.19	9.22	9.26
V_o	8.99	8.99	9.00	9.07	9.11

Me THF = methyl tetrahydrofuran; Me Formate = methyl formate; DMA = dimethyl acetamide; DEF = diethyl formamide.

Approximate values for V_t were determined using 16 compounds, some of which are included in Table I, with molecular weights of about 100 or less and containing a variety of functional groups. In THF the retention volumes of these compounds fell within a fairly narrow range within about ±5% of V_t. By comparison, the reproducibility of retention volume for toluene in THF was 0.01 mL and in DMF was 0.14 mL. As exemplified by the retention times of hexane, toluene and dioxane in Table I, increasing the solvent polarity substantially increased the retention volumes of the more non-polar compounds to as much as 50% more than the estimated V_t. This is almost certainly an adsorption effect, with the relatively non-polar column packing increasingly retaining the non-polar solutes as the solvent polarity increases. The best estimate of V_t was obtained by following the retention volumes of compounds chemically similar to the solvent being used. These were detected using the refractive index detector. For THF, V_t was estimated using methyl tetrahydrofuran. For the DMF containing eluants, dimethyl acetamide, diethyl formamide, etc., were used. The narrow range of elution volumes for all the THF and DMF analogues when using the mixed THF/DMF solvent appears to justify this approach. The increase in V_t of about 1 mL with the more polar solvents represents about a 10% increase in the pore volume of the packing material. The small increase in V_o and larger increase in V_t may or may not be consistent with swelling of the polymer beads. Because of the change in column parameters, comparison of solute elution with change in solvent was subsequently made after calculation of solute distribution coefficients ($K_D = (V_r - V_o)/(V_t - V_o)$ where V_r is the solute retention volume) (19).

Figure 4 shows the change in elution of the polystyrene standards with a change in solvent. The retention volumes of the PS polymers substantially increased as the solvent polarity increased such that in DMF and DMF modified with LiBr and formic acid, elution beyond V_t was observed. Similarly, but to a lesser extent, the elution volumes of the other polymer standards (Table II) also increased. An increase in K_D may be explained by either a solute-solvent or a solute-column gel interaction. It is certainly conceivable that a change in solvent could produce a change in the hydrodynamic volume of a solute. Large molecules are considered to be primarily random coils in solution. The coiled orientation of the molecule, and therefore its hydrodynamic volume, is likely to be strongly affected by the solvent being used. Small molecules have less flexibility so this mechanism of size change should be less important for small molecules. The formation of solute-solvent complexes via mechanisms such as hydrogen bonding would increase molecular size resulting in smaller K_D. A decrease in molecular size will increase the K_D of a solute; however, K_D should not be greater than 1 unless solute adsorption to the column packing takes place.

In this study the lower molecular weight PS were much more strongly affected by the changes in solvent with K_D values up to 1.5 being observed. The PMMA were affected to a much lesser extent and the Igepals the least of all by the solvent changes. The PS are clearly being retained by the column gel by adsorption as solvent polarity is increased. Since the column gel is a copolymer of styrene and divinyl benzene and is chemically similar to the PS standards, affinity of the PS for the packing material should increase as solvent polarity increases. The lower molecular weight PS are more strongly affected probably because they permeate a larger fraction of the total pore volume. Any effect due to change in size of the PS polymers is obscured by the strong adsorption effect. The smaller increases in K_D for the PMMA and Igepals are probably the result of a lower affinity of these polymers for the packing material although the influence of increased molecular size as a counterbalance to adsorption may also be important. As has been noted before (3), PS are very poor standards for HPSEC when using DMF-based solvents, as they exhibit substantial non-size exclusion

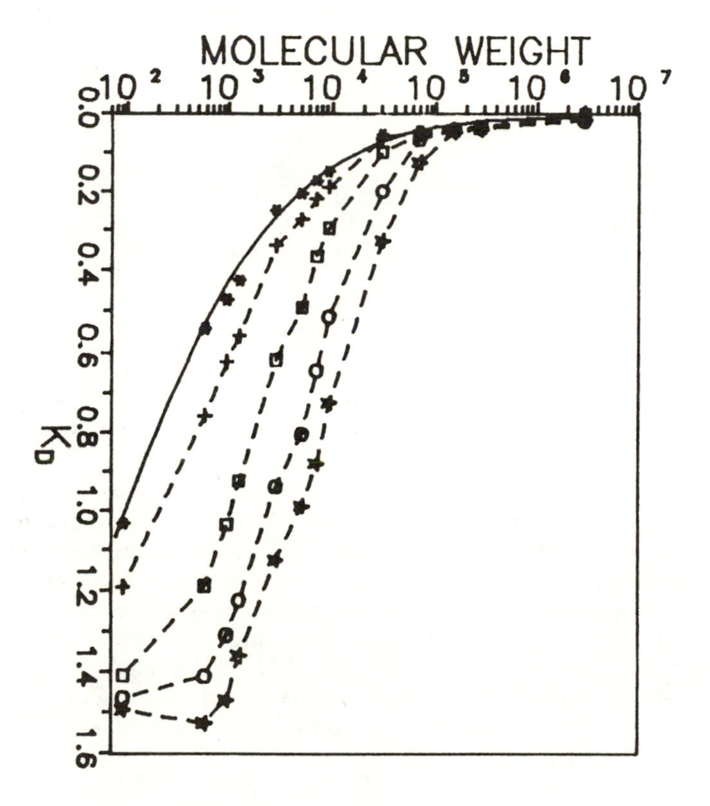

Figure 4. Elution of polystyrene standards with different eluants. ∗ = THF; + = THF/DMF; □ = DMF; ○ = DMF/LiBr; ⋆ = DMF/HCO_2H.

Table II. Effect of solvent changes on elution of PMMA and Igepals

	Solute Distribution Coefficient (K_D)				
Mol. Wt.	THF	THF/DMF	DMF	DMF/LiBr	DMF/HCO_2H
PMMA					
1300000	0.02	0.01	0.02	0.02	0.02
330000	0.03	0.04	0.04	0.03	0.03
107000	0.04	0.04	0.06	0.04	0.04
60000		0.05		0.05	0.05
27000	0.06	0.07	0.08	0.09	0.09
10300	0.15	0.16	0.18	0.20	0.20
3000	0.30	0.32	0.39	0.47	0.44
Igepals					
4626	0.19	0.20	0.22	0.23	0.23
1983	0.31	0.33	0.37	0.39	0.39
749	0.47	0.52	0.59	0.64	0.66
441	0.58	0.65	0.77	0.81	0.84
308	0.68	0.76	0.94	0.96	1.03

behavior. The other polymer standards could possibly be used, but only if the polymers being analyzed experienced very similar non-size exclusion effects.

Figure 5 shows a typical example of HPSEC of an acetylated lignin in the various solvent systems. Addition of DMF to the mobile phase resulted in a multimodal chromatogram with a substantial fraction of the lignin polymer being excluded from the pores of the column gel. In fact, elution of material up to 1 ml before V_o was observed. This type of behavior has been reported by several researchers and is generally attributed to association of lignin molecules. Connors *et al.* (8), using dimethyl sulfoxide and μ-styragel columns, observed association of both underivatized and acetylated kraft lignins, indicating that the mechanism of association was not through hydrogen bonding. In non-aqueous solvents, Ekman and Lindberg (6), studying exhaustively methylated lignins, and Chum *et al.* (3), with acetylated lignins and DMF as solvent, also found association taking place. Sarkanen (9) reported that underivatized organosolv lignins also exhibited association during gel permeation chromatography on dextran gels using aqueous sodium hydroxide as solvent. He proposed that interactions of the highest occupied molecular orbital- lowest unoccupied molecular orbital (HOMO-LUMO) type could be governing these associative processes. A number of researchers (3,8) have shown that addition of an electrolyte (LiCl, LiBr, etc.) to the solvent disrupts molecular association so that unimodal elution of lignins is again obtained. From Figure 5 it appears that neither formic acid (5 wt%) nor LiBr (0.1 M) could break all the associative interactions, with formic acid being slightly less effective. These associative interactions have been observed with other polymers and it has been suggested (3) that addition of LiBr could eliminate association by shielding the dipoles of individual molecules. By using model compounds a better understanding of these interactions should be obtained.

Table III describes the effect of solvent change on the lignin model compounds. None of the model compounds exhibited evidence of association; all had unimodal elution in the different solvents. The K_D of the fully derivatized model compounds tended to increase as solvent polarity was increased, as had that of the polymer standards. As these are fully derivatized, relatively small molecules, the possibility for size change through interaction with the solvents is small. Increasing affinity for the column gel as the solvent polarity increased is the most probable explanation for their greater retention.

Table III. Effect of Solvent Changes on Elution of Lignin Model Compounds and Polymers

		Solute Distribution Coefficients (K_D)				
	Mol.Wt.	THF	THF/DMF	DMF	DMF/LiBr	DMF/HCO_2H
Fully Derivatized Models[1]						
IA	1078	0.46	0.47	0.51	0.54	0.54
IIA	1034	0.49	0.52	0.58	0.62	0.62
IIIA	446	0.64	0.64	0.68	0.71	0.72
VI	378	0.70	0.68	0.75	0.77	0.81
VII	256	0.80	0.80	0.91	0.94	0.97
Partially or Underivatized Models						
I	784	0.49	0.44	0.45	0.42	0.46
V	408	0.66	0.62	0.67	0.65	0.70
III	320	0.66	0.59	0.59	0.57	0.61
IV	304	0.68	0.64	0.66	0.65	0.69
VIII	196	0.82	0.74	0.75	0.76	0.77
IX	166	0.83	0.76	0.79	0.77	0.82
α,β-bis(0-4-aryl) Ether Bonded Model Polymers[2]						
X	6100	0.18	0.19	0.21	0.22	0.23
XI	4500	0.21	0.22	0.24	0.26	0.28
XII	1200	0.40	0.40	0.45	0.46	0.49

[1] The A indicates peracetylated derivatives of model compounds I, II and III.

[2] Apparent molecular weights for the lignin model polymers shown here were determined by HPSEC.

The partially derivatized and underivatized model compounds had K_D that appeared to vary quite differently. The K_D for these compounds were generally lower in the mixed THF/DMF solvent than in the pure solvents. Addition of LiBr in DMF generally resulted in lower K_D, whereas addition of formic acid resulted in higher K_D than were observed using the pure solvent. As these compounds have free phenolic and alcoholic hydroxy groups, it is likely that their size was substantially affected because of hydrogen bonding with the solvents. However, the complex variability of the behavior of these underivatized model compounds is probably due to the combination of several interactions with the HPSEC system.

The effect of solvent change on the elution of the lignin model polymers is also shown in Table III. All the polymers were increasingly retained by the column gel as eluant polarity was increased. However, as can be seen in the example in Figure 6, addition of DMF to the solvent produced multimodel elution similar to that observed with acetylated lignins which has been attributed to association. As observed with lignins, addition of LiBr or formic acid prevented association with only a slight shoulder in the chromatogram obtained in DMF/formic acid indicating that any association remained. Comparing the chromatograms of the acetylated lignins and lignin model polymers, it appears that the lignins associated to a greater degree than did the model polymers.

The question posed by these results is: Why do the model polymers appear to behave like lignins while the model compounds do not? Chemically the model compounds (e.g., IA and IIA) are more similar to the acetylated lignins than are the model polymers. The linkages between C_9 units in the model compounds are either β-0-4 or biphenyl which are two important interunit linkages found in organosolv lignins. The linkages between C_9 units in the model polymers are α-0-4, but they do have a guaiacyl group bonded to the β- carbon of the propyl sidechain. While the model polymers are unacetylated, they have very few free hydroxyl groups. The model polymers are produced by anion initiated polymerization of a quinonemethide using hydroxide or methoxide. Chain propagation occurs by reaction of the resulting phenoxide with the unsaturated α-C of the quinonemethide, with termination occurring when the phenoxide reacts with a proton instead. Consequently, the model polymers should only have one free phenolic hydroxyl and one free aliphatic hydroxyl (or a methoxyl) per molecule independent of their molecular weight. Despite the differences between the acetylated lignins and the model polymers, their HPSEC behavior in the various solvents is similar and different from that of the acetylated model compounds. There is one property in which the model polymers are similar and the model compounds are dissimilar to lignins, and that is their dispersity. The model polymers and lignins are polydisperse, while the model compounds are monodisperse. Connors *et al.* (3) observed multimodal elution from gel permeation chromatography of a synthetic DHP lignin when using DMF as solvent. Similar behavior was found for a kraft lignin in the same system. Tentative conclusions that may be drawn from these results are that high and low molecular weight components must be present for association to occur with lignin-like molecules, such that complexes sufficiently large to be excluded from the pores of the column gel are produced, or that the complex three-dimensional structures of these molecules somehow enhance associative interactions.

Experiments to obtain evidence of associative interactions between low

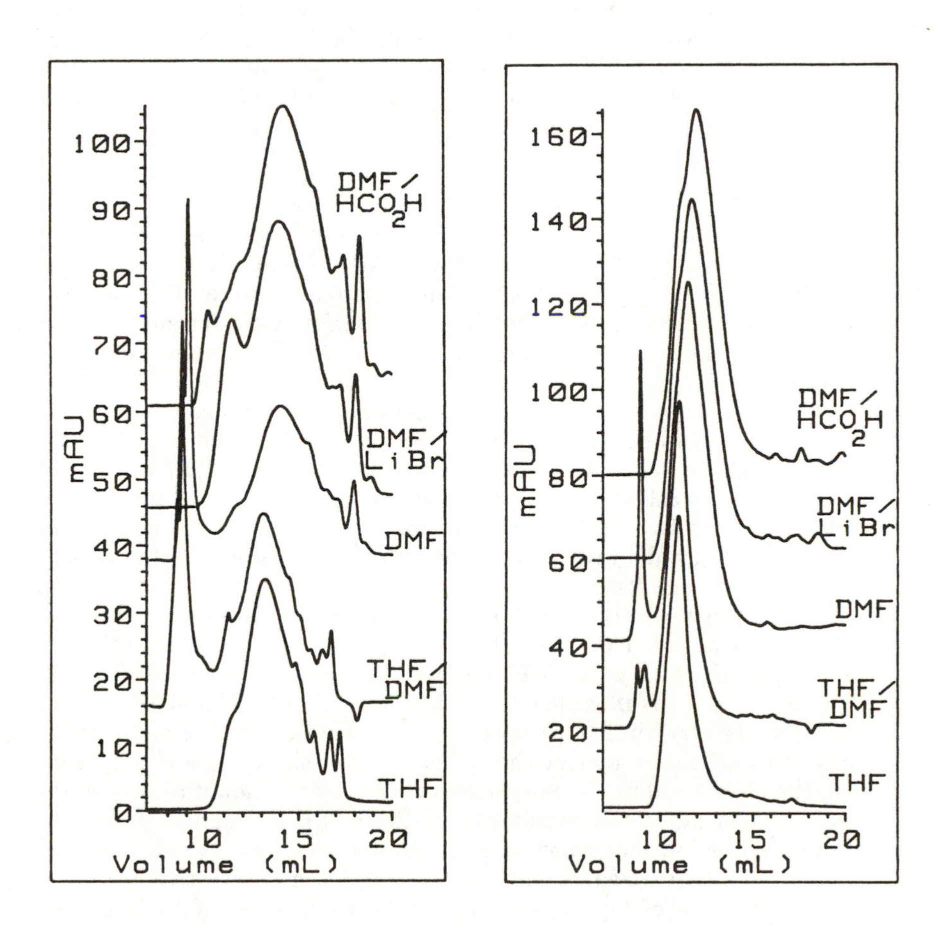

Figure 5. HPSEC of an acetylated organosolv aspen lignin in the various eluants.

Figure 6. HPSEC of an α, β-bis(o-4-aryl) ether bonded model polymer (XI) in the various eluants.

and high molecular weight lignin-like components have been attempted. HPSEC of mixtures of acetylated model compounds (IA, IIA and IIIA) with an acetylated organosolv lignin were obtained using THF and DMF as eluants. In all cases, no additional associative behavior was observed. The HPSEC of the mixtures appeared to be simply the addition of the HPSEC of the model compounds to that of the lignin. In THF no excluded peaks were observed and in DMF the excluded peaks of the lignin were reproduced without modification. There was no clearly discernible interaction between these lignin model tetramers and dimer and the lignin molecules.

Conclusions

1. A physical change in the column gel was found when using the high fractional polarity eluants made with DMF. There was apparently a small increase in the void volume and approximately a 10% increase in the total pore volume. The observed changes were reproducible as the column set was cycled through eluants of high and low fractional polarity.
2. As shown previously in the literature, polystyrenes are strongly affected by adsorption when using solvents containing DMF and polystyrene-divinylbenzene gels. The PMMA, Igepals, derivatized lignin model compounds, and lignin model polymers also appear to be affected by adsorption but to a lesser extent. Unless a different column gel is used for chromatography with DMF, this non-size exclusion effect is always likely to be a problem.
3. The α,β-bis(0-4-aryl) ether bonded lignin model polymers exhibited associative behavior on going to the high fractional polarity solvents such as DMF/THF and DMF. The association was decreased by addition of 0.1 M LiBr more than by addition of 5 wt% formic acid. This behavior is very similar to that observed for derivatized and underivatized organosolv and other lignins. These simple linear polymers are therefore good model polymers for lignins and should receive more attention as calibration standards for the HPSEC of lignins.
4. None of the low-molecular-weight lignin model compounds used in this study, derivatized or not, clearly exhibited associative behavior in the presence of higher fractional polarity eluants. This indicates that the self-association constants for these compounds in these media are small, at least for the concentration range studied.
5. The chromatograms of mixtures of well-defined low-molecular-weight lignin model compounds and α,β-bis(0-4-aryl) ether bonded lignin model polymers and acetylated lignins in high fractional polarity solvents appeared to be merely additive. In the concentration range investigated, these mixtures did not exhibit evidence of association between the low-molecular-weight compounds and the polymers, in addition to that already exhibited by the polymers themselves within the experimental errors of these measurements.

Acknowledgments

The authors wish to thank Dr. M. E. Himmel and Dr. S. Sarkanen for many profitable discussions and suggestions. The support of the Biomass Energy Technology Division of the U.S. Department of Energy through FWP BF82 is gratefully acknowledged.

Literature Cited

1. Yau, W. W.; Kirkland, J. J.; Bly, D. D. *Modern Size Exclusion Chromatography*; Wiley: New York, 1979.
2. Provder, T. In *Size Exclusion Chromatography: Methodology and Characterization of Polymers and Related Materials*; ACS Symp. Ser. No. 245; American Chemical Society: Washington, DC, 1984.
3. Chum, H. L.; Johnson, D. K.; Tucker, M. P.; Himmel, M. E. *Holzforschung* 1987, **41**, 97-108.
4. Faix, O.; Lange, W.; Beinhoff, O. *Holzforschung* 1980, **34**, 174-76.
5. Faix, O.; Lange, W.; Salud, E. C. *Holzforschung* 1981, **35**, 3-9.
6. Ekman, K. H.; Lindberg, J. J. *Paperi ja Puu* 1966, **46**, 241-44.
7. Pellinen, J.; Salkinoja-Salonen, M. *J. Chromatogr.* 1985, **328**, 299-308.
8. Connors, W. J.; Sarkanen, S.; McCarthy, J. L. *Holzforschung* 1980, **34**, 80-85.
9. Sarkanen, S.; Teller, D. C.; Hall, J.; McCarthy, J. L. *Macromolecules* 1981, **14**, 426-34.
10. Philip, C. V.; Anthony, R. G. In *Size Exclusion Chromatography*; Provder, T., Ed.; ACS Symp. Ser. No. 245; American Chemical Society: Washington, DC, 1984; pp. 257-72.
11. Johnson, D. K.; Chum, H. L. In *Pyrolysis Oils from Biomass: Producing, Analyzing and Upgrading*; Soltes, E.; Milne, T., Eds.; ACS Symp. Ser. No. 376; American Chemical Society: Washington, DC, 1988; pp. 156-66.
12. Hyatt, J. A. *Holzforschung* 1987, **41**, 363-70.
13. Chum, H. L.; Johnson, D. K.; Palasz, P.; Smith, C. Z.; Utley, J. H. P. *Macromolecules* 1987, **20**, 2698-702.
14. Himmel, M. E.; Tatsumoto, K.; Oh, K. K.; Grohmann, K.; Johnson, D. K.; Chum, H. L. In *Lignin: Properties and Materials*; Glasser, W. G.; Sarkanen, S., Eds.; ACS Symp. Ser.
15. Somerville, G. R.; Lopez, J. A. In *Solvents: Theory and Practice*; Tess, R. W., Ed.; ACS Symp. Ser. No. 124; American Chemical Society: Washington, DC, 1973; pp. 175-85.
16. Himmel, M. E.; Oh, K. K.; Sopher, D. W.; Chum, H. L. *J. Chromatogr.* 1983, **267**, 249-65.
17. Chum, H. L.; Johnson, D. K.; Black, S.; Baker, J.; Grohmann, K. Wallace, K.; Schroeder, H. A. *Biotechnol. Bioeng.* 1988, **31**, 643-49.
18. Gierer, J.; Lindeberg, O. *Acta Chem. Scand.* 1980, **B34**, 161-70.
19. In *Sephadex Filtration in Theory and Practice*; Pharmacia Fine Chemicals: Uppsala, 1979; p. 32.

RECEIVED March 17, 1989

Chapter 9

Determination of the Molar Mass Distribution of Lignins by Gel Permeation Chromatography

Kaj Forss, Raimo Kokkonen, and Pehr-Erik Sågfors

The Finnish Pulp and Paper Research Institute, P.O. Box 136, 00101 Helsinki, Finland

This paper describes the fractionation of lignin sulfonates on elution through Sephadex G-50, G-75 and Sephacryl S-300 using water or a 0.5M sodium chloride solution buffered to pH 8 as the eluent. Fractionation of kraft lignin on elution through Sephadex G-50 with 0.5M sodium hydroxide as well as the influence of the sodium hydroxide concentration is also described. By comparing the retention volumes of proteins and lignin sulfonate fractions with known molar masses, it is shown that several commercially available proteins can be used for calibration of the columns. It is shown that on elution through Sephadex G-25 with 0.5M sodium hydroxide the retention volume of monomeric compounds is influenced more by their functional groups than by their molecular size.

A method is needed for the determination of the molar mass and molar mass distributions of lignins. The method should give reproducible results when used in different laboratories. The procedure should not be complicated, and the calibration components must be readily available and reasonably priced.

The present paper describes the fractionation of lignin sulfonates and kraft lignin by gel permeation chromatography (GPC) and the method developed and used for several years at the Finnish Pulp and Paper Research Institute.

Sulfonated Lignins

Elution of Lignin Sulfonates with Water. As can be seen from Figures 1, 2, 3, and 4, the fractionation of lignin sulfonates on elution with water through Sephadex G-50 and G-75 takes place in such a way that the logarithms of

0097–6156/89/0397–0124$06.00/0

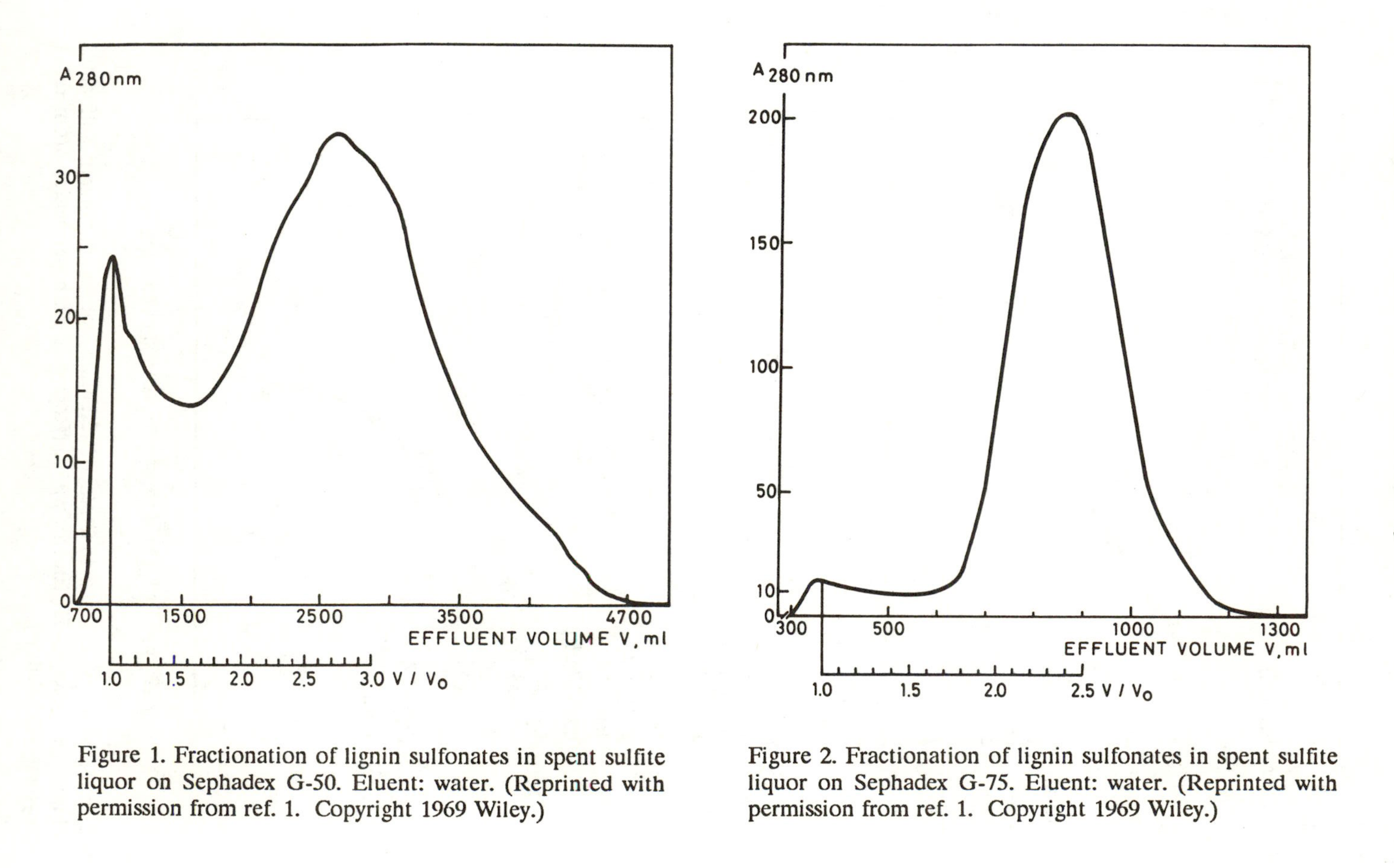

Figure 1. Fractionation of lignin sulfonates in spent sulfite liquor on Sephadex G-50. Eluent: water. (Reprinted with permission from ref. 1. Copyright 1969 Wiley.)

Figure 2. Fractionation of lignin sulfonates in spent sulfite liquor on Sephadex G-75. Eluent: water. (Reprinted with permission from ref. 1. Copyright 1969 Wiley.)

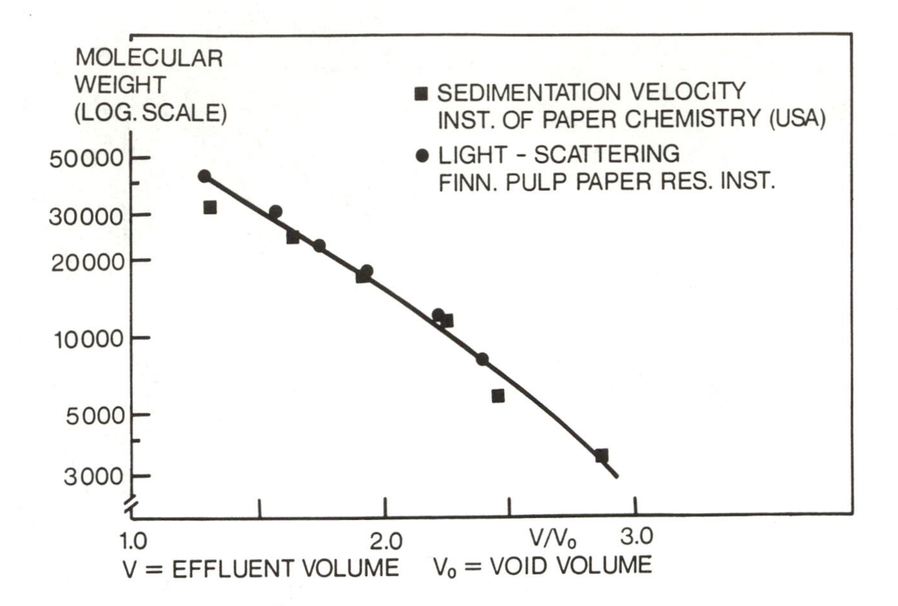

Figure 3. Molar mass as a function of retention volume for lignin sulfonates. Column: Sephadex G-50. Eluent: water. (Reprinted with permission from ref. 1. Copyright 1969 Wiley.)

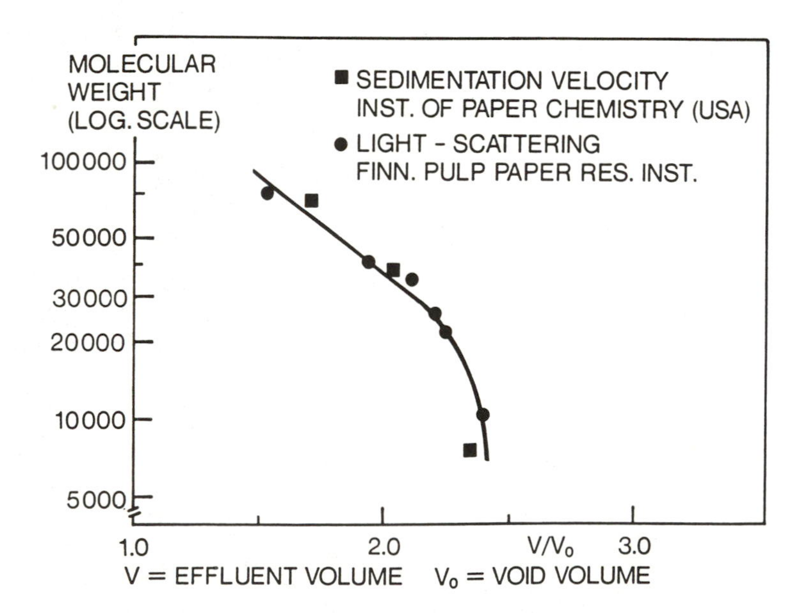

Figure 4. Molar mass as a function of retention volume for lignin sulfonates. Column: Sephadex G-75. Eluent: water. (Reprinted with permission from ref. 1. Copyright 1969 Wiley.)

the molar masses of the lignin sulfonates show a straightline relationship with the retention volumes (1).

In these experiments Sephadex G-50 effected fractionation of the lignin sulfonates with molar masses of 5 000–50 000 dalton and Sephadex G-75 those with molar masses of 20 000–100 000 dalton. The molar mass determinations carried out using the light-scattering and ultracentrifugal sedimentation velocity methods gave identical results. However, calibrating the columns using light-scattering or ultracentrifugation is too awkward for routine determinations. There is thus a need for readily available calibration standards of known molar mass that elute at the same retention volume as lignin sulfonates of the same molar mass.

It should be noted that the relationships between molar mass and retention volume for lignin sulfonates shown in Figures 3 and 4 are strictly only valid for the samples studied in these experiments because lignin sulfonates are polyelectrolytes and thus interact with each other and with the gel matrix of the column. The shape of the calibration curve is thus affected by, among other things, the size and concentration of the sample (2). Interactions between molecular species can be eliminated by eluting with a suitable electrolyte.

Elution of Lignin Sulfonates with Electrolyte Solution and Calibration of Columns

The effects caused by the electrolyte nature of lignin sulfonates are eliminated by using a 0.5M sodium chloride solution as eluent. This eluent is made 0.1M with respect to Tris-HCl and buffered to pH 8 with hydrochloric acid in order to dissolve the proteins used as calibration standards (Fig. 5).

The column is calibrated using proteins of known molar mass. The relative retention volumes 0.0 and 1.0 are defined by the elution of Blue Dextran (molecular weight 2 000 000) and sulfosalicylic acid (molecular weight 218), respectively.

In the experiment described in Figure 6, four lignin sulfonate fractions with known molar masses were eluted.

Figure 6 shows that the proteins used for calibration elute in the same way as lignin sulfonates, which justifies the use of proteins as calibration standards. A comparison between Figures 4 and 6 shows that elution with an electrolyte solution fractionates lignin sulfonates in the range 3 000–80 000 dalton, but that elution with water fractionates those in the range 20 000–100 000 dalton.

Because of the very high molar masses of enzymically polymerized lignin sulfonates, Sephacryl S-300 is used as the gel matrix. The fractionation range is 10 000 to 1 000 000 dalton. Even in this case proteins of known molar mass can be used as calibration standards (Figs. 7 and 8).

One disadvantage of using salt solution as eluent is that the lignin sulfonates tend to adsorb onto the gel matrix, resulting in a resolution inferior to that obtained by elution with water. On the other hand, elution behavior with water is adversely affected by the polyelectrolyte properties of the lignin sulfonates. Adsorption, which is caused by the phenolic hydroxyl

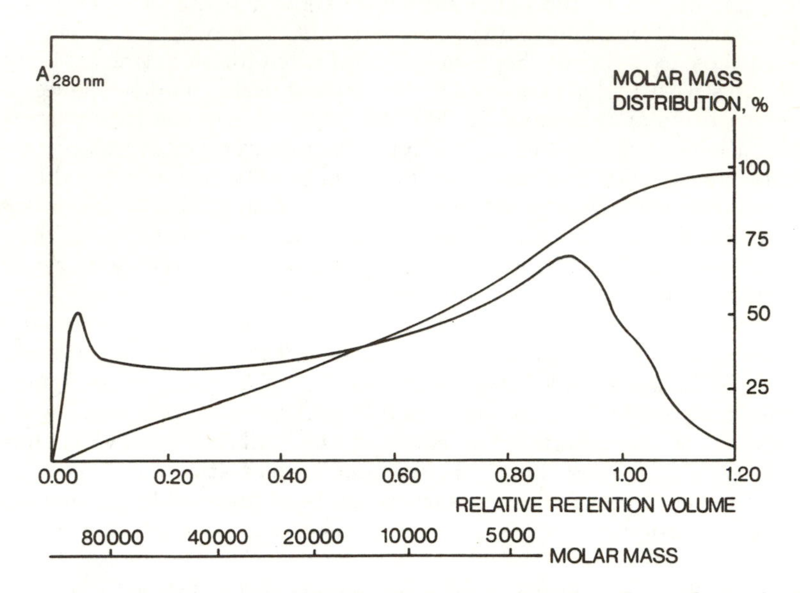

Figure 5. Fractionation of lignin sulfonates on elution through Sephadex G-75. Eluent: 0.5M NaCl, 0.1M Tris-HCl (pH 8).

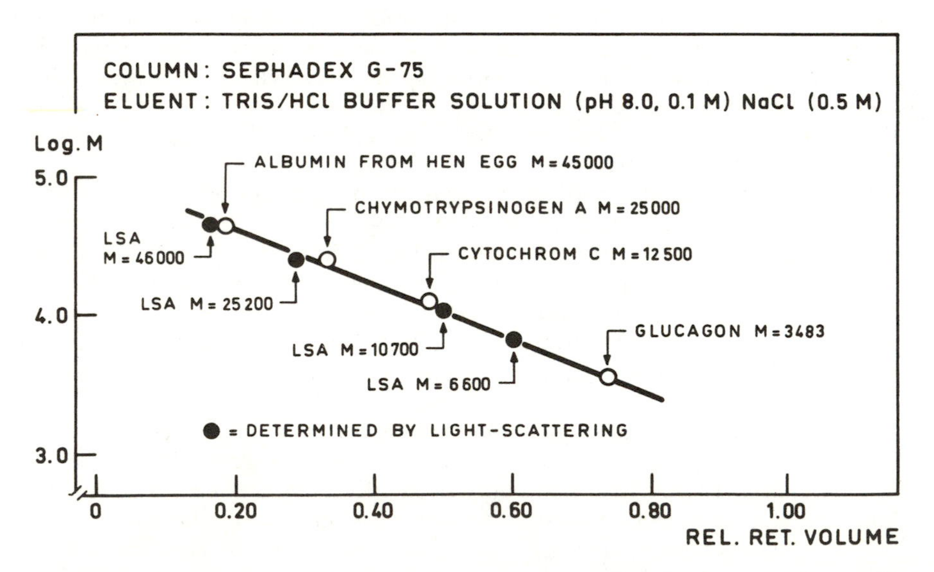

Figure 6. Calibration of Sephadex G-75. Eluent: 0.5M NaCl, 0.1M Tris-HCl, pH 8 (4).

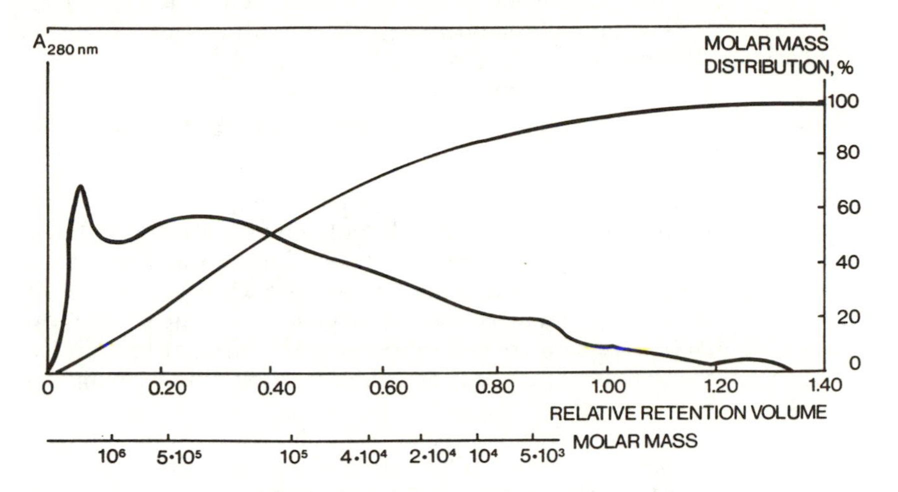

Figure 7. Fractionation of enzymically polymerized lignin sulfonates through Sephacryl S-300. Eluent: 0.5M NaCl, 0.1M Tris-HCl (pH 8).

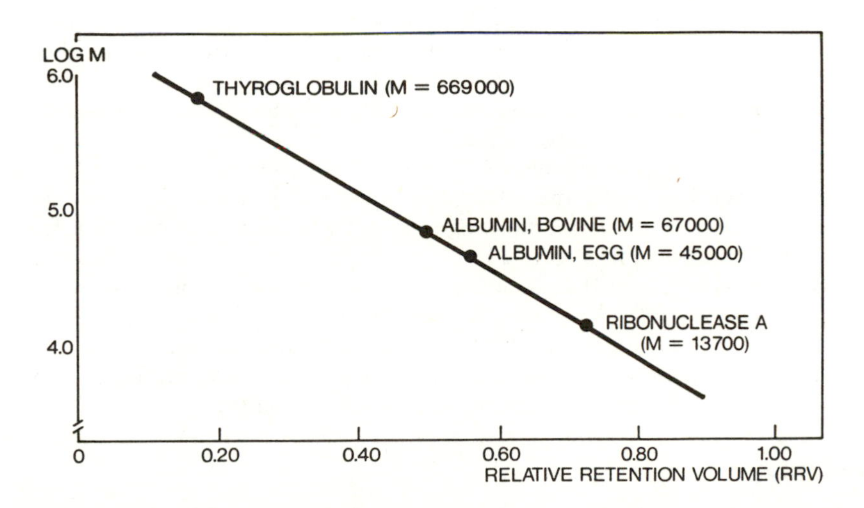

Figure 8. Calibration of Sephacryl S-300. Eluent: 0.5M NaCl, 0.1M Tris-HCl (pH 8).

groups of the lignin sulfonates, can be avoided by using a strongly alkaline eluent to ionize the phenolic hydroxyl groups. However, the Sephadex G-75 used for the fractionation of lignin sulfonates in the molar mass range 3 000–80 000 dalton is rapidly degraded by strong alkaline solutions. Sephacryl is stable enough, but the high molar mass calibration proteins cannot be used in strong alkaline solution.

Elution of Non-Sulfonated Lignins with Sodium Hydroxide Solution as Eluent

Non-sulfonated lignins such as those from alkaline pulping processes are insoluble in water but easily soluble in sodium hydroxide solutions. When dissolved in and eluted with a sodium hydroxide solution, they show polyelectrolyte properties, i.e., the molecular species interact. As revealed by Figure 9, the fractionation result is strongly dependent on the sodium hydroxide concentration up to a concentration of 0.4M. A 0.5M sodium hydroxide solution is thus an appropriate eluent for fractionation on Sephadex G-50 (3).

With 0.5M sodium hydroxide as eluent, Sephadex G-50 effects fractionation in the molar mass range 1000–15000 dalton and can be used for a period of 3-4 weeks with a single calibration carried out with proteins and polypeptides of known molar mass, as revealed by Figure 10. Relative retention volumes 0.0 and 1.0 are defined with Blue Dextran and phenol, respectively.

Due to the dark color of alkali lignins, their molar masses cannot be determined by means of the light-scattering method. However, as shown by Figure 10, elution with sodium hydroxide also brings about a consistent elution pattern of lignin sulfonates and polypeptides. It is assumed that this also applies to the kraft lignins.

Fractionation on Sephadex G-25 using 0.5M sodium hydroxide as eluent causes the low molar mass lignin components in black liquor to elute in the relative retention volume range 0.3-1.3 with partial separation from each other, as shown in Figure 11.

It should be noted that in this relative retention volume range elution does not necessarily take place in order of decreasing molecular size, because, as seen from Figure 12, functional groups may have a greater effect on elution behavior than molecular size (Forss, K.; Talka, E., The Finnish Pulp and Paper Research Institute, unpublished results).

Summary

It has been shown that the molar mass distributions of lignin sulfonates and kraft lignin can be determined by gel permeation chromatography. Calibration of the columns with lignin sulfonates of known molar mass or, alternatively, with commercially available proteins and polypeptides has been shown to give the same result.

Because of the polyelectrolyte properties of lignins, elution is performed with electrolyte solutions. If the lignins are water soluble and the column

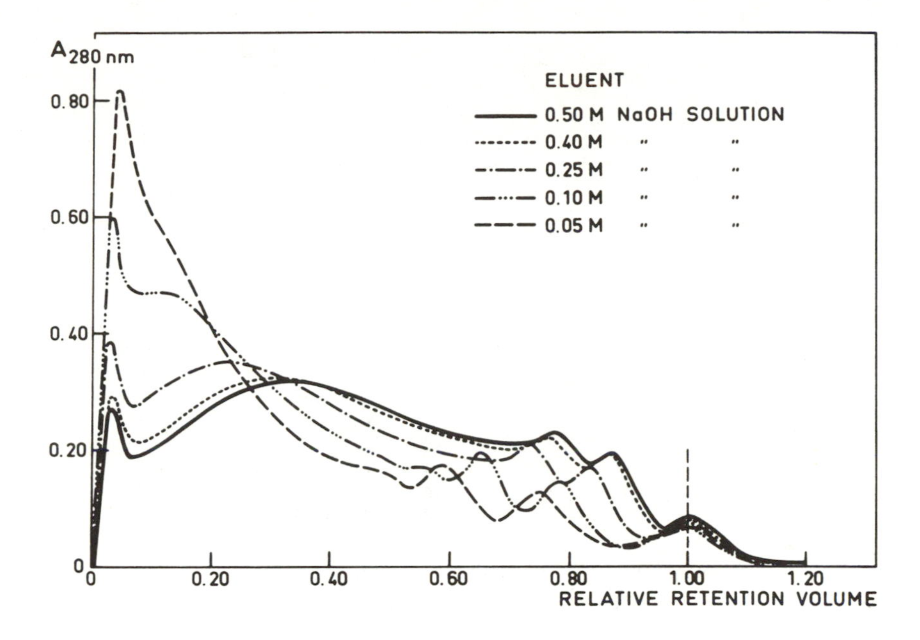

Figure 9. Influence of sodium hydroxide concentrations in the eluent on fractionation of lignins in draft black liquor. Column: Sephadex G-50. (Reprinted with permission from ref. 3. Copyright 1976 Wiley.)

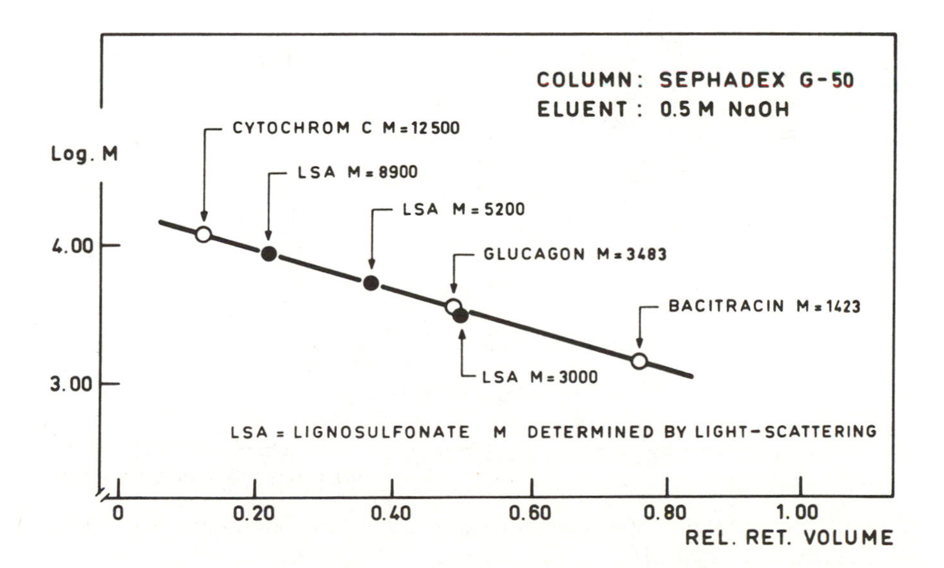

Figure 10. Calibration of Sephadex G-50. Eluent: 0.5M NaOH (4).

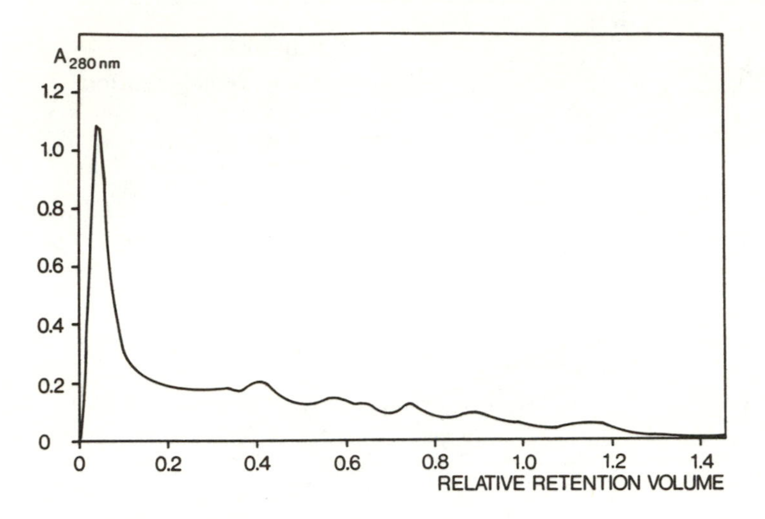

Figure 11. Fractionation of low molar mass lignin components in kraft black liquor. Column: Sephadex G-25. Eluent: 0.5M NaOH.

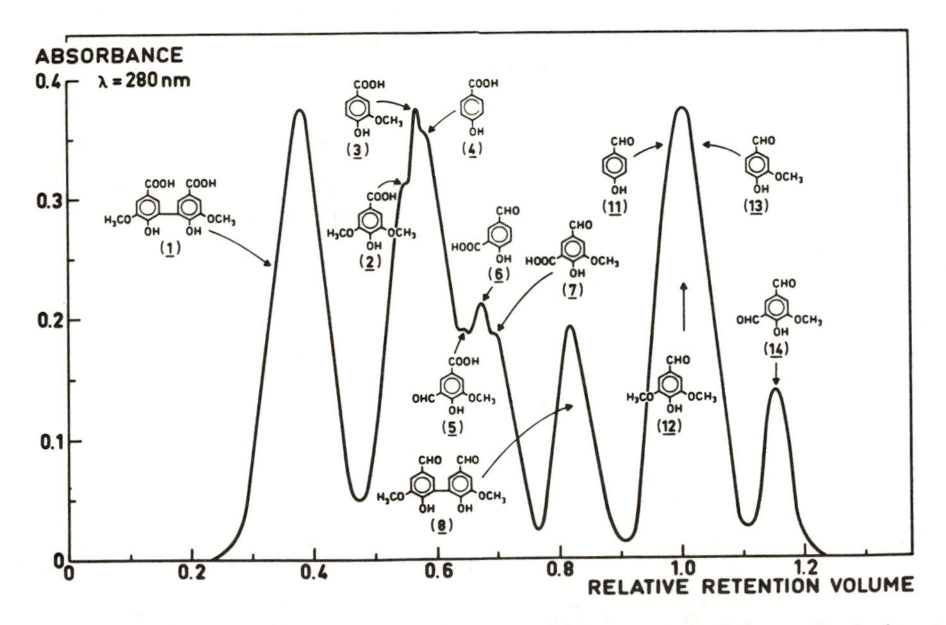

Figure 12. Fractionation of low molar mass model compounds. Column: Sephadex G-25. Eluent: 0.5M NaOH.

gel matrix is not stable enough in the presence of an alkaline eluent, the elution is performed with 0.5M sodium chloride solution adjusted to pH 8 with 0.1M Tris-HCl buffer. If the lignins are soluble only in alkaline solution the elution is performed with 0.5M sodium hydroxide solution.

Literature Cited

1. Forss, K.; Stenlund, B. *Pap. Puu* 1969, **51**, 93-105.
2. Stenlund, B. *Pap. Puu* 1970, **52**, 55-62; 121-30; 197-206; 333-39; 671-79.
3. Forss, K.; Stenlund, B.; Sågfors, P.-E. *Appl. Polym. Symp.* 1976, **28**, 1185-94.
4. Forss, K. G.; Furmann, A. G. M. U.S. Patent 4 105 606, 1978.

RECEIVED April 3, 1989

Chapter 10

Determinations of Average Molecular Weights and Molecular Weight Distributions of Lignin

P. Froment and F. Pla

Laboratoire de Génie des Procédés Papetiers Associé au Centre National de la Recherche Scientifique, UA 1100, INPG, Ecole Française de Papeterie B.P. 65 F–38402 St. Martin d'Hères Cedex, France

Three methods of molecular weight determination were investigated and their application to lignins is discussed. VPO gives suitable results for $\overline{M}_n$ if adequate thermistor beads are used. Reliable $\overline{M}_w$ values are obtained by LALLS if anisotropy, fluorescence and light absorption are taken into account. Preliminary experiments by on-line SEC-LALLS are promising but need more investigation.

The understanding of the macromolecular properties of lignins requires information on number- and weight-average molecular weights ($\overline{M}_n$, $\overline{M}_w$) and their distributions (MWD). These physico-chemical parameters are very useful in the study of the hydrodynamic behavior of macromolecules in solution, as well as of their conformation and size (1). They also help in the determination of some important structural properties such as functionality, average number of multifunctional monomer units per molecule (2, 3), branching coefficients and crosslink density (4, 5).

Number-average molecular weights are mainly determined by colligative methods, viz. cryoscopy, ebulliometry, vapor pressure osmometry and membrane osmometry. Among these methods, membrane osmometry can only be used for molecular weights higher than 25,000. In the case of lignins, vapor pressure osmometry (VPO) seems to be the most suitable and practical method in spite of some experimental difficulties. However, it only allows determinations of $\overline{M}_n$ in the range of 100 to 10,000. It follows that there is a molecular weight range of 10,000 to 25,000 which cannot be explored by using these two absolute methods. Size exclusion chromatography (SEC) covers this range and is more and more used to characterize lignins, although much remains to be done in order to achieve comparable absolute results from various laboratories. Indeed, the application of this technique to lignins has until now only given qualitative results owing to problems related to molecular associations and calibration procedures.

0097–6156/89/0397–0134$06.00/0

Sedimentation equilibrium and light scattering are the two absolute methods that have been used to determine $\overline{M}_w$ of lignins and lignin derivatives. These methods, together with SEC, cover a wide range of molecular weights. Light scattering is an excellent technique, particularly with the advent of more sophisticated instruments as discussed below (6,7).

This paper deals with a detailed analysis concerning the determination of $\overline{M}_n$, $\overline{M}_w$ and MWD of lignins using respectively VPO, low angle laser light scattering (LALLS) and SEC. In addition, recent results obtained by on-line LALLS-SEC are presented.

Results and Discussion

Vapor Pressure Osmometry. VPO is a very practical method for determining $\overline{M}_n$ values in a wide range of solvents and temperatures. Recently, results obtained with classical pendant-drop instruments showed a significant dependence of the calibration constant upon the molecular weight of the standards (8,9). On the other hand, with an apparatus equipped with thermistors allowing the volume of the drops to be kept constant, this anomaly is not observed (10,11).

The thermistors used in this study were of the latter type. As shown in Tables I and II, the corresponding K values appeared clearly to be a function of the solvent and temperature but not of the molecular weight of the standards.

Table I. Influence of Solvent and Temperature on the Calibration Constant with Benzil as Calibrating Compound

	Temperature (°C)				
Solvent	25	37	45	60	90
Tetrahydrofuran	–	–	3775	–	–
Dioxane	–	–	2680	4765	–
2-methoxyethanol	435	950	1445	2520	5690

Table II. Influence of the Molecular Weight of the Calibrating Compound on the Calibration Constant in THF at 45°C

Calibrating Compound	Molecular Weight	K(ohm.mole^{-1}.kg)
Benzil	210	3775
POE	310	3750
POE	420	3790
Polystyrene	1790	3780

Given the sensitivity of colligative methods to the presence of low molecular weight impurities, particular care was taken to isolate lignin samples free of such foreign contaminants. Thus, for example, the extraction of a sample of spruce dioxane lignin for eight hours with diethyl ether produced an increase of $\overline{M}_n$ from 1200 to 3000, the corresponding weight loss being about 5%.

Conversely, the presence of macromolecular aggregates are not detected adequately by VPO. A suitable change of solvent or a temperature increase can lead to dissociation and a corresponding effect on $\overline{M}_n$. Tables III and IV clearly show that this associative phenomenon is not relevant in the case of two typical samples.

Table III. $\overline{M}_n$ of Narrow Fractions of Spruce Dioxane Lignin

Fractions	D1	D2	D3	D4	D5	D6	D7	D8	D9
$\overline{M}_n$ (THF)	820	970	1250	1650	2250	3100	4000	5000	6050
$\overline{M}_n$ (Dioxane)	800	970	1300	1700	2250	3100	4000	5000	6050

Table IV. $\overline{M}_n$ of Black Cottonwood Alkali Lignin Fractions in Different Solvents and at Several Temperatures

		$\overline{M}_n$				
		THF	2-methoxyethanol			
Sample	$\overline{M}_w$	45°C	25°C	37°C	45°C	60°C
F-1	17600	3650	3700	3750	3700	3700
F-2	24000	5100	5100	5000	5050	5000
F-3	43500	6000	5900	5950	5850	5900
F-4	55000	6300	6400	6400	6400	6450

In order to overcome the solubility limitation typical of lignin fractions, chemical modifications have been envisaged. Obviously only those methods giving nearly quantitative recovery are adequate for the purpose of measuring $\overline{M}_n$. Table V shows results related to the acetylation technique where only a slight increase in $\overline{M}_n$ is observed as expected.

Table V. $\overline{M}_n$ of Spruce Alkali Lignin Fractions Before and After Acetylation (in 2-methoxyethanol at 60°C)

	$\overline{M}_n$	
Sample	Before Acetylation	After Acetylation
SA-1	2000	2350
SA-2	2850	3150
SA-3	3850	4450

Light Scattering Photometry. We will limit our discussion to LALLS because it constitutes a net improvement as compared to other methods of $\overline{M}_w$ determination. It has the advantage of providing absolute values of the Rayleigh ratio by direct comparison of scattered and transmitted light.

$$R(\theta) = G(\theta)G(0)^{-1}D(\sigma l)^{-1} \tag{1}$$

where $R(\theta)$ is the Rayleigh ratio, $G(\theta)$ and $G(0)$ are respectively the photomultiplier readings for scattered and transmitted light, D is the attenuation used for the measure of the transmitted light, and the product σl is an instrumental factor. Another advantage is that, owing to the small angle between the incident and scattered beam, the form factor $P(\theta)$ is equal to unity. The equation relating the Rayleigh ratio to the molecular weight thus reduces to:

$$Kc/\Delta R(\theta) = 1/\overline{M}_w + 2A_2c \tag{2}$$

where c is the concentration (g.ml^{-1}), A_2 is the second virial coefficient, $\Delta R(\theta)$ is the difference between the Rayleigh ratios of the solvent and the solution, and K is the light scattering constant which, for vertically polarized incident light, is given by:

$$K = 4\pi^2 n^2 (dn/dc)^2 \lambda^{-4} N^{-1} \tag{3}$$

where n is the solvent refractive index, dn/dc is the refractive index increment of the solution, λ is the wavelength, and N Avogadro's number. $\overline{M}_w$ is obtained by extrapolating $Kc/\Delta R(\theta)$ to zero concentration.

Nevertheless, in order to obtain reliable results with lignins, fluorescence, light absorption, and anisotropy must be taken into account. Fluorescence is easily eliminated by the use of an adequate interference filter. It is well known that lignins exhibit an absorption spectrum with a maximum in the near ultra-violet tailing all the way into the visible.

The correction of absorption depends on the geometry of the cell and of the scattering volume. Fortunately, in the KMX-6 instrument, the scattering volume is at the center of the cell and the incident and scattered beams are equally attenuated. Consequently the correction of absorption is reduced to the measure of the intensity of the transmitted beam for each concentration.

The net effect of anisotropy is an enhancement of the scattered light. This excess scattering can be eliminated by measuring the vertical and horizontal components of the scattered light with an analyzing polarizer. These two components are complex functions of $\sin\theta$ and $\cos\theta$ (12). For small values of θ setting $\sin\theta = 0 (\cos\theta = 1)$ results in a negligible error. The Rayleigh ratio corrected for anisotropy is then given by the usual expression:

$$\Delta R(\theta) = \Delta Rv(\theta) - (4/3)\Delta Rh(\theta) \tag{4}$$

where $\Delta R(\theta)$ and $\Delta Rh(\theta)$ stand respectively for the vertical and horizontal excess Rayleigh ratios.

Tables VI and VII give results corresponding to two series of lignin fractions obtained with a flow-through reactor (3). (The units for dn/dc and A_2 are respectively $ml.g^{-1}$ and $mole.ml.g^{-2}$). These results show that LALLS allows the determination of low $\overline{M}_w$ values. The dn/dc values differ from sample to sample but vary only slightly for a given set of fractions. The second virial coefficient exhibits no definite trend. Negative values indicate perhaps some association effects but light scattering alone is not sufficient to ascertain this point.

Table VI. LALLS Results on Acidic Organosolv Lignin Fractions from Black Cottonwood

	Fraction							
Parameter	1	2	3	4	5	6	7	8
$\overline{M}_w$	1500	2550	4000	9000	19000	29000	59500	74000
dn/dc	0.168	0.159	0.157	0.157	0.156	0.157	0.159	0.157
$A_2.10^2$	2.1	1.5	2.8	1.7	0.15	0.11	0.18	0.19

Table VII. LALLS Results on Alkali Lignin Fractions from Black Cottonwood

	Fraction							
Parameter	1	2	3	4	5	6	7	8
$\overline{M}_w$	4700	10500	11500	17500	24000	36000	43500	55000
dn/dc_2	0.187	0.184	0.188	0.190	0.192	0.191	0.189	0.193
$A_2.10^2$	0	-1.2	2.1	1.4	1.3	0.9	1.7	1.1

Size Exclusion Chromatography. One can define three levels of interpretation in SEC. The first is a visual inspection of the weight fraction versus elution volume curve. Knowing that for a pure size exclusion mechanism the molecular weight decreases when the elution volume increases, this gives a picture of the distribution and allows a qualitative comparison of samples. The calibration and subsequent calculation of average molecular weights constitute the second level. The third level refers to the correction of column dispersion. Because the aim of this paper is the determination of molecular weight averages, attention will be focused on the second level, assuming that all the problems associated with the first one have been solved. This is an ambitious assumption but indispensable for proceeding to the second level.

From the Q factor (13) to Benoit's universal calibration (14) and the more recent "southern calibration" (15), a lot of papers have been devoted to this question. Unfortunately, the application of these procedures to lignins has been unsuccessful. The best approach is the calibration of the columns with well characterized lignin samples of broad or narrow distribution (16). Another practice is the direct use of the calibration established with polystyrene samples. When the exponents of the Mark-Houwink equation for both polystyrene and the sample under study are the same, this method gives "relative molecular weights." If the condition above is not fulfilled, as for lignins, there is no simple relation between the calculated molecular weights and the real ones.

Figure 1 shows the calibration curves for polystyrene and lignin in the low molecular weight region. Even in this region where the effects of branching are reduced, there is no coincidence between lignin and polystyrene. When the molecular weight increases the two curves diverge.

On-Line SEC-LALLS. The possibility afforded by on-line SEC-LALLS to continuously calculate the molecular weight of the molecules eluting from a set of columns allows one to overcome the calibration problem. In Figure 2 are shown the recorder traces for the vertical (Vv) and horizontal (Hv) components of the scattered light, the transmitted light (GO) and the concentration of the effluent (DRI) corresponding to an acetylated organosolv hornbeam lignin. Note that these recordings need, at least, three sample injections. The molecular weight averages are calculated by summing over the elution volume range spanned by the sample, the expression:

$$\overline{M}_w(i)c(i) = 1/[K/\Delta R(\theta, i) - 2A_2] \tag{5}$$

where $\overline{M}_w(i)$ and $c(i)$ are respectively the weight average molecular weight and the concentration of the species eluting at v(i). K, A_2 and $\Delta R(\theta, i)$ have been defined above. The summation leads to:

$$\overline{M}_w = (\Delta v/p)\Sigma\{1/[K/\Delta R(\theta, i) - 2A_2]\} \tag{6}$$

$$\overline{M}_n = \Sigma c(i)/\Sigma[c(i)/\overline{M}_w(i)] \tag{7}$$

where p is the injected weight and Δv is the elution volume increment. In practice A_2 is neglected (17). Equation (6) shows that $\overline{M}_w$ depends only on the intensity of the light scattering signals. The value of $\overline{M}_n$ calculated according to equation (7) is always higher than that obtained by absolute methods.

A glance at curve Vv reveals an interesting feature. In the high molecular weight region (small elution volumes) there is a small peak located near the void volume. The absence of equivalent peaks at the same elution volume on the concentration detector and the horizontal component curves indicates the presence of a small amount of high molecular weight species in the sample. It remains to be ascertained if this high molecular weight component is a fundamental characteristic of the sample or an artifact. Some experiments carried out in our laboratory show a tendency for this peak to become reduced with storage time. Further experiments are needed to elucidate the exact nature of this phenomenon. Fortunately, the effect on the calculation of molecular weight averages is modest. For

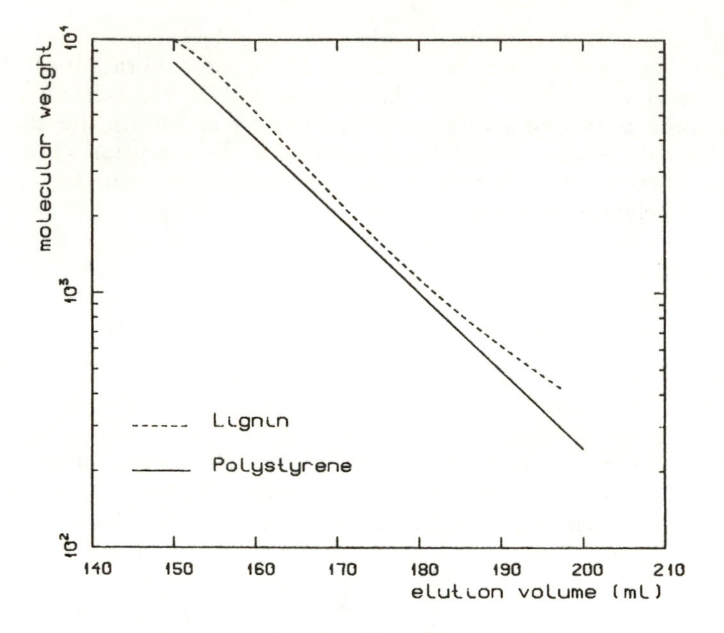

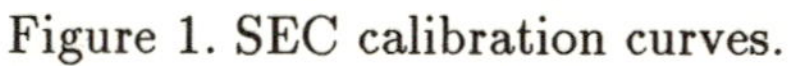
Figure 1. SEC calibration curves.

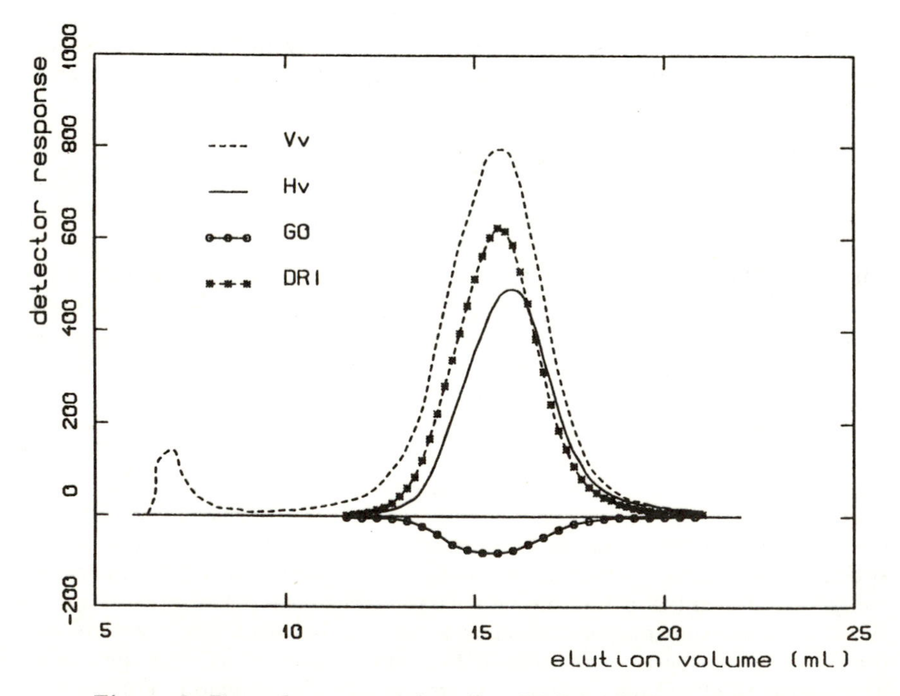

Figure 2. Recorder traces of on-line SEC-LALLS experiment.

the sample studied the $\overline{M}_w$ value is 15000 instead of 12000 when the high molecular weight component is neglected. The $\overline{M}_n$ value is 4500 instead of 4000 determined by VPO.

Light scattering is very sensitive to high molecular weight compounds but, on the other hand, low molecular weight components are not detected and must be eliminated in order to avoid difficulties and to obtain reliable number-average molecular weights as stated previously with VPO.

Conclusion

Lignins are polydisperse macromolecules whose average molecular weights are rather low. Their determination by means of absolute methods becomes now easy owing to improvements in techniques, and this in spite of many problems inherent in the polymer itself.

VPO appears to be the most practical colligative method for determining $\overline{M}_n$. Nevertheless, care must be taken concerning lignin purification and drying, the purity of the solvent and instrument calibration. On the other hand, reliable $\overline{M}_w$ values can be obtained using LALLS if fluorescence, light absorption and anisotropy are accounted for.

SEC alone allows only comparison between lignin samples. The calibration of the columns still remains a real problem. The association of SEC with LALLS, following the procedure described here, is certainly the best approach to determining molecular weight averages and the MWD of lignin derivatives. However, there remains to be explained the appearance of small amounts of high molecular weight fractions. This topic is presently under investigation.

Experimental

Dioxane Lignin. Spruce wood meal was extracted with a boiling solution composed of 1000 ml dioxane and 10 ml HCl (sp. gr. = 1.19) (18). The resulting lignin was then Soxhlet-extracted with diethyl ether and fractionated by SEC (4,18) to obtain narrow fractions.

Alkali Lignin. Black cottonwood platelets were cooked in a flow-through reactor with 1.0N NaOH at 160°C flowing at a steady rate of about 17.5 $ml.min^{-1}$ (3). The effluent was collected as several successive fractions from which lignin was precipitated and purified (3). The same procedure was applied, at 170°C, to spruce matchsticks.

Organosolv Lignins. Acidic organosolv lignin was obtained from black cottonwood using again a flow-through reactor (3). The delignification conditions were methanol/water (70/30 v/v) in the presence of 0.01M H_2SO_4 at 150°C. The successive lignin fractions were recovered as described in Ref. 3.

Hornbeam chips were cooked in a batch reactor under the following conditions: ethanol/water (50/50 v/v); catalyst: $MgCl_2$ 0.05M on o.d. wood; liquor-to-wood ratio: 10; maximum temperature: 170°C for 3 hours. The resulting organosolv lignin was recovered as described elsewhere (19).

Quantitative Acetylation of Lignins. To avoid fractionation of the lignins and loss of low molecular weight components, which occur in the usual acetylation procedures, the quantitative acetylation described by H. Chum *et al.* (20,21) was employed.

Vapor Pressure Osmometry. Number-average molecular weights were evaluated with a vapor pressure osmometer (Knauer) following a previously described method (18).

Low Angle Laser Light Scattering Photometry. Weight-average molecular weights were determined with a KMX-6 photometer (LDC/Milton Roy). The light source was a 2 mW vertically polarized helium-neon laser (λ = 632.8 nm). The instrument was equipped with an analyzing polarizer to measure both vertical and horizontal components of the scattered light. Fluorescence was eliminated by the use of an interference filter centered at 632.8 nm and with a band with of 4 nm. The solvent was 2-methoxyethanol at room temperature. Solvent and solutions were clarified by filtration through a 0.2 μm pore size teflon filter. The refractive index increments were evaluated at $20 \pm 0.01°C$ and $\lambda = 632.8$ nm with a BP 2000 differential refractometer (Brice Phoenix). The refractive index of the solvent was determined under the same conditions with an Abbe refractometer. For on-line SEC-LALLS, a flow-through cell was used.

High Performance Size Exclusion Chromatography. SEC was carried out on polystyrene-divinylbenzene gels with porosites ranging from 100 to 10^4Å. The solvent, THF, had a flow rate of 1 ml/min. The detector was a differential refractometer.

Literature Cited

1. Goring, D. A. I. In *Lignins: Occurrence, Formation, Structure and Reactions*; Sarkanen, K. V.; Ludwig, C. H., Eds.; Wiley-Interscience: New York, 1971; Chapter 17.
2. Dolk, M.; Pla, F.; Yan, J. F.; McCarthy, J. L. *Macromolecules* 1986, **19**, 1464-70.
3. Pla, F.; Dolk, M.; Yan, J. F.; McCarthy, J. L. *Macromolecules* 1986, **19**, 1471-77.
4. Pla, F.; Robert, A. *Cell. Chem. Technol.* 1974, **8**, 3-10.
5. Pla, F.; Yan, J. F. *J. Wood Chem. Technol.* 1984, **4**, 285-99.
6. Plastre, D. Thèse Docteur-Ingenieur, INPG, Grenoble, 1983.
7. Pla, F.; Froment, P.; Capitini, R.; Tistchenko, A. M.; Robert, A. *Cell. Chem. Technol.* 1977, **11**, 711-18.
8. Brzezinsky, J.; Glowala, H.; Kornas-Calka, A. *Eur. Polym. J.* 1973, **9**, 1251-53.
9. Marx-Figini, M.; Figini, R. V. *Makromol. Chem.* 1980, **181**, 2401-7.
10. Kamide, K.; Terakawa, T.; Uchiki, H. *Makromol. Chem.* 1976, **177**, 1447-64.
11. Burge, D. E. *J. Appl. Polym. Sci.* 1979, **24**, 293-99.
12. Russo, P. S.; Bishop, M.; Langley, K. H.; Karasz, F. E. *Macromolecules* 1984, **17**, 1289-91.

13. Cazes, J. *J. Chem. Educ.* 1966, **43**, A567-76.
14. Grubisic, Z.; Rempp, P.; Benoit, H. *J. Polym. Sci.* 1967, **5**, 753-59.
15. Hester, R. D.; Mitchell, P. H. *J. Polym. Sci. Polym. Chem.* 1980, **18**, 1727-38.
16. Froment, P.; Robert, A. *Cell. Chem. Technol.* 1977, **11**, 691-96.
17. Kim, C. J.; Hamielec, A. E.; Benedek, A. *J. Liq. Chromatogr.* 1982, **5**, 425-41.
18. Froment, P.; Pla, F.; Robert, A. *J. Chim. Phys.* 1971, **68**, 203-206.
19. Ivanow, T. Thèse, INPG, Grenoble, 1987.
20. Chum, H. L.; Johnson, D. K.; Ratcliff, M.; Black, S.; Schroeder, H. A.; Wallace, K. *Proc. 3rd ISWPC* 1985, p. 223.
21. Chum, H. L.; Johnson, D. K.; Tucker, M. P.; Himmel, M. E. *Holzforschung* 1987, **41**, 97-108.

RECEIVED February 27, 1989

Chapter 11

Association of Kraft Lignin in Aqueous Solution

Sri Rudatin, Yasar L. Sen, and Douglas L. Woerner

Department of Chemical Engineering, University of Maine, Orono, ME 04469

A technique has been developed to study the self-association of kraft lignins in alkaline solution using ultrafiltration membranes. The effects of alkalinity, lignin concentration and molecular weight distribution, ionic strength and some additives have all been explored. The association process accompanies the neutralization of the phenoxide groups. Neutralization can be accomplished by protonation or chelation. Large molecules show a strong affinity to associate with small molecules. High lignin concentrations increase the extent of association. Association in a 5.0 M urea solution was greatly reduced at alkalinities less than pH 13.0, indicating that hydrogen bonding may play an important role.

Utilization of kraft lignins has been stymied by several factors: the difficulty in producing purified material, inability to separate the kraft lignin by molecular weight, and a general lack of understanding of kraft lignin. In general, the phenomenon of association has greatly retarded the understanding and utilization of kraft lignins. Disagreement on the molecular weights and molecular weight distributions of lignin preparations is widespread, undoubtedly due to different degrees of association under many different conditions (1). When the underlying mechanism and the conditions affecting association are well understood, many of the difficulties in characterization and utilization may be overcome. The viscosity of kraft lignin solutions can be modeled as colloidal phenomena. The origin of the colloidal particles may be association.

Association of kraft lignin is mainly based on internal and external factors. The composition and the functional groups within the kraft lignin structure are important internal factors in determining the thermodynamic behavior of kraft lignin. The major groups are the aromatic ring (1/C9),

0097–6156/89/0397–0144$06.00/0

carboxyl groups (1/C9), phenolic hydroxyl groups (0.6/C9), aliphatic hydroxyl groups (0.48/C9), and ether groups (1/C9). The extent of association in kraft lignin solution will also change with external factors such as solvent, alkalinity, concentration, ionic composition, organic additives, time, and temperature.

There are four different mechanisms of molecular association: hydrogen bonding (H-bonding), stereoregular association, lyophobic bonding, and charge transfer bonding. H-bonding is a dipole-dipole interaction between two electronegative atoms on which one has a hydrogen atom covalently bonded. The high strength of these bonds (20 kJ) arises because the dipoles are only separated by the very small hydrogen atom. Stereoregular association occurs between two highly ordered polymers which share many van der Waals interactions along the polymer chains. Many biological polymers demonstrate stereoregular bonding, the most notable being the DNA double helix. Lyophobic bonding is a misnomer for the exclusion of solutes from solution because of the strong intermolecular attraction between the solvent molecules. The major requirement for lyophobic bonding is the development of a structure of the solvent molecules such as water and benzene show. Charge transfer bonding is the sharing of positive and negative charges as exhibited between cationic and anionic polymers.

A long history of association of kraft lignins in solution exists in the literature, almost all of it based on the determination of molecular weight. In 1958 Gross *et al.* (2) discussed H-bonding with respect to irreproducibility of MW measurements by cryoscopy. Benko in 1964 (3) suggested that H-bonding or hydrophobic bonding was responsible for the changes in diffusivity and viscosity of kraft lignins. In 1967 Brown (4) suggested that H-bonding was responsible for molecular association in vapor pressure osmometry experiments for the determination of molecular weight. Lindström in 1979 and 1980 (5) and Yaropolov and Tishchenko in 1970 (7,8) researched the changes in viscosity of kraft lignin solutions and ascribed the increasing viscosity with decreasing alkalinity to be due to intermolecular bonding. Milled wood lignins have been shown to have intramolecular hydrogen bonding primarily between the phenolic, aliphatic and ether groups. Some intermolecular hydrogen bonding was observed with the gamma aliphatic hydroxy groups (9).

The most recent work in this field has been performed by Sarkanen and co-workers. A series of papers (10-13) propose that kraft lignin undergoes stereoregular association between highly ordered fragments of the native lignin which remain after pulping. Bonding is predominantly HOMO-LUMO interactions of the pi-orbitals of the benzene rings between large polymers and small oligomers and is stoichiometrically constrained. The major tool used in this work is size exclusion chromatography of precipitated lignins, with a variety of supporting evidence. The major findings of this work are: (1) complexes in a variety of solvents can be broken down with ionic additives; (2) the molecular weight distribution of precipitated kraft lignins is very similar below 4000 daltons; (3) dissociation is favored by working at high alkalinity and low solute concentration.

A simple model of the degree of association has been presented (Woerner, D.L.; McCarthy, J. L. *Macromolecules*, in press) in which the degree of association is related to the solubility of the protonated lignin in neutral water and the degree of ionization of the phenolic groups. The assumptions are: (1) all ionized phenolic ions are in solution; (2) a certain fraction of the protonated phenolic groups is soluble; (3) a single K_a and solubility limit, S, are applicable to all lignin species of all molecular weights. The fundamental reaction was ionization (Equation 1) and the degree of association is given in Equation 2.

$$LOH = LO^- + H^+ \tag{1}$$

$$1 - \%ass = S + (1 - S) \times X \text{ where } 1/X = 1 + \frac{K_w}{K_a[OH]} \tag{2}$$

Here S is the solubility of the lignin in neutral water, X is the fraction of phenolic groups that have been ionized, and K_a is the dissociation constant of the phenolic groups. The pK_a was determined to be approximately 11.4 from light scattering results and the solubility limit was measured to be about 0.16 mass fraction. A reasonable fit of the permeate concentration from ultrafilter experiments with a 10,000 molecular weight cutoff (MWCO) was obtained.

Experimental Procedures

Two lignin preparations were used in this work. The primary kraft lignin was isolated from a northern softwood mill liquor by precipitation and washing following the procedures developed by Kim (14). The precipitated lignins were separated into three groups: FAM (fully associated molecules) were the as-precipitated molecules, SAM (small associating molecules) were molecules from FAM which were able to pass a 10,000 MWCO membrane, and LAM (large associating molecules) were the FAM molecules not able to pass the 10,000 MWCO membrane. The LAM solution contained some low molecular weight material. A fourth lignin sample was obtained from the non-precipitated lignins and called NASM (non-associating small molecules). The second kraft lignin was Indulin AT obtained from Westvaco Corp., Charleston, S.C., and was used as obtained.

The lignin was analyzed for concentration and molecular weight distribution. The concentration was obtained by UV absorption at 280 nm assuming the validity of Beer's law with an absorption coefficient of 20 cm mL/g. The molecular weight distributions were obtained following Sarkanen (10) using Sephadex G-100 in a 25 mm by 70 cm column (Pharmacia). All data were collected and analyzed by Fortran computer programs written at the University of Maine following Yau (15).

The membrane sampling technique was used extensively to get samples which were representative of the solution in equilibrium with the associated complexes. Briefly, the sample is equilibriated in a sealed polyethylene container under nitrogen at least overnight. A 200 mL portion of the sample is placed in the ultrafilter cell (Amicon 8200) with an XM300 membrane (Amicon 300,000 MWCO). Earlier studies (Woerner, D. L.; McCarthy, J.

L., *Macromolecules*, in press) had shown that this membrane did not allow associated complexes to pass through the membrane. The cell is pressurized with nitrogen. The first 5 mL of permeate are discarded, and then a 2 mL sample is taken for analysis. The ultrafilter conditions were adjusted to minimize concentration polarization, CP (16), by using a stirrer rate of 390 rpm and 60 kPa pressure drop across the membrane. At high lignin concentrations CP could not be avoided and many different stirrer rates and pressures were used to provide a correction for CP.

The results are often presented in terms of the rejection coefficient which is a dimensionless number defined as

$$R = 1 - C_p/C_b \tag{3}$$

which allows direct comparison of results at different parent solution concentrations. The rejection coefficient is primarily a function of the membrane pore size and the solute hydrodynamic radius. A change in the rejection coefficient when the solution characteristics are changed is indicative of a change in the degree of association. An increase in the rejection coefficient indicates a decrease in the number of molecules available to penetrate the membrane presumably as a result of an increase in the degree of association.

Factors Affecting Association

Alkalinity. The FAM solution was divided into several aliquots and the pH was adjusted with H_2SO_4 to a pH ranging from 8.5 to 13.8, each at a concentration of about 11 g/L. Each solution was individually ultrafiltered and a permeate sample obtained. The pH 8.5 solution was titrated with 0.1M NaOH to several alkalinities and ultrafiltered. The permeate concentrations and rejection coefficients are presented in Table I, and the permeate molecular weight distributions are shown in Figure 1.

The presence of kraft lignin association was indicated at all alkalinities and all molecules participated. The association was strongly pH dependent and completely reversible. However, the extent of the association for the small molecules and the large molecules was different. The high molecular weight molecules show considerable association in the pH 13.5 to pH 12.0 range, with little or no further association at lower alkalinities. The small molecules showed no association above pH 13.0 and the major association was in the pH 13.0 to 11.0 region. Below 10.0 there was no further association.

These results indicate that protonation of the phenoxide ion is a necessary step in the association process, and the average K_a for ionization of the phenolic hydroxyl group increases with molecular weight.

Molecular Weight. The four KL solutions and 50:50 mixtures of the SAM, NASM and LAM with FAM were ultrafiltered at pH 13.8 and 8.5 to explore the effect of the parent solution molecular weight distribution on the association process. The results at pH 13.8 are presented as rejection coefficient distributions in Figure 2. The rejection coefficient distribution was

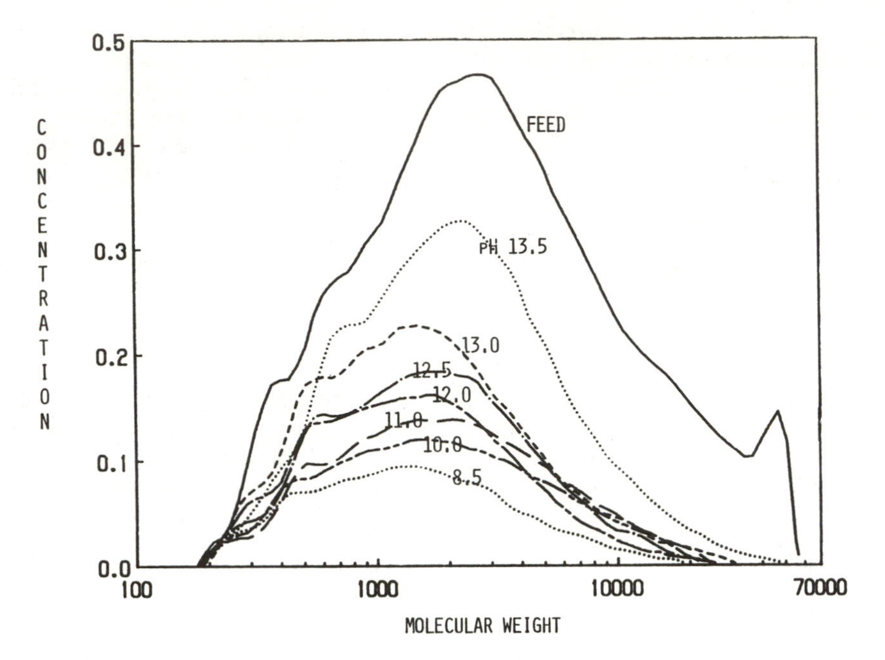

Figure 1. Molecular weight distributions of the permeate obtained from the titration of FAM demonstrating the association of large molecules in the pH 13.8 to 12.0 region and small molecules in the pH 13.0 to 10.0 range. Sephadex G-100/0.10 M NaOH.

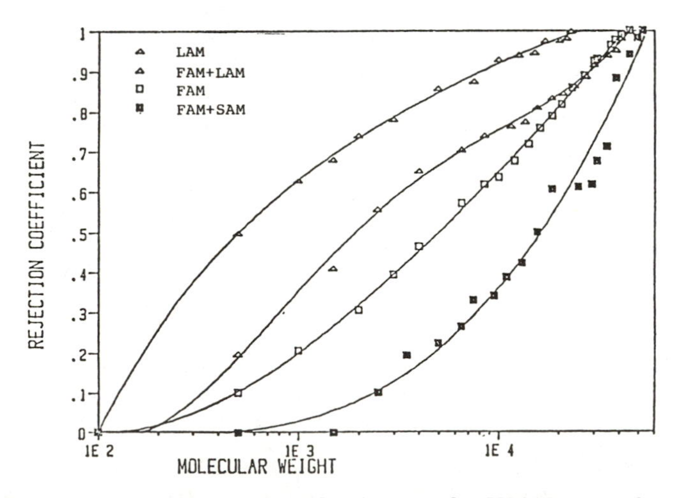

Figure 2. Rejection coefficient distribution on the XM300 membrane showing the association caused by an increase in the concentration of large molecules at constant total lignin concentration.

Table I. Titration of Kraft Lignin Using XM300 Membrane

Run	pH	C feed, g/L	C permeate, g/L	Rej. Coeff.
1	13.5	12.10	7.00	0.42
2	13.0	10.80	5.35	0.50
3	12.5	10.75	4.50	0.58
4	12.0	10.65	4.10	0.61
5	11.5	10.70	4.00	0.62
6	11.0	10.80	3.85	0.64
7	10.5	11.10	3.85	0.65
8	10.0	11.10	3.50	0.68
9	9.5	11.00	3.00	0.72
10	9.0	11.00	2.75	0.75
11	8.5	11.00	2.50	0.77
12[1]	10.0	11.00	3.25	0.70
13[1]	12.0	10.50	4.35	0.58
14[1]	13.5	10.10	7.00	0.31

[1] Back titration results.

obtained by dividing the permeate concentration by the parent solution concentration at every molecular weight. The rejection coefficient should be identical except for changes imparted by association.

The association of kraft lignin was strongly influenced by its molecular weight distribution (MWD) at pH 13.8. The feed MWD and permeate MWD profile of these solutions should coincide up to a certain value of molecular weight regardless of the MWD of the parent solution if association is not present. The feed and permeate MWD profile were identical, i.e., rejection coefficient = 0, up to 3000 for NASM, SAM, FAM-NASM, and FAM-SAM. For the solutions having more large molecules (FAM, LAM and FAM-LAM), the feed MWD and permeate MWD coincided up to 600. Apparently when an excess of small molecules is present (NASM, SAM, NASM+FAM and SAM+FAM) the associated complexes appeared to become saturated. On the other hand more large molecules in the solution provided more sites for association and small molecules were effectively removed from the solution. This result indicates that there is considerable interaction between the small molecules and the large molecules at high alkalinity.

The amount of small molecules in the permeate at low alkalinity increased with the relative concentration of small molecules (Figure 3). If hydrophobic bonding or simple solubility was the dominant mechanism for the association, the solubility of lignin molecules should be solely a function of pH regardless of the MWD of the parent solution. Consequently, the permeate MWD should be approximately the same and not a function of the MWD of the parent solution. Since the experimental permeate MWD varied with the parent MWD, the solubility or hydrophobic interaction mechanism was no longer accepted.

Lignin Concentration. Lignin concentrations have been examined in the range of 10 to 100 g/L. These concentrations represent concentrations normally present in black liquors and therefore are of industrial relevance. Typical lignin concentrations in black liquor range from 50 g/L for 15% solids to over 300 g/L at firing.

The membrane sampling technique is much more complicated at these concentrations because of the effects of CP at the membrane surface. The theory of rejection by the CP layer (17) suggests that a plot of $ln(C_p)$ as a function of the reduced flux rate Jw/J−1 at several stirrer rates and transmembrane pressures should yield a straight line. The intercept is $ln(C_{b,app} \times (1 - R_{mem}))$. At each concentration approximately 20 ultrafilter experiments are performed at several stir bar rates and pressures and plotted as shown in Figure 4. The intercept values from these plots are divided by $(1 - R_{mem})$ and the parent lignin solution concentration to give the degree of association. The results are shown in Table II. The effect of lignin concentration on the molecular weight distribution of solutes in equilibrium with the associated complexes has not yet been ascertained because we need knowledge of the diffusion coefficient of kraft lignin as a function of molecular weight and the rejection coefficient distribution without association or concentration polarization.

Table II. Association at High Lignin Concentrations[a]

Parent Solution Concentration	pH			
	13.8	13.0	12.0	10.0
10	–	1.0	.479	.77
25	.940	.848	.604	.364
50	.846	.748	.740	.696
100	.744	.834[c]	.911[c]	–
50 + I[b]	.440	–	–	–

[a] The numbers are the permeate concentrations divided by the bulk concentration. All the corrections to determine the degree of association are not yet known, but a decrease in this number is evidence of an increase in the degree of association. The solute is Indulin AT. The permeate data are obtained from the intercept of figures similar to Figure 4.

[b] 50 g/L solution, ionic strength 3.0 M with the addition of NaCl.

[c] This apparent decrease in the degree of association is under verification.

The results based on bulk lignin concentration demonstrate that the degree of association increases as the concentration of the solute increases. This result is observed at all alkalinities. The degree of association increases as the alkalinity decreases at any solute concentration.

The simple solubility model would predict that the concentration in the permeate would remain constant as the total lignin concentration increases.

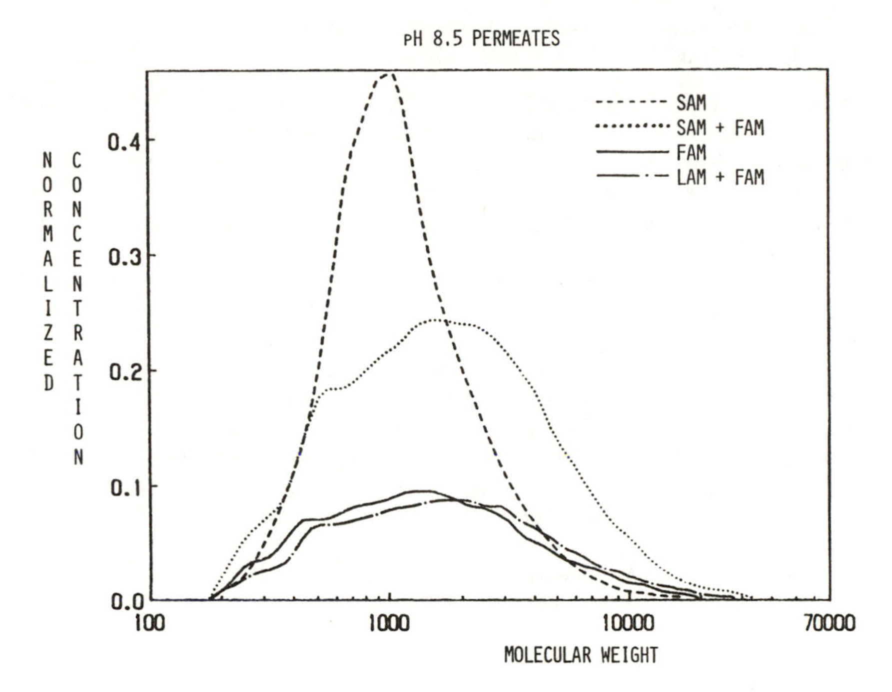

Figure 3. Molecular weight distributions of permeates obtained from parent solutions of differing molecular weight distributions at pH 8.5. Sephadex G-100/0.10 M NaOH.

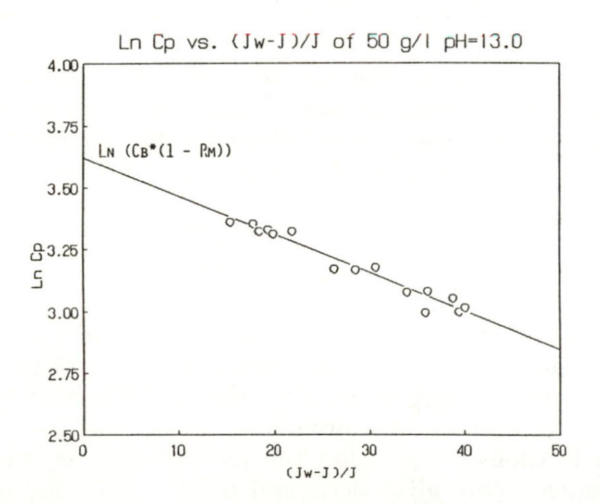

Figure 4. The correction for concentration polarization effects. Parent solution was 50 g/L at pH 13.0. Stirrer rates varied from 350 to 510 rpm, transmembrane pressure ranged from 34 to 207 kP_a.

This concept clearly overpredicts the degree of association at low alkalinities and underpredicts that at high alkalinities.

Additives. Sodium dodecyl sulfate (SDS), 5.0 M urea and 0.1 M betaine were added to the FAM solution to disrupt the association process to help understand the mechanism. Betaine had previously shown some dissociative properties, and urea is effective in preventing association caused by H-bonding. Solutions at different alkalinities were produced as before and ultrafiltered.

The rejection coefficients listed in Table III showed the reduction of association from FAM solution expressed by the reduction of R at various additives usage. In general the extent of association of FAM solution decreased moderately with the addition of betaine at any pH. Urea increased the degree of dissociation at high alkalinity but more importantly completely stopped association between pH 13.0 and 10.0, the region in which the majority of the phenoxide group protonation occurs. 0.14 M SDS showed an initial decrease in the degree of association at high alkalinity, but then increased the degree of association at all lower pH's.

Table III. Rejection Coefficients of FAM Solution and Additives

	pH			
Additive	13.7	13.0	10.0	8.5
None	0.47	0.53	0.64	0.68
Urea	0.41	0.45	0.46	0.55
Betaine	0.36	0.44	0.51	0.56
SDS	0.33	0.58	0.79	–

The effectiveness of betaine can be explained better by its blocking proton uptake from solution. Betaine lessened the amount of protonated phenolic hydroxyl groups at any pH and consequently the intermolecular association between kraft lignin molecules. The chemical reactions can be shown as follows:

$$LOH \longrightarrow LO^- + H^+$$

$$LO^- + {}^-B^+ \longrightarrow LO^- + B^-$$

The equilibrium of the ionization reaction was shifted to a smaller mole fraction by betaine addition. However, as the pH was dropped, protonation of ionized groups took place to maintain the equilibrium of both reactions and association complexes were formed.

These indications suggest that the predominant mechanism of kraft lignin association as the pH is decreased is hydrogen bonding. Hydrogen bonding could be formed by phenolic OH-phenolic OH linkages or phenolic OH-ether linkages, which could be present abundantly in kraft lignin solute components. Carboxyl groups have pK_a's less than 5 and would be fully ionized while the aliphatic hydroxyl groups of the lignin molecules would not ionize in the alkaline region and therefore should not contribute to the above results. Another possibility is that the hemicellulose attached to lignin forms hydrogen bonds, and protonation of the phenoxide is necessary only to reduce electrostatic repulsions.

Ionic Strength. The role of ionic strength may be pivotal in understanding the thermodynamics of black liquor solutions which have ionic strengths of 2.0 M for 15% solids liquor and about 7.0 M at firing. Preliminary results show that the ionic strength has a major effect on the degree of association at high alkalinities as shown in Table II. When the ionic strength of a 50 g/L lignin solution at pH 13.8 was raised from 1.0 M to 3.0 M, the extent of association increased substantially. As of this time the reason for this is not known. A somewhat related effect is demonstrated in the importance of ionic strength in adsorption of lignins on Sephadex G25 (13). Research in this area is continuing to verify the single result.

Conclusions

1. The membrane technique being developed at the University of Maine is a powerful tool to study intermolecular interactions when one of the solutes is a macromolecule or in a separate phase.
2. Protonation of phenoxide is a key step in the association process.
3. The K_a of kraft lignins is a function of molecular weight.
4. The simple solubility model is not adequate to explain the variation of observed effects with molecular weight, concentration and ionic strength. However, its simplicity and relative ease of use warrant further modification.
5. The additive studies indicate that hydrogen bonding plays an important role in the association process at low alkalinity (less than pH 13.0). It is not clear if hydrogen bonding occurs between the phenolic hydroxyl groups or if bonded hemicellulose is responsible.
6. Higher lignin concentrations and higher ionic strengths increase the degree of association.
7. The mechanism of association at high alkalinity may be different from that at low alkalinity.

Literature Cited

1. Chum, H. L.; Johnson, D. K.; Tucker, M. P.; Himmel, M. E. *Holzforschung* 1980, **41**, 97.
2. Gross, S. K.; Sarkanen, K. V.; Schuerch, C. *Anal. Chem.* 1958, **30**, 518.
3. Benko, J. *Tappi* 1964, **47**, 508.
4. Brown, W. *J. Appl. Polym. Sci.* 1967, **11**, 2381.
5. Lindström, T. *Colloid Polym. Sci.* 1979, **257**, 277.
6. Lindström, T. *Colloid Polym. Sci.* 1980, **258**, 168.
7. Yaropolov, N. S.; Tishchenko, D. V. *Zh. Prikl. Khim.* 1970, **43**, 1120.
8. Yaropolov, N. S.; Tishchenko, D. V. *Zh. Prikl. Khim.* 1970, **43**, 1351.
9. Michell, A. J. *Cell. Chem. Technol.* 1982, **16**, 87.

10. Connors, W. J.; Sarkanen, S.; McCarthy, J. L. *Holzforschung* 1980, **34**, 80.
11. Sarkanen, S.; Teller, D. C.; Hall, J.; McCarthy, J. L. *Macromolecules* 1981, **17**, 426.
12. Sarkanen, S.; Teller, D. C.; Stevens, C. R.; McCarthy, J. L. *Macromolecules* 1984, **17**, 2588.
13. Garver, T. M.; Sarkanen, S. *Holzforschung* 1986, **40**, 93.
14. Kim, H. K. Ph.D. Thesis, University of Maine, Orono, ME, 1985.
15. Yau, W. W.; Kirkland, J. J.; Bly, D. D. *Modern Size-Exclusion Chromatography*; Wiley: New York, 1979.
16. Blatt, F.; Dravid, A.; Michaels, A. S.; Nelson, L. In *Membrane Science and Technology*; Flinn, J. E., Ed.; Plenum Press: New York, 1970.
17. Woerner, D. L.; McCarthy, J. L. *AIChE Symp. Ser.* 1986, **82**, 77.

RECEIVED February 27, 1989

Chapter 12

Modes of Association Between Kraft Lignin Components

Sunil Dutta, Theodore M. Garver, Jr., and Simo Sarkanen[1]

Department of Forest Products, University of Minnesota, St. Paul, MN 55108

Ultraviolet-visible spectral concomitants have been identified for the associative/dissociative processes which affect the apparent molecular weight distributions of kraft lignin preparations in aqueous alkaline solution. Their kinetic time course has a form which appears to be the same as that of the weight-average molecular weight itself. The results indicate that the velocities for both dissociation *and* association of the individual molecular species are independent of molecular weight. These findings suggest that the kraft lignin components interact productively with the associated complexes in a particular order. Protonation of the kraft lignin phenoxide moieties facilitates, in nonaqueous polar solvents such as DMF, much more extensive association; here the underlying intermolecular forces are different from, yet in their selectivity complementary to, those operative in aqueous alkaline solution. The severe restrictions prevailing upon associative/dissociative behavior imply that the macromolecular kraft lignin complexes embody a well-defined regular structure derived, presumably, from the original configuration of the native biopolymer.

It is now quite well established that the behavior of the molecular species comprising various lignin derivatives is strongly affected by attractive intermolecular interactions between them (1-9). There is nothing inherently

[1]Address correspondence to this author.

0097-6156/89/0397-0155$06.25/0

unique about this, except perhaps in the current divergence of opinion among workers in the field regarding the extent, importance and implications of such phenomena. Indeed a comparison of reported findings suggests that at least two kinds of intermolecular forces exert their respective influences under rather different circumstances (6,9).

However, a declared intention to embark upon detailed studies of non-covalent interactions between kraft lignin components is almost certain to cause some surprise. The majority of chemists would find it difficult to entertain a more unattractive prospect than the investigation of complicated macromolecular structures constituted from (*p*-hydroxyphenyl)propane units by *random* proportionate distributions of ten different linkages. Their concerns would hardly be mitigated by the belief (7) that such structures might adequately describe native lignins, the second most abundant group of biopolymers. But to bedevil the situation further by exposing these macromolecules to the industrial kraft process (developed, after all, for converting wood into pulp) would surely confound even the most dedicated analyst. Two-hour contact at 170°C with aqueous solutions containing 45 gL^{-1} NaOH and 12 gL^{-1} Na_2S can engender quite drastic modifications through interunit covalent bond cleavage and formation, redox reactions, and numerous other transformations, both intelligible and obscure (10,11).

Yet sometimes productive insights can appear from quite unexpected quarters, and it is the contention of this chapter that the associative interactions prevailing between kraft lignin components in solution afford such an occasion. The ensuing account embodies selected excerpts from recent work carried out at the University of Minnesota (12,13) that will be published in complete detail elsewhere. The kraft lignin adopted for investigation was derived from Douglas fir (*Pseudotsuga menziesii* (Mirb.) Franco) and recovered from an industrial black liquor sample donated by the Weyerhaeuser Co. from their mill in Everett, Washington, U.S.A. (8).

Molecular Weight Distributions of Kraft Lignins

The experience of prior work has repeatedly confirmed that the molecular weight distributions of kraft lignin preparations can be reliably deduced from Sephadex G100/aqueous 0.10 *M* NaOH elution profiles by ultracentrifuge sedimentation equilibrium techniques (5, 6, 8). Although their resolution does not approach that obtained under HPSEC conditions, such profiles have a particular advantage in depicting the populations of macromolecular complexes and individual components prevailing at pH's greater than 11.5: below this point the effects of intermolecular hydrogen bonding are no longer counteracted by the negative charge densities arising from unprotonated phenoxide groups on the species involved (9). Indeed well-defined associative processes may be readily detected in aqueous alkaline solutions containing kraft lignin concentrations greater than 100 gL^{-1} at pH's in a fairly narrow range immediately above 11.5; conversely dissociation occurs at kraft lignin concentrations less than 1 gL^{-1} in aqueous 0.10 *M* NaOH (6).

Ultraviolet-Visible Spectral Concomitants of Associative Interactions between Kraft Lignin Species

Figure 1 illustrates the ultraviolet-visible spectral changes accompanying the dissociation of a Douglas fir kraft lignin (8) during incubation at 0.047 gL^{-1} in aqueous 0.10 *M* NaOH. In the regions around 230 and 290 nm, and above 450 nm, the marked reduction in absorptivity is characterized by a time course which is reasonably well described by superimposed first and pseudo-zero order kinetic rate processes. Owing to the fact that the respective velocities are comparable in magnitude, through the Levenberg-Marquardt nonlinear regression procedure (14) it is difficult to determine the corresponding first order rate coefficients accurately from data in these spectral regions.

Around 350 nm, however, the absorbance increases with time in a first order process superimposed upon a much smaller zero order velocity (Figure 2). Indeed, appropriate subtraction of the linear asymptote to the change reveals that the predominant rate process obeys first order kinetics quite accurately for at least two half-lives (Figure 3). Interestingly, the net extent of this first order process increases appreciably with decreasing kraft lignin concentration in solution (Table I). Despite the accentuated errors inherent in their numerical analyses, the spectral changes around 230 nm and above 450 nm are not inconsistent with their first order components' being characterized by rate coefficients of the same magnitude as that deduced at 350 nm.

Table I. Variation in Extent of Kraft Lignin Dissociation with Concentration in Aqueous 0.10 *M* NaOH at 25.0°C[a]

Concentration gL^{-1}	k_1 h^{-1} [b]	Spectrophotometric Pathlength, cm	$A_\infty - A_0$ [c] at 350 nm
4.62×10^{-1}	3.9×10^{-3}	0.1	0.059
4.68×10^{-2}	4.2×10^{-3}	1.0	0.064
4.72×10^{-3}	4.1×10^{-3}	10.0	0.118

[a] Determined spectrophotometrically at 350 nm.
[b] Rate coefficient for first order physicochemical process.
[c] Overall change in absorbance arising from first order process.

Kinetic Changes in Weight-Average Molecular Weight

The foregoing ultraviolet-visible spectral changes are directly correlated with the variation in molecular weight distribution of the kraft lignin sample under comparable conditions in aqueous 0.10 *M* NaOH (Figure 4). Phenomenologically, the decrease in weight-average molecular weight, $\overline{M}_w$, with time also kinetically conforms to a superposition of first and pseudo-zero order rate processes (Figure 5). Employing the Levenberg-Marquardt procedure (14) to enable subtraction of the pseudo-zero order contribution, $\overline{M}_{w,\infty,t}$, from the overall change in $\overline{M}_{w,t}$, the remaining plot of

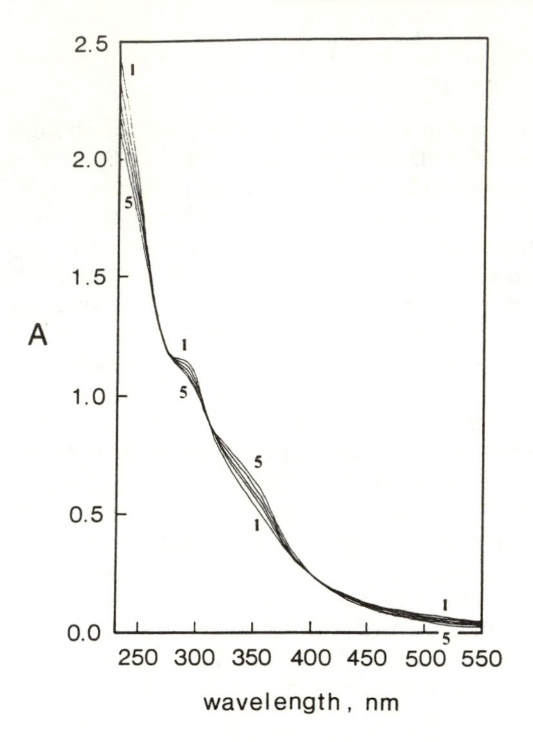

Figure 1. Ultraviolet-visible spectral concomitants of dissociation and oxidative cleavage of kraft lignin preparation during incubation at 0.047 gL^{-1} in aqueous 0.10 *M* NaOH between (1) 0 h and (5) 1340 h.

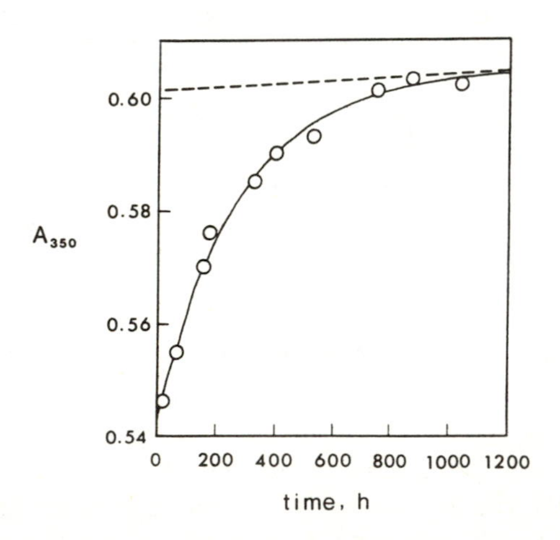

Figure 2. First and pseudo-zero order kinetic components of change in absorbance at 350 nm with time for kraft lignin preparation during incubation at 0.46 gL^{-1} in aqueous 0.10 *M* NaOH at 25.0°C; k_1 = 0.0039 h^{-1}, v_0 = 0.0000026 $A_{350}h^{-1}$ (0.1 cm pathlength).

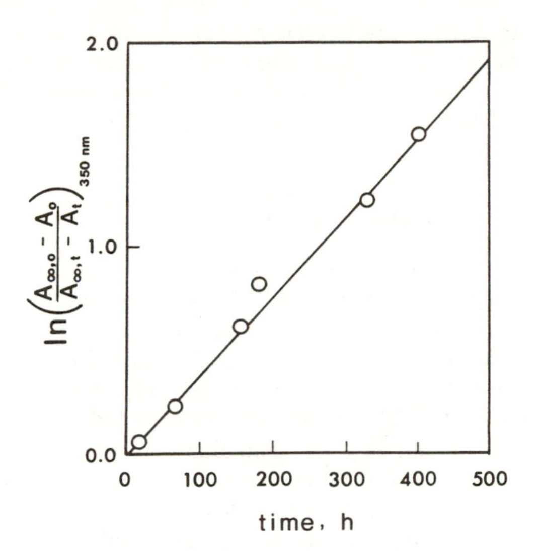

Figure 3. Rate plot for first order kinetic component of change in absorbance at 350 nm with time for kraft lignin preparation during incubation at 0.46 gL^{-1} in aqueous 0.10 *M* NaOH at 25.0° C; k_1 = 0.0039 h^{-1}.

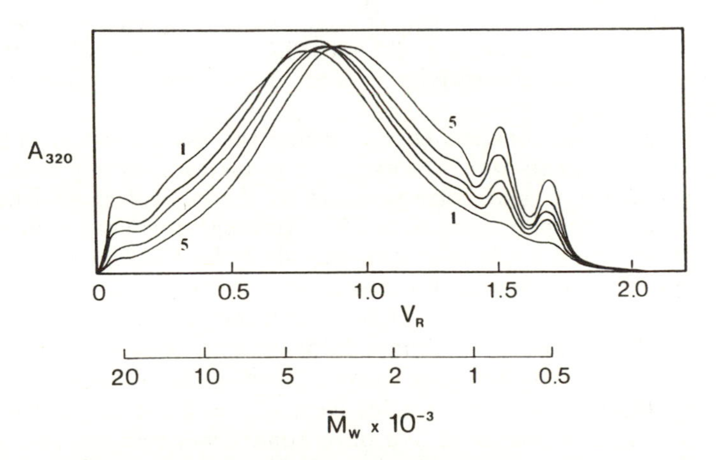

Figure 4. Dissociation and oxidative cleavage of kraft lignin preparation during incubation at 0.50 gL^{-1} in aqueous 0.10 *M* NaOH for (1) 48 h, (2) 168 h, (3) 336 h, (4) 840 h and (5) 1780 h. (Sephadex G100/aqueous 0.10 *M* NaOH elution profiles monitored at 320 nm.)

$\ln(\overline{M}_{w,t} - \overline{M}_{w,\infty,t})$ against time, t, reveals that the estimated first order rate coefficient, $5.0 \pm 0.3 \times 10^{-3}$ h^{-1}, is quite close in magnitude to that ($4.05 \pm 0.15 \times 10^{-3}$ h^{-1}) characterizing the variation in A_{350}. In a similar manner, the phenomenological rate coefficient describing the first order contribution to the change in $\overline{M}_w$ during the dissociation at 0.50 gL^{-1} of a preassociated kraft lignin sample in aqueous 0.10 *M* NaOH (Figure 6) was found to be $3.9 \pm 0.4 \times 10^{-3}$ h^{-1}.

Implications at the Molecular Level

A number of structures thought to be present in kraft lignins, including stilbenes (15), styryl aryl ethers (16, 17) and aromatic rings of phenolic moieties (18), are susceptible to cleavage by oxygen in aqueous alkaline solution. Indeed the oxygen content of kraft lignin fractions increases significantly during incubation for extended periods at 0.5 gL^{-1} in aqueous 0.10 *M* NaOH solution (12). If the pseudo-zero order components of the changes in absorptivity and $\overline{M}_w$ are identified with the results of such oxidative cleavage reactions, the first order processes must reflect the dissociation of the macromolecular kraft lignin complexes.

Indeed the characteristics of the ultraviolet-visible spectral changes for the kraft lignin solution in aqueous 0.10 *M* NaOH during the course of dissociation and covalent cleavage differ according to which process contributes to the greater extent (Iwen, M.L., Sarkanen, S., University of Minnesota, unpublished results). Initially, when the decrease in molecular weight originates primarily from dissociation, quasi-isosbestic points reside near 305 and 455 nm; subsequently, when covalent cleavage predominates, quasi-isosbestic points appear around 330 and 385 nm. The overall effect is one where the change in absorptivity is least in the regions about 315 and 400 nm (Figure 1).

The equilibrium constant characterizing the association of individual kraft lignin components would be expected to increase rapidly with molecular weight above a value corresponding to the critical chain length for macromolecular complex formation (19). Although this change in equilibrium constant would not be dictated exclusively by differences in the phenomenological rate coefficients for dissociation, it is reasonable to anticipate that kraft lignin components of higher molecular weight would dissociate more slowly than those of lower molecular weight. Such an effect may be exemplified by the reported interaction between poly(*N*-vinylpyrrolidone) and dansyl labeled poly(acrylic acid) with $\overline{M}_\eta = 5.9 \times 10^5$, the data for which support the existence of hydrogen bond donor and acceptor groups in equivalent concentrations (20): introduction of an over five-fold molar excess of unlabeled poly(acrylic acid) with $\overline{M}_\eta = 6.9 \times 10^5$ facilitated a rate of interchange that is ten times slower for 5.7×10^4 than for 1.0×10^4 molecular weight poly(*N*-vinylpyrrolidone) (21).

Consequently any model proposed to account for the dissociation of individual components from associated kraft lignin complexes should pay particular attention to restrictions that would rationalize the concomitant first

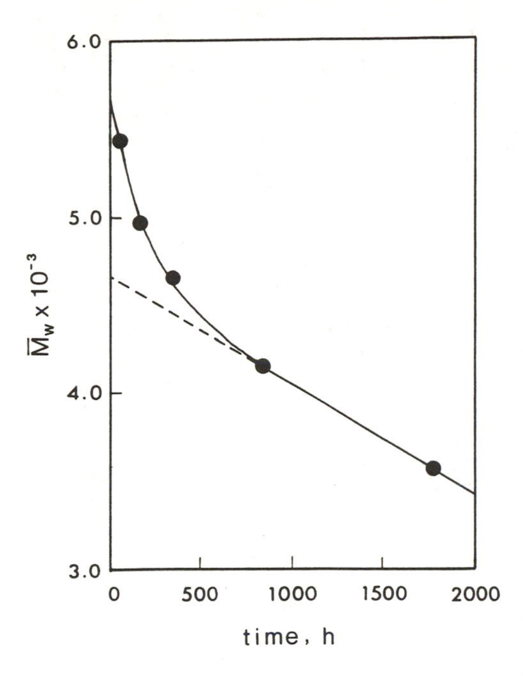

Figure 5. First and pseudo-zero order kinetic components of change in $\overline{M}_w$ with time for kraft lignin preparation during incubation at 0.50 gL^{-1} in aqueous 0.10 *M* NaOH (calculated from Sephadex G100/aqueous 0.10 *M* NaOH elution profiles monitored at 320 nm.)

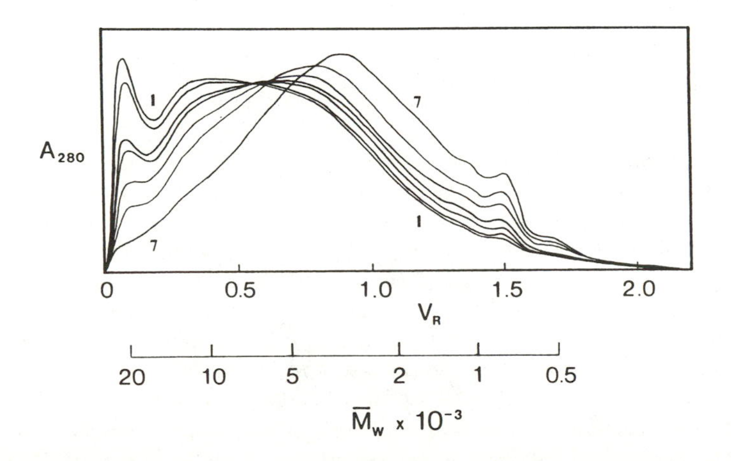

Figure 6. Dissociation of preassociated kraft lignin sample during incubation at 0.50 gL^{-1} in aqueous 0.10 *M* NaOH for (1) 0 h, (2) 25 h, (3) 118 h, (4) 171 h, (5) 401 h, (6) 575 h and (7) 1340 h. (Sephadex G100/aqueous 0.10 *M* NaOH elution profiles monitored at 280 nm.)

order behavior of both absorptivity and weight-average molecular weight. The following scheme involving the consecutive release of components L_l from complexes C_{c+1} in a specific order could successfully meet these requirements:

$$\begin{array}{llll} \ldots\; C_{c+2} & \rightleftharpoons & C_{c+1} & + L_{l+1} \\ C_{c+1} & \rightleftharpoons & C_c & + L_l \\ C_c & \rightleftharpoons & C_{c-1} & + L_{l-1} \\ C_{c-1} & \rightleftharpoons & C_{c-2} & + L_{l-2} \;\ldots \end{array}$$

Stoichiometry implies $[C_{c+1}] - [C_{c+1}]_0 = [L_{l+1}] - [L_{l+1}]_0 - ([L_l] - [L_l]_0)$, where $[C_{c+1}]_0$, $[L_{l+1}]_0$ and $[L_l]_0$ are the molar concentrations $[C_{c+1}]$, $[L_{l+1}]$ and $[L_l]$, respectively, of the complex C_{c+1} and components L_{l+1} and L_l at arbitrary time $t = 0$. Adopting m to represent molecular weight, since $m_{c+1} = m_c + m_l$,

$$\begin{aligned} \sum_{c,l}(m_c^2[C_c] + m_l^2[L_l]) = \sum_{c,l}(m_{c+1}^2[C_{c+1}]_0 &+ m_{c+1}^2([L_{l+1}] - [L_{l+1}]_0) \\ &- m_c^2([L_l] - [L_l]_0) - 2m_c m_l([L_l] - [L_l]_0) \\ &- m_l^2([L_l] - [L_l]_0) + m_{l+1}^2[L_{l+1}]) \\ = \sum_{c,l}(m_{c+1}^2[C_{c+1}]_0 &+ m_l^2[L_l]_0 \\ &- 2m_c m_l([L_l] - [L_l]_0)). \end{aligned}$$

Accordingly, the weight-average molecular weight, $\overline{M}_w$, for the system of interacting species at any time during the associative/dissociative process can be written as

$$\overline{M}_w = \frac{\sum_{c,l}(m_{c+1}^2[C_{c+1}]_0 + m_l^2[L_l]_0 - 2 < m_c m_l > ([L_l] - [L_l]_0))}{\sum_{c,l}(m_c[C_c] + m_l[L_l])},$$

where $< m_c m_l >= \sum_l m_c m_l([L_l] - [L_l]_0)/(\sum_l([L_l] - [L_l]_0))$ and it is understood that a unique relationship exists between c and l.

The number-average (with respect to the individual components released) product of the molecular weights, $< m_c m_l >$, of interacting species, $\{C_c, L_l\}$, will remain constant if the evolution with time of $([L_l] - [L_l]_0)$ follows exactly the same kinetic course for each L_l. This is reminiscent of earlier findings that the relative ratios of kraft lignin components with molecular weights below 3500 do not vary with the degree of association for the sample as a whole (6). An appropriately positioned rate-determining step in the sequence of dissociative events would impose such a restriction upon the system. If the step in question were to exhibit first order behavior, so also would the changes in $\overline{M}_w$ and absorptivity accompanying the overall dissociative process.

Associative/Dissociative Homology among Kraft Lignin Samples

The associative/dissociative processes in which kraft lignin species participate are reversible: association and dissociation take place respectively at high and low kraft lignin concentrations in aqueous alkaline solution. For example, a kraft lignin sample that has been preassociated (during incubation at 190 gL^{-1} in 1.0 *M* ionic strength aqueous 0.6 *M* NaOH) spontaneously dissociates when diluted in aqueous 0.10 *M* NaOH to 0.5 gL^{-1} (Figure 6). Conversely, a kraft lignin sample that has been predissociated (through incubation at 0.5 gL^{-1} in aqueous 0.10 *M* NaOH and subsequent recovery by ultrafiltration with a nominally 500 molecular weight cutoff membrane) associates spontaneously when dissolved to a level of 160 gL^{-1} in 1.0 *M* ionic strength aqueous 0.40 *M* NaOH (Figure 7).

Onto all such associative/dissociative processes, the effects of oxidative covalent cleavage reactions will be appended to varying extents depending upon the prevailing circumstances. Consequently the molecular weight distributions depicted by the corresponding elution profiles exhibit points of intersection that may vary quite widely (Figures 4 and 6). However, the kraft lignin species capable of interacting with one another can be readily separated from those components that, owing to covalent modification, cannot. This may be accomplished through the straightforward expedient of partly relaxing the impediments to association imposed by solution conditions; appropriately elaborated chromatographic fractionation of the macromolecular complexes will then complete the task.

To this end kraft lignin samples possessing different degrees of association in aqueous sodium hydroxide solutions can be secured by acidification to pH 7.5 and freeze-drying. Thereupon dissolution in a portion of eluent containing 0.325 *M* NaOH and subsequent fractionation through Sephadex LH20 with aqueous 35% dioxane separates the sample into two groups of species: a subset of associated complexes appearing at the void volume is segregated from the remaining components eluting just after the salt band (Figure 8; *cf.* ref. 8). After freeze-drying, refractionation of the leading peak through Sephadex LH20 with aqueous 35% dioxane allows a net 60-65% of the original kraft lignin to be recovered as an associated salt-free specimen.

The molecular weight distributions depicted by the Sephadex G100/aqueous 0.10 *M* NaOH elution profiles (Figure 9) of these recovered kraft lignin preparations reflect the degrees of association for the constituent subsets of species in the original samples. They exhibit a common intersection point, and as such represent the molecular weight distributions of preparations that are homologous from an associative/dissociative point of view: the relationship between successive members of the series is uniformly confined to systematic differences in their degrees of association without perturbations arising from the effects of covalent cleavage reactions.

This harmonious outcome from a simple procedure has far-reaching implications. That the species appearing at the void volume in aqueous 35% dioxane from Sephadex LH20 are associated is readily confirmed by the corresponding size-exclusion chromatographic profiles from a 10^6Å pore-size

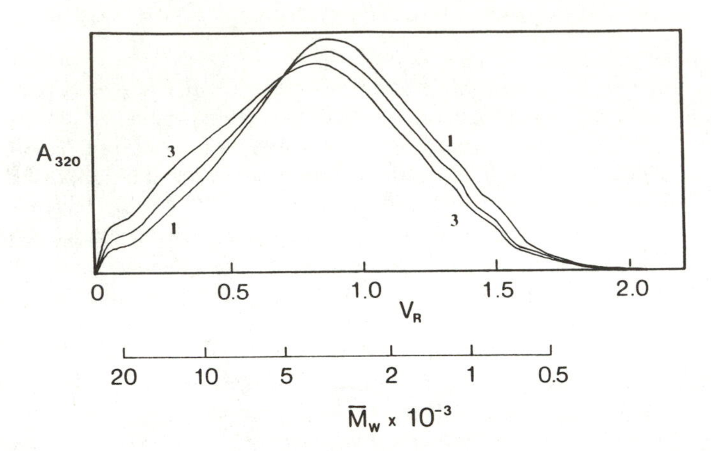

Figure 7. Association of predissociated kraft lignin sample (92% retained by nominally 500 molecular weight cutoff ultrafiltration membrane) during incubation at 160 gL^{-1} in 1.0 *M* ionic strength aqueous 0.40 *M* NaOH for (1) 0 h, (2) 50 h and (3) 480 h.

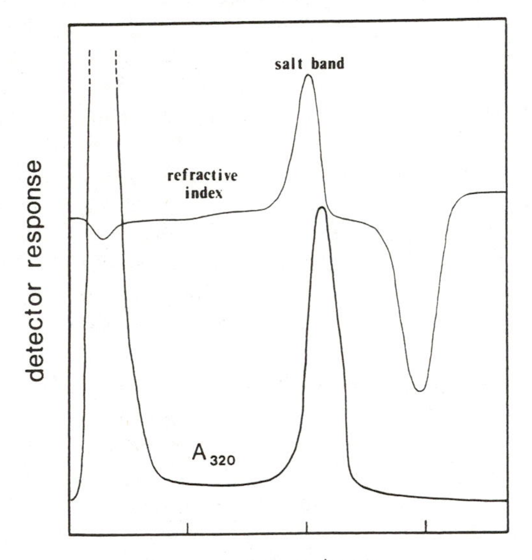

Figure 8. Kraft lignin fractionation into two subsets of species during desalting by elution with aqueous 35% dioxane through Sephadex LH20 of an initially 4.5 gL^{-1} sample solution at 2.3 *M* ionic strength containing 0.32 *M* aqueous NaOH.

poly(styrene divinylbenzene) column. While the intermolecular forces causing association in aqueous dioxane necessarily differ from those in aqueous alkaline solution, their respective stoichiometric selectivities towards the individual kraft lignin components must be directly complementary. The relatively rapid association facilitated by lowering the pH is presumably mediated by hydrogen bonding (9) and/or dipolar interactions; the slower associative processes encountered in aqueous alkaline solutions are probably caused by nonbonded orbital interactions, the concomitants of which contribute to the long-wavelength absorbance changes in the ultraviolet-visible spectrum (Figure 1).

The Behavior of $< m_c m_l >$ during Association and Dissociation

It has not been possible experimentally to detect systematic differences in the ultracentrifuge sedimentation equilibrium weight-average molecular weight calibration curves for the Sephadex G100/ aqueous 0.10 M NaOH elution profiles of kraft lignin preparations with different degrees of association (Figure 10). Consequently the same relationship has been employed to calculate the overall weight- and number-average molecular weights, $\overline{M}_w$ and $\overline{M}_n$, of the preparations as associative or dissociative processes are underway.

Accordingly, the average product of the molecular weights, $< m_c m_l >$, of interacting kraft lignin species (*vide supra: Implications at the Molecular Level*) can be evaluated numerically by plotting $\overline{M}_w$ *versus* $1/\overline{M}_n$: the slope at any point on the resulting curve is given by $-2 < m_c m_l >$ (4). It is hereby evident that the value, 1.03×10^7, of $< m_c m_l >$ characterizing the relationship between the members of the homologous series of kraft lignin preparations is, within experimental error, identical in magnitude to those encountered during association of the original kraft lignin sample at 195 gL^{-1} in 1.0 M ionic strength aqueous 0.4 M NaOH (Figure 11), dissociation of the preassociated kraft lignin preparation (Figures 6 and 11), and association of the predissociated kraft lignin preparation (Figures 7 and 11).

During dissociation of the original kraft lignin sample at 0.46 gL^{-1} in aqueous 0.10 M NaOH (Figure 4), however, the apparent value, 5.1×10^6, of $< m_c m_l >$ is considerably lower (Figure 11). Indeed the manner in which $\overline{M}_w$ varies with $1/\overline{M}_n$ during the dissociation of the preassociated kraft lignin preparation (Figures 6 and 11) suggests that $< m_c m_l >$ similarly becomes smaller when a comparable range of degree of association has been reached. Presumably the magnitude of $< m_c m_l >$ under these conditions is influenced by additional contributions arising from covalent cleavage of individual kraft lignin components, the effects of which would be more extensive when the degree of association is smaller, the pH higher, and the solution more dilute (whereupon the stoichiometric ratio of dissolved oxygen to individual kraft lignin components is larger). In this connection it should be pointed out that, in aqueous alkaline solutions containing the highest kraft lignin concentrations where association is favored, the pH tends to approach a value around 11.6, just above the region where the

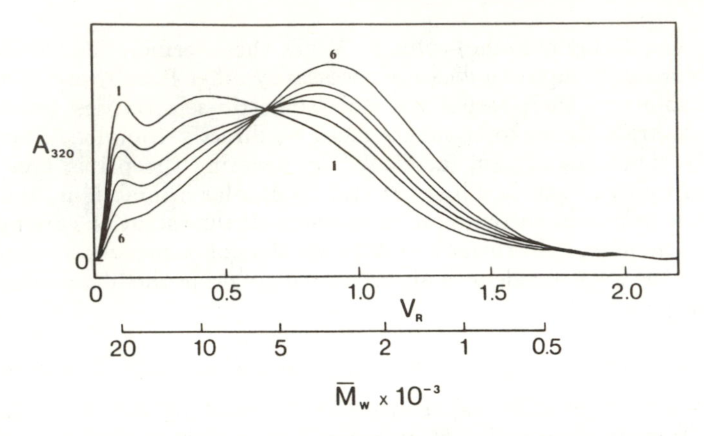

Figure 9. Apparent molecular weight distributions representing a homologous series of kraft lignin samples secured by desalting after association and dissociation in aqueous alkaline solutions for: (1) 300 h, (2) 144 h and (3) 48 h at 170 gL^{-1} in 1.0 *M* ionic strength aqueous 0.40 *M* NaOH; (4) 0 h; (5) 144 h and (6) 644 h at 0.50 gL^{-1} in aqueous 0.10 *M* NaOH. (Sephadex G100/aqueous 0.10 *M* NaOH elution profiles monitored at 320 nm.)

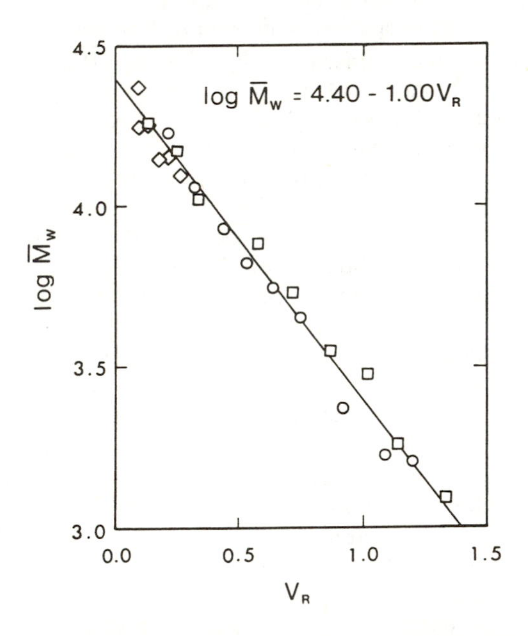

Figure 10. Molecular weight calibration curves for kraft lignin samples differing in degree of association eluted from Sephadex G100 with aqueous 0.10 *M* NaOH: (◇) associated sample after 385 h at 180 gL^{-1} in 1.0 *M* ionic strength aqueous 0.40 *M* NaOH; (○) original preparation; (□) dissociated sample precipitated upon acidification to pH 3.0 after 2000 h at 0.50 gL^{-1} in aqueous 0.10 *M* NaOH.

molecular weights of the macromolecular complexes fall most rapidly during titration of the component phenolic groups with hydroxide (9).

Remarkably, the magnitude of $< m_c m_l >$ delineating the associative or dissociative influences upon molecular weight distribution does not appear to be affected by the small variations in component composition resulting from differences in the preparative histories of the samples. Indeed the same constant value of $< m_c m_l >$ prevails during association and dissociation despite marked changes in the relative populations of the constituent species. Both processes thus embody restrictions whereby the individual components are captured by or released from the macromolecular complexes at respective rates of which the kinetic forms are independent of molecular weight. In the context imposed by the dissolved kraft lignin preparation, therefore, the equilibrium constants for all pairs of interacting species are operationally the same as one another.

Contrary to *a priori* expectations (19), then, the equilibrium constant for the association of kraft lignin components does not increase with the degree of polymerization. Rather the behavior encountered indicates that the rates of association and dissociation are governed by a particular rate determining step. Such a restriction strongly suggests that the individual molecular species interact with complementary loci on the corresponding macromolecular complexes in a specific order. It is difficult to conceive how selectivity on this scale could be sustained in a system without a high level of structural regularity in, and molecular organization among, the species involved.

Further Evidence for Nonrandom Interactions between Kraft Lignin Species

There are other quite independent indications that the associative processes taking place in lignin samples are not random. This is exemplified by the composition of a partially dissociated kraft lignin preparation secured by precipitation (59% of original sample) upon acidification to pH 3.0 of a 0.50 gL^{-1} solution in aqueous 0.10 *M* NaOH that had been incubated for 2000 h. Paucidisperse fractions selected from the Sephadex G100/aqueous 0.10 *M* NaOH elution profile of the preparation can be separated into two subsets of components B and C by eluting with water through Sephadex G25 in the presence of a high (initially 2.0 M) ionic strength salt band (8). The components in subsets B and C respectively contribute primarily to the higher and lower molecular weight regions of the Sephadex G100/aqueous 0.10 *M* NaOH elution profile (Figure 12).

The behavior of the kraft lignin species during elution through Sephadex G25 is determined by three coupled physicochemical processes, *viz.* adsorption at high solution ionic strengths of the lower molecular weight components onto the gel, intermolecular association, and diffusion of the higher molecular weight entities through the salt band. The overall effect facilitates more complete separation between the smaller and larger kraft lignin species when the differences in their molecular weights are greater. The weight-average molecular weights of the component subsets

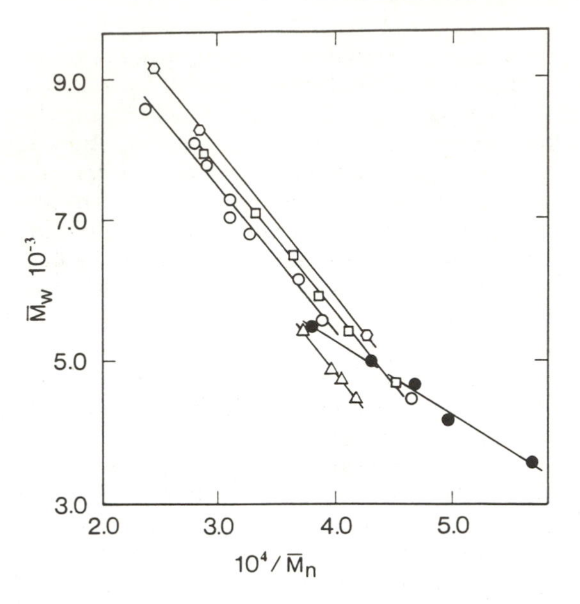

Figure 11. Relationship between $\overline{M}_w$ and $\overline{M}_n$ during associative/dissociative processes between molecular kraft lignin species in aqueous alkaline solutions: (□) desalted homologous series of samples with different degrees of association; association of (⬡) preparation at 180 gL^{-1} and (△) predissociated sample at 160 gL^{-1}, both in 1.0 *M* ionic strength aqueous 0.40 *M* NaOH; (○) dissociation of preassociated sample and (●) dissociation and covalent cleavage of preparation, both at 0.50 gL^{-1} in aqueous 0.10 *M* NaOH.

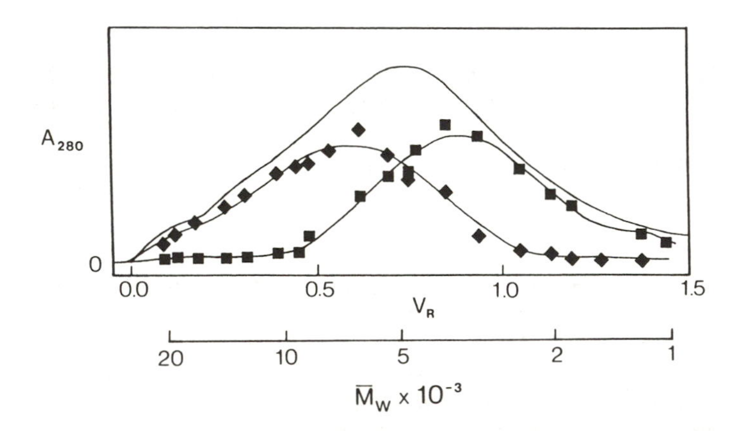

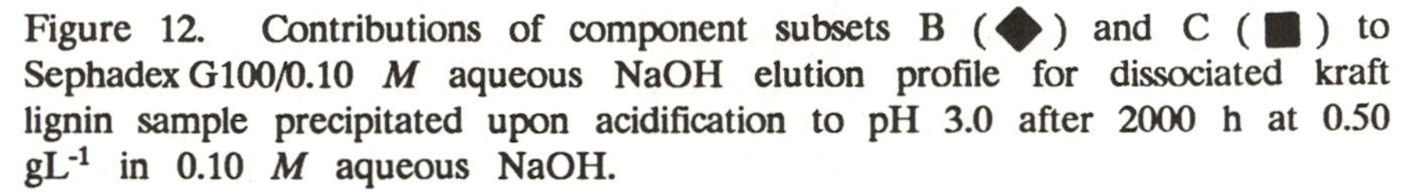
Figure 12. Contributions of component subsets B (◆) and C (■) to Sephadex G100/0.10 *M* aqueous NaOH elution profile for dissociated kraft lignin sample precipitated upon acidification to pH 3.0 after 2000 h at 0.50 gL^{-1} in 0.10 *M* aqueous NaOH.

B and C are juxtaposed in Figure 13 to those of the parent paucidisperse fractions from the partially dissociated kraft lignin preparation. This comparison establishes that a particular subset of components characterized by a weight-average molecular weight of 3500 can be spontaneously liberated from the entire range of kraft lignin species with molecular weights initially falling between 10,000 and 23,000.

An important physicochemical concomitant of the associative/dissociative processes in which kraft lignin components participate can be detected in the ultraviolet-visible spectra of paucidisperse fractions in aqueous alkaline solution derived from kraft lignin preparations characterized by different degrees of association. The ratio of the absorptivity at 230 nm to that at 300 nm generally decreases with increasing frequency of phenoxide moieties among the species present (22). Except in the very lowest reaches of the molecular weight range, this ratio for those components with molecular weights less than 3500 is independent of degree of association for the entire kraft lignin preparation (Figure 14). On the other hand, the frequencies of unprotonated phenoxide groups among the higher molecular weight species clearly increases with degree of association for the sample as a whole. These observations are consistent with the supposition (6) that the associative processes involve preferential interactions between the subset of components with molecular weights below 3500 and the complementary higher molecular weight entities; accompanying protonation of the phenoxide moieties, which tends to reduce the charge density on the associated complexes, evidently need not be complete.

Association in Nonaqueous Polar Solvents

When the phenoxide moieties of the kraft lignin components participating in associated complex formation are fully protonated, far more extensive association can occur in accommodating solvents. Such an outcome may be readily demonstrated for the series of kraft lignin preparations differing homologously with respect to their degrees of association in aqueous 0.10 *M* NaOH (Figure 9). When calibrated with paucidisperse polystyrene standards, the corresponding elution profiles (Figure 15) in DMF from 10^6Å pore-size poly(styrene-divinylbenzene) exhibit apparent molecular weight distributions characterized by weight-average molecular weights 1.5 to 1.9×10^4 times greater than those in aqueous 0.10 *M* NaOH (Table II).

When measured through a 26 nm bandwidth interference filter (rejecting most fluorescence), Rayleigh scattered light intensities (multiplied by reciprocal solution transmittance to correct for absorbance) furnish weight-average molecular weights for the same series of kraft lignin preparations in DMF that are between 190 and 1100 times larger than those in aqueous 0.10 *M* NaOH (Table II). The trends in, rather than the absolute values of, the reported numbers have the greater significance. The forms of the respective Zimm plots (Figure 16) are typical of associating macromolecular species (23): the apparent second virial coefficients, calculated from the

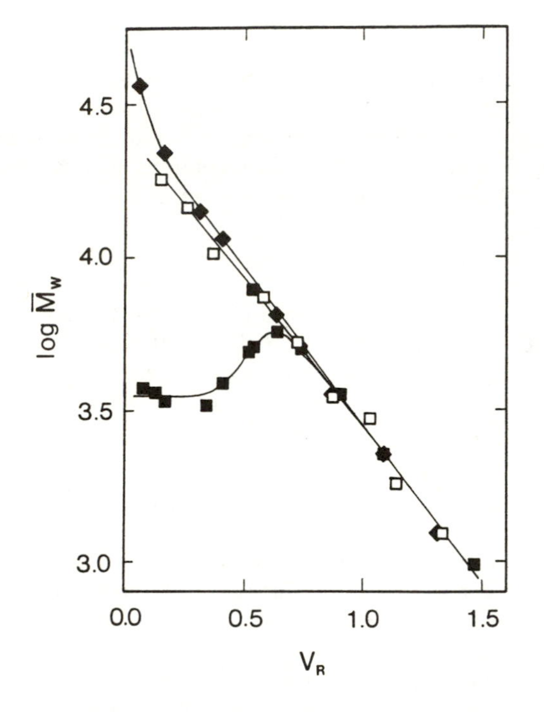

Figure 13. Comparison of weight-average molecular weights for component subsets B (◆) and C (■) with those for complete paucidisperse fractions (□) isolated from Sephadex G100/0.10 *M* aqueous NaOH elution profile of the dissociated kraft lignin sample precipitated upon acidification to pH 3.0 after 2000 h at 0.50 gL^{-1} in 0.10 *M* aqueous NaOH.

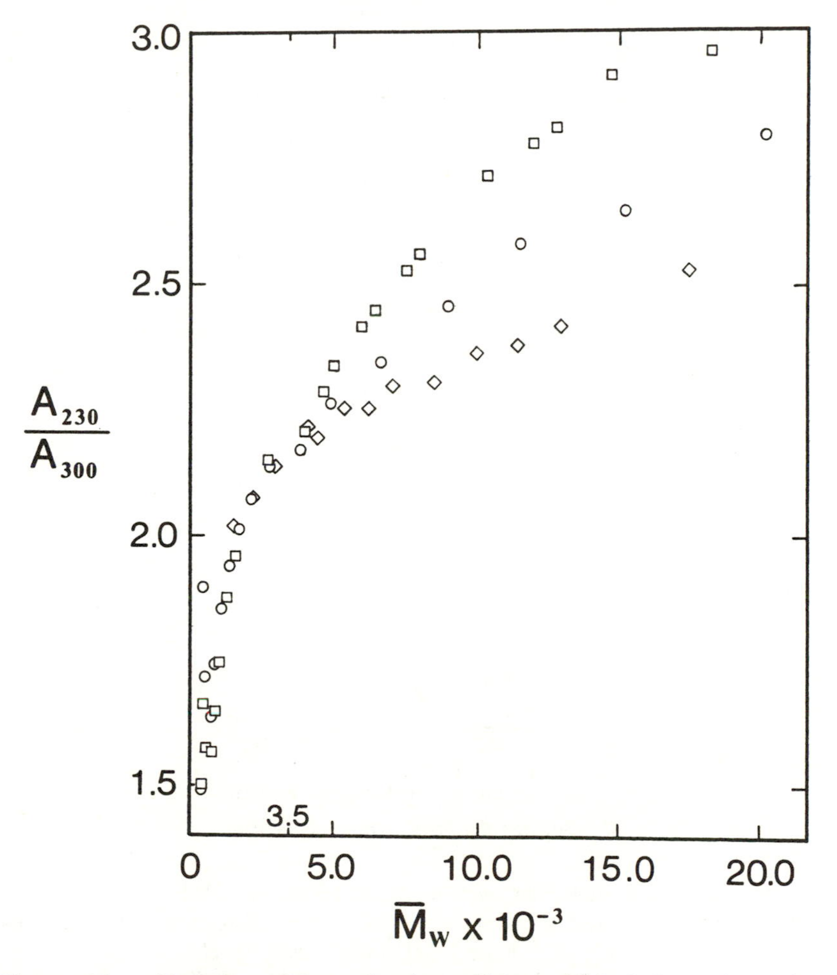

Figure 14. Variation with molecular weight exhibited by ratios of absorbance values at 230 and 300 nm for paucidisperse kraft lignin fractions in aqueous 0.10 *M* NaOH from (◇) associated sample after 385 h at 180 gL^{-1} in 1.0 *M* ionic strength aqueous 0.40 *M* NaOH; (○) original preparation; (□) dissociated sample precipitated upon acidification to pH 3.0 after 2000 h at 0.50 gL^{-1} in aqueous 0.10 *M* NaOH.

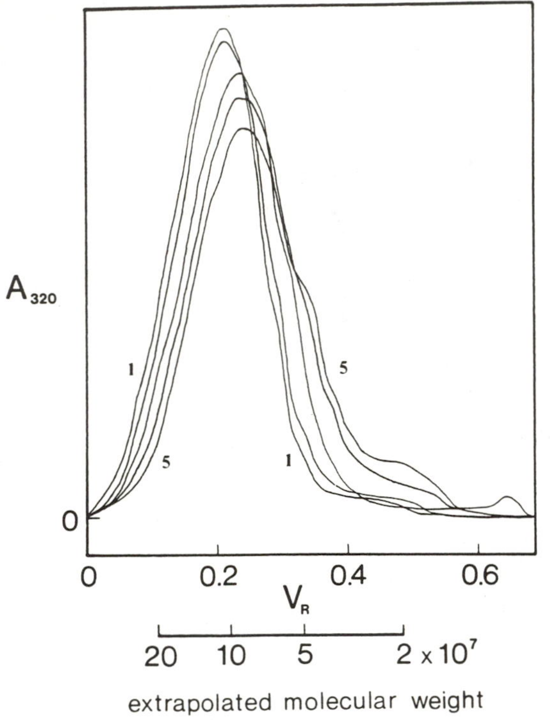

Figure 15. Apparent molecular weight distributions in DMF of kraft lignin samples secured by desalting after association and dissociation in aqueous alkaline solutions for: (1) 300 h and (2) 144 h at 170 gL^{-1} in 1.0 *M* ionic strength aqueous 0.40 *M* NaOH; (3) 0 h; (4) 144 h and (5) 644 h at 0.50 gL^{-1} in aqueous 0.10 *M* NaOH. (Profiles from 10^6 Å pore-size 20 μ particle poly(styrene-divinylbenzene) column monitored at 320 nm.)

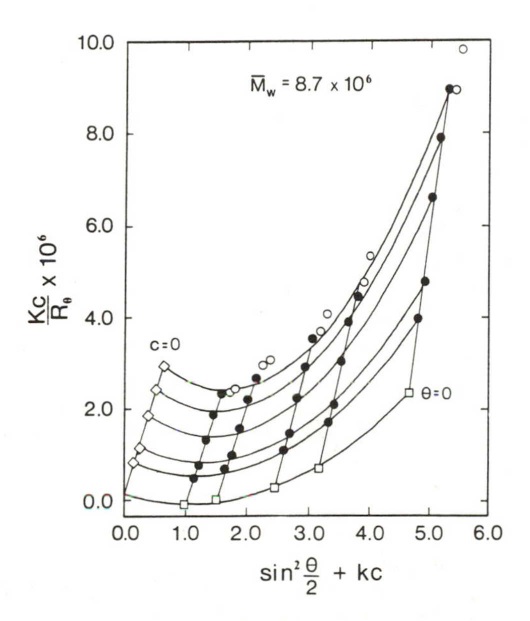

Figure 16. Zimm plot of 514.5 nm light-scattering data characterizing associated kraft lignin sample in DMF following desalting after incubation at 170 gL^{-1} for 300 h in 1.0 *M* ionic strength aqueous 0.40 *M* NaOH.

Table II. Comparison of Homologous Kraft Lignin Samples in DMF and Aqueous 0.10 *M* NaOH[a]

Parameter	Dissociated Kraft Lignin[b]	Original Kraft Lignin[c]	Associated Kraft Lignin[d]
$\overline{M}_{w,NaOH}$ [e]	4.67×10^3	5.89×10^3	7.91×10^3
$\overline{M}_{w,DMF}$ [f]	8.7×10^5	1.3×10^6	8.7×10^6
$\frac{\overline{M}_{w,DMF}}{\overline{M}_{w,NaOH}}$ [e,f]	190	220	1100
$A_{2,app}$ [f] $cm^3 mol\ g^{-2}$	-8.4×10^{-4}	-4.2×10^{-4}	-3.9×10^{-4}
R_G nm[f]	74	130	310
$\overline{M}_{w,DMF}$ [g] from HPLC	8.9×10^7	1.0×10^8	1.2×10^8

[a] Desalted by elution with aqueous 35% dioxane through Sephadex LH20 after incubation in aqueous NaOH solutions.
[b] After 644 h at 0.50 gL^{-1} in aqueous 0.10 *M* NaOH.
[c] Without prior incubation in aqueous NaOH solution.
[d] After 300 h at 170 gL^{-1} in 1.0 *M* ionic strength aqueous 0.40 *M* NaOH.
[e] Calculated from Sephadex G100/aqueous 0.10 *M* NaOH elution profiles monitored at 320 nm.
[f] Deduced from Zimm plots of 514.5 nm light-scattering data from solutions containing between 0.35 and 2.3 gL^{-1} kraft lignin.
[g] Apparent values calculated on the basis of the extrapolated polystyrene calibration curve for 10^6Å pore-size 20μ particle poly(styrene-divinylbenzene) column.

initial slopes of the curves incorporating the points extrapolated to zero angle, are negative (Table II).

Dissolution in, and subsequent recovery from, DMF does not alter the molecular weight distributions of the kraft lignin preparations in aqueous 0.10 *M* NaOH. Consequently the associative processes occurring under these two solution conditions must be governed by *different* intermolecular forces—presumably hydrogen bonding and nonbonded orbital interactions, respectively. Yet the effect of the latter influences the extent of the former: preassociation of the kraft lignin preparation in aqueous NaOH engenders a greater degree of cooperativity in the pronounced associative processes occurring in DMF (Table II). This may be of more than passing interest if the stoichiometric selectivities of the putative intermolecular nonbonded orbital and hydrogen bonded interactions with respect to the individual kraft lignin components are directly complementary (*vide supra: Associative/Dissociative Homology among Kraft Lignin Samples*).

Concluding Remarks

The kinetics of macromolecular kraft lignin complex dissociation in aqueous alkaline solution are successfully accounted for through a simple scheme whereby the individual components are released in a specific order. Owing to concomitant changes in solution viscosity, the association of kraft lignin components cannot be kinetically delineated in a straightforward way. Nevertheless, the average product of the molecular weights of the interacting species remains fixed at the same constant value as during dissociation. This confirms that the principle of microscopic reversibility is rigorously operative upon both facets of the system. Such drastic restrictions upon associative/dissociative behavior suggest that the kraft lignin complexes embody the memory of a *regular* native macromolecular structure. As the two distinct intermolecular forces putatively governing association, nonbonded orbital interactions and hydrogen bonding exhibit selectivities towards the individual kraft lignin components that are strictly complementary to one another.

The experimental observations summarized in this chapter resist integration with prevailing views about kraft lignins. Inevitably, their present interpretation will be judged as over-simplified from the perspective of further insights yet to be divulged in future work. But when a remarkable harmony emerges from the cacophony of kraft pulping, is it not time to listen for the echo of a new theme in lignin chemistry?

Acknowledgments

Acknowledgment for support of this research is made to the National Science Foundation (Grant CBT 8412604), the Blandin Foundation, the Graduate School of the University of Minnesota, the Minnesota Agricultural Experiment Station, and the University of Minnesota Computer Center. The authors are indebted to Professor V.A. Bloomfield, Department of Biochemistry at the University of Minnesota, in whose laboratories the light scattering measurements were carried out.

Paper No. 16,650 of the Scientific Journal Series of the Minnesota Agricultural Experiment Station funded through Minnesota Agricultural Experiment Station Project No. 43-68, supported by Hatch funds.

Literature Cited

1. Benko, J. *Tappi* 1964, **47**, 508-14.
2. Brown, W. *J. Appl. Polym. Sci.* 1967, **11**, 2381-96.
3. Connors, W. J.; Sarkanen, S.; McCarthy, J. L. *Holzforschung* 1980, **34**, 80-5.
4. Sarkanen, S.; Teller, D. C.; Hall, J.; McCarthy, J. L. *Macromolecules* 1981, **14**, 426-34.
5. Sarkanen, S.; Teller, D. C.; Abramowski, E.; McCarthy, J. L. *Macromolecules* 1982, **15**, 1098-104.
6. Sarkanen, S.; Teller, D. C.; Stevens, C. R.; McCarthy, J. L. *Macromolecules* 1984, **17**, 2588-97.

7. Garver, T. M., Jr.; Sarkanen, S. In *Renewable-Resource Materials: New Polymer Sources*; Carraher, C. E., Jr., Sperling, L. H., Eds.; Plenum: New York, 1986; pp 287-303.
8. Garver, T. M., Jr.; Sarkanen, S. *Holzforschung* 1986, **40** (Suppl.), 93-100.
9. Woerner, D. L.; McCarthy, J. L. *4th International Symposium on Wood and Pulping Chemistry* Preprints: Paris, 1987; vol 1, pp 71-6.
10. Gierer, J. *Wood Sci. Technol.* 1980, **14**, 241-66.
11. Gierer, J.; Wännström, S. *Holzforschung* 1984, **38**, 181-4.
12. Garver, T. M., Jr. Ph.D. Thesis, University of Minnesota, 1988.
13. Dutta, S. M.S. Thesis, University of Minnesota, 1988.
14. Marquardt, D. W. *J. Soc. Indust. Appl. Math.* 1963, **11**, 431-41.
15. Gierer, J.; Pettersson, I.; Szabo-Lin, I. *Acta Chem. Scand. B* 1974, **28**, 1129-35.
16. Gierer, J.; Imsgard, F.; Norén, I. *Acta Chem. Scand. B* 1977, **31**, 561-72.
17. Gellerstedt, G.; Lindfors, E. L.; Lapierre, C.; Monties, B. *Svensk Papperstidn.* 1984, **87**, R61-7.
18. Chang, H.-M.; Gratzl, J. S. In *Chemistry of Delignification with Oxygen, Ozone and Peroxides*; Gratzl, J. S., Nakano, J., Singh, R. P., Eds.; Uni Publishers: Tokyo, 1980; pp 151-63.
19. Tsuchida, E.; Abe, K. *Adv. Polym. Sci.* 1982, **45**, 77-85.
20. Bednár, B.; Li, Z.; Huang, Y.; Chang, L.-C. P.; Morawetz, H. *Macromolecules* 1985, **18**, 1829-33.
21. Chen, H.-L.; Morawetz, H. *Eur. Polym. J.* 1983, **19**, 923-8.
22. Aulin-Erdtman, G.; Sandén, R. *Acta Chem. Scand.* 1968, **22**, 1187-209.
23. Elias, H.-G. In *Light Scattering from Polymer Solutions*; Huglin, M. B., Ed.; Academic Press: New York, 1972; Chapter 9, pp 397-457.

RECEIVED February 27, 1989

Chapter 13

Reversed-Phase Chromatography of Lignin Derivatives

Kaj Forss, Raimo Kokkonen, and Pehr-Erik Sågfors

The Finnish Pulp and Paper Research Institute, P.O. Box 136, 00101 Helsinki, Finland

The paper shows that lignosulfonates and kraft lignin can be fractionated according to their polarities by reversed-phase liquid chromatography. The high molar mass lignin derivatives are fractionated in such a way that those with highest molar mass are eluted last. Lignin-carbohydrate compounds can be separated from virtually carbohydrate-free lignin.

In our opinion, lignin in wood consists of high molar mass glycolignin bound to carbohydrates and of a group of low molar mass lignins collectively referred to as hemilignins. The term lignin(s) is used collectively for glycolignin and hemilignins. In spruce wood, hemilignins representing 15-20% of the total lignin consist of monomeric, dimeric and oligomeric molecules. During the acid bisulfite and kraft pulping processes, the hemilignins and glycolignin are rendered soluble; as glycolignin undergoes both depolymerization and polymerization during the cook, the result is a complex mixture of molecules of different sizes and characteristics (1-4).

The complexity of this mixture is in no way reduced by the fact that small fragments peel off the glycolignin during the delignification and that some of the dissolved lignins are probably bound to carbohydrates as lignin-carbohydrate compounds. It is interesting to note in the present context that these lignin-carbohydrate compounds are much more polar than the other lignin compounds.

A study of dissolved lignin derivatives first requires their separation from each other. For this purpose gel permeation chromatography (GPC) is widely used. In this technique separation of lignin derivatives is based largely on the size and shape of the molecule, namely its hydrodynamic volume.

However, it is possible for compounds with the same molecular size to have different chemical structures. Such compounds may not be separated

0097–6156/89/0397–0177$06.00/0

by GPC. There is thus a need in experimental lignin research for fractionation techniques that separate molecules on the basis of properties other than size. One such fractionation method, which is based on the polarity of the components, is reversed-phase liquid chromatography (RPC). The purpose of this paper is to describe the fractionation of lignin derivatives by means of this method.

The hydrophobic stationary phase used in reversed-phase chromatography is a silica gel or polymeric matrix to which hydrocarbon chains have been attached by silylation. The most commonly used are C_{18}, C_8, C_6 and C_2 chains.

Elution in reversed-phase chromatography is often carried out using a gradient, produced from water and some water-miscible organic solvent. The solute components are thus distributed between the stationary and mobile phases mainly on the basis of their polarities. In reversed-phase chromatography hydrophilic compounds elute before hydrophobic ones.

Fractionation of Lignosulfonates

In order to study birch lignosulfonates, spent sulfite liquor, from which monosaccharides had been removed by ion exclusion chromatography, was fractionated on the basis of molecular size by preparative GPC (Fig. 1).

It can be seen from the figure that almost half of the birch lignosulfonates have a molar mass greater than 1000 g/mol.

The fractions in the region 700-1230 mL in Figure 1 were combined in order to study the structure of the polymeric portion of the birch lignosulfonates. The combined solution was then refractionated by preparative RPC into five fractions (Fig. 2).

Figure 2 shows that the lignosulfonates are fractionated into two portions. The lignosulfonates eluted in the retention time range 0-15 minutes are strongly polar, whereas those eluted in the range 15-40 minutes behave as less polar compounds with polarity decreasing with increasing retention time.

It must be noted that lignosulfonates are strong polyelectrolytes and thus polar components. However, part of the high molar mass molecule is non-polar in character, and this part of the molecule causes high molar mass lignosulfonates to elute as non-polar compounds.

The reason why lignosulfonates elute in the retention time range 3-9 minutes could be because they are, in fact, strongly polar lignin-carbohydrate compounds. To investigate this possibility, fractions I-V were subjected to acid hydrolysis and the monosaccharide content and composition of the resulting mixture determined by liquid chromatography. The carbohydrate and lignosulfonate contents are shown in Table I.

Table I shows that carbohydrates account for about one-third of the solids in fraction I. Fractions III-V contained considerably less carbohydrates.

After hydrolysis, fraction I contained xylose and arabinose in the ratio 10:1. Hydrolysis of fractions III-V yielded very small amounts of xylose. The other monosaccharides present in fractions III-V were arabinose and

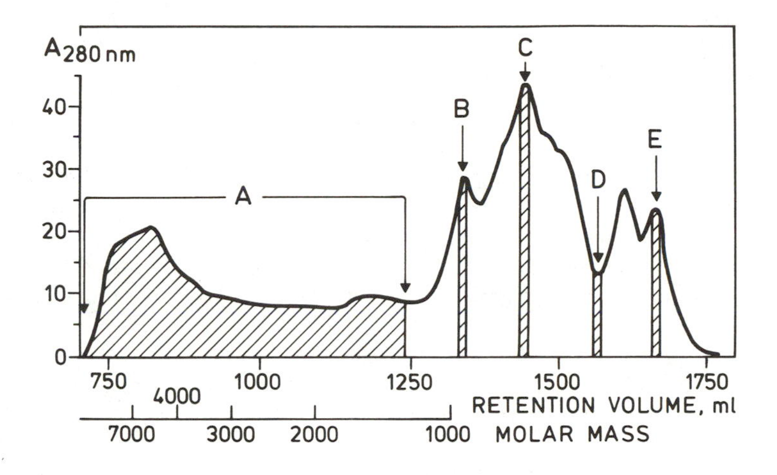

Figure 1. Fractionation of birch lignosulfonates by preparative gel permeation chromatography.

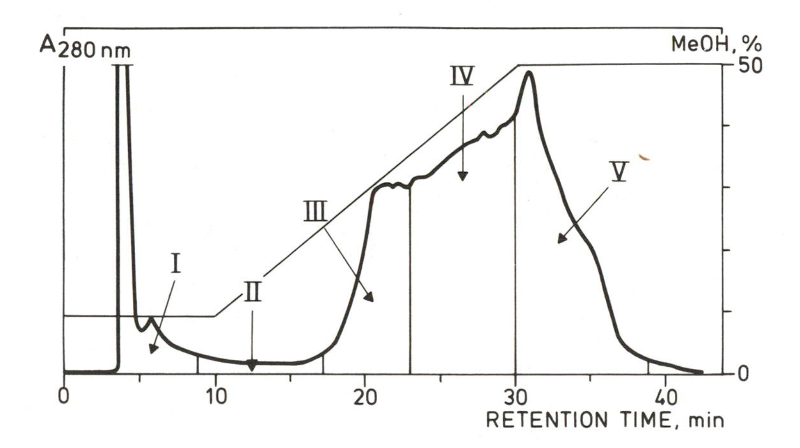

Figure 2. Fractionation of high molar mass birch lignosulfonates (fraction A in Figure 1) by preparative reversed-phase chromatography.

Table I. Carbohydrate and Lignosulfonate Contents of the Hydrolyzed Fractions I-V

	Fraction				
Compounds	I	II	III	IV	V
Monosaccharides, % (w/w)	30	–	5	2	6
Lignosulfonates, % (w/w)	70	–	95	98	94

rhamnose. These results suggest that fraction I contains lignosulfonate-xylan compounds.

Figure 2 shows that fraction I elutes over a very narrow retention time range, whereas fractions II-V are spread over a wide range. To determine the reason for this, fractions I-V were fractionated by analytical GPC (Fig. 3).

It can be seen from Figures 2 and 3 that the virtually carbohydrate-free fractions III-V are eluted by RPC in order of increasing molar mass and that fraction V contains the highest molar mass lignosulfonates. A large part of fraction I is eluted by GPC in the same region as fraction V. This supports the supposition that the lignosulfonates of fraction I are bound to carbohydrates. Otherwise they would have been eluted by RPC in fraction V.

It should be noted that the broad molar mass distribution of fraction I in Figure 3 reflects the molar mass distribution of the lignin-carbohydrate compounds and not that of the lignin portion of the lignin-carbohydrate compounds.

It was shown that high molar mass lignosulfonate compounds can be fractionated by RPC into hydrophilic and hydrophobic compounds. It can be seen from Figures 4 and 5 that the birch lignosulfonates with low molar mass (fractions B and C in Figure 1) were also fractionated into hydrophilic and hydrophobic portions with no clearly resolved peaks.

On the other hand, fractions D and E, which elute later in preparative GPC (Fig. 1), show clearly separated peaks in both the hydrophobic and hydrophilic zones when fractionated by RPC (Figs. 6 and 7).

It can be seen from Figure 3 that the molar mass distribution of the hydrophilic compounds (fraction I) is broad although they are eluted by RPC in a narrow zone (Fig. 2). Their reversed-phase chromatographic fractionation is thus based almost exclusively on their polarity, molecular size having no effect on the process.

The results show that RPC will also separate low molar mass lignosulfonates into hydrophilic and hydrophobic fractions as well as into a far greater number of individual components than obtained by fractionation with GPC.

Fractionation of Kraft Lignin

In the same way as with birch lignosulfonates, preparative RPC can be used

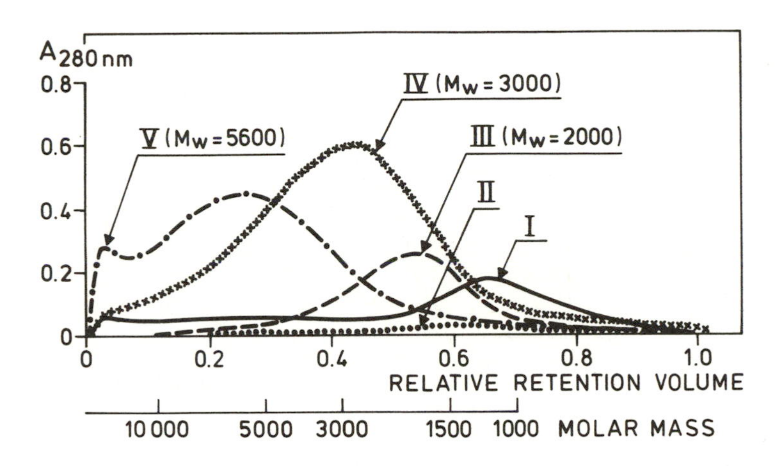

Figure 3. Molar mass distribution of high molar mass birch lignosulfonates (fractions I-V in Figure 2).

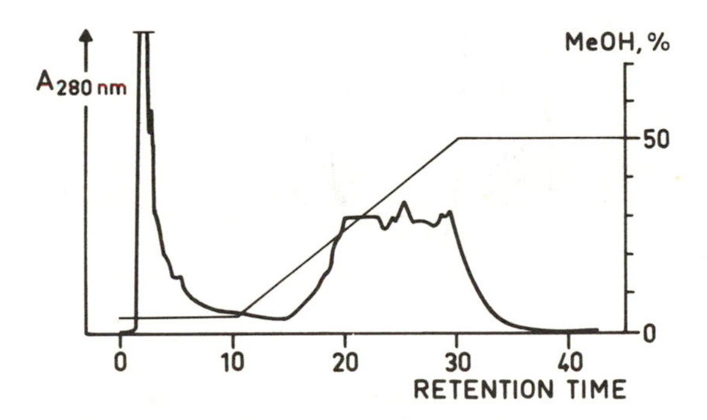

Figure 4. Fractionation of low molar mass birch lignosulfonates (fraction B in Figure 1) by reversed-phase chromatography.

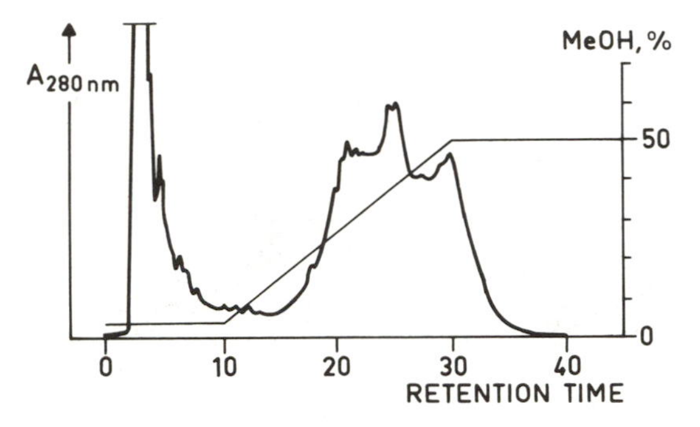

Figure 5. Fractionation of low molar mass birch lignosulfonates (fraction C in Figure 1) by reversed-phase chromatography.

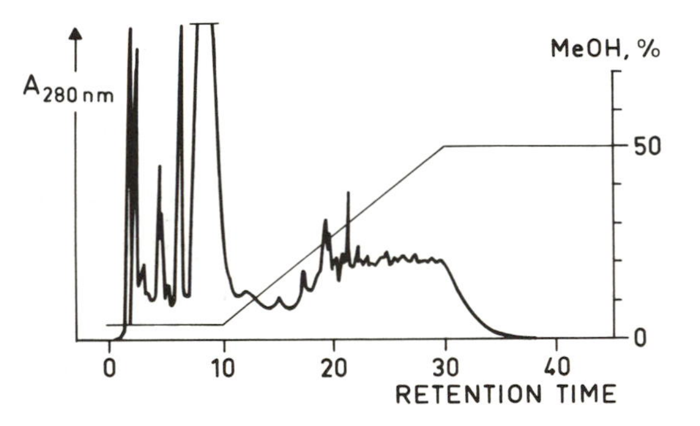

Figure 6. Fractionation of low molar mass birch lignosulfonates (fraction D in Figure 1) by reversed-phase chromatography.

to fractionate kraft lignin into hydrophilic and less hydrophilic compounds (Fig. 8).

The figure shows that hydrophilic lignin derivatives (fraction I) elute in the retention time range 30-80 minutes, compared with 90-140 minutes for the hydrophobic compounds (fractions II-IV). Refractionation by analytical GPC of fractions I-IV (Fig. 9) shows that these fractions encompass wide molar mass ranges.

Fraction I, which is less hydrophobic than fractions III and V (Fig. 8), contains lower molar mass compounds than fraction III, which in turn contains lower molar mass compounds than the most hydrophobic fraction (fraction IV).

Fraction I, which consists mainly of low molar mass compounds, also contains a small amount of high molar mass lignin derivatives eluting with relative retention volumes of 0-0.1. These derivatives are polar and some may be bound to carbohydrates, or otherwise they would have been eluted by RPC along with the hydrophobic fractions II-IV.

The analytical reversed-phase chromatograms in Figure 10 show that high molar mass hydrophobic kraft lignin in kraft black liquor elutes in the retention time range 60-90 minutes. The corresponding lignosulfonates elute sooner in the retention time range 30-60 minutes because of their sulfonate groups and consequently their more highly hydrophilic nature. It can also be seen from Figure 10 that the hydrophilic sulfonated hemilignins in the spent sulfite liquor elute in the retention time range 0-15 minutes.

Figure 11 shows that the elution of monomeric benzene derivatives in RPC is closely connected with the structure of their functional groups and thus with their polar properties. The figure shows that the strongly hydrophilic sulfonate is the first of the model compounds to be eluted. It is also seen that in RPC monomeric acids elute before the corresponding alcohols, which elute before the aldehydes. Guaiacyl compounds elute before the corresponding syringyl compounds, which in turn elute before the veratryl compounds.

Conclusions

It has been shown that RPC can be used to fractionate both lignosulfonates and kraft lignin on the basis of polarity. Strongly hydrophilic lignin-carbohydrate compounds can be separated from virtually carbohydrate-free lignin. High molar mass lignosulfonates and kraft lignin are fractionated on the basis of molar mass, with the highest molar mass compounds eluted last.

Preparative and analytical reversed-phase chromatography combined with GPC is a useful tool in experimental lignin research.

Experimental

Lignosulfonates. Samples of birch and spruce wood meal extracted with ethanol-cyclohexane (1:3) were heated from 20°C to 135°C during 1 h and then cooked for 6 h at 135°C in 150 mL reactors with sodium bisulfite liquor

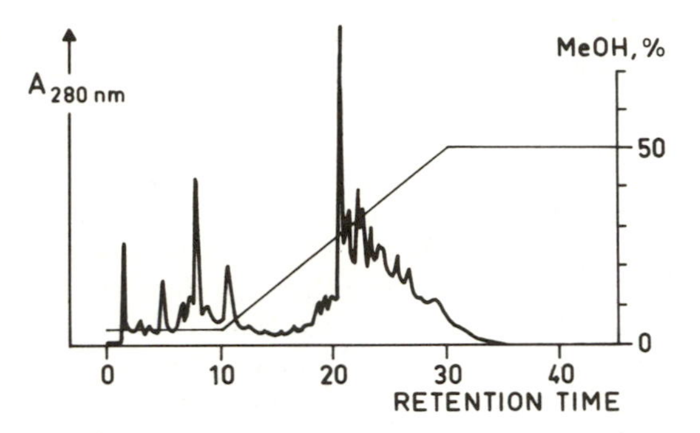

Figure 7. Fractionation of low molar mass birch lignosulfonates (fraction E in Figure 1) by reversed-phase chromatography.

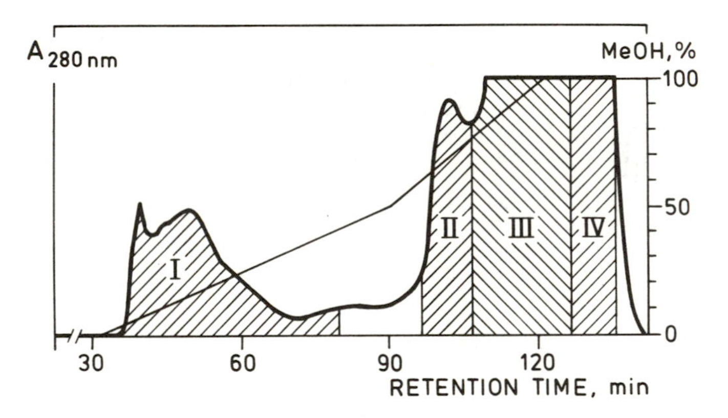

Figure 8. Fractionation of pine kraft lignin by preparative reversed-phase liquid chromatography.

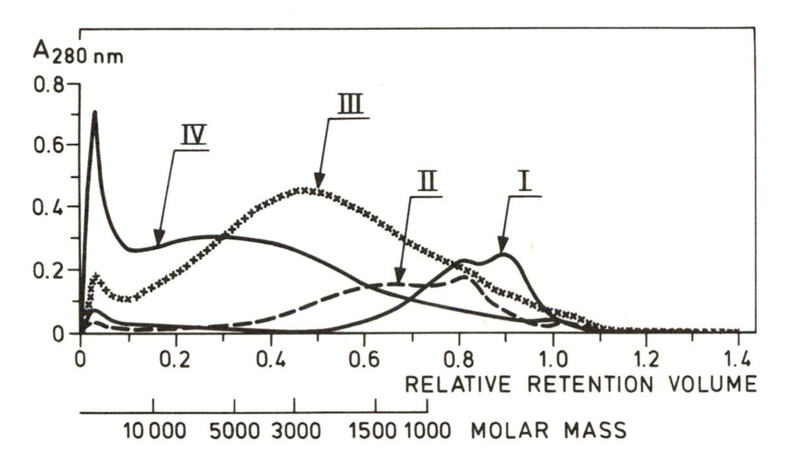

Figure 9. Molar mass distribution of pine kraft lignin (fractions I-IV in Figure 8).

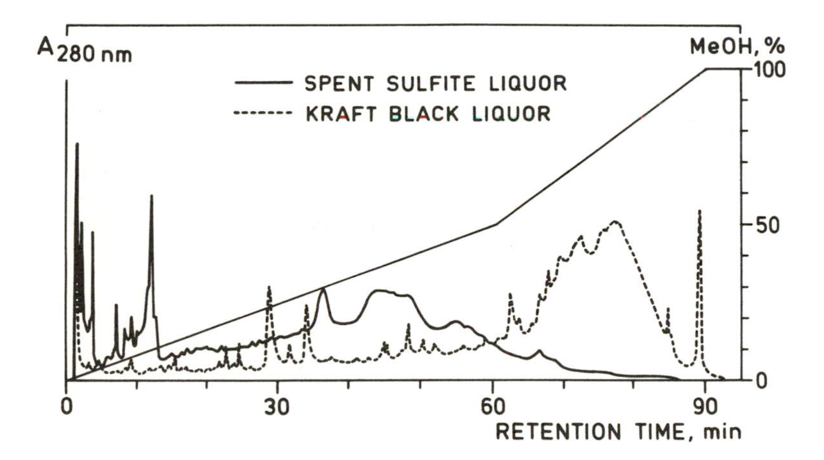

Figure 10. Fractionation of spruce spent sulfite liquor and pine kraft black liquor.

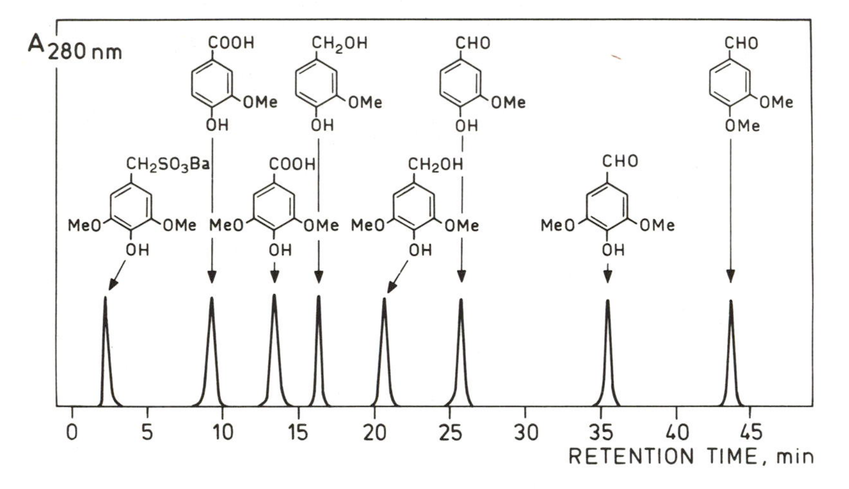

Figure 11. Influence of functional groups on retention time.

containing 7% SO_2 and 1% Na_2O. The spent and washing liquors from the pulp were combined and evaporated.

Sulfite and sulfate ions were precipitated from the spent birch liquor (Fig. 1) with barium hydroxide. Monosaccharides and other low molar mass non-electrolytes and weak electrolytes were separated quantitatively from the lignosulfonates by means of ion exclusion chromatography (5).

Kraft Lignin. The industrial pine black liquor was diluted with water (1:10) before analytical RPC.

RPC System and Conditions.

Reversed-phase columns:

Analytical:	Figures 4,5,6,7	Spherisorb C_6, 5 μm, 140/6.4 mm (Phase Separation, UK)
	Figures 10,11	Spherisorb ODS 2, 5 μm, 140/4 mm (Phase Separation, UK)
Preparative:	Figure 2	LiChroprep C_8, 40-60 μm, 250/10 mm (E. Merck, FRG)
	Figure 8	Sepralyte C_{18}, 40 μm, 300/25 mm (Analytichem, USA)

Solvent delivery system: Model LC-5060 (Varian, USA)

Mobile phase:	Figures 2,4,5, 6,7,10,11	A) KH_2PO_4 50 mM/L + KOH 1.15 mM/L B) MeOH
	Figure 8	A) H_2O B) MeOH

Gradient:

Figures 2,4,5, 6,7	Time (min)	0	10	30	
	v/v, % (A)	96	96	50	
	v/v,% (B)	4	4	50	
Figures 10,11	Time (min)	0	60	90	
	v/v, % (A)	100	50	0	
	v/v, % (B)	0	50	100	
Figure 8	Time (min)	0	30	90	120
	v/v, % (A)	100	100	50	0
	v/v, % (B)	0	0	50	100

Flow rate:	Figures 2,8	2.0 mL/min
	Figures 4,5,6,7	1.5 mL/min
	Figures 10,11	1.0 mL/min

Detection of A_{280nm}: Spectrophotometric Detector Model LC-75 (Perkin-Elmer, USA)

Injection: Syringe Loading Sample Injector Model 7125 (Rheodyne, USA)
with analytical 20 μL loop and preparative 2 mL loop.

GPC System and Conditions.

Gel permeation columns:

Analytical:	Figures 3,9	Sephadex G-50, fine, 1500/10 mm (Pharmacia, Sweden)
Preparative:	Figure 1	Sephadex G-50, fine, 1400/40 mm (Pharmacia, Sweden)

Solvent delivery system: STA-multipurpose peristaltic pump 13 19 00, (Desaga, FRG)

Mobile phase:	Figures 3,9	H_2O
	Figure 1	0.5 *M* NaOH
Flow rate:	Figures 3,9	90 mL/h
	Figure 1	20 mL/h

Detection of A_{280nm}:

Analytical:	Figures 3,9	UV-detector UVICORD S Model 2138 (LKB, Sweden)
Preparative:	Figure 1	Collected fractions measured with a spectrophotometer Model PMQ2 (C. Zeiss, FRG)

Injection: Syringe injection, analytical volume 0.5 mL and preparative volume 100 mL.

Literature Cited

1. Forss, K.; Fremer, K.-E. *Tappi* 1964, **47**, 485-93.
2. Forss, K.; Fremer, K.-E. *Pap. Puu* 1965, **47**, 443-54.
3. Forss, K.; Fremer, K.-E.; Stenlund, B. *Pap. Puu* 1966, **48**, 565-74, 669-76.
4. Forss, K.; Fremer, K.-E. *Appl. Polym. Symp.* 1983, **37**, 531-47.
5. Jensen, W.; Fremer, K.-E.; Forss, K. *Tappi* 1962, **45**, 122-7.

RECEIVED March 17, 1989

GENERAL MATERIALS

Chapter 14

Specialty Polymers from Lignin

J. Johan Lindberg, Tuula A. Kuusela, and Kalle Levon[1]

Department of Wood and Polymer Chemistry, University of Helsinki, SF–00170 Helsinki, Finland

The chemical and technological aspects of lignin modification and utilization are discussed. Competition with petrochemicals does not, for a number of reasons, permit the commercial utilization of non-oil based products made from lignin. However, pressure from governments and the public for a cleaner environment and new pulping processes (organosolv processes) may inevitably result in full scale industrial use of lignin other than fuel within the not too distant future.

The yearly growth of the world forests is about 7-9 billion cubic meters of biomass (FAO 1966). During the production of 140 million tons of cellulose and pulp from a part of this biomass, about 50 million tons of lignin are formed. Over 95% of the lignin residue is used as an energy source for the recovery of inorganic pulping chemicals or disposed of as a waste (1, 2). Although lignin has given its technological developers a lot of trouble and economic disappointments, there is a clear and positive trend favoring lignin modification and use for purposes other than as a valuable industrial fuel.

The difficulties involved in the chemical utilization of lignin and the factors which promote such developments can, in the authors' opinion, be summarized as follows:

Negative or retarding factors:

- chemical and molecular weight inhomogeneity causes high fractionation and modification costs;
- the three-dimensional structure involving carbon-carbon bonds that are difficult to break and which resist degradation to low molecular weight compounds;
- high oxygen content and hygroscopicity;
- comparatively high basic cost as a raw material owing to its use as an important industrial fuel;

[1]Current address: Department of Chemistry, Polytechnic University, 333 Jay Street, Brooklyn, NY 11201

0097–6156/89/0397–0190$06.00/0

- if spent liquors are combined with effluents from bleaching processes based on chlorine or reactions with halogen containing chemicals, a range of harmful compounds are formed through thermal and/or chemical treatment (23).

Positive or promoting factors:
- readily available in huge amounts;
- if disposed of as waste, black liquor can be a serious environmental pollution hazard;
- high energy content owing to the aromatic nuclei;
- a number of reactive points are present on the carbon skeleton which can be used for a wide range of substitution and addition reactions;
- good compatibility with several important basic chemicals;
- excellent colloidal and rheological properties, especially in the case of lignosulfonic acids;
- good adsorbent and ion exchange and adhesive properties;
- a direct source of various kinds of phenolic and aromatic compounds.

The cost of one kilogram of lignin is equivalent to about 0.6 kilograms of heavy oil. This fact, coupled with the points listed above, indicate why oil is a significant competitor in most common industrial applications.

Some basic facts are given below to support the following discussion. The tentative structure of softwood lignin is seen in Figure 1 (3), and some important data regarding its present uses are given in Table I (4). After cellulose and agricultural products, lignin is the largest source of organic matter in the plant kingdom that is readily available for industrial purposes.

As far as the utilization of lignin for polymers and other chemicals is concerned, we can proceed along two principal routes: either by using the whole unfractionated lignocellulosic material obtained from the spent liquor for the manufacture of simple and cheap bulk products, or by extracting certain fractions or low-molecular weight degradation products and modifying them into speciality polymers or high quality chemicals. In the next chapters we shall discuss various aspects of these processes.

Chemical and Physical Properties

Data concerning the elemental composition and methods for isolating lignin from the wood structure, as well as the mean molar masses and distributions of the obtained fragments, are given in Table II (1,5).

Figure 1 and Tables I and II indicate the importance of carbon-carbon and ether bonds. The former bonds can only be disrupted using strong reaction conditions, whereas the latter bonds are broken by much milder treatment. The methoxyl groups in the ortho position in softwoods, and in the ortho and para positions with relation to the phenolic hydroxyl group in hardwoods, have a marked influence on the reactivity and solubility behavior. They can be chemically blocked or removed.

The double bonds, carbonyl groups, carboxyl groups and large number of hydroxyl groups determine the physical and reaction behavior. In the case of lignosulfonates, the strongly acidic and polar sulfonic acid groups

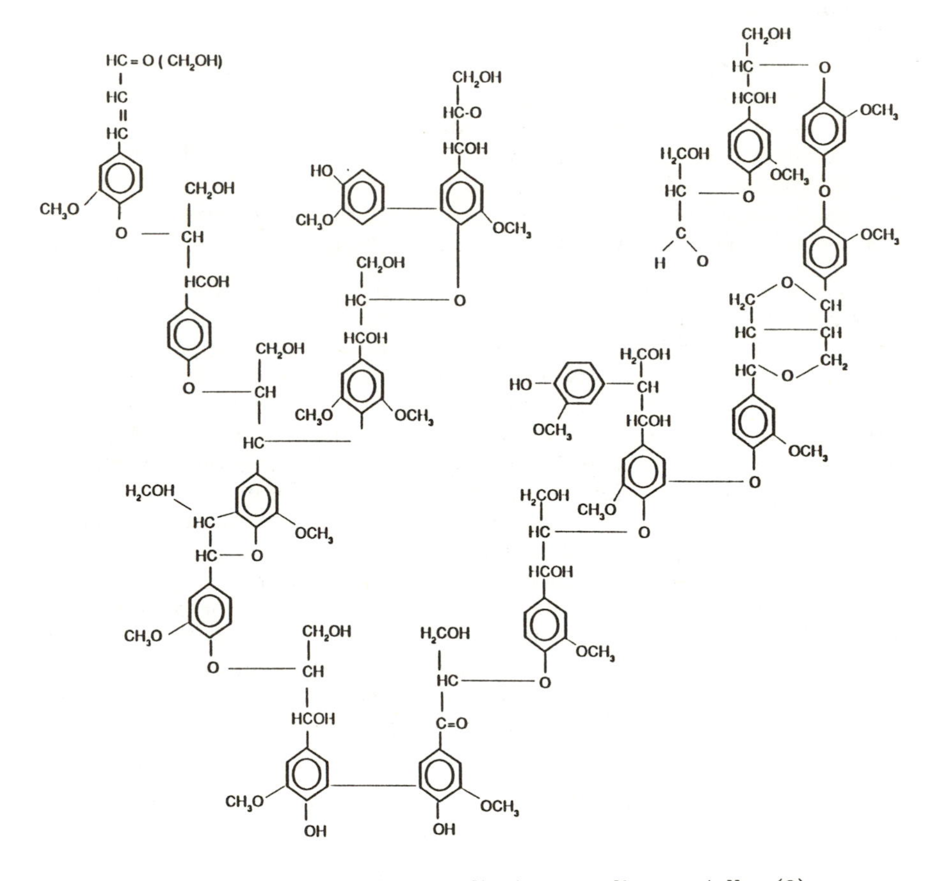

Figure 1. Structure of native lignin according to Adler (3).

Table I. Technical applications of spray dried spent sulfite liquors (SSL) and kraft lignin or its modifications (KL)

Binder and adhesive:
Pelletizing animal fodder (SSL)
Road dust (SSL)
Substitute for phenolic resins (SSL, KL)
Carbon black and rubber filler (KL)
Ore and mold binders (SSL)
Grinding aid and additive for cement (SSL)
Dispersants:
Dyestuffs (SSL, KL)
Clays and ceramics (SSL, KL)
Paints (SSL)
Pesticides (SSL)
Emulsifier and stabilizer for emulsions (SSL)
Oil well drilling mud (SSL)
Tanning and sequestering agent (SSL)
Electrical uses:
Storage battery plates (KL)
Electrolytic refining (KL)

primarily determine the solubility and colloidal behavior. The sulfonic acid groups are easily split off by thermal treatment to give a chemically active intermediate (6).

Raw Materials for Polymers by Degradation

Some monomeric and oligomeric products derived from lignin by degradation are given in Table III (3). Of hundreds of lignin degradation products, only guaiacol, vanillin, dimethyl sulfide and dimethylsulfoxide (DMSO) have been produced on a small industrial scale. As the yields are generally low and costs are high, the markets for these speciality chemicals are somewhat limited, especially as sources for monomers for bulk polymers.

On the other hand, there are potentially large markets for purified phenols derived from lignin. Quite high theoretical yields have been claimed for both wastes from sulfur consuming pulping processes (4) (kraft lignin and lignosulfonates) and soda lignin and especially non-sulfur organosolv processes. However, the heterogeneity of the phenolic fractions obtained is especially a drawback in the synthesis of high-quality homogeneous polymers.

A more promising way of obtaining more uniform degradation products is pyrolysis at high temperature. Synthesis gas is obtained at ever higher temperatures (700-1000°C), but according to Nimz, at the present it is not competitive with the corresponding processes based on coal (2).

Another way is to produce acetylene from sulfur-free lignin by flash

Table II. Some Basic Data of Lignins. Milled Wood Lignin = MWL

Elemental Composition (MWL) and Functional Groups (units/100 C_9):

	Spruce	Pine	Birch
Carbon %	63.15	63.70	58.8
Hydrogen%	6.21	6.29	6.5
Methoxyl%	15.90	15.50	21.5
Methoxyl groups	92-96		139-158
Phenolic groups	15-30		9-13
Benzyl alcohol groups	15-20		
Benzyl ether	6-8	6-8	
Carbonyl	20		
2- or 6-condensation	2.5-3	1.5-2.5	

Molar Mass and Polydispersity:

Wood Species Method of Isolation	MW	Mw/Mn
Eastern spruce MWL (22)	20600	2.6
Western hemlock MWL (22)	22700	2.4
Western hemlock Sulfonation (1)	400/150000	7.1
Norway spruce Sulfonation (1)	5300/131000	3.1
Birch Ethanolysis (77)	1750	2.5
Spruce Ethanolysis (77)	1750	2.5
Poplar Ethanolysis (76) (*Populus tremuloides*)	3540	3.5

Note: The content of functional groups depends on the origin of the lignin in the cell wall. Great variations in molar masses and Mw/Mn-values are reported for technical lignins.

pyrolysis by the Crown Zellerbach process (2). Bench scale experiments (68) indicate that lignins also may be used as an agent in liquefaction of coal to obtain oligomers or low molecular weight hydrocarbons. Both methods, and also methods based on kraft lignin (4), lead to well-known raw materials for polymer syntheses.

Table III. Main products obtained on pyrolysis of hardwood

Gases:	Water, carbon monoxide, carbon dioxide, methane, hydrogen, hydrocarbons
Distillates, tars:	Acetic acid, methanol, acetone, hydrocarbons, phenols, levoglucosan, furfural
Char and charcoal → activated carbon	
Ash (g/100 g): wood: 2.5; lignin: 1.9; holocellulose: 1.4.	The values vary with source of wood and growing conditions.
Calorific value (MJ/kg):	Wood: 20; lignin: 26.4; holocellulose: 16.2.

Organosolv Lignins as Raw Materials for Polymers

The partially degraded lignin can easily be separated from other wood components by refluxing wood chips at moderately elevated temperatures in weakly acid media of organic solvents, e.g., acetosolv, alcohol, peroxyformic acid, and similar pulping processes (7). In contrast to the normal technical processes, the lignin in this case is free from sulfur and chlorine, and is not burned to recover the inorganic pulping chemicals. However, the development of any organosolv pulping method is a difficult and complex matter which has been solved on a pilot plant scale in a few cases only (7, 69). At present it would appear that hardwoods are especially suitable for organosolv processes, whereas softwoods require higher cooking temperatures and greater amounts of additives.

Non-woody materials have been used comparatively little for organosolv cooking. Apart from silicate problems, they may be easier to deal with than wood (8).

Chemical Modification

In combination with the above discussed functional groups, the aromatic nature of the polymer provides a potential for further reaction, cf. (69):

- Alkylation (9) and dealkylation (14, 22)
- Oxyalkylation (71, 72)
- Amination (20, 21, 25)
- Carboxylation and acylation
- Halogenation and nitration
- Hydrogenolysis
- Methylolation
- Oxidation and reduction
    - ⋆ Chemical
    - ⋆ Electrochemical
    - ⋆ Microbiological (63)
- Polymerization (28, 36-42)

- Sulfomethylation (50)
- Sulfonation (62)
- Silylation (27,67)
- Phosphorylation (26,67)
- Nitroxide formation (59)
- Grafting
- Composite formation

Polymers from Degradation Products and Lignin Fragments

A wide range of polymers can be synthesized using the aforementioned reactions as well as some reactions which are discussed in later chapters on engineering plastics, and divalent reactants. It is thus possible to obtain various types of homopolymers and copolymers from the phenolic degradation products mentioned before, as well as from macromolecular fragments, by modification reactions. A proposed reaction scheme for some routes to polymers and oligomeric derivatives is given in Figure 2.

The routes give, using well-known condensation and radical reactions, bakelites (I), polyazophenylenes (II), polyimides (III), polyurethanes (IV), nitro compounds and polyamides (V), aromatic polyethers and polyesters (VI), polychalcones (VII), polyphenylene sulfides (IX), ammonia lignin (X), carbon fibers (XI), silicones (XII), and phosphorus esters (XIII). In addition, radiation and chemical grafting can be used to obtain polymers of theoretical interest and practical use. Although the literature on the above subject is very large, there are comprehensive summaries available (1,28,69).

Aromatic Units and Phenols

The acid condensation reaction of the aromatic and phenolic units is a typical reaction of lignin. The presence of acids results in resonance stabilized carbonium ion structures formed in the lignin macromolecule. These carbonium ion structures react further, e.g., with unsubstituted positions in the lignin macromolecule. Thus, thermal treatment of powdered wood in acidic conditions causes condensation, the coniferyl aldehyde and coniferyl alcohol groups being especially reactive. In addition, other inter- and/or intramolecular condensations may occur.

Chen (29) found that the amount of sulfuric acid directly determines the hardening time in the acid condensation of spent sulfite liquors used in plywood and veneers. However, in general the adhesives based purely on acid condensed lignins have often been found to be an uneconomic and qualitatively inferior alternative to adhesives based on synthetic polymers and phenol or lignin-formaldehyde resins.

If lignin is heated with phenol, the phenol condenses with lignin in the α-position of the side chain. Phenol generally couples in its para- position (15). Under optimum conditions, from 2.5 to 3 moles of phenol or phenolic derivative per phenylpropane unit are added to the protolignin (18).

One of the most widely used condensation reactions is between lignin and phenol using formaldehyde as a coupling agent. This reaction depends

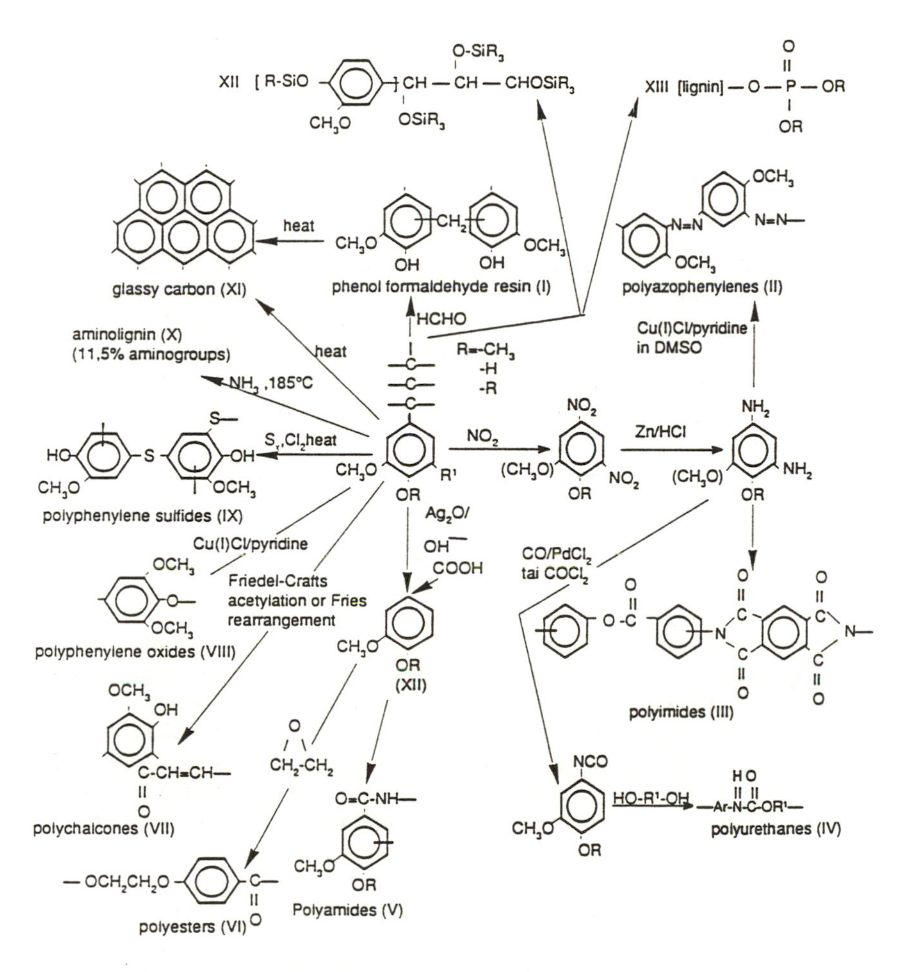

Figure 2. A general scheme of various ways to modify lignin and its low molecular weight degradation products.

on the well known bakelite reaction, which since the 1960's has been used to synthesize lignin-phenol based adhesives. One of the best known processes of this type is the base catalyzed one-step Karatex process, in which can be used either lignosulfonates or kraft lignin as raw material (19). Lignin may also react in a comparable manner with nitrogen containing reagents such as hexamethylene tetramine, urea, and melamine (20).

Ammonia and ammonium persulfate can also be used as reagents (21). However, as pointed out by Nimz (48) and other authors, owing to the non-uniformity of the product, only 15% of urea-formaldehyde and 25% phenol-formaldehyde resin binder may be replaced by calcium base spent sulfite liquors.

Thermal Stability and Antioxidant Effect

It is evident from model studies on the pyrolysis of sulfonated aromatic ion exchangers (30) and the synthesis of sulfur lignin, as well as from studies on the coextrusion of lignosulfonates with vinyl polymers (6), that only part of the sulfur is evolved as sulfur dioxide. A considerable amount of the residual sulfur is converted during the pyrolysis to sulfonyl and sulfur bridges between the base polymers. Model studies (30) indicate that these bridges make the base polymer thermally stable. The above reactions also permit the use of lignin containing sulfonic acid groups as stabilizers for vinyl polymers such as polyethylene, polypropylene and polyvinyl chloride.

Gul *et al.* (31) found that lignin derivatives increase the cold and UV light resistance of polyphenylene. According to Levon *et al.* (6), a blending of lignosulfonate with polyethylene in a plasticorder increases the stability, as indicated by rheological behavior and spin density, if the activation temperature is higher than 463 K. A high carbohydrate content in the lignin gives poorer results. The antioxidant effect of hydroxylated lignin on polyethylene, measured as oxygen uptake, has been reported to be only 10 times less than for usual commercial antioxidants (32).

From model studies on guaiacol derivatives and thioglycolic acid lignin it is known that stable nitroxide radicals can easily be introduced by Schotten-Baumann reactions in the lignin polymer (59). It is evident from the general chemical and photochemical properties of nitroxide radicals (60) that nitroxide derivatives of lignins could be used as preservative agents and radical scavengers as well as processing stabilizers for polymers.

Electrically Conducting Polymers

Lignin has long been used together with carbon black in master batches for rubbers with good success (33). It is to be expected that intermixing with carbon would also give it properties enabling it to be used as an extender in conductive compounding materials.

It is evident from the studies carried out by Wesseling (34) on other polymer compounding systems that the electrical conductivity in such cases is of a complex percolation type which is primarily confined to the thin surface layer of the graphite coating the large, globular insulating polymer particles. The macromolecular spheres may agglomerate to chain-like linear structures which build up networks, where the driving percolating force is the interfacial energy.

As already shown by Chupka and Rykova (65), other mechanisms are involved when semiconducting polymers are obtained from lignins by doping them with Lewis acids. Lignosulfonates and kraft lignins give generally poor results, but a derivative, sulfur lignin, has been found to be better suited for this purpose (35). The mechanism of conduction is definitely very similar to that of other doped semiconductor polymers, although a chemical complex formation mechanism cannot naturally be excluded (58). Under certain conditions we have also noted a reversible charging effect ("battery effect") (73), cf. also the theory in (74). The long term stability of the materials is quite good, but their potential uses have still not been explored.

The work of Chupka (65) and preliminary work in our laboratory indicate that lignin can under certain conditions be a photoconductor.

Engineering Plastics from Lignin

A considerable amount of data has accumulated regarding the modification of lignins to engineering plastics. Unfortunately, the incorporation of various monomers and polymers, such as di- and polyvalent epoxyphenols, esters and isocyanates, in the lignin structure in most cases resulted in brittle or tarry materials whose properties designated them as potential adhesives, lacquers, dispersants and films, but not as structural materials (36-40).

Recent systematic studies on the relation between network structure and substituents in kraft lignin, steam exploded, have shown that the lignin containing networks can be modified in new ways, cf. e.g. (80). Also the toughening of glassy, structural thermosets can be achieved by incorporating a variety of polyether and rubber-type soft segment components in the polymer network structure.

Glasser and co-workers (41) applied this principle to lignins. They found that toughening elements can also be built into these materials to form elastic polyurethanes through hydroxyl functionality, acrylates from vinyl functionality, and epoxies from amine functionality. It is thus possible to completely abolish the generally noted brittleness of lignin caused by the globular structure of the lignin fragments. Also Kringstad and co-workers have been working on similar questions (42).

According to Glasser (79), the low glass transition temperature and the decrease in brittleness can be explained by introduction of soft molecular segments capable of a plastic response to mechanical deformation. The introduction of hydrophilic polyether segments results predominantly in single phase morphology, and less polar rubber segments seem to favor two-phase morphology.

The use of lignins in polymer blends, e.g., with polyvinyl alcohol (81) with no sign of phase separation, is a new promising technique. It is found that blends prepared by injection molding have consistently better material properties than blends prepared by usual solution casting.

To conclude the above methods of incorporation of modified lignin in polymer networks and blends opens new promising possibilities for the technical use of lignins and makes it competitive with other raw materials for engineering plastics.

Gelation and Crosslinking

It has been widely known that a water-insoluble gel is obtained when spent sulfite liquors are treated with dichromate solutions. This property has been used, among other things, for increasing the tanning effect of lignosulfonates and for making oil well drilling muds.

The lignosulfonates react in the gelation process with di- or polyvalent metal ions and form covalent coordinative bonds. The hydroxyl, carbonyl and carboxyl groups in the lignin structure seem to be effective in the complexation process. According to Hayashi and Goring (43), the catechol groups formed by demethylation during the pulping process are operative when dichromate complexes are formed with lignosulfonate. Further investigations indicate that the greatest part of the hydroxyl groups disappear and a small amount of carboxyl groups are formed during the process (44).

Many other agents promoting oxidative coupling catalyze the gelation process. It has therefore been proposed that an oxidative coupling process could be involved (45). The gel formation process is in all circumstances a three-stage one: complex formation→intermolecular bridge formation→gel formation.

Grafting of Lignin

It is also possible to modify lignin by grafting using, e.g., styrene or acrylic monomers. The grafting process is a free radical reaction which can be initiated either by radiation or by peroxide, as used in ordinary polymerization processes. Thus, acrylic monomers can be grafted onto lignosulfonates in aqueous solution using hydrogen peroxide/iron (II) catalyst. The radicals add probably to the non-substitute positions in the aromatic nucleus (46,47).

Experimental evidence has indicated that γ-radiation can also be used to graft styrene onto kraft lignin to give a product with similar solubility properties as polystyrene. The reaction seems to be promoted by methanol (56).

Also, acrylonitrile may be grafted by radiation to lignosulfonates with grafting yields of about 20%. However, the compatibility of the grafted product as a filler in SBR seems to be inferior to that of the original lignosulfonate (57).

Intermolecular Association

It is well known from very early studies on lignin reactivity that hydrogen bond formation and other secondary valence forces strongly affect the solubility and colloid behavior as well as reactivity and general rheological behavior of lignins (52-54). As a consequence, water and other polar solvents and foreign matter are removed only with difficulty.

The application of lignosulfonate-chromium derivatives in oil well drilling mud has already been mentioned as a typical example of the above effects in combination with crosslinking. The use of lignosulfonates as active extenders in concretes is a further example of the applicability of the above effects (61). Another is the use of lignins as extenders for asphalt cement (70).

Recently Bogolytsin and co-authors (62) found that the associations of sulfur dioxide and other inorganic sulfur derivatives interact with the aromatic nuclei of lignins, and strongly influence their reaction and redox behavior. This association effect also forms a selective pre-association state in the sulfonation reaction in wood pulping. It is evident that a wide range of similar association effects may be present, but they remain to be detected and studied.

Materials of Future

When a new method is reported for synthesis of polymers, in a short time it is also tried on lignins. New lignin-based raw materials are also constantly appearing: the use of steam exploded hardwood lignin for making plywood adhesives has recently been explored by Gardner and Sellers (64) and found promising in this intensively competitive area.

Enzymatically degraded lignin also seems to be a potential source of chemicals, as can be seen from the extensive review of Harvey and coauthors (63).

Conclusions

It is evident from the above discussion that lignin is still a material of the future in the areas of structural plastics and as a source of chemicals. However, the ever-increasing demand for a clean environment hastens the day when lignin is used on a large industrial scale. This day is much nearer than was predicted only a few years ago. From this point of view, it is evident that research should continue and more funds should be allocated to research work on lignin modification.

Literature Cited

1. Sarkanen, K. V.; Ludwig, C. H. *Lignins*; Wiley-Interscience: New York, 1971.
2. Nimz, H. In *Fourth International Symposium on Wood and Pulping Chemistry*; Paris, 1987, pp. IIIA-IIIK.
3. Adler, E. *Wood Sci. Technol.* 1977, **11**, 169-218.
4. Lindors, T.; Enkvist, T. *Finska Kemists. Medd.* 1965, **74**, 29.
5. Sjostrom, E. *Wood Chemistry, Fundamentals and Applications*; Academic Press: New York, 1981.

6. Levon, K.; Huhtala, J.; Malm, B.; Lindberg, J. J. *Polymer* 1987, **28**, 745-750.
7. Laamanen, L.; Poppius, K. *Paper and Timber* 1988, 143-148.
8. Judt, M. F. In *Fourth International Symposium on Wood and Pulping Chemistry*; Paris, 1987, IIA-IIE.
9. Ishikawa, H.; Oki, T.; Fujita, F. *Japan Wood Res. Soc.* 1961, **7**, 85.
10. Falkehag, S. I. U.S. Patent 3 672 817, 1972.
11. Fukuzumi, T.; Sakuma, S.; Takahashi, H.; Yonita, K.; Fujihara, K.; Isome, Y.; Shibamato, T. *Holzforschung* 1966, **20**, 51.
12. Glasser, W. G.; Gratzl, J. S.; Collins, J. J.; Forss, K.; McCarthy, J. L. *Macromolecules* 1979, **8**, 565.
13. Struszczyk, H.; Allan, G. G.; Balaba, W. *Kem. Kemi* 1980, **7**, 492.
14. Goheen, D. W. *For. Prod. J.* 1967, **12**, 471.
15. Allan, G. G. In *Lignins*; Sarkanen, K. V.; Ludwig, C. H., Eds.; Wiley-Interscience: New York, 1971; Ch. 13.
16. Arseneau, D. F.; Pepper, J. *Pulp and Paper Mag. Can.* 1965, **66**, 415.
17. Marton, J.; Adler, E. *Tappi* 1963, **46**, 92.
18. Domburg, G. E.; Sharapova, T. E. *Khim. Drev.* 1973, 47.
19. Forss, K.; Fuhrmann, A. *Paper and Timber* 1976, **58**, 817.
20. Falkehag, S. I. U.S. Patent 3 697 497, 1972.
21. Gelfand, E. O.; Tushina, L. F. *Arkhang. Lesotekn. Inst.* 1973, 49. Balcere, D. *Khim. Drev.* 1973, 72.
22. Sjostrom, E. *Wood Chemistry, Fundamentals and Applications*; Academic Press: New York, 1981; 153-154.
23. Kuehl, D. W.; Butterworth, B. C.; DeVita, W. M.; Sauer, C. P. *Biomedical and Environmental Mass-Spectrometry* 1987, **14**, 443-447.
24. Chudakov, M. I.; Kusina, N. A.; Kirpicheva, L. M.; Mironava, Ya. Va. *Lesnoi Zhurnal* 1977, **20**, 125.
25. Nagaty, A.; Mansour, O. Y. *Am. Dyestuff Reptr.* 1979, **68**, 64.
26. Doughty, J. B. U.S. Patent 3 081 293, 1963. Mansour, O. Y.; Nagah, A.; Nagieb, Z. A.; Nossier, M. *Am. Dyestuff Reptr.* 1983, **72**, 28-32, 34-35.
27. Blount, D. H. U.S. Patent 4 051 115, 1977.
28. Lindberg, J. J.; Erä, V. A.; Jauhiainen, T. P. *Appl. Polym. Symp.* 1975, **28**, 269-275.
    Lindberg, J. J.; Melartin, J. *Kem.-Kemi* 1982, **9**, 736-744.
    Lindberg, J. J.; Hortling, B. *Chemia Stosowana* 1983, **XXVII**, 3-12.
    Lindberg, J. J.; Levon, K.; Kuusela, T. *Acta Polymerica* 1988, **39**, 47-50.
29. Chen, K. C. *Adhesive Age* 1978, **21**, 31.
30. Matsuda, M.; Funabashi, K. *J. Polym. Sci.* 1987, **25**, 669-673.
31. Gul, V. E.; Lyubeshkina, E. G.; Shargodskii, A. M. *Polym. Mech.* 1965, 1.
32. Bronovitskij, V. E.; Sharipdzhanov, A. *Plast. Massy* 1977, **1**, 34.
33. West Virginia Pulp and Paper Co., Polychemicals Div., Technical Bull. 300.
34. Wessling, B. *Kunststoffe* 1986, **76**, 930-936.
    Wessling, H. *Makromol. Chem.* 1984, **18**, 1265-1275.

35. Levon, K.; Kinanen, A.; Lindberg, J. J. *Polym. Bull.* 1986, **16**, 433. Lindberg, J. J.; Turunen, J.; Hortling, B. U.S. Patent 410 711, 1982.
36. Moores, H. H.; Dougherty, W. K.; Ball, F. Y. U.S. Patent 3 519 581, 1970.
37. Tai, S.; Sawanobori, T.; Nakano, J.; Migita, N. *J. Japan Wood Res. Soc.* 1968, **14**, 46.
38. Hsu, O. H.; Glasser, W. G. *Wood Sci.* 1976, **9**, 97.
39. Mihailov, M.; Gerdjikova, S. *Compt. Rend. Acad. Bulg. Sci.* 1965, **18**, 43.
40. Brown, S. U.S. Patent 4 131 573, 1978.
41. Glasser, W. G. *Fourth International Symposium on Wood and Pulping Chemistry*; Paris 1987, Vol. I, 45.
42. Kringstad, K.; Mörck, R.; Reimann, A.; Yoshida, H. *Fourth International Symposium on Wood and Pulping Chemistry*; Paris, 1987, Vol. I, 67.
43. Hayashi, A.; Goring, D. A. I. *Pulp and Paper Mag. Can.* 1965, **66**, T154.
44. James, A. N.; Tice, P. A. *Tappi* 1964, **47**, 43.
45. Tanaka, H.; Senju, R. *Tappi* 1970, **53**, 1657.
46. Sadakata, M.; Takahashi, K.; Saito, M.; Sakai, T. *Fuel* 1967, **66**, 1667-1672.
47. Beck. S. R.; Wang, M. *Ind. Eng. Chem. Process Des. Dev.* 1980, **19**, 312.
48. Nimz, H. In *Wood Adhesives*; Pizzi, A., Ed.; Marcel Dekker: New York, 1983; 248-288.
49. Palenius, I. *Papier* 1982, **36**, 13-19.
50. Oita, O.; Nakano, J.; Migita, N. *J. Japan Wood Res. Soc.* 1966, **12**, 239.
Falkehag, S. I. U.S. Patent 3 763 139, 1973.
Griggs, B. F.; Zhao, L.-W.; Chen, C.-L.; Gratzl, J. S. *TAPPI Res. Devt. Conf.* 1984, 235-247.
51. Faix, O.; Meier, D.; Grobe, I. *J. Anal. Appl. Pyrolysis* 1987, **11**, 403-416.
52. Lindberg, J. J. *Paper and Timber* 1955, **37**, 206; 1960, **42**, 193.
53. Lindström, T. Ph.D. Thesis, Royal Univ. of Technol., Stockholm, 1979.
54. Sarkanen, S.; Teller, D. C.; Stevens, C. R.; McCarthy, J. L. *Macromolecules* 1984, **17**, 2588-97.
55. Namet, C. *et al. J. Polym. Sci.* 1971, **9**, 855.
56. Hatakeyama, H. Japan Patent 119 091, 1975.
57. Phillips, R. B. *et al. J. Appl. Polym. Sci.* 1975, **17**, 443.
58. Kuusela, T.; Lindberg, J. J.; Levon, K.; Osterholm, J.-E.; Am. Chem. Soc. Symp. 1988, in press.
59. Törmälä, P.; Lindberg, J. J.; Koivu, L. *Paper and Timber* 1972, **54**, 158.
Lindberg, J. J.; Bulla, I.; Törmälä, P. *J. Polym. Sci. Symp.* 1975, **33**, 167-171.

60. Allen, N. S. In *New Trends in the Photochemistry of Polymers*; Allen, N. S.; Rabek, J. F., Eds.; Elsevier Appl. Sci. Publ.: London, 1985; 209-246.
61. Bialski, A. M. *CPPA/TAPPI/Int. Sulfite Conf.* Preprint 1986, 131-146.
62. Bogolytsin, K.; Lindberg, J. J. *Cell. Chem. Technol.* 1983, **17**, 19; 1985, **19**, 437.
62. Harvey, P. J.; Shoemaker, H. E.; Palmer, J. M. In *Plant Products and the New Technology*; Fuller and Gallon, Eds.; Oxford Univ. Press: Oxford, 1985; 249-266.
64. Gardner, D. J.; Sellers, T., Jr. *For. Prod. J.* 1986, **36**, 61-67.
65. Chupka, E. I.; Rykova, T. M. *Khim. Prirod. Soed.* 1983, 82-85.
66. Telysheva, G. M.; Sergeyeva, V. N. *Int. Symp. Wood and Pulping Chem.*; Japan, 1983, **4**, 225-227.
67. Holmberg, K.; Johansson, J. A. *Svensk Papperstidn.* 1983, **86**, 152-158.
68. Coughlin, R. W.; Davoudzadeh, F. *Nature* 1983, **303**, 5920:789-791.
69. Boye, F. *Utilization of Lignins and Lignin Derivatives*; Bibliography Ser. 292, I-II: Appleton, WI, 1984; Suppl. I, 1985.
70. Sundstrom, D. W.; Keel, H. E.; Daubenspeck, T. H. *I. & E. C. Prod. Res.* 1983, **22**, 496-500.
71. Dilling, P.; Sarjeant, P. T. U.S. Patent 4 454 066, 1984.
72. Lin, S. Y. U.S. Patent 4 184 845, 1980.
73. Kuusela, T. A.; Laantera, M. L. A.; Lindberg, J. J., to be published.
74. Rantner, M. A.; Shriver, D. F. *Chem. Rev.* 1988, **88**, 109-124.
75. Glasser, W. G.; Kelley, S. S. *Lignin*, In *Encyclopedia of Polymer Science and Engineering*; Vol. 8, 2nd Ed.; John Wiley & Sons, Inc.: New York, 1987; 795-852.
76. Glasser, W. G.; Barnett, C. A.; Sano, Y. *J. Appl. Polym. Sci., Appl. Polym. Symp.* 1983, **37**, 441.
77. Lange, W.; Faix, O.; Beinhoff, O. *Holzforschung* 1983, **37**, 163.
78. Wu, L. C.-F.; Glasser, W. G. *J. Appl. Polym. Sci.* 1985, **29**, 1111-1123.
79. Saraf, V. P.; Glasser, W. G.; Wilkes, G. L. *J. Appl. Polym. Sci.* 1985, **30**, 2207-2224.
80. Muller, P. C.; Kelley, S. S.; Glasser, W. G. *J. Adhesion* 1984, **17**, 185-206.
Glasser, W. G.; Wu, L. C.-F.; Selin, J.-F.; *Wood. Agricult. Residues* 1983, 149-165.
81. Ciemniecki, S. L.; Glasser, W. G. *Polymer* 1988, **29**, 1021-1029; 1030-1036.

RECEIVED February 27, 1989

Chapter 15

High-Performance Polymers from Lignin Degradation Products

Hyoe Hatakeyama[1], Shigeo Hirose[1], and Tatsuko Hatakeyama[2]

[1]Industrial Products Research Institute, 1–1–4 Higashi, Tsukuba, Ibaraki 305, Japan
[2]Research Institute for Polymers and Textiles, 1–1–4 Higashi, Tsukuba, Ibaraki 305, Japan

High-performance polymers having 4-hydroxyphenyl, guaiacyl and syringyl groups were synthesized from lignin degradation products. Physical properties of the polymers were investigated by differential scanning calorimetry (DSC), thermogravimetry (TG), gel permeation chromatography (GPC), viscosity measurement, etc. The relationship between the chemical structure and physical properties of polymers was analyzed from the viewpoint of molecular design. It was found that physical properties such as molecular weight, solubility for solvents, crystallinity, relaxation in glassy state, thermal decomposition temperature, etc., could be controlled by the appropriate arrangements of chemical bonds and functional groups such as phenylene group, methoxyl group, alkylene group, etc.

Lignin is widely found in nature and exists abundantly next to cellulose in higher plants. Technically most of it is obtained as a by-product of pulping process and is used as fuel to obtain energy to operate pulping mills.

In spite of extensive research to expand the use of lignins into industrial materials, the results obtained have not been very successful. To a large extent, this difficulty of efficient utilization of lignin is believed to be dependent on its heterogeneous and complex nature. It is generally accepted that lignin molecules consist of three basic units such as 4-hydroxyphenyl, guaiacyl and/or syringyl groups which are considered to link each other in statistically different ways. Therefore, most research results concerning the degradation of lignin have shown that mononuclear phenols having the above basic units are major degradation products (1).

The purpose of this paper is to describe, from the standpoint of molecular design, the relationship between the chemical structures and physical

0097–6156/89/0397–0205$06.00/0

properties of new types of high-performance polymers which have recently been synthesized from lignin degradation products at our laboratory.

Polyhydroxystyrene Derivatives

Poly(4-hydroxystyrene) [I], poly(4-hydroxy-3-methoxystyrene) [II], poly(4-hydroxy-3,5-dimethoxystyrene) [III] and their acetates were synthesized from 4-hydroxybenzaldehyde, vanillin and syringaldehyde (2), which were synthesized as shown in Scheme 1.

Molecular weight and molecular weight distributions of the samples were measured by gel permeation chromatography (GPC). The molecular weight of the polymers was controlled by changing polymerization conditions: from 6.6×10^3 to 3.7×10^5 (2). The molecular weight distribution (M_w/M_n) was from 2.2 to 4.3 (2).

Glass transition temperatures (T_g's) of styrene derivatives [I], [II], and their acetates were measured by differential scanning calorimetry (DSC) (2-4). Figure 1 shows the relationship between the T_g and the molecular weight of the styrene derivatives. The influence of molecular weight on molecular motion of the polymers is clearly recognizable. At the same time, the effect of substituent groups is also noteworthy. As seen from the figure, T_g values for [I] and [II] are 10-60 K higher than those of polystyrene and the acetylated samples. This fact seems to indicate that the introduction of a hydroxyl group at the 4-position of the aromatic ring forms hydrogen bonds and restricts the molecular motion of the main chain (2-4). On the other hand, if the methoxyl group is introduced in the 3-position of the aromatic ring, adjacent to hydroxyl or acetoxyl group at the 4-position, T_g decreases due to the steric hindrance caused by the methoxyl group. The above results indicate that the effect of substituent groups such as hydroxyl and methoxyl groups on T_g is more prominent than that of molecular weight (2,3).

Polyesters Having Spiro-Dioxane Rings

As illustrated in Scheme 2, 3,9-bis (4-hydroxy-3-methoxyphenyl)- 2,4,8,10-tetraoxa-spiro [5,5] undecane, designated bisphenol [IV], was synthesized from vanillin and pentaerythritol. Polyesters were obtained by the reaction of [IV] with terephthaloyll chloride or sebacoyl chloride (5).

The thermal stability of the obtained polyesters, polyterephthalate (PTS) (inherent viscosity, η_{inh} = 1.30 dl/g) and polysebacate (PSS) (η_{inh} = 0.89 dl/g) having spiro-dioxane rings, was analyzed by thermogravimetry (TG). PTS started to decompose at 568 K and PSS at 527 K. This shows that the former is thermally more stable than the latter. DSC studies of the above polymers were carried out in an atmosphere of nitrogen at the heating rate of 10 K/min from room temperature to a temperature a few K's below the thermal decomposition temperature (T_d) determined by TG. Although no transition was detected in the DSC curve of PTS, a glass transition at 363 K was seen in a DSC curve of PSS. The X-ray diffractogram of PTS showed a crystalline pattern, while that of PSS showed an amorphous halo pattern.

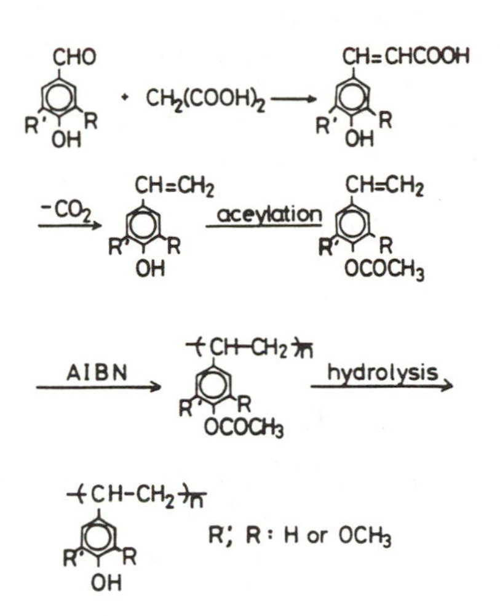

[I]: R, R' = H

[II]: R = OCH_3, R' = H

[III]: R, R' = OCH_3

Scheme 1

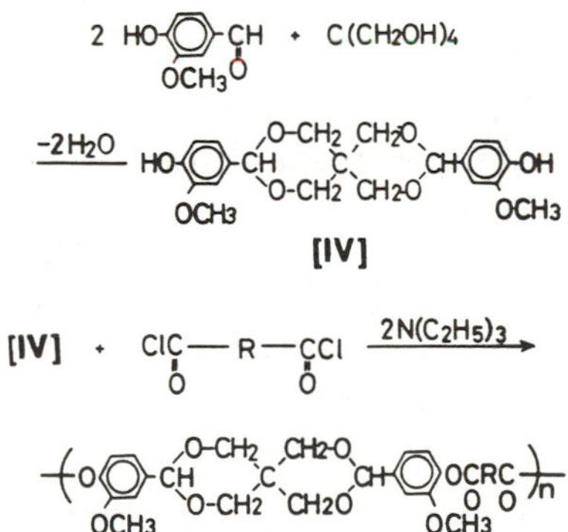

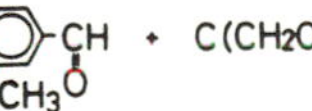

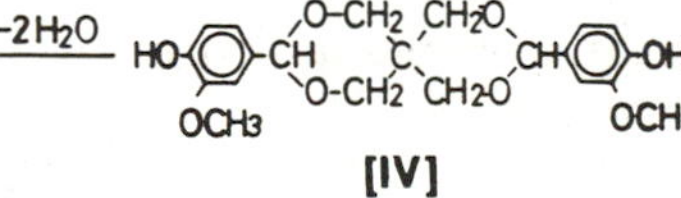

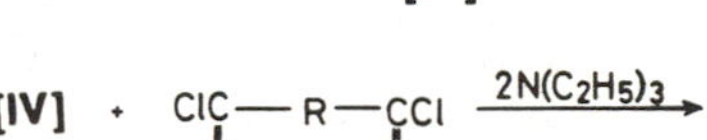

Scheme 2

One of the purposes of the present study is to investigate the relationship between the properties and structure of the polymers synthesized. Generally speaking, as T_g values of a polymer depend on its rigidity, it should be possible to estimate the rigidity of the spiro-dioxane rings in PSS by comparing the T_g with those of polysebacates from other bisphenolic compounds. Among the polyesters from bisphenolic compounds, the thermal properties of polyesters from disubstituted bis(4-hydroxyphenyl)methane (DBHM) (6) have been studied in detail. Therefore, we attempted to compare T_g of PSS with those of a series of polyesters obtained from DBHM. The result of the comparison showed that the rigidity of the spiro-dioxane ring is almost similar to DBHM.

Polyethers and Polyesters Having Methoxybenzalazine Units

Polyethers and polyesters having methoxybenzalazine units with various alkylene groups (C_4, C_6 and C_8) in the main chain were synthesized from vanillin (7,8). The condensation reaction of 4,4′-alkylenedioxybis (3-methoxybenzaldehyde) [VI] with hydrazine monohydrate was applied to the synthesis of polyethers [VII] (M_n, 7.4×10^3 for C_4, 7.3×10^3 for C_6 and 4.1×10^3 for C_8 derivatives), as shown in Scheme 3. Polyesters [IX] (η_{inh}, 0.35 dl/g for C_4, 0.38 dl/g for C_6 and 0.43 dl/g for C_8 derivatives) were synthesized from 4,4′-dihydroxy-3,3′-dimethoxybenzalazine [VIII] and dicarboxylic acid chlorides by conventional low temperature solution polycondensation, as shown in Scheme 4.

The thermal stability of the samples was studied by TG. As shown in Table I, the decomposition temperatures (T_d) of polyethers were higher than those of polyesters. However, it was found that T_d did not depend on chain length of the alkylene groups in both the polyethers and polyesters.

Table I. Starting temperatures of decomposition (T_d's) of polyethers [VII] and polyesters [IX] having methoxybenzalazine units with various alkylene groups

Alkylene groups	Polyethers T_d (K)	Polyesters T_d (K)
C_4	574	533
C_6	570	541
C_8	571	539

C_4, C_6, C_8: tetra, hexa, and octamethylene, respectively.
T_d: Starting temperature of weight loss. Heating rate: 20 K/min.

Figure 2 shows the DSC curves of poly(oxy-2-methoxyl, 4-phenyleneoxyoctamethylene) [VII-C_8]. In Figure 2, an endothermic peak of melting is observed in the heating curve of a sample preheated at 490 K. In the cooling curve, an exothermic peak of crystallization is observed at around 415 K. As shown in Figure 2, the polyethers crystallized during cooling from

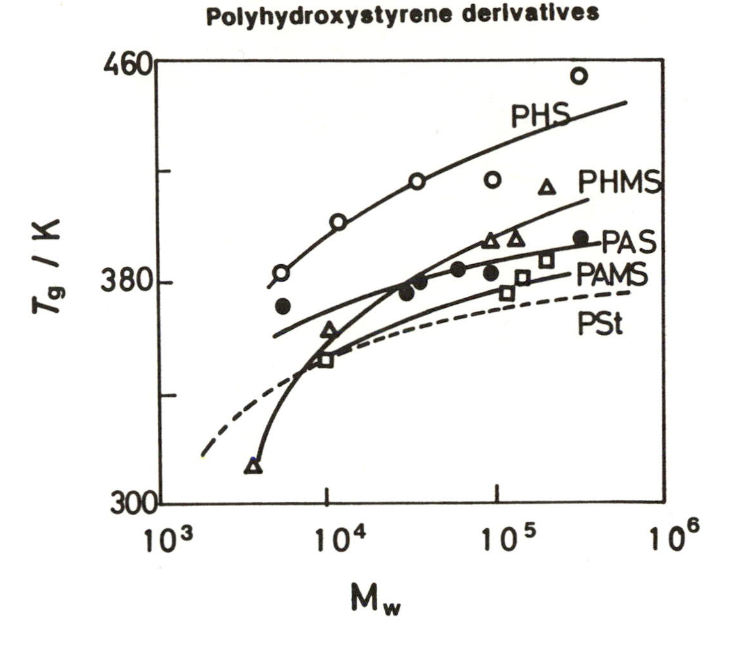

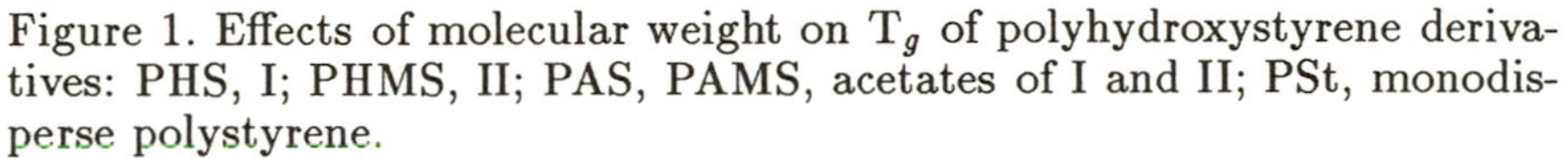
Figure 1. Effects of molecular weight on T_g of polyhydroxystyrene derivatives: PHS, I; PHMS, II; PAS, PAMS, acetates of I and II; PSt, monodisperse polystyrene.

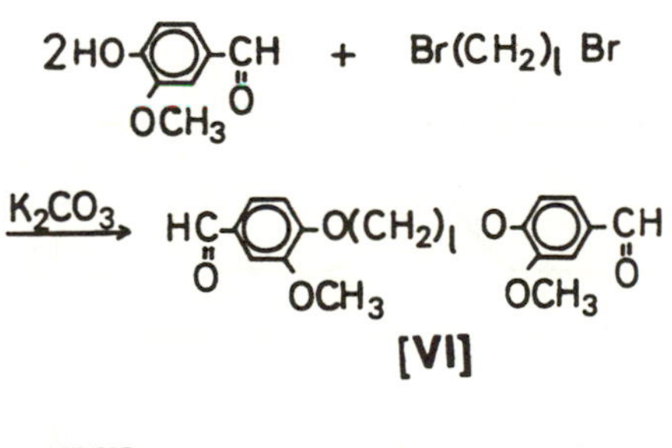

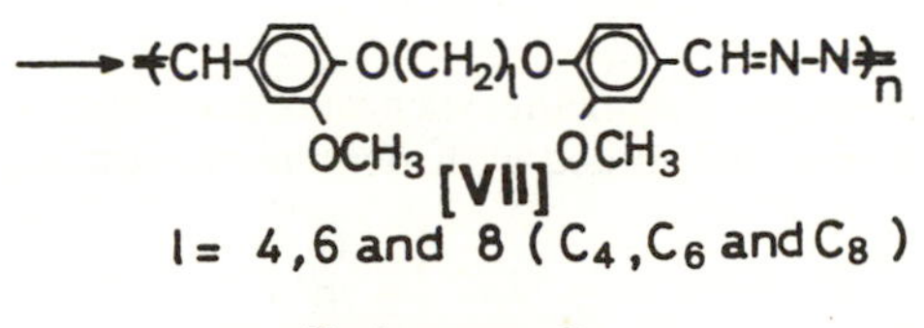

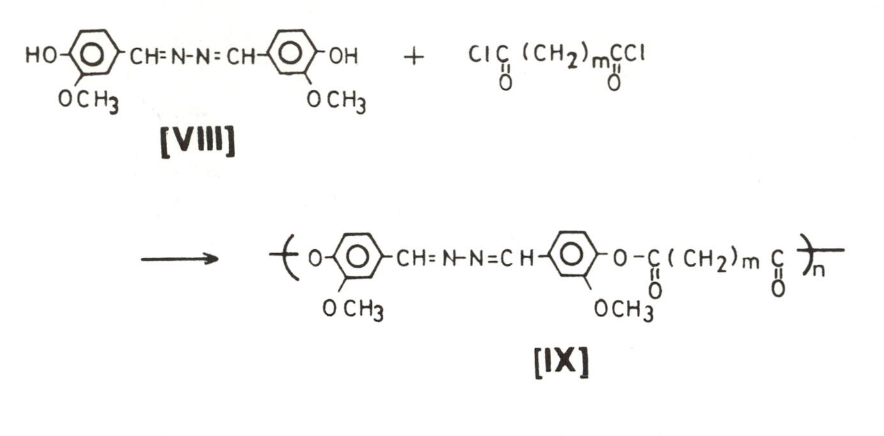

m = 4, 6 and 8 (C_4, C_6 and C_8)

Polymethoxybenzalazine-polyether

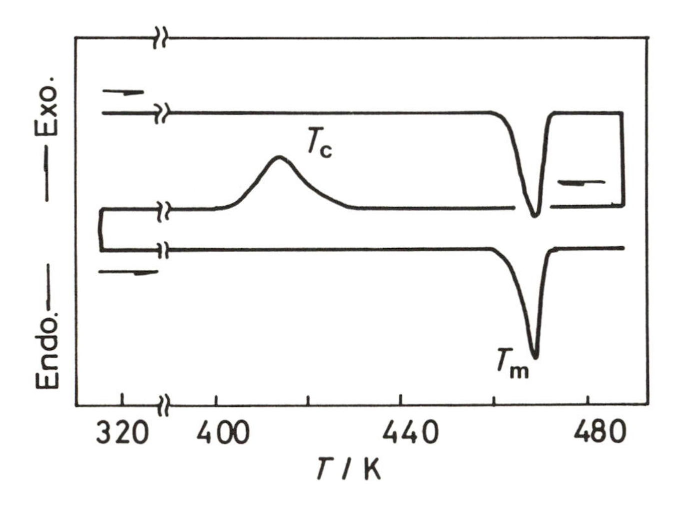

Figure 2. DSC curves of polymethoxybenzalazine ether, VII-C_8: heating and cooling rates, 10 K/min. (Reprinted with permission from ref. 8. Copyright 1986 Sen-i Gakkaishi.)

the molten state. Therefore, it can be said that the polyethers crystallize isothermally from the molten state.

Figure 3 shows the isothermal DSC curves of polyether VII-C_8 which was allowed to stand at each of the predetermined temperatures.

In order to analyze the isothermal crystallization of polyether VII-C_8 in more detail, Avrami's equation (9), $lnX = -kt^n$, was applied to the results shown in Figure 3. In this equation, X is the fraction of material which has not yet been transformed into a crystalline state at time t, and k and n are constants. Avrami's index n calculated were between 3.7 and 3.8, i.e., n = ca. 4. The value of $n = 4$ suggests that the crystals grow three-dimensionally if the nucleation process is homogeneous and the growth process is linear. However, it is known that Avrami's index does not always show the direct reflection of the crystallization process. Therefore, in order to obtain information about the growth of crystals, the polarizing microscopic measurements were made. The measurement under crossed nicols indicated a Maltese cross for each crystal which is characteristic of a spherulitic structure, thus supporting the obtained Avrami's indices.

Figure 4 shows the DSC curve for the as-polymerized sample of poly (oxy-2-methoxyl, 4- phenylene- methylidenenitrilonitrilo-methylidene-3- methoxy-1, 4-phenyleneoxysebacoyl) [IX-C_8]. As shown in Figure 4, an endothermic peak of melting was observed. The cooling curve did not show any exothermic peak of crystallization. The reheating curve showed glass transition and an exothermic peak of cold-crystallization followed by an endothermic peak of melting.

The data for the isothermal crystallization of [IX-C_8] from the glassy state were analyzed by the same method as that for the polyethers. Avrami's index obtained was between 2.1 and 2.2, i.e., n = ca. 2. This value, $n = 2$, suggests that the crystals grow two-dimensionally, if the nucleation process is heterogeneous and the growth process is diffusion-controlled.

Polyacylhydrazones Having Guaiacyl Units with Alkylene Groups

Polyacylhydrazones [X] having two guaiacyl units with the alkylene groups in each repeating unit were synthesized using vanillin and dibromoalkanes as starting materials as shown in Scheme 5 (10).

The phase transition and the thermal stability of the six polymers synthesized were studied by DSC and TG. Inherent viscosities, T_g's and T_d's are listed in Table II. The T_d does not very much differ among the polymers.

The polymers synthesized were crystalline, when they were obtained as the precipitates from DMSO solutions. Melting peaks were observed in the as-polymerized samples at around 460-500K in DSC curves. However, an exothermic peak attributable to the crystallization was not found in a DSC curve. The X-ray diffractograms of the samples cooled from the molten state to room temperature showed typical amorphous patterns. Glassy samples did not crystallize, although they were annealed at temperatures below melting for a long time. This suggests that the molecular

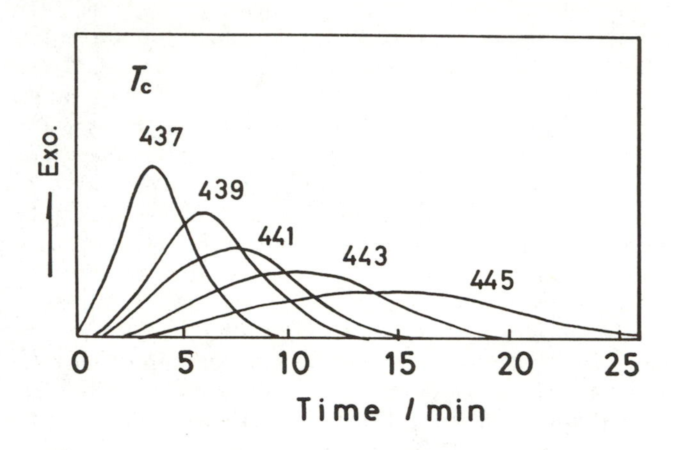

Figure 3. Isothermal DSC curves showing crystallization of polymethoxybenzalazine ether, VII-C_8: crystallization temperature is indicated for each curve. (Reprinted with permission from ref. 8. Copyright 1986 Sen-i Gakkaishi.)

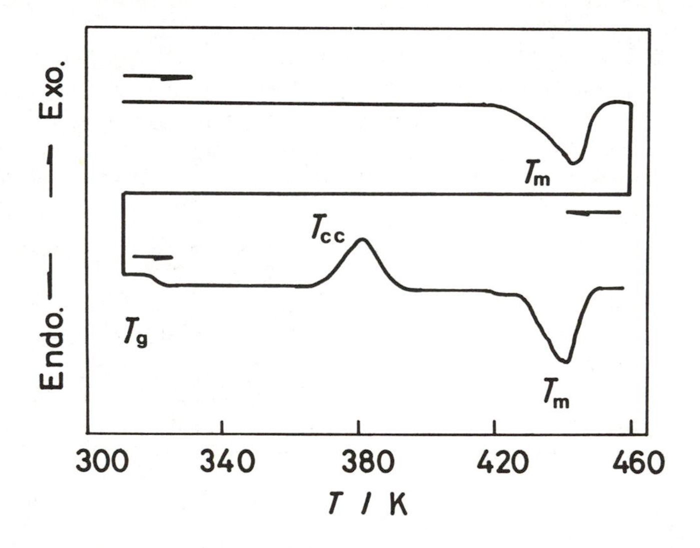

Figure 4. DSC curves of polymethoxybenzalazine ester, IX-C_8: heating and cooling rates, 10 K/min. (Reprinted with permission from ref. 8. Copyright 1986 Sen-i Gakkaishi.)

Table II. Inherent viscosities (η_{inh}'s), glass transition temperatures (T_g's) and starting temperatures of decomposition (T_d's) of polyacylhydrazones (X)

Polyacylhydrazones $-(CH_2)_m-$, m	-R-	η_{inh}(dl/g)	T_g (K)	T_d (K)
2	$(CH_2)_4$	0.56	394	609
4	$(CH_2)_4$	0.60	368	609
6	$(CH_2)_4$	0.53	363	610
2	m-C_6H_4	–	460	622
4	m-C_6H_4	0.92	450	618
6	m-C_6H_4	0.95	430	615

rearrangement from the glassy state to a more ordered state may be disturbed by the presence of guaiacyl units in the main chain in the case of the polyacylhydrazones synthesized in this study.

Figure 5 shows the heat capacities of quenched polyacylhydrazones at around the glass transition temperatures (T_g's). T_g values, estimated from DSC curves, are listed in Table II. Figure 5 and Table II show that the T_g decreases with the increase of m numbers in $-(CH_2)_m-$ of polymers having the same R shown in Scheme 5. This fact shows that the flexibility of the polymers increases with the length of the alkylene group of each repeating unit. On the other hand, polymers having the m-phenylene group showed higher T_g's than those having the tetramethylene group if the m number in $-(CH_2)_m-$ group was the same. This shows that the aromatic group is more rigid than the alkylene group.

Polyesters Having Syringyl-Type Biphenyl Units

In this study, polyesters [XII] having syringyl-type biphenyl units were synthesized from 4,4′-dihydroxy-3,3′,5,5′-tetramethoxybiphenyl (XI) which was prepared from 2,6-dimethoxyphenol (11). As shown in Scheme 6, polyesterification of XI with terephthaloyl, isophthaloyl and sebacoyl chloride were carried out by the low temperature solution polycondensation and by the interfacial polycondensation. The polyterephthalate with $\eta_{inh} = 1.42$ dl/g was obtained by the interfacial polycondensation. The polyisophthalate with $\eta_{inh} = 0.73$ dl/g and the polysebacate with $\eta_{inh} = 0.43$ dl/g were obtained by the low temperature solution polycondensation.

Thermal properties of the polyesters obtained were studied by TG and DSC. The starting temperature of decomposition (T_d) of each polymer in air and nitrogen are shown in Table III. As seen from Table III, T_d of each polymer measured in air is lower than that measured in nitrogen. T_d's of polyterephthalate and isophthalate are higher than T_d of polysebacate.

A DSC curve of polyterephthalate obtained in the temperature range between room temperature and that just below T_d did not show any phase transition. However, an X-ray diffractogram of this polymer showed a crystalline pattern. Accordingly, it was considered that the molecular chain of

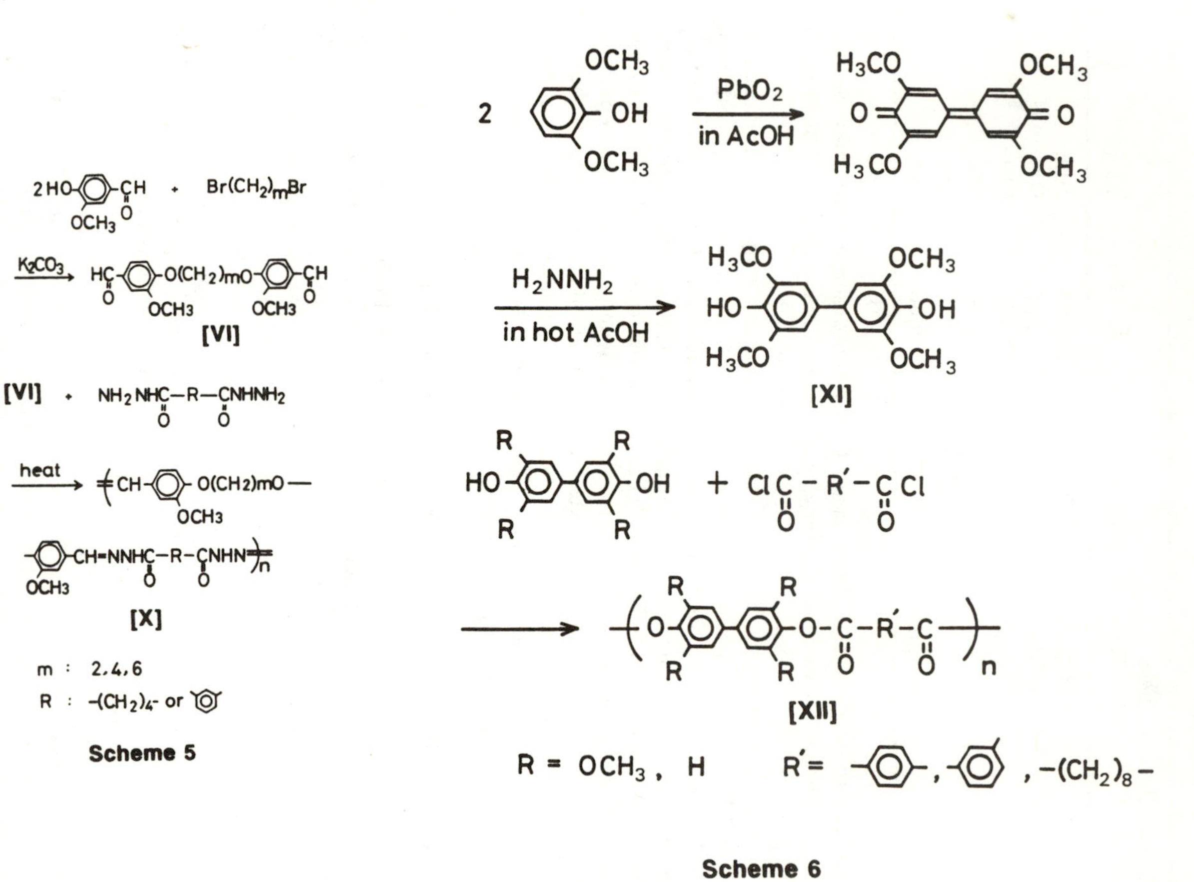

Scheme 5

Scheme 6

Table III. Starting temperatures of decomposition (T_d's) of polyesters (XII) with syringyl-type biphenyl units

Polymer	T_d (K) in N_2	in Air
Polyterephthalate	614	599
Polyisophthalate	618	591
Polysebacate	588	562

the polymer was rigid and no phase transition was detected in the temperature range studied by DSC. In the case of polyisophtahalate and polysebacate, the glass transition was observed on DSC curves. The glass transition temperatures (T_g's) for polyisophthalate and polysebacate were 525 K and 367 K, respectively. The low T_g for polysebacate may be attributed to flexibility of the sebacoyl group of the polymer.

Aromatic Polyethers Having Phosphine Oxide Groups

As shown in Scheme 7, the polymerization of bis(4-fluorophenyl) phenylphosphine oxide (BFPO) with bisphenols [XIII] was carried out, and thermal properties of the obtained polyethers [XIV-A (η_{inh} = 0.63 dl/g) and XIV-B (η_{inh} = 0.48 dl/g)] were studied by DSC and TG (12). In this study, 2, 2-bis(4-hydroxyphenyl) propane and 4, 4′-dihydroxybiphenyl were used as compounds XIII's.

TG measurements of the polyethers XIV-A and -B in nitrogen showed that the polymers started to decompose at 778 and 808 K, respectively. The polyethers started to decompose at 713 and 763 K in air. The obtained results are listed in Table IV.

Table IV. Glass transition temperatures (T_g's) and starting temperatures of decomposition (T_d's) of polyethers XIV-A and -B

Polyether	R in bisphenol	T_g (K)	T_d (K) in N_2	in Air
XIV-A	$-C(CH_3)_2-$	470	778	713
XIV-B	–	498	808	763

The phase transition of obtained polyethers, XIV-A and XIV-B, was studied by DSC in the temperature range lower than 700K. DSC heating and cooling curves of the polymers did not show any first order transition. However, glass transition of the polymers was found in the DSC heating curves. The X-ray diffractograms of as-polymerized samples of the polymers showed the typical halo pattern indicating that the samples were amorphous. The glass transition temperatures (T_g's) of the polymers are listed in Table IV.

The T_g of polyether B is 28 K higher than that of polyether A. This suggests that the biphenyl units in polyether B are more rigid than

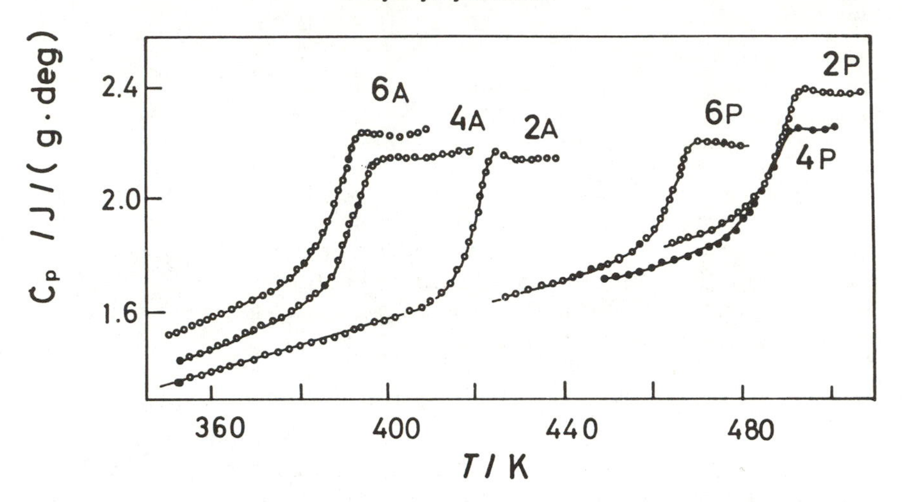

Figure 5. Heat capacities of polyacylhydrazones, X's, quenched at around T_g: 2A, 4A, 6A; m=2, 4, 6 and R=$(CH_2)_4$, respectively. (Reprinted with permission from ref. 10. Copyright 1983 Sen-i Gakkaishi.)

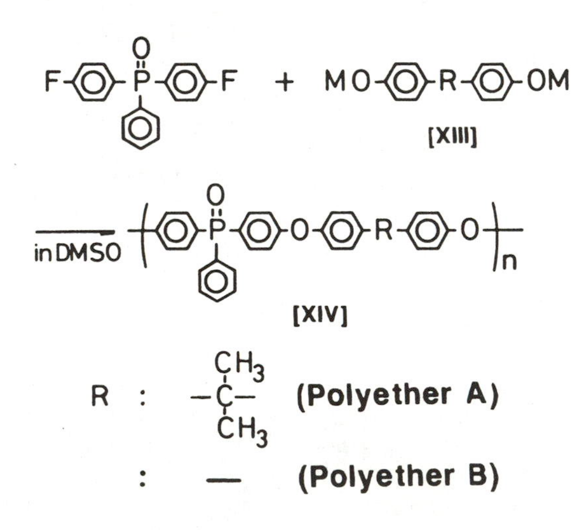

Scheme 7

2,2-diphenyl-propane units in polyether A. It is natural that highly heat-resistant polymers with high T_g are amorphous, since it is difficult for polymers having strong intermolecular force and rigid non-linear backbone chains to form the regular lamellae structure.

It is known that the dissociation energies (D's) of the bonds in the main chain contribute considerably to the thermal stability of polymers, as thermal degradation of condensation polymers proceeds via random scission of polymer chains in radical mechanism (13). Figure 6 shows the chemical structures of polyethers A and B, and also the values of D's which are indicated for each bond in kJ/mol. The D's were calculated using D values of organic compounds with low molecular weight (14, 15). As shown in Figure 6, 322 kJ/mol of D_{c-p} for the bonds in the triphenylphosphine oxide unit is the smallest value among the bonds in the main chains of polyethers A and B. This suggests that the scission of C-P bonds is mainly related to the degradation of the polyethers. In the case of polyether A, the scission of C-C bonds in isopropylidene units is also associated with the degradation and the reduction of the stability of the polymer, since D_{c-c} of the bonds is only 302 kJ/mol. Activation energies (E's) of degradation of the polyethers A and B were calculated according to the method reported by Ozawa (16). The calculated E's were 164 kJ/mol for polyether A and 217 kJ/mol for polyether B. The difference between E and D suggests that the degradation of polyether A and B proceeds not only through the homolytic random scission of bonds in polymer chains but also through other reactions such as the chain transfer reaction of polymer radicals which reduces the E value.

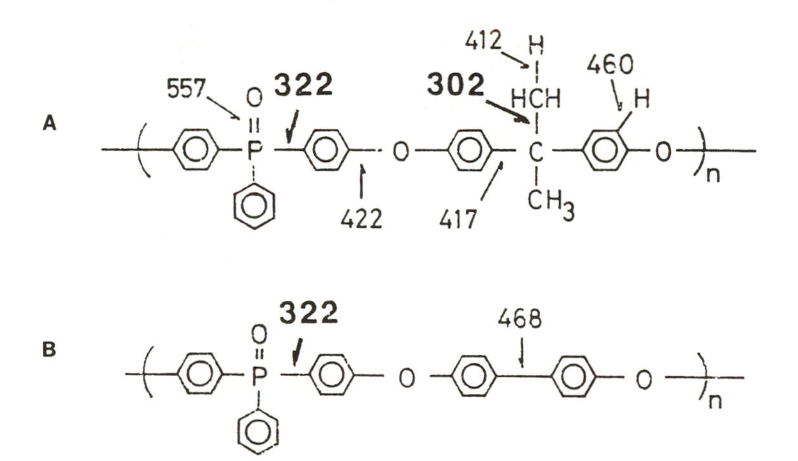

Figure 6. Chemical structures and dissociation energies (D's) of the bonds in polyphenylphosphine oxides, XIV-A and XIV-B. The value of D in kJ/mol is indicated for each bond. (Reprinted with permission from ref. 12. Copyright 1987 Sen-i Gakkaishi.)

Literature Cited

1. Goldstein, I. S. *J. Appl. Polym. Sci., Appl. Polym. Symp.* 1975, **28**, 259.
2. Hatakeyama, T.; Nakamura, K.; Hatakeyama, H. *Polymer*, 1978, **19**, 593.
3. Nakamura, K.; Hatakeyama, T.; Hatakeyama, H. *Polymer* 1981, **22**, 473.
4. Nakamura, K.; Hatakeyama, T.; Hatakeyama, H. *Kobunshi Ronbunsyu* 1982, **39**, 53.
5. Hirose, S.; Hatakeyama, T.; Hatakeyama, H. *Sen-i Gakkaishi* 1982, **38**, T-507.
6. Morgan, P. W. *Macromolecules* 1970, **3**, 536.
7. Hirose, S.; Hatakeyama, T.; Hatakeyama, H. *Kobunshi Ronbunshu* 1982, **39**, 733.
8. Hirose, S.; Hatakeyama, H.; Hatakeyama, T. *Sen-i Gakkaishi* 1986, **42**, T-49.
9. Avrami, M. *J. Chem. Phys.* 1939, **7**, 1103.
10. Hirose, S.; Hatakeyama, H.; Hatakeyama, T. *Sen-i Gakkaishi* 1983, **39**, T-496.
11. Hirose, S.; Hatakeyama, H.; Hatakeyama, T. *Sen-i Gakkaishi* 1985, **41**, T-432.
12. Hirose, S.; Nakamura, K.; Hatakeyama, T.; Hatakeyama, H. *Sen-i Gakkaishi* 1987, **43**, 595.
13. Mita, I. In *Aspects of Degradation and Stability of Polymers*; Gellinek, H. H. G., Ed.; Elsevier: New York, 1978; p. 247.
14. Chem. Soc. Japan, Ed. *Kagaku Binran*; Maruzen: Tokyo, 1984; p. 324.
15. Hays, H. R.; Peterson, D. J. In *Organic Phosphorous Compounds*; Kolosapoff, G. M.; Maier, L., Eds.; Wiley-Interscience: New York, 1972; Vol. 3, p. 418.
16. Ozawa, T. *Bull. Chem. Soc. Japan* 1965, **38**, 1881.

RECEIVED February 27, 1989

Chapter 16

Modification of Lignin to Electrically Conducting Polymers

Tuula A. Kuusela[1], J. Johan Lindberg[1], Kiran Levon[1], and J. E. Österholm[2]

[1]Department of Wood and Polymer Chemistry, University of Helsinki, SF–00170 Helsinki, Finland
[2]Neste Oy, Research Centre, Kulloo, 06850 Finland

Sodium lignosulfonate (NaLS) and sulfur lignin (SL) have been investigated as components in electrically conducting polymers. A thermostable and probably non-toxic polymer, sulfur lignin, has been obtained by modifying lignin-based sodium lignosulfonate through reactions with elemental sulfur in an autoclave at 473-513K. Sodium lignosulfonate and sulfur lignin are normally electrical insulators, but their conductivity can be increased for instance by compounding with graphite or doping with various electron acceptors or donors. We have investigated the combination of compounding with graphite followed by doping with bromine. The conducting properties were measured on samples ground and pressed to the size of IR-pellets. The measurements were made with standard four-point probe or two-point probe techniques. The doping was followed by infrared (IR) and electron spin resonance (ESR) spectroscopy. The percolation phenomenon in the combination of sulfur lignin and graphite was determined by measuring conductivities. The combination of sulfur lignin, graphite, and bromine was studied by measuring conductivities.

Lignosulfonate, a water soluble polymer, is isolated from sulfite spent liquor, a by-product from the sulfite pulping process. It contains extraneous substances, such as sugars, alcohols, terpenes, and sulfite or sulfate salts (1). The polyphenol-like material is separated by ultrafiltration to remove these residues. The lignosulfonate contains such functional groups as methoxyl, carbonyl, phenolic and aliphatic hydroxyl, and sulfonic acid groups. The methoxyl group is the most characteristic group for all lignins. Since sulfonate groups in lignosulfonate macromolecules are ionized in neutral solution, lignosulfonates are anionic polyelectrolytes (2). Lignosulfonates are

0097–6156/89/0397–0219$06.00/0

non-linear, highly branched polymers with sulfonate groups along the backbone as shown in Figure 1.

We have investigated ultrafiltrated sodium lignosulfonate which is a three dimensional polyelectrolytic macromolecule. Rauma-Repola Corporation, Finland, produced ultrafiltrated sodium lignosulfonate in 95% purity, which contained only 5% impurities, mostly sugars.

It was found (3) that the black residual material on the strainer of the sulfite digester was a polymeric sulfur-containing substance. This led us to investigate the reaction between lignosulfonate and sulfur, simulating sulfite pulping process conditions. Under these conditions, a similar polymeric material was formed (3-5).

Preparation of Sulfur Lignin

The synthesis of sulfur lignin is relatively simple, and on a large scale its preparation becomes less expensive than the preparation of synthetic polymers currently used in the semiconductor industry.

The thermal degradation of lignin occurs over a wide temperature range, from about 420-773 K. Several practical ways exist for the preparation of sulfur lignin. These are summarized in Table I.

Table I. Some of the methods to prepare sulfur lignin at the temperature of 473-513 K

Sodium lignosulfonate with:	1. N,N-dimethyl acetamide and sulfur
	2. Water and sulfur in autoclave
	3. Water and sodium sulfide
	4. Sodium carbonate, water, and sulfur

We have used the first method and got a raw material, which was treated with acetone and diethyl ether in a Soxhlet extractor for several hours. After drying in vacuum we determined some details of its structure. The reaction of sodium lignosulfonate with sulfur in N,N-dimethyl acetamide cannot be fully described. We do not know all the reactions which are proceeding simultaneously in the autoclave during the process. Anyhow, differential scanning calorimetry (DSC) measurements of some model compounds have revealed that sulfur partially sublimes, and that N,N-dimethyl acetamide evaporates. In addition, the reaction is slow, and endothermic reaction is observed at 363-423 K indicating the reaction of sulfur with the aliphatic chains. The exothermic reaction at higher temperatures indicates the formation of aromatic carbon-sulfur linkages while simultaneous reactions like desulfonation, demethylation and condensation might also occur (6,7).

Samples of the final synthesis products were investigated using the techniques listed in Table II.

Synthesis conditions and analytical results (Table II) are consistent with the hypothetical structures of sulfur lignin shown in Figure 2. Sulfur lignin is an aromatic thioether and it resembles polyphenylene sulfide derivatives (8).

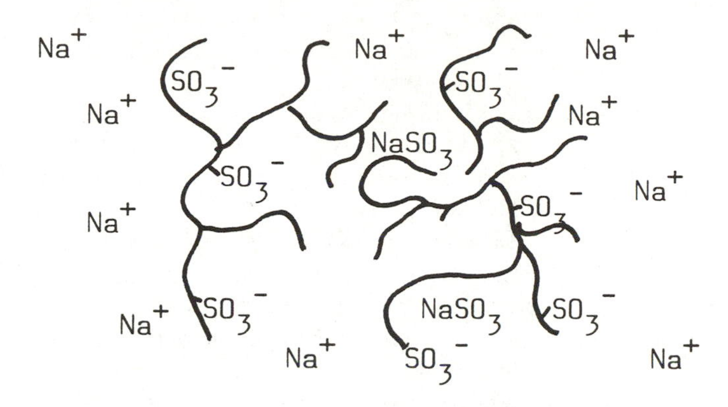

Figure 1. Schematic representation of a lignosulfonate macromolecule.

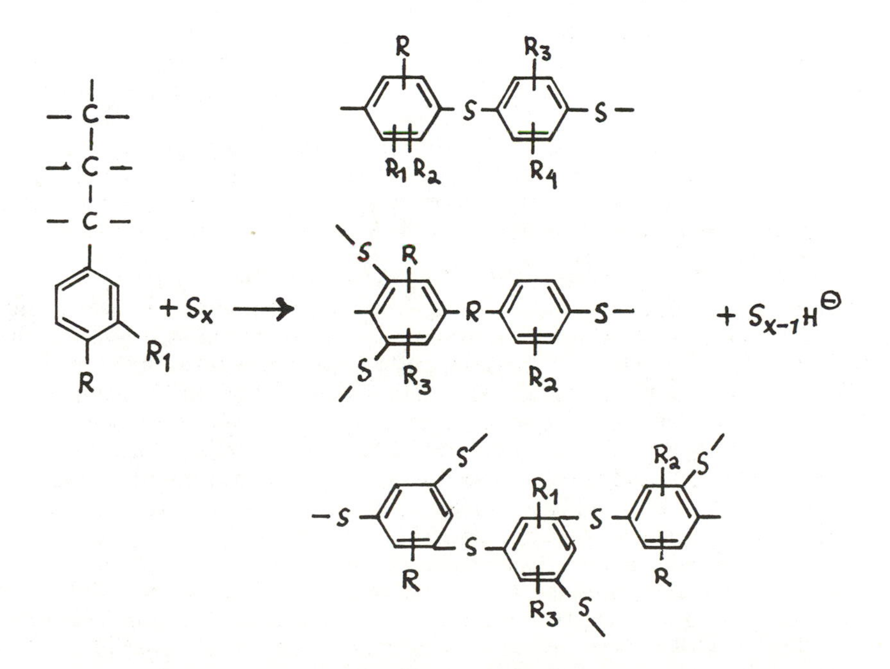

Figure 2. The reaction and possible structures of sulfur lignin.

Table II. Analysis of the structure of sulfur lignin

Methods	Results
X-ray	Sulfur lignin is a totally amorphous material
SEM-microscopy	Amorphous structure, particle sizes are between 1000-40000 nm
Combustion gases	Sulfur content in samples from different experiments varies between 10-26 mass-%
Solid state NMR	Stable sulfur linkages between the aromatic units are formed
IR	Material is nearly completely desulfonated
Elemental analysis	Material is demethoxylated and the propyl group is split off. Polymer is branched.
Conductivity	Sulfur lignin can be made conducting by compounding and by doping
ESR-spectroscopy	Paramagnetic structures and radicals are found

Conducting Polymers

Conducting polymeric materials have become an area of interest to academic and industrial research groups, not only due to their interesting properties, but also due to their technologically promising future in a wide range of applications. Such applications include, for instance, organic batteries, photovoltaic devices, switching and memory devices, electromagnetic interference shielding, and many others which require conducting or semiconducting materials (9). There is every prospect that amorphous, three-dimensional materials hold promise for future investigations.

Future Prospects for Modified Lignin Materials

Sulfur lignin (SL) is seemingly nontoxic and very stable in air. Pure sulfur lignin is almost an insulator. Its intrinsic conductivity varies between 2 and 200 pS/cm ($2\text{-}200\times10^{-12}$ S/cm) depending on the moisture content (10). The dry material has a conductivity of 2 pS/cm. Sulfur lignin seems to be a good adsorbent, too. We have determined the moisture content of our materials, and they vary between 2 and 6 mass-%. Dried starting materials were not doped equally well as those with original atmospheric moisture.

Conductivities of sulfur lignin (SL), sodium lignosulfonate (NaLS), and some common materials are represented in Table III.

Polyaromatic sodium lignosulfonate and sulfur lignin resin can be made conducting by doping. The conductivity increases by several decades through doping with electron acceptors and donors as shown in Table IV.

The majority of the common polyaromatic compounds are donors. For these compounds doping with acceptors is facile.

We have doped sulfur lignin and sodium lignosulfonate in vapor-phase (iodine, bromine, and ammonia) and in liquid phase (sodium and ferrichloride). The conductivity mainly depends on the nature of the dopant ion and the doping degree. Doping can be monitored by IR-spectroscopy. The intensities of the peaks decrease, and the fine structure vanishes, when the

Table III. Conductivities of SL, NaLS, and some common materials (S/cm)

METALS					SEMICONDUCTORS						INSULATORS
		INORGANIC				ORGANIC					
6	4	2	0	-2	-4	-6	-8	-10	-12	-14	-16
exponent 10		(relative scale)									
Ag	Bi		Ge	polyphenylenes							teflon
Cu			Si					SL			quartz
											polystyrene
		graphite									polyethene
				doped SL							anthracene
				doped NaLS							polyethylene

Table IV. Conductivities of sodium lignosulfonate (NaLS) and sulfur lignin (SL) with some dopants

Material	Doping with	Conductivity	Maximum Doping	Stability in Air
NaLS	undoped	140.0 pS/cm	–	stable
NaLS	iodine	2.0 uS/cm	52.0 mass-%	stable
NaLS	bromine	0.3 mS/cm	49.3 mass-%	stable
NaLS	ammonia	1.0 mS/cm	32.1 mass-%	unstable
NaLS	sodium	40.1 mS/cm	21.0 mass-%	unstable
SL	undoped	120.0 pS/cm	–	stable
SL	iodine	280.0 uS/cm	59.4 mass-%	stable
SL	bromine	340.0 uS/cm	66.0 mass-%	stable
SL	ammonia	5.4 mS/cm	49.8 mass-%	unstable
SL	ferrichloride	6.0 mS/cm	32.0 mass-%	stable
SL	sodium	10.0 mS/cm	21.9 mass-%	unstable

doping degree increases. Conductivity increases as well. The ESR-spectra also indicate the doping degree which coincides with conductivity. The Dysonian lineshape is assumed throughout, and A/B-ratios are measured (Table V) (10-12).

Table V. A/B-ratios of sulfur lignin doped with bromine

Doping %	A/B	Conductivity
0	1.1	120.0 pS/cm
14	1.02	2.3 uS/cm
25	0.93	21.2 uS/cm
41	0.90	428.9 uS/cm
66	0.85	6.2 mS/cm

The effect of doping sulfur lignin with bromine on the IR spectra is shown in Figure 3a and 3b.

The Compounding of Sulfur Lignin with Graphite

We have made samples from sulfur lignin with increasing amounts of graphite (Merck, particle size < 50000 nm). This system produces increasing conductivity at a certain threshold graphite content. The conduction of electricity through a sulfur lignin-graphite system, where only graphite is conducting, depends on the concentration of the graphite phase. A current may only flow if a graphite path exists through the solid sulfur lignin. When a path is formed, the conductivity of the sample shows a large increase. This critical concentration is called the *percolation threshold*. The percolation occurs at a percentage value of about 5 mass-% graphite in sulfur lignin as shown in Figure 4a. The curve of these samples has the common percolation shape.

The Compounding with Bromine

A new series of sulfur lignin-graphite samples with known graphite content were treated with bromine in the vapor phase (doped). After measuring the conductivities, we recorded the curve shown in Figure 4b (Kuusela, T. A., The University of Helsinki, Finland, to be published).

It was a previously discovered phenomenon that graphite can bind bromine into its aromatic structure (13). Both of these components, graphite (conductivity 0.77 S/cm) and sulfur lignin (conductivity 120 pS/cm), can be doped simultaneously with bromine (conductivity < 10 nS/cm). After bromine doping, measured conductivities have increased noticeably, and this is attributed to the change in the percentage of graphite in the total system. The conductivity increases up to 23 S/cm (which is an intercalation compound of graphite with bromine without sulfur lignin), as shown in Figure 4b. The difference between lower and higher conductivity values is smaller than in the sulfur lignin-graphite system (14). Bromine content is about 50 mass-% in all sulfur lignin-graphite-bromine samples. Pure graphite can bind about 9 mass-% bromine into its structure.

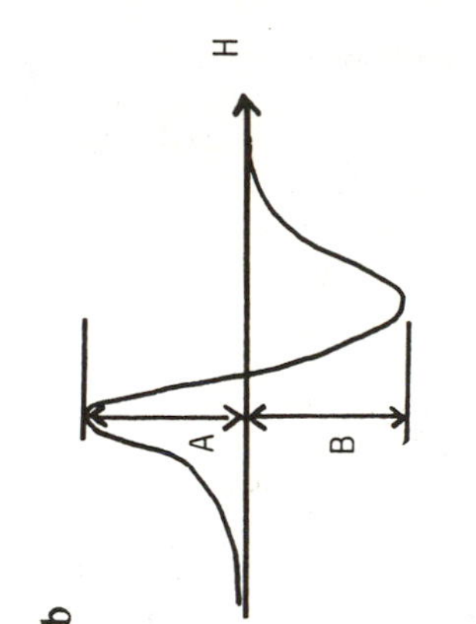

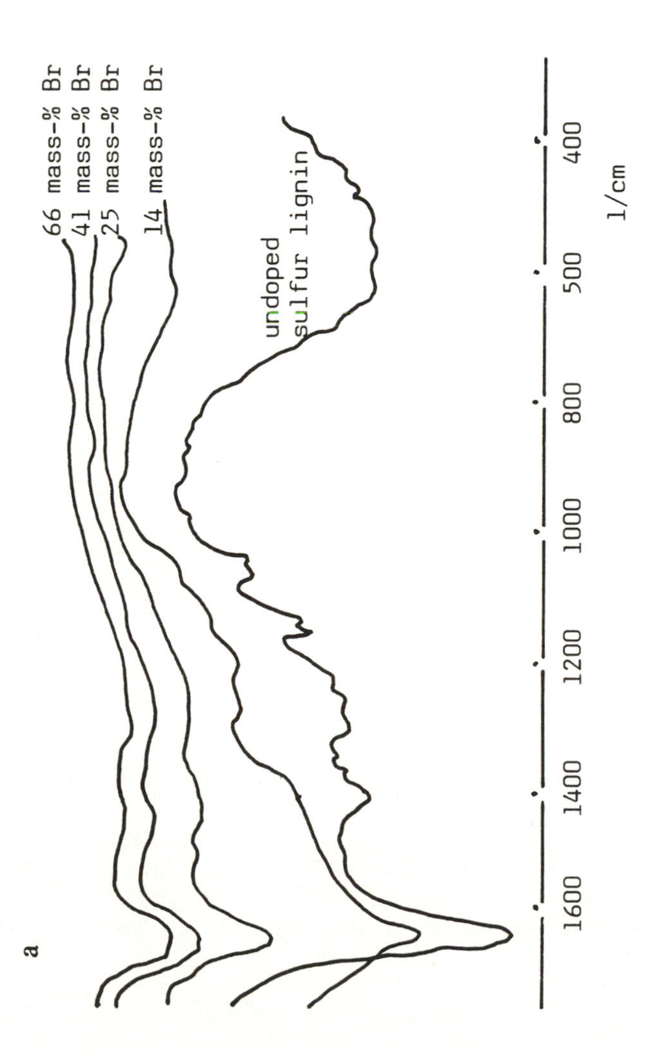

Figure 3. a, the effect of doping degree on IR spectra (bromine); b, ESR-spectra of sulfur lignin doped with bromine.

Graphites with larger surface areas or greater porosities have a distinctly lower percolation threshold. It is assumed that the conductivity of a compound depends upon the structured agglomerates being sufficiently close to each other, or in direct contact above the percolation point, and on the continuous current pathways created thereby (14-15).

Mechanism of Conductivity

When sodium lignosulfonate or sulfur lignin are compounded, for instance, with iodine or bromine, complexes supposedly form (16-17). These systems are conductors with mixed ionic and electronic nature. Presumably they are charge transfer complexes, since the electronic conductivity predominates (18-19). These compounded materials form charge transfer structures (20). Water is supposed to introduce ionic conductivity to the system. Impurities affect conductivity, too (21). In any case, the main models of conductivity are probably based on the *band model* and/or the *hopping model*.

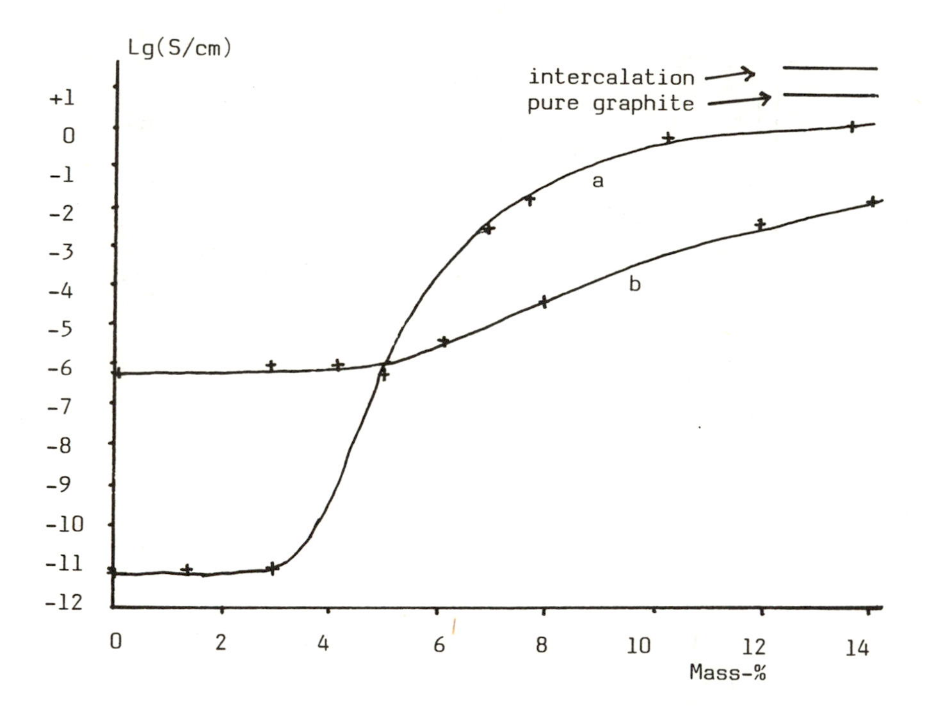

Figure 4. Compounding sulfur lignin with graphite (a) and with graphite and bromine (b).

Conclusions

There has long been an interest in understanding the mechanism of conductivity. In graphite, the electrons are the main current carriers, but intercalation with bromine increases the conductivity of this combination.

Modified lignin materials may serve as components in (semi)conducting systems. All heterogeneous and amorphous polymeric complexes are now of interest for possible future employment in electrically conducting materials.

Literature Cited

1. Hrutfiord, B. F.; McCarthy, J. L. *Tappi* 1964, **47**, 381.
2. Rezanowich, A.; Goring, D. A. I. *J. Colloid Sci.* 1960, **15**, 452.
3. Turunen, J. Ph.D. Thesis, University of Helsinki, Helsinki, 1963.
4. Lindberg, J. J.; Turunen, J.; Hortling, B. Finnish Patent No. 50 998, 1982.
5. Lindberg, J. J.; Turunen, J.; Hortling, B. U.S. Patent No. 410 711, 1982.
6. Levon, K. Ph.D. Thesis, University of Tokyo and University of Helsinki, Tokyo, 1986.
7. Hortling, B. Ph.D. Thesis, University of Helsinki, Helsinki, 1979.
8. Laakso, J.; Hortling, B.; Levon, K.; Lindberg, J. J. *Polymer Bull.* 1985, **14**, 138.
9. Aldssi, M. *Polym.-Plast. Technol. Eng.* 1987, **26**(1), 45.
10. Kaila, E.; Kinanen, A.; Levon, K.; Turunen, J.; Österholm, J.-E.; Lindberg, J. J. U.S. Patent No. 4 610 809, 1986.
11. Lindberg, J. J.; Levon, K.; Kuusela, T. A. *Acta Polymerica* 1988, **39**(1/2), 47.
12. Levon, K.; Kinanen, A.; Lindberg, J. J. *Polym. Bull.* 1986, **16**, 433.
13. Kagan, H. B. *Pure and Appl. Chem.* 1976, **46**, 177.
14. Wessling, B.; Volk, H. *Synthetic Metals* 1986, **16**, 127.
15. Wessling, B. *Macromol. Chem.* 1984, **185**, 1265.
16. Neoh, K. G.; Tan, T. C.; Kang, E. T. *Polymer* 1988, **29**, 553.
17. Kang, E. T.; Tan, T. C.; Neoh, K. G. *Eur. Polym. J.* 1988, **24**, 371.
18. Ratner, M. A.; Schriver, D. F. *Chem. Rev.* 1988, **88**, 109.
19. Hardy, L. C.; Schriver, D. F. *J. Am. Chem. Soc.* 1985, **107**, 3823.
20. Stranks, D. R.; Heffernan, M. L.; Lee Dow, K. C.; McTigue, P. T.; Withers, G. R. A. In *Chemistry: A Structural View*; Cambridge University Press, 1970; Chapter 25, p. 422.
21. Jernigan, J. T.; Chidsey, C. E. D.; Murray, R. W. *J. Am. Chem. Soc.* 1985, **107**, 2824.

RECEIVED May 29, 1989

Chapter 17

Production and Hydrolytic Depolymerization of Ethylene Glycol Lignin

R. W. Thring[1], E. Chornet[1], R. P. Overend[1,2], and M. Heitz[1]

[1]Département de Génie Chimique, Faculté des Sciences Appliquées, Université de Sherbrooke, Sherbrooke, Québec J1K 2R1, Canada
[2]Conseil National de Recherches du Canada, Ottawa, Ontario K1A 0R6, Canada

A prototype solvolytic lignin has been isolated from *Populus deltoides* using ethylene glycol as solvent in a process development unit. The characteristics of this lignin by elemental analysis, methoxyl content, thermogravimetric analysis, gel permeation chromatography, FTIR, and ^{13}C NMR, suggest that this is a native-type lignin. Depolymerization of this lignin by alkaline hydrolysis produces reasonable yields of monomeric compounds: at optimum conditions of 2% sodium hydroxide, and a treatment severity corresponding to 300°C and 1 h reaction conditions, a maximum of 11% of identifiable monomers, based on the initial lignin, are produced, of which 9% are catechol and its methyl and ethyl derivatives.

Hydrolysis of lignin in acidic and basic media has received attention due to the rather few and simple degradation products obtained. Acid-catalyzed hydrolysis reactions applied to isolated lignin have been studied by a number of workers. Lundquist (1), for example, subjected Bjorkman lignin to acidolysis and obtained significant yields of monomeric products. A review of the work prior to 1971 has been made by Wallis (2).

Alkaline degradations, which also involve hydrolytic reactions, can split the lignin and yield useful low molecular fragments. The OH^- group has long been known to cleave lignin bonds and is one of the two primary catalytic agents ($S^=$ is the other) responsible for the dissolution of lignin in the kraft process.

Alkaline hydrolysis, as compared to other methods of lignin degradation, has certain advantages. The catalytic reagents are inexpensive and available commercially, and require no elaborate method of preparation.

0097–6156/89/0397–0228$06.00/0

Also, the yields of the monomeric products can be significant and the product spectrum as a whole is not extensive. It is established (3) that these hydroxyl groups will only cleave certain C-O-C bonds, namely the following:

1. α-aryl ether bonds provided they contain a free phenolic OH^- group in the para position of the α-aryl ether group or a free alcoholic OH^- group on the β-carbon atom.
2. β-aryl ethers if the phenolic OH^- group in the para position to the β-aryl ether side chain is etherified and if there is a free alcoholic OH^- group bound to the α- and/or γ-position(s) of the propane side chain.

The behavior of milled wood lignin and its various modifications towards alkali (sodium hydroxide) treatments agreed well with the results of model experiments (4).

Clark and Green (5) studied the production of phenols from alkaline hydrolysis of kraft lignin. The lignin was cooked at 260-310°C in solutions of sodium hydroxide and sodium sulfide. Principal products identified and quantified were guaiacol, catechol, methyl- and ethyl- guaiacols, methyl- and ethyl- catechols, and phenol. A maximum quantity of these, amounting to 11% of the lignin, occurred when the lignin was cooked in 4% NaOH at 300°C for 30 minutes. Catechol was found to be the most abundant monomeric product, that is, about 5.3% of the lignin at these optimum conditions. Demethylation of guaiacyl compounds and degradation of others took place when sodium sulphide was included during cooking, which resulted in reduced yields of total phenolic compounds identified.

Hagglund and Enkvist (6) developed a laboratory scale method for manufacturing methyl sulfide from kraft black liquor by pressure heating after addition of sodium sulfide. This process was later taken over by Crown-Zellerbach in the United States and developed in pilot plant and full scale. However, the yield is only about 7% of the initial lignin utilized in the process.

Enkvist and co-workers (7) pressure heated kraft black liquors with additions of small amounts of Na_2S and NaOH for 10-20 minutes at 250-290°C in batch as well as in simple continuous apparatus. The addition of sodium sulfide was found to increase the formation of ether-soluble material. The highest yields of ether-soluble phenols occurred with pressure heating at about 291°C of spent kraft liquor containing 20.4% of organic substance after addition of 3.2% Na_2S and 1.6% NaOH. The maximum yield of ether-soluble material was 33% of the organic substance, with catechol and its nearest homologues comprising 5%.

Another product that is produced commercially from the alkaline hydrolysis of lignin is vanillin. It was found that the yield could be improved by the presence of oxygen during the alkaline hydrolysis reaction. A number of commercial ventures have been based on this procedure (8, 9).

The focus of our work is to produce a prototype solvolytic lignin, characterize it, and degrade it to lower molecular weight chemicals by hydrolysis with sodium hydroxide. The breakage of aromatic ether linkages has been shown to be a dominant reaction in alkaline delignification processes. It is therefore of interest to use this apparently simple and promising approach to investigate the chemical utilization of our lignin.

The efficiency of ethylene glycol-water as a delignifying solvent has been demonstrated by Gast and Puls (10). Results showed that sufficiently delignified pulps could be obtained. Also, the lignins produced showed promising results as extenders in phenolic resin adhesives.

Ethylene glycol pulping has been previously studied in our laboratories (11,12). When applied to ***Populus deltoides***, it acts as a "protective" solvent for cellulose which is only slightly depolymerized even at temperatures as high as 220-230°C. Under these conditions the lignin can be scavenged from the wood matrix and dissolved in the ethylene glycol. A fractionation of the main constitutive polymers can thus be achieved. The present work aims at using the lignin fraction as feedstock for alkaline depolymerization to monomeric products.

This paper, then, describes the preparation and characterization of a solvolysis lignin prepared under conditions that are very different to previous solvolytic lignin preparations. In particular, no water, acid or base is used in this process during the delignification step. The lignin is then acid precipitated and hydrolyzed in dilute aqueous solution to produce monomeric substituted phenols.

Experimental

Materials. Air dried wood (*Populus deltoides*) was ground to pass through a screen of 0.5 mm mesh size. These "fines" were used in the subsequent solvolytic studies. The chemicals used were: ethylene glycol (practical grade); sodium hydroxide (practical grade); ethanol (95%); diethyl ether. Calibration-phenols were purchased from Aldrich Chemicals Limited and Alfa Chemicals Limited.

Set-up for Ethylene Glycol Lignin Production. A process development unit (PDU), previously described by Chornet and co-workers (11), was used for the experiments. A typical preparation consists of initially mixing 1-1.2 kg of wood meal with 10 l of ethylene glycol. The mixture is allowed to stand overnight for imbibition to take place. To enhance solvent to substrate penetration, the slurry is homogenized at 200°C in the pretreatment section of the PDU. It is then pumped through the treatment section which consists of a tubular reactor at 220°C. The product slurry is collected in a receiver. The detailed procedure and choice of conditions above have been published elsewhere (11,12).

The product slurry is suction filtered, whilst hot, and the residue washed with hot ethylene glycol (at 100°C). The liquid product and first washing are combined and stored in the refrigerator to be used later. For isolating the lignin from the glycol/hemicellulose solution, it was necessary to use dilute aqueous HCl as a precipitating catalyst. The best conditions found were for 0.05% aqueous HCl (acid : black liquor = 3:1) at T =50-60°C. By this procedure 70-80% of the lignin from the wood is recovered.

Lignin Characterization. Elemental analysis, TGA, methoxyl content, GPC, FTIR, and ^{13}C NMR were used to characterize our prototype solvolytic lignin.

Typical of most isolated lignins, an elemental analysis of glycol lignin in our laboratories using a Perkin Elmer 240C instrument showed it to contain 61.85% C, 6.25% H, 0.16% N, and 33.83% O (by difference). Methoxyl content determinations were made according to the method initially proposed by Zeisel (13) and later modified by Haluk (14). Methoxyl and ash content were 20% and 0.11%, respectively. The presence of sugars was determined to be less than 1%. A comparison between the thermal degradation rate of glycol lignin and kraft Indulin AT by TGA showed that the former degraded at a faster rate than the latter. The maximum rate of weight loss was about twice that of kraft lignin and occurred at about 360°C, compared to 400°C for kraft.

GPC analysis of acetylated glycol lignin was carried out on a Varian 5000 Liquid Chromatograph equipped with two PL gel columns (50Åand 500Å) connected in series. Tetrahydrofuran was used as eluent. The column set was calibrated with monodisperse polystyrene standards for molecular weight determination. Molecular weight averages for derivatized glycol lignin were calculated to be: $\overline{M}_n = 986$ and $\overline{M}_w = 4762$.

The infrared spectrum of glycol lignin is shown in Figure 1. A 5DXB Nicolet FTIR spectrometer in diffuse reflectence mode was used. The sample was prepared by the KBr disk method. Assignment of absorption bands is based on information from Herget (15) and Winston (16). Infrared spectra of glycol and milled wood lignin from the same wood appeared to be similar. One notable difference was the large carbonyl band at 1738 cm^{-1} due to carbohydrates present in the MWL. The sample of MWL used was known to contain 6-8% sugars.

The ^{13}C NMR spectrum (90 MHz, 8.4 T, 2 seconds delay) of glycol lignin is presented in Figure 2. DMSO-d_6 was used as solvent. Assignment of the major signals is based on the work of Lapierre *et al.* (17). The spectrum is very similar to that of milled wood lignin except that at the low field there appear to be more pronounced aliphatic peaks. Identification of these were not undertaken because the glycol lignin was isolated from as-received hardwood which was not extractives-free.

Depolymerization. Subsequent depolymerization of this lignin was carried out in a 500 ml magnetically stirred autoclave. A typical procedure for the experiments was to load the autoclave with 5 g of dry lignin (dried at 60°C overnight), 100 ml solvent, and 0-6 g sodium hydroxide. The bomb was sealed, secured onto its support frame, then the gas inlet, outlet, and pressure gauge were connected. After purging the reactor with nitrogen to remove air, the stirrer was set at 500 rpm and switched on.

To start a run, the reactor was immersed in a preheated salt bath and the temperature and time were recorded by computer for subsequent calculation of the reaction ordinate, to be defined later. Typically, the desired reaction temperature (±3°C experimental error) was reached within the first ten minutes of heating.

After the desired treatment was attained, the reactor was quenched in a cold water bath. The stirrer and computer recording were stopped

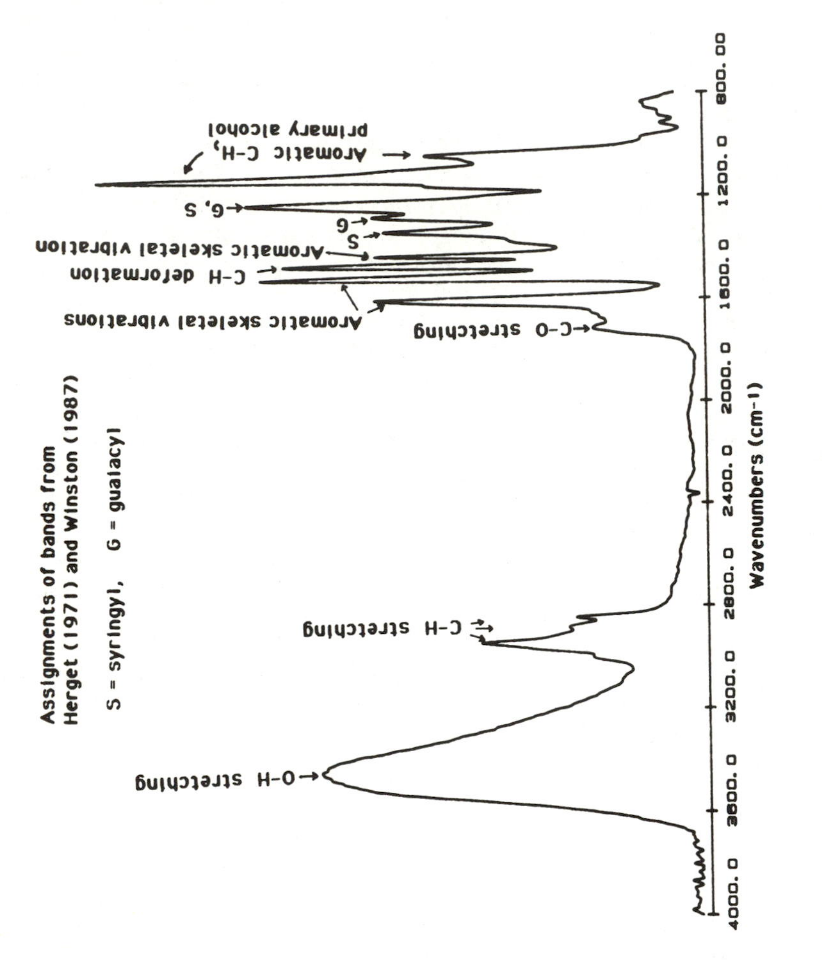

Figure 1. Infrared spectrum of glycol lignin.

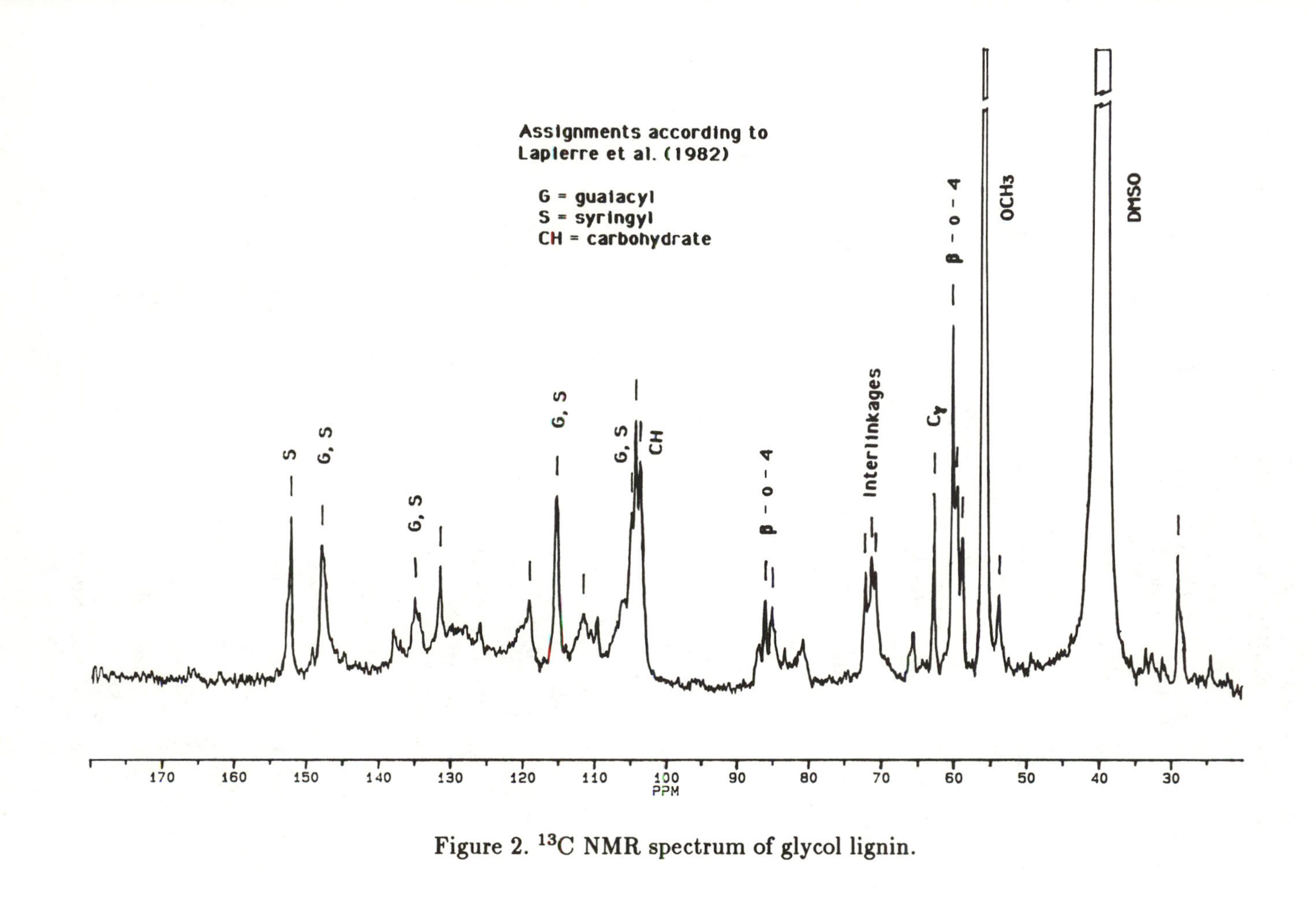

Figure 2. ^{13}C NMR spectrum of glycol lignin.

when the reactor temperature reached its initial value. The bomb was depressurized, if necessary, removed from its frame, opened, and drained of its contents.

After reaction, the slurry product consisted of a liquid mixture together with some insoluble material. This slurry was filtered and the residue thoroughly washed with water. Further separation of the reaction products was carried out as shown in Figure 3. An aliquot of the liquid was acidified, after ethanol removal (if necessary) with 10% (v/v) HCl to a pH of 1-2, then filtered. The filtrate was extracted with diethyl ether till the aqueous layer was judged colorless by eye. After removal of the ether by rotary evaporation, the extract was analyzed by capillary gas chromatography.

The solvolytic treatment and the recovery procedures employed yielded about 75% of the original Klason lignin present in the wood. This percentage could be further optimized but was not attempted in the present work. The recovered lignin was subjected to alkaline depolymerization. Yields presented throughout the paper are expressed as percentages of the recovered lignin.

Material balances during the depolymerization step were within 5% closure for all the experiments reported.

Gas Chromatographic Analysis. Ether-soluble fractions from the hydrolysis runs were analyzed by capillary GC (Hewlett-Packard Model 5890A gas chromatograph); DB-5 column of 30 m length, 0.25 mm I.D.; flame ionization detector; 1.5 ml He/min; split injection 1:100; oven temperature programmed from 65°C to 220°C at 3°C/min (hold 140°C for 5 min and 220°C for 10 min). Products were acetylated by adding 1-2 ml acetic anhydride and 2 drops of pyridine to the sample and heating at 60°C for 1 h. The peaks assigned to phenol, o- and p- cresol, guaiacol, 4-methyl, 4-ethyl- , 4-propylguaiacol, catechol, 4-methyl-, 4-ethylcatechol, syringol, vanillin, acetovanillone, syringaldehyde, and acetosyringone, were identical (retention time) to those of pure compounds. Identification of peaks was further confirmed by comparing mass spectral fragmentation patterns of products in the sample to pure compounds by GC/MS.

For quantitative estimation of identified monomers, response factors were calculated, using 4-ethyl-resorcinol as an internal standard. The relative error in the determination of all the compounds was ±4%.

Results and Discussion

Most of the results obtained from the hydrolysis experiments were analyzed and represented as a function of R_o, a reaction ordinate, previously defined and used (18). It is defined as

$$R_o = \int_O^t \exp\left[(T - T_O)/\lambda\right] \cdot dt \qquad (1)$$

where T_O =constant (taken to be 100°C)
λ =constant (taken to be 14.75°C
t =time (in min.).

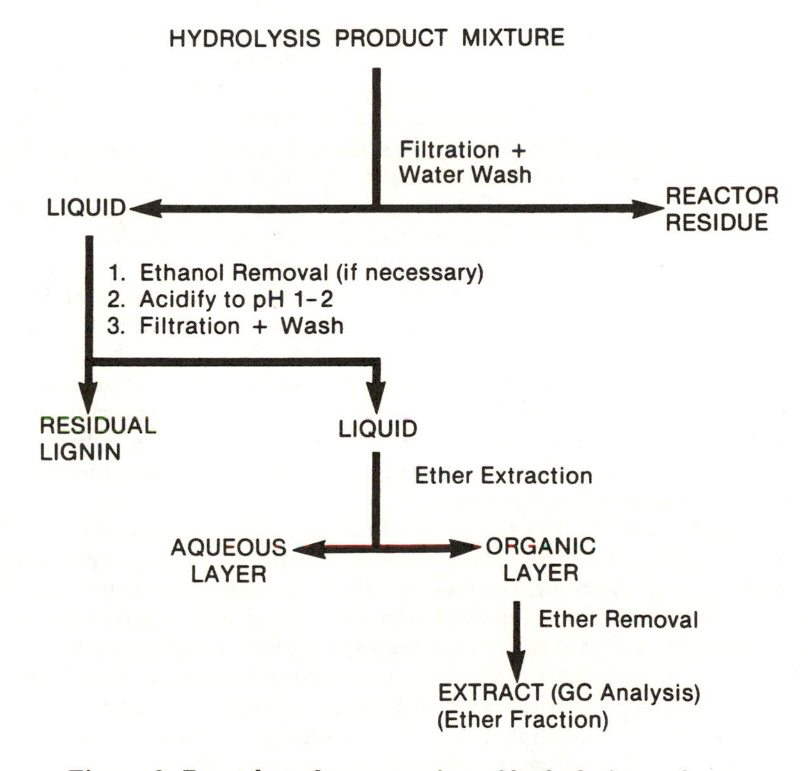

Figure 3. Procedure for separation of hydrolysis products.

The reaction ordinate is utilized here as a means of combining the effects of temperature and time. In presenting and discussing our results, the data is presented on plots of the derived data versus $\log_{10} R_o$.

Conversion of lignin to liquid and gaseous products was determined as follows:

$$\% \text{ conversion} = (W_i - (W_1 + W_2)/W_i) \times 100 \quad (2)$$

where W_i = weight of initial dry lignin (g)
W_1 = weight of reactor residue (g)
W_2 = weight of residual lignin (g)

Effect of Solvent. Figure 4 shows the hydrolysis of glycol lignin in 3% sodium hydroxide using water only and an ethanol/water mixture as solvent. As seen, conversion can be approximated by a linear increase with severity of treatment, for both cases. A similar result was found in the alkaline hydrolysis of kraft lignin in 2% sodium hydroxide solution by Clark and Green (5). When water only is used, the maximum yield of ether soluble material is about 20% of the original lignin, with most of the lignin decomposing to volatile products at higher treatment severities.

However, the quantity of ether soluble material obtained using the solvent mixture was higher and reached a maximum of approximately 25% of the original lignin. This suggests that the presence of ethanol enhances the degradation of lignin in alkaline media. Also, it appears that for either solvent, there is a critical value of reaction temperature and time (and obviously of R_o) beyond which the quantity of ether soluble material remains essentially the same. Thus, it can be said that in using either solvent, only a fraction of the glycol lignin is hydrolyzed into an ether soluble material containing monomeric compounds.

The spectra of monomeric products in the ether extract, when either solvent was used in the reaction, were very similar. Compounds identified were: phenol, o-cresol, p-cresol, guaiacol, 4-methyl-, 4-ethyl-, and 4-n-propyl-guaiacol, catechol, 4-methyl- and 4-ethyl-catechol, vanillin, syringol, syringaldehyde, acetovanillone and acetosyringone. Confirmation of their identity was carried out by comparing retention times and mass spectral fragmentation patterns with those of authentic compounds. Figure 5 shows plots of the yields of phenol, guaiacol and syringol versus R_o using water only and a 50/50 ethanol-water mixture, respectively. For the same reaction conditions, the yields of each of the compounds, especially guaiacol and syringol, are higher when water only is used. It is apparent that the presence of ethanol inhibits the depolymerization of glycol lignin to the monomers cited above. The complete set of data points is shown as a function of temperature in Figure 6. The experiments reported below were carried out using water as the solvent.

Effect of Alkali Concentration. Figure 7A depicts that the variation of the conversion of glycol lignin with sodium hydroxide concentration reaches a plateau at about 60%. Also, the ether soluble material remains constant

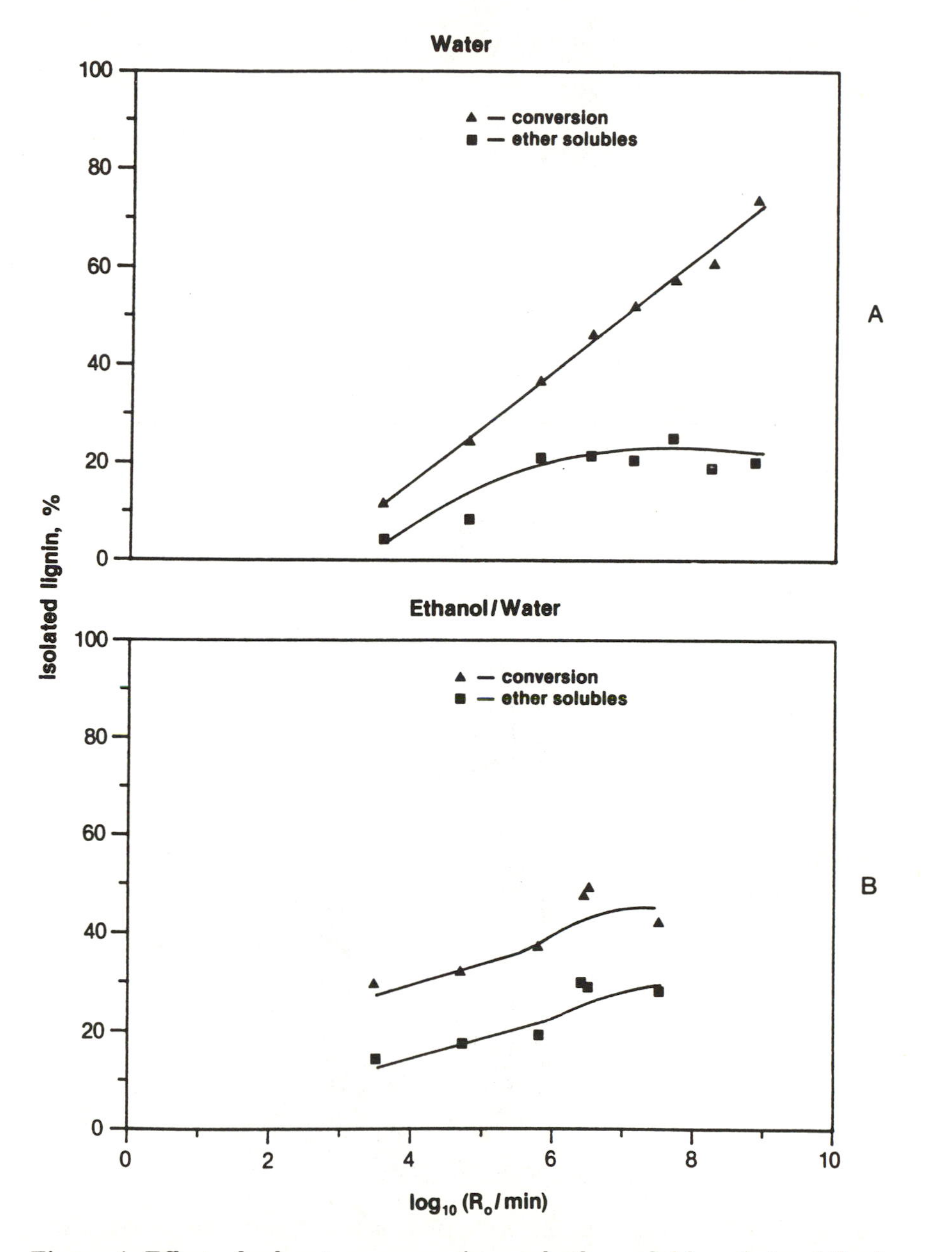

Figure 4. Effect of solvent on conversion and ether solubles of glycol lignin depolymerized in 3% NaOH.

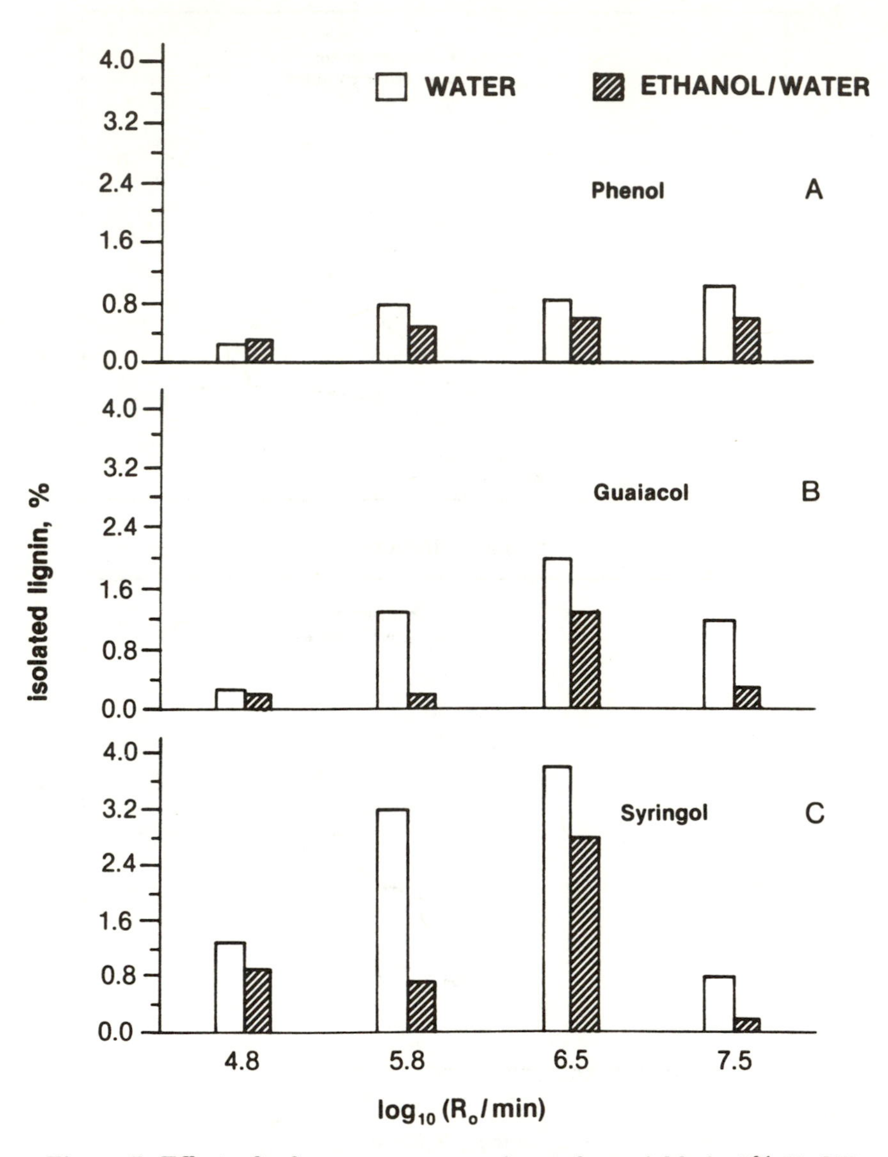

Figure 5. Effect of solvent on monomeric product yields in 3% NaOH.

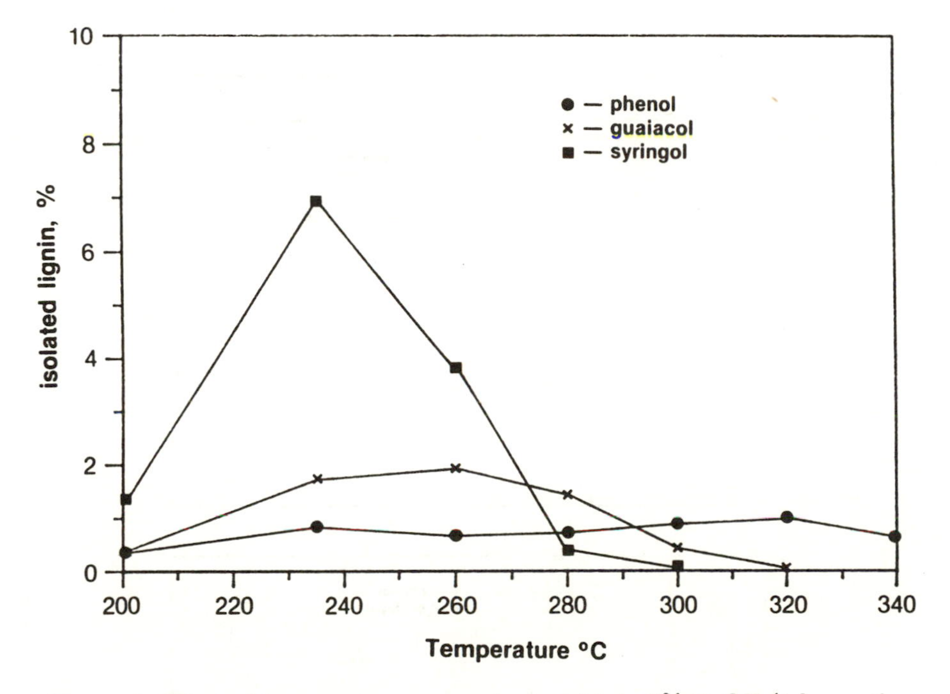

Figure 6. Effect of temperature on product yields in 3% NaOH (1 h reaction time).

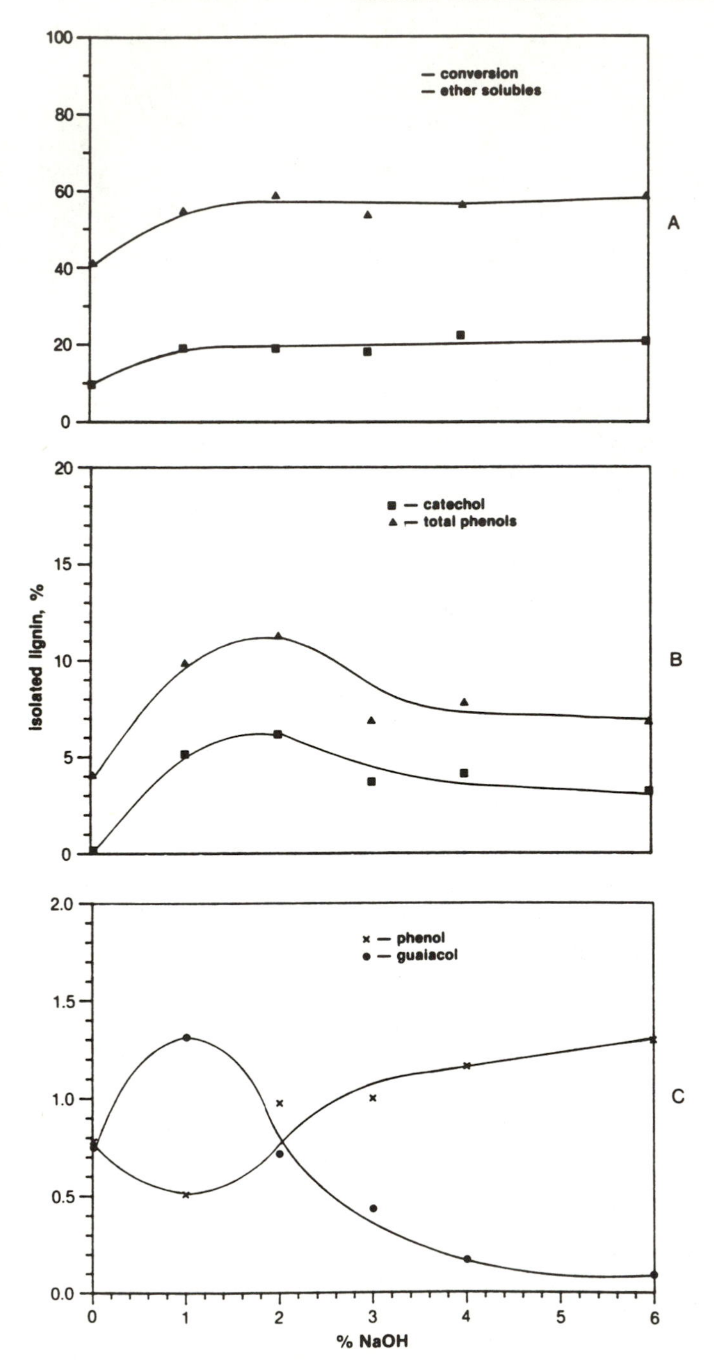

Figure 7. Effect of alkali concentration on conversion and product yields ($\log_{10}R_o = 7.7$).

at a maximum of 20% of the initial lignin with increasing concentration of sodium hydroxide. It is interesting to note that an aqueous treatment of glycol lignin without using sodium hydroxide under the same conditions reported in Figure 7A yields an ether soluble fraction of about 10%.

The variation of catechol and total monomers identified with NaOH concentration is shown in Figure 7B. A maximum of approximately 11% total phenols and 6% catechol is reached at 2% NaOH with increasing alkaline concentration for the same treatment severity. It should be noted that Clark and Green (5) also obtained a maximum of 11% phenolic compounds when they cooked kraft lignin in 4% NaOH at 300°C for 30 minutes. The quantity of catechol was found to be 5.3% of the initial lignin under these conditions.

Figure 7C shows that phenol reaches a minimum of 0.5% and stays constant around 1% with increasing concentration of sodium hydroxide. This indicates that the production of phenol from glycol lignin is relatively unaffected by alkaline hydrolysis. Guaiacol, on the other hand, reaches a maximum at 1% NaOH and then rapidly decreases with increasing concentration of sodium hydroxide. This increase in demethoxylation and/or demethylation of guaiacol with alkaline concentration has also been ascertained by Sarkanen and co-workers (19), who investigated the rate of hydrolysis by sodium hydroxide of methoxyl groups in lignin models and lignin.

Catechol Production at Optimum Conditions. It has already been demonstrated that the production of monomeric compounds reaches a maximum when glycol lignin is cooked in 2% NaOH at a treatment severity corresponding to a reaction temperature and time of 300°C and 1 h, respectively. It was then necessary to see if higher yields could be achieved when the severity of treatment (R_o) is varied.

Figure 8A indicates that conversion and ether soluble material vary in the same manner as when 3% NaOH was used. However, the ether solubles only reach a maximum of 20% at higher treatment severities. The distribution of the identified total phenols and total catechols obtained with increasing reaction ordinate is shown in Figure 8B. The increasing amount of catechols in the total phenols identified implies that these are secondary lignin degradation products, probably originating from primary products such as vanillin, syringol, and syringaldehyde. These were found to occur at lower treatment severities in our experiments. As R_o is increased, reduction in the yields of all monomeric products occurs, demonstrating the increasing influence of pyrolytic reactions in the lignin to produce volatile liquid and gaseous products.

As indicated in the separation scheme in Figure 6, a solid residue was recovered from the liquid hydrolysis products. The carbon-to-oxygen ratio of this material was found to increase rather dramatically with increasing R_o, as seen in Figure 8C. This ratio can either be taken as a measure of the degree of condensation, i.e., a highly condensed lignin means that it has a large number of interunit carbon-carbon bonds, or it can be interpreted as being the result of extensive demethoxylation. The differing levels of

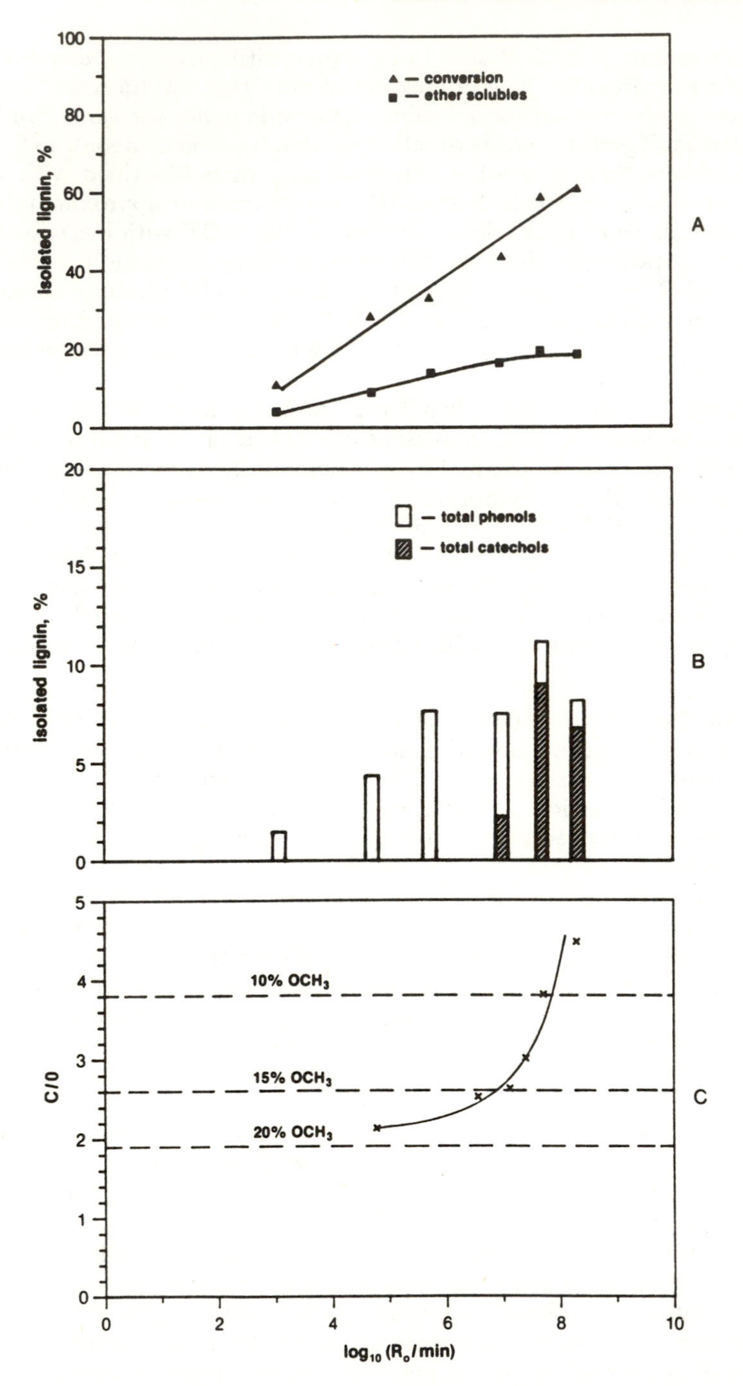

Figure 8. Effect of treatment severity on glycol lignin depolymerization in 2% NaOH. C/O = carbon/oxygen ratio in residual lignins.

demethoxylation are shown as parallel lines in Figure 8C where the original lignin (20% OCH_3) appears as being progressively demethoxylated with increasing treatment severity. A qualitative test on the solutions indicated the presence of methanol which provides further evidence of demethoxylation.

Conclusions

This study has again demonstrated the effectiveness of alkaline hydrolysis as a prime method of lignin degradation to readily identifiable phenolic products.

If, as a criterion of value to the study of lignin depolymerization by alkaline hydrolysis, the maximum yield of oxygen-bearing, phenylpropanoid derivatives is chosen, then the conditions of such a study have been optimized here at a treatment severity corresponding to a reaction temperature of 300°C for 1 hour. Under these conditions, 20% of the lignin is recovered as ether-solubles of which 55% is identifiable as monomeric derivatives. The rest of this material probably consists of dimeric-type compounds not identified by capillary gas chromatography.

From this work it is evident, then, that the hydrolysis of glycol lignin with sodium hydroxide does yield a rather significant amount of catechol and its methyl and ethyl derivatives. A maximum quantity of these, which amounted to 9% of the original lignin, was obtained at the conditions cited above.

The above results appear consistent with a model for lignin proposed by Pepper *et al.* (20) which consists of a core structure to which are attached the more readily accessible side-chain units. These units comprise that fragment of lignin which is easily degraded and released as low molecular weight identifiable products by any degradation procedure.

In view of our results, a possible route for lignin depolymerization could consist of producing an organosolv lignin by methods analogous to those used in our work. The lignin thus produced can be directly treated under alkali conditions at mild severities ($\log_{10} R_o \simeq 7.7$) to recover a low molecular weight fraction in which monomeric products are predominant.

As far as the ethylene glycol lignin is concerned, it has been shown to be a native-like lignin which can be produced and recovered by direct solvolytic treatment of the initial lignocellulosic substrate. It would also be possible to remove the hemicelluloses via an aqueous/steam treatment prior to solvolytic separation of the lignin and cellulose. Such an option would facilitate the recovery of the three main constitutive fractions of lignocellulosics in significant yields. Work in this direction is now underway.

Acknowledgments

The authors are indebted to J. Bureau, J. P. Lemonnier, J. Bouchard, and P. Vidal for their technical support. Financial contributions of NSERC, NRCC and FCAR are gratefully acknowledged.

Literature Cited

1. Lundquist, K. *Acta Chem. Scand.* 1973, **21**, 2597.
2. Wallis, A. F. A. In *Lignins: Occurrence, Formation, Structure and Reactions*; Sarkanen, K. V., Ludwig, C. F., Eds.; Wiley-Interscience: New York, 1971; pp. 345-372.
3. Gierer, J.; Noren, I. *Acta Chem. Scand.* 1962, **16**, 1713.
4. Gierer, J.; Lenz, B.; Noren, I.; Soderberg, S. *Tappi* 1964, **47**(4), 233.
5. Clark. I. T.; Green, J. *Tappi* 1968, **51**(1), 44.
6. Hagglund, E.; Enkvist, T. U.S. Patent 2 711 430, 1955.
7. Enkvist, T.; Turunen, J.; Ashorn, T. *Tappi* 1962, **45**(2), 128.
8. Tomlinson, G. H.; Hibbert, H. *Am. Chem. Soc.* 1936, **58**, 345.
9. Salvesen, J. R.; Brink, D. L.; Diddaus, D. G.; Owzarski, P. U.S. Patent 2 434 626, 1948.
10. Gast, D.; Puls, J. *EEC Proc.* 1985, p. 949.
11. Chornet, E.; Vanasse, C.; Overend, R. P. *Entropie* 1986, **130/131**, 89.
12. Vanasse, C. M.Sc. Thesis, Univ. of Sherbrooke, Québec, Canada, 1986.
13. Zeisel, S. *Monatsh Chem.* 1985, **6**, 989.
14. Haluk, J. P.; Metche, M. *Cell. Chem. Technol.* 1986, **20**, 31.
15. Herget, H. L. In *Lignins: Occurrence, Formation, Structure, and Reactions*; Sarkanen, K. V., Ludwig, C. H., Eds.; Wiley-Interscience: New York, 1971; pp. 267-293.
16. Winston, M. H. Ph.D. Thesis, North Carolina State University, Raleigh, NC, 1987.
17. Lapierre, C.; Lallemand, J. Y.; Monties, B. I. *Holzforschung* 1982, **36**(6), 275.
18. Heitz, M.; Carrasco, F.; Rubio, M.; Brown, A.; Chornet, E.; Overend, R. P. *Biomass* 1987, **13**, 255.
19. Sarkanen, K. V.; Chirkin, G.; Hrutfiord, B. F. *Tappi* 1963, **46**(6), 375.
20. Pepper, J. M.; Steck, W. F.; Swoboda, R.; Karapally, J. C. *Adv. Chem. Ser.* 1966, **59**, 238.

RECEIVED May 29, 1989

Chapter 18

New Trends in Modification of Lignins

Henryk Struszczyk

Institute of Chemical Fibres, 19 C. Sklodowska Str., 90–570 Lodz, Poland

New trends in the modification of lignins related to the formation of polymeric materials with such special properties as thermal stability, fire resistance and use as carriers for controlled release, bioactive compounds are discussed. Several properties of new polymeric materials, especially their thermal behavior, were studied. The activation energy of the thermal degradation process was found to be in the range of 21-176 kJ/mol for several derivatives. Practical agricultural tests with lignin-based carriers containing chemically bound 2, 4-D demonstrated that lignin is capable of providing herbicide release over prolonged time periods.

The worldwide increase in the price of petroleum and coal has created an interest in alternative sources of raw materials. Biomass is an attractive renewable raw material comprising all types of agricultural and silvicultural vegetation. These renewable resources have recently been considered major alternative raw materials for the chemical industry.

Lignins represent the second most abundant polymeric component of biomass. Lignins in spent pulping liquors of chemical pulping processes have so far been used as an energy source, where they do not always live up to their full economic potential. Lignins, owing to their reactive sites which are mainly aromatic as well as aliphatic hydroxyl groups, have been potential raw materials for the manufacture of new polymers (1-7).

This paper presents two different directions of lignin modification research:

- modification by at least difunctional reactive modifiers to form special types of polymeric materials characterized by, among others, thermal stability, fire resistance, and chemical resistance; and

0097–6156/89/0397–0245$06.00/0

- modification of lignins or of lignin derivatives to form polymeric carriers for controlled release bioactive substances.

Special Lignin-Based Polymeric Materials

Two main lignin types were used for these investigations: kraft lignin (Indulin AT, Westvaco Co., USA), and (a) lignin sulfonates (Ultra B002, Rauma Repola Oy, Finland) with an intrinsic viscosity of 8.0 mL g^{-1}, a total hydroxyl content of 14.5% and a phenolic hydroxyl content of 4.7%; (b) lignin sulfonates (Borresperse NA) with an intrinsic viscosity of 5.0 mL g^{-1}, a total hydroxyl group content of 14.9% and a phenylic hydroxyl content of 3.7%; and (c) lignin sulfonates (Ultrazine NAS) with an intrinsic viscosity of 12.1 mL g^{-1}, a total hydroxyl group content of 15.1%, and a phenylic hydroxyl content of 4.2% (both produced by Borregaard Inc., Norway).

Chlorophosphazenes ($NPCl_2$), in the form of hexachlorocyclotriphosphazene (m.p. 112-113°C, Inabate Co., Japan) as well as cyclic oligomers (m.p. 87-91°C, Poland) and terephthaloyl chloride (Merck, FRG), served as reactive modifiers.

The modification of lignins by at least difunctional reactive modifiers was carried out in three systems: in solution (A), in suspension (B), and in soild state (C) in the presence of hydrogen chloride acceptors (8-10). The modification in solution (A) was carried out by dissolving lignins (1.0 g) and a corresponding amount of hydrogen chloride acceptor in a suitable solvent. Difunctional reactive modifier (chlorophosphazenes or terephthaloyl chloride) dissolved in a suitable solvent was next added dropwise during continuous agitation. The mixture was allowed to react at the boiling point for 3 h. The reaction mixture was poured into ice water, and the solid precipitate was centrifuged at 66.6 rps for 15 min. It was washed several times with dioxane and water, or with 5% sodium bicarbonate solution and water, to obtain a neutral reaction product, and it was dried (8, 10-13).

The lignin modifications in suspension (B) were carried out using lignins (1.0 g) dispersed in a suitable medium containing also the hydrogen chloride acceptor. Difunctional reactive modifier dissolved in a suitable solvent was next added dropwise during continuous agitation. The reaction mixture was allowed to react at the boiling point for 3 h. The purification process was carried out as described before (8, 10-12).

The lignin modifications in solid (C) were carried out using lignins (1.0 g) mixed with corresponding amounts of hydrogen chloride acceptor and chlorophosphazenes. The reaction was carried out by heating at a temperature of 100°C, i.e., above the melting point of chlorophosphazene. The reaction mixture was poured into ice water, and then the solid product was purified as described before (9).

The properties of the derivatives obtained were determined by the usual analytical methods (8-17).

Differential thermal analysis (DTA) was performed in air with an OD-102 Derivatograph (MOM, Hungary) using 100 mg samples at a heating rate of 10°C/min. Differential thermogravimetry (DTG) and thermogravimetry

(TG) were carried out with a DuPont Thermal Analyzer, Type 990, with a 10 mg sample in air. The sample weight was recorded against temperature in a range of 20-600°C at a heating rate of 10°C/min.

The hydrolytic resistance of the products was investigated quantitatively in 0.4 N potassium hydroxide and 0.5 N sulfuric acid solutions by mixing powdered materials (40 mg) in a suitable solvent (20 mL) at 50 ± 0.1°C for 24 h.

Modification of Lignins with Chlorophosphazenes

The modification of lignins with chlorophosphazenes allows the manufacture of products characterized by flame resistance and thermal stability. This can be attributed to the aromatic structure of the lignin-phosphazene polymer as well as to the presence of such flame inhibiting elements as phosphorous, nitrogen and sulfur. Other useful properties may also result from this combination. It has previously been reported (8-13) that the modification provides crosslinked products with suitably low chlorine content. This is a consequence of incomplete substitution of the phosphazenes cycles. Additional modification of the reaction products by chemical compounds with reactive hydroxyl or amine groups reduces the unreacted chlorine content and improves product properties (8-13). Some properties of the derivatives obtained are presented in Table I.

A novel modification method, in solid state, has recently been tested (9). Simplicity as well as effectiveness seem to hold promise for this technique (9). Some experimental results are summarized in Table II.

An additional modification of lignins with hydroxyl or amine group-containing compounds was found to further improve the product properties (Table III).

The results reveal that Ultrazine NAS lignosulfonates have the highest reactivity toward chlorophosphazenes. This must probably be explained with their structure, their higher purity, their higher molecular weight, and their higher dispersing ability in contrast to other lignosulfonates.

The modification of lignins by chlorophosphazenes allows the formulation of polymeric materials characterized by:

- High flame resistance: The derivatives are distinguished by high flame resistance and failure to glow completely, whereas the unmodified lignins ignited easily and sustained a flame (kraft) or glowed rapidly (lignosulfonates);
- High hydrolytic resistance against aqueous acid and base: The modified lignins were found not to consume more alkali than the parent lignin (i.e., 50-80 mg KOH/g of derivative as compared to 85 mg KOH/g of parent lignin);
- Chemical resistance to the action of organic solvents;
- Low manufacturing costs in comparison to other types of phosphazene polymers; and
- Continuous improvements of chlorophosphazene-modified lignins, allowing the evaluation of a wide range of applications.

Table I. Properties of the Lignin Modification Products

Symbol of Sample	Type of		Reagents amount, mmol		Yield of Product[a] (g)	Chemical Content, %			Infrared Frequency of Units, cm^{-1}	
	Lignin	$NPCl_2$	$NPCl_2$	Pyridine		P	S	Cl	P=N	POC
1/A	Indulin	olig.	12.208	24.364	1.033	6.76	--	1.35	1260	1158
2/A	Indulin	trimer	3.052	6.091	0.600	1.20	--	--	1210	1025
3/B	Ultra	olig.	8.584	19.392	1.003	2.55	--	0.44	1260 1210	1155 1030
4/B	Ultra	trimer	2.146	4.848	0.035	1.01	--	--	1212	1165 1030
5/B	Borresperse NA	olig.	12.160	7.492	0.185	5.90	3.90	1.46	1210 1030	1170
6/B	Ultrazine NAS	olig.	12.160	27.492	0.192	8.76	3.43	1.31	1250 1200	1160 1025
6a/B	Ultrazine NAS	trimer	6.085	13.746	0.038	3.86	2.60	1.10	1200	1170 1045

[a]All reactions were carried out on lignins (1.0 g) at the boiling point in dioxane as a solvent (A) or in suspension (B).

Table II. Modification Parameters and Derivative Properties Using Solid State Reaction

Symbol of Sample	Type of Lignin	Acceptor Type	Reaction Time, h	Reagent Amount, mmol		Yield of Product[a], g	Chemical Content, %		
				$NPCl_2$	Acceptor		P	S	Cl
7	Borresperse NA	P	3	12.16	27.49	0.79	4.40	3.96	2.64
8	Borresperse NA	P	6	12.16	27.49	0.81	6.83	3.43	2.19
9	Borresperse NA	NaOH	3	12.16	182.20	0.15	5.62	4.10	4.90
10	Ultrazine NAS	P	3	6.09	13.75	0.99	5.61	4.40	3.06

P = pyridine.
[a]All reactions were carried out on lignins (1.0 g) using oligomers of $NPCl_2$.

Table III. Properties of Additionally Modified Products

Symbol of Sample	Symbol of Initial Product	Type of Lignin Used	Type of Modifying Compound	Yield of Product[a], g	Chemical Content, %			Infrared Frequency of P=N Unit, cm^{-1}
					P	S	Cl	
11	1	Indulin AT	n-propanol	1.097	5.82	--	1.22	1255,1210
12	1	Indulin AT	diethylamine	0.612	7.42	--	1.00	1260,1210
13	5	Borresperse NA	diethylamine	0.142	4.10	4.73	0.65	1240,1220

[a]Reactions were carried out using 0.145 mol of diethylamine or 0.0213 mol of n-propanol.

The modification of lignins by chlorophosphazenes distinctly increases hydrogen bonding energy of the products obtained, i.e., 17-23 kJ/mol as compared to 14-20 kJ/mol for the parent lignin. This phenomenon, which was additionally confirmed by X-ray data (10, 12), shows the augmentation of the probable formation of regions with increased degree of supermolecular order.

Thermal Behavior of Lignins Modified by Chlorophosphazenes

The thermogravimetric (TG) as well as difference thermogravimetric (DTG) analysis data are summarized in Table IV. The thermograms of selected derivatives are presented in Figure 1.

The TG as well as DTG data (Table IV) show a higher thermal stability of lignosulfonates. The derivatives obtained are distinguished by higher thermal stability in comparison with the parent lignins, particularly in the case of additional modifications by n-proponol or diethylamine (Table IV, Samples 11 and 13). The increase in thermal stability must be explained with the cross-linked structure as well as the saturation with such aromatic groups as phenylpropane and phosphazene. The increased phosphorous content increases at the same time the flame resistance. In all cases, low amounts of volatile products formed in the exothermic degradation process effectively stifled flame development (Table IV). Thermal behavior studies of the lignin-phosphazene derivatives show that their degradation process results in volatile and non-volatile products. Assuming first-order reaction kinetics for the thermal degradation, reaction constant (k) and activation energy (E) can be calculated by dynamic TGA (18, 19). The E-value can thus be calculated from the leastsquare slope of the plot of log k versus reciprocal absolute temperature as shown in Figure 2. The activation energy values for the thermal degradation of selected derivatives as well as unmodified lignins are summarized in Table V.

As can be seen from Figure 2, a dual segment approximation was necessary in the case of some lignins as well as derivatives. This phenomenon indicates that pyrolysis takes place by two mechanisms in the temperature range studied while the degradation of, for example, Ultrazine NAS could be explained by a single mechanism. The kraft lignin as well as Ultra B002 have a higher value of E as compared to other lignin sulfonates. This may result from differences in the lignin structure.

The modification of lignins with chlorophosphazenes results in changes of E that correspond to the degree of substitution and the phosphorous content (Table V).

Lignin modification with chlorophosphazenes is an example of how this renewable resource may be utilized in special polymeric materials.

Lignin Modification with Terephthaloyl Chloride

Another potential lignin modification method concerns the reaction with difunctional acid chlorides, especially terephthaloyl chloride, to form cross-linked polymeric materials (1, 10, 14, 20). Several studies have dealt with

Table IV. TG and DTG Analysis of the Products of Modification by Chlorophosphazenes

Symbol of Sample	Type of lignin	P Content, %	Temperature of Maximum Rate of Mass Loss, °C	Range of Temp. of Max. Rate of Mass Loss, °C	Percentage of Mass Loss at Different Temperatures					
					100°C	200°C	300°C	400°C	500°C	600°C
1	Indulin AT	1.20	---	---	3.5	7.0	12.5	22.2	30.8	43.5
2	Indulin AT	6.76	492	350	6.7	9.3	18.0	26.1	45.2	61.3
11[a]	Indulin AT	5.82	---	---	5.8	8.5	12.8	22.4	32.4	40.7
13[b]	Indulin AT	7.42	305	235-350	2.0	3.3	11.8	22.0	28.4	34.1
3	Ultra B002	2.55	250,570	225-300	4.2	5.5	16.2	24.0	32.8	50.1
7	Borresperse NA	4.40	275	270-310	11.0	15.0	25.0	30.0	41.0	48.1
9	Borresperse NA	5.02	300	270-330	8.0	14.0	20.0	45.0	56.0	75.0
5	Borresperse NA	5.90	---	---	6.0	9.0	10.0	13.0	16.0	20.0
8	Borresperse NA	6.83	280	250-300	12.0	20.0	27.0	33.0	42.0	51.0
10	Ultrazine NAS	5.61	280,289	250-300	4.0	8.0	29.0	35.0	46.0	56.0
--	Indulin AT	--	393	300-398	1.8	4.1	13.0	76.3	86.1	88.1
--	Ultra B002	--	292,489	250-495 472-495	3.4	4.0	22.6	29.8	63.1	64.4
--	Borresperse NA	--	325	250-350	4.5	7.1	28.2	44.9	51.3	--
--	Ultrazine NAS	--	280,370	240-430	3.5	5.0	12.1	26.1	31.5	45.0

[a]Modified by n-propanol.
[b]Modified by diethylamine.

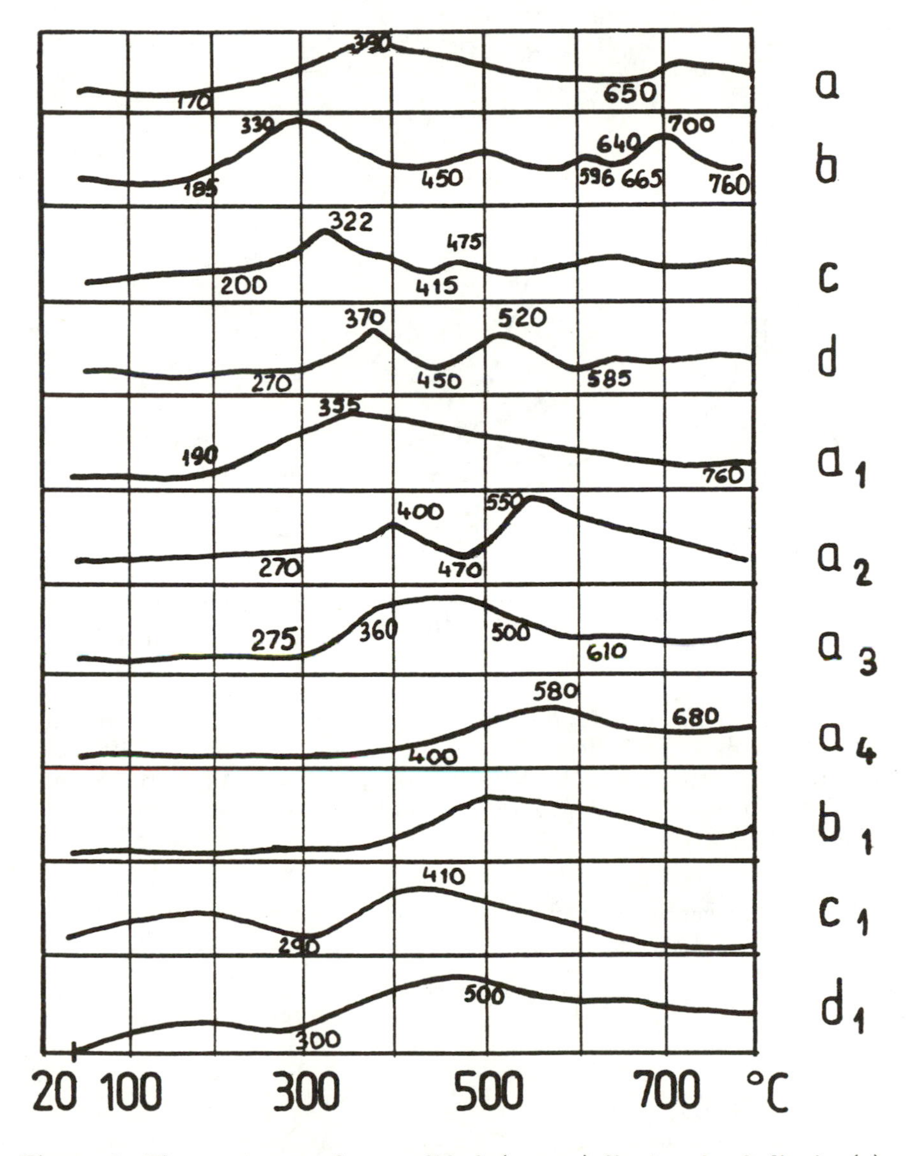

Figure 1. Thermograms of unmodified (parent) lignins: kraft lignin (a); Ultra B002 (b); Borresperse NA (c); and Ultrazine NAS (d), as well as their derivatives: 1(a_1); 2(a_2); 11(a_3); 12(a_4); 3(b_1); 13(c_1); 10(d_1).

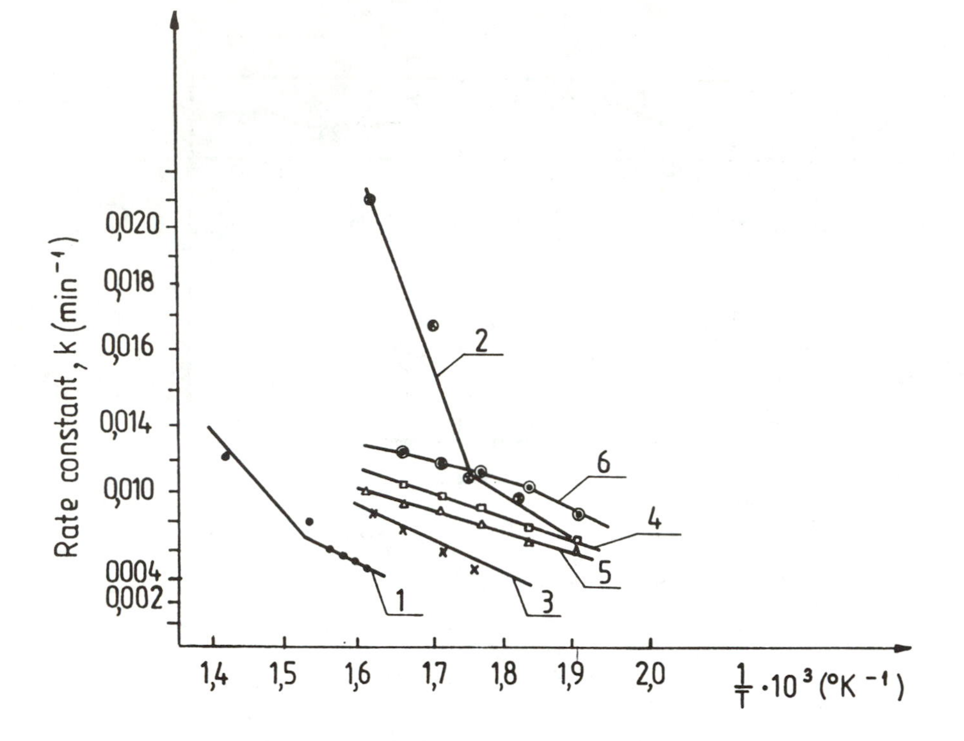

Figure 2. Arrhenius plot for selected lignins and their derivatives from dynamic TGA data: kraft lignin (1, ●); Borresperse NA (2, ⊗); Ultrazine NAS (3, ×); Borresperse NA derivatives of symbol #7 (4,□); Borresperse NA derivatives of symbol # 9 (5, △); and Ultrazine NAS derivatives of symbol #10 (6, ⊙).

Table V. Activation Energy of the Thermal Degradation Process.

Sample	Type of Lignins	P Content, %	kJ/mol	Temperature Range, °C
--	Borresperse NA	---	46.0	250-350
--	Ultrazine NAS	---	33.3	250-350
--	Ultra B002	---	162.4	250-350
--	Indulin AT	---	176.7	350-400
7	Borresperse NA	4.40	70.0	250-350
9	Borresperse NA	5.62	44.0	250-350
8	Borresperse NA	6.83	34.8	250-350
10	Ultrazine NAS	5.61	41.0	250-350

this reaction and reported on limited success. A recent reinvestigation of this reaction has produced meltable aromatic ester-like polymeric materials. The modification of lignin sulfonates with terephthaloyl chloride in solvent or suspension (10) permits the formation of lignin-based polymers with ester groups. The products are characterized by several advantages: melting point is observed between 290 and 330°C; hydrolytic and chemical resistance is good; thermal stability is increased; color is white to yellow; the product can be mixed in the melt with other polymers; the presence of such functional groups as sulfonates permits the application as additives to modify other polymer properties.

Some results of the modification of lignin sulfonate Ultra B002 by reaction with terephthaloyl chloride are summarized in Table VI. The total hydroxyl content of the lignosulfonates as well as their derivatives are presented in Table VII. The hydrolytic resistance of selected products is evaluated in Table VIII. The results presented in Tables VI-VIII stress several advantages of the derivatives with terephthaloyl chloride. The modified lignin sulfonates were insoluble, or only very slightly soluble, in organic solvents. They were, however, soluble in dimethyl sulfoxide. Ordered structures were identified by X-ray studies (16, 17).

The modification of lignins with terephthaloyl chloride is of interest to the thermal properties of the derivatives (15). These are presented in Table IX. The results reveal that an exothermic temperature event is shifted to higher temperature as a consequence of modification. The endothermic temperature event can be attributed to the glass transition and the melting process of the derivatives. The activation energy of the degradation process was calculated for several samples based on the results of dynamic TG studies (Table X). A distinct decrease in activation energy is observed in the case of Ultra B002. At the same time, the E values seem to correspond with the higher temperature of the modification.

The modification of lignin sulfonates with terephthaloyl chloride produces new polymeric materials containing ester groups. This modification can be used to utilize lignins also for improvement of chemical fiber properties. This is presently under investigation.

Modified Lignins as the Carriers for Controlled Release Preparations

In the hope of finding large markets for lignins, a review of potential agricultural applications has been conducted (1, 6, 14). Modification of lignins with bioactive compounds containing suitable reactive groups permits the formation of various derivatives characterized by a wide range of release rate of active component.

2, 4-dichlorophenoxyacetic acid (2, 4-D) as a standard herbicide containing reactive carboxylic functionality was used in the present study. Additionally, the lignin sulfonates of "Klutan" (Niedomice, Poland) and the reaction residues from vanillin production (LS-W, Wloclawek, Poland) were also used. The carriers containing chemical bound 2, 4-D used the following preparations:

Table VI. Effect of Modification with Terephthaloyl Chloride on the Properties of the Ultra B002 Product.

Reaction Medium	Reagent Molar Ratio	Color of Product	Yield of Product[a]	Chemical Content, %			M.p. °C
				C	H	S	
Dioxane	1:1.10	yellow	0.419	53.8	3.95	0.95	310-316
	1:1.65	yellow	0.359	55.3	4.17	0.84	318-321
	1:2.20	yellow	0.399	58.5	4.20	0.75	327-331
Water	1:0.55	lt.yellow	0.250	54.0	4.09	0.76	298-311
	1:1.10	lt.yellow	0.475	45.3	3.95	1.36	306-314
	1:1.65	lt.yellow	0.576	51.5	3.87	1.00	311-320
	1:2.20	lt.yellow	1.038	56.8	4.10	0.79	309-322

[a]All reactions were carried out on a lignin (1.0 g) at 30°C for 15 min.

Table VII. Total Hydroxyl Content of Lignin Sulfonates and Their Derivatives with Terephthaloyl Chloride.

Symbol of Sample	Type of Lignin	Total Hydroxyl Content, %
--	Borresperse NA	14.9
NA-2-20	Borresperse NA	1.9
NA-2-70	Borresperse NA	1.5
--	Ultra B002	14.5
R-2-10	Ultra B002	0.1
R-2-70	Ultra B002	0.1

LD: lignin sulfonates modified by 2,4-D (21)
LSD: lignin sulfonates modified by terephthaloyl chloride and subsequently by 2,4-D (22)
LF: lignin sulfonates modified by phenol and formaldehyde and then by 2,4-D (23).

Table VIII. Hydrolytic Resistance of the Lignins Modified with Terephthaloyl Chloride

Type of Product	Average Amount of H_2SO_4 (g) Consumed by 1 g of Product ($x\ 10^{-3}$)	Average Amount of KOH Consumed by 1 g of Product ($x\ 10^{-3}$)
Modified Borresperse NA	50-200	10-50
Modified Ultra B002	50-300	10-50
Borresperse NA	450	85
Ultra B002	450	85

Application of several types of lignin sulfonates as carriers in connection with reactive herbicides (21-23) permits the development of effective controlled release (CR) preparations. The properties of chemically bound lignin 2,4-D-based herbicides are summarized in Table XI.

Several preparations were subjected to standard release analysis in an aqueous medium (14). The release data are shown in Figure 3. These data suggest that the lignin sulfonate 2,4-D-based preparations have the capacity for releasing herbicide activity over a prolonged time period.

The most important verification of CR preparations is their practical test under agricultural conditions. Such a performance test was carried out at the Plant Protection Institute (Poznań, Poland) using two doses of *Pielik*, a Polish 2,4-D preparation in 85% purity: "a" dose of 1 kg of *Pielik* per 1 ha, and "b" dose of 2 kg of *Pielik* per 1 ha. Epiphyfic preparations were applied at three stages: directly after germination of three types of test plants (sunflower, wetch and white mustard) (I); one week after germination (II); and two weeks after germination (III). The weight loss of the plants was determined until 30 days following application. The test results are summarized in Table XII.

These results reveal that LSD-6 (W) was the most effective preparation when applied in the higher dose (b). The other preparations were characterized by a lower practical action within the test period. At the same time these preparations have effectively acted for a longer period than that studied in the test of Table XII.

It can be concluded that lignin sulfonates modified by reaction with herbicide functionality (21-23) can be used as potential carriers for controlled release herbicides. This application of lignin sulfonates offers several advantages both for agriculture and for the environment.

Table IX. Thermal Properties of Lignosulfonates Modified by Terephthaloyl Chloride.

Symbol of Sample	Type of Lignin	Temperature of exothermic °C			Temperature of endothermic °C		
		Event Start	Max.	End	Event Start	Max.	End
--	Borresperse NA	200	322	425	---	---	---
--	Ultra B002	185	330	450	---	---	---
NA-2-20	Borresperse NA	380	405	425	100	122	130
		435	462	470	250	358	367
Na-2-70	Borresperse NA	355	375	400	100	122	130
					250	340	350
R-2-10	Ultra B002	348	370	400	100	120	125
		420	450	470	250	330	345
R-2-70	Ultra B002	355	370	400	100	120	130
		410	450	470	250	342	355

Table X. Activation Energy of the Degradation Process of Selected Lignin Sulfonates as well as their Derivatives with Terephthaloyl Chloride

Type of Sample	Type of Lignins	E kJ/mol	Temperature Range, °C
--	Borresperse NA	46.0	250-350
--	Ultra B002	162.4	250-350
NA-2-10	Borresperse NA	49.0	250-350
NA-2-70	Borresperse NA	95.0	250-350
R-2-10	Ultra B002	21.4	250-350
R-2-70	Ultra B002	48.1	250-350

Table XI. Properties of 2,4-D Modified Lignin Sulfonates [a]

Symbol of Sample	Type of Lignin	Chemical Content, %				2,4-D Content, %
		C	H	S	Cl	
LD-3	Borresperse NA	57.1	4.38	4.14	6.38	19.2
LD-3(W)	LS-W	46.5	3.43	4.52	7.97	24.8
LD-3(K)	Klutan	41.2	3.28	5.35	7.14	22.2
LSD-6(W)	LS-W	52.3	4.11	2.38	0.73	2.24
LSD-6	Borresperse NA	57.1	3.99	1.43	0.74	2.31
LF-2	Borresperse NA	--	--	--	4.53	14.1

[a]The reaction conditions were reported previously (21-23).

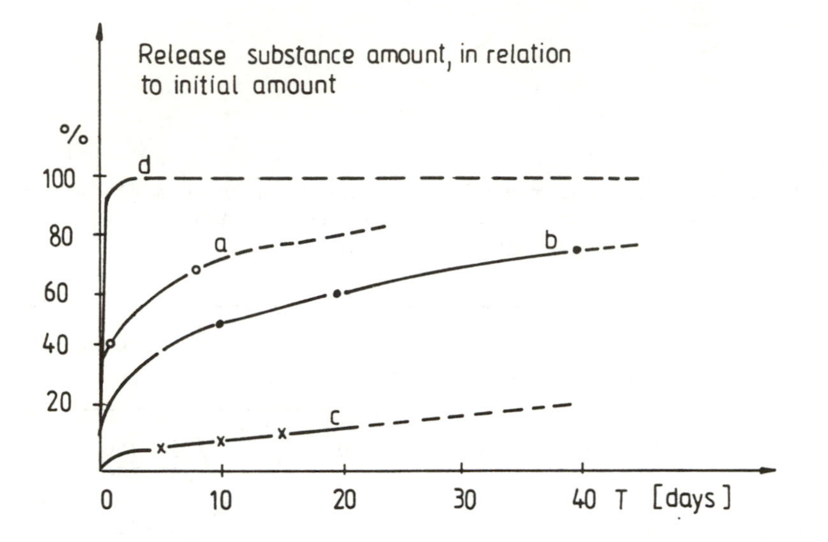

Figure 3. Release curves of 2, 4-D from preparations of: LSD-6 (a); LF-2 (b); LD-3 (c); and standard 2, 4-D (d).

Table XII. Test Results of the Effect of 2,4-D Containing Preparations on Plant Growth.

Type of Preparation	Dose	Average Wt. Loss, % I	II	III	Average Weight Loss from 3 Stages, %
LD-3	a	41.8	60.8	43.4	48.7
	b	53.7	57.7	63.7	58.4
LD-3(W)	a	36.8	50.5	35.7	41.0
	b	70.0	51.7	61.2	60.9
LD-3(K)	a	61.5	49.8	58.5	56.6
	b	55.8	59.8	79.3	64.9
LSD-6(W)	a	54.0	56.0	59.2	56.4
	b	63.7	82.5	73.7	73.3
LSD-6	a	55.8	63.2	49.7	56.2
	b	59.0	60.5	65.3	61.6
Standard 2,4-D (Pielik)	a	70.3	81.2	49.3	66.9
	b	72.7	69.3	72.7	71.6

Lignin utilization studies have been conducted in several fields. However, no practical, commercial process has yet emerged for the manufacture of lignin-based polymeric materials. This is undoubtedly related to such factors as product inhomogeneity, reactivity differences, and also with lack of familiarity with this type of raw material. It should be also stressed that lignins are only by-products from pulp manufacture. However, the utilization of lignins in polymeric materials seems to be possible.

Acknowledgment

It is a pleasure to acknowledge the contribution of my assistant, Mrs. K. Wrześniewska-Tosik, for her continued cooperation.

Literature Cited

1. Sarkanen, K. V.; Ludwig, C. H., Eds. *Lignins*; Wiley-Interscience: New York, 1971.
2. Simonescu, C. I. *Cell. Chem. Technol.* 1978, **12**, 577.
3. Naraian, H. *Indian Chem. Manuf.* 1981, **19**(6), 11.
4. Lindberg, J. J.; Melartin, J. *Kem. Kemi* 1982, **9**(11), 736.
5. Glasser, W. G.; Hsu, O. H.; Reed, D. L.; Forte, R. C. *1979 Canadian Wood Chemistry Symposium Proceedings*; Harrison Hot Springs, Canada.
6. Allan, G. G.; Balaba, W.; Dutkiewicz, J.; Struszczyk, H. *Chemicals from Western Hardwoods and Agricultural Residues*; Semi-Annual Report, April 1979, NSE-7708979; Univ. of Washington, Seattle, USA.
7. Chen, R.; Kokta, B. V.; Valade, J. L. *J. Appl. Polym. Sci.* 1979, **24**, 1609.
8. Struszczyk, H.; Laine, J. E. Polish Patent 125877, 1981.
9. Struszczyk, H. Polish Patent Appl. P-265167, 1987.
10. Struszczyk, H.; Krajewski, K. Polish Patent 134256, 1982.
11. Struszczyk, H.; Laine, J. E. *J. Macromol. Sci.-Chem.* 1982, **A17**(8), 1193.
12. Struszczyk, H. *Fire and Materials* 1982, **6**(1), 7.
13. Struszczyk, H. *J. Macromol. Sci.-Chem.* 1986, **A23**(8), 973.
14. Struszczyk, H.; Wrześniewska-Tosik, K. *Proc. Inter. Symp. on Fibre Sci. and Technol.*; Hakone, Japan, 1985, p. 336.
15. Struszczyk, H.; Allan, G. G.; Balaba, W. *Proc. of Euchem '80 Conf.*; Helsinki, Finland, 1980.
16. Struszczyk, H. *Proc. Hungarian Symp. on Thermal Analysis*; Budapest, Hungary, 1981.
17. Struszczyk, H. *Proc. of 31st IUPAC Conf. Macro '87*; Merseburg, Ger. Dem. Rep., 1987.
18. Ramian, M. V. *J. Appl. Polym. Sci.* 1970, **14**, 1323.
19. Tang, W. G. *U.S. For. Serv. Res. Pap. FPL 71*, 1967.
20. Van der Klashort, G. H.; Forbes, C. P.; Psotta, K. *Holzforschung* 1983, **37**(6), 279.
21. Struszczyk, H.; Wrześniewska-Tosik, K. Polish Patent 141253, 1985.
22. Struszczyk, H.; Wrześniewska-Tosik, K. Polish Patent 141254, 1985.
23. Struszczyk, H.; Wrześniewska-Tosik, K. Polish Patent 141255, 1985.

RECEIVED April 18, 1989

Chapter 19

Application of Computational Methods to the Chemistry of Lignin

Thomas Elder

School of Forestry, Auburn University, Auburn, AL 36849

Computational and theoretical techniques have been used to describe a wide range of compound classes, but have been only sparingly utilized in studies on the properties and reactions of lignin. A brief summary of the capabilities and limitations of molecular mechanics and molecular orbital calculations is presented, along with a survey of specific applications to lignin that have been reported in the literature.

An examination of the other articles in this text serves as an excellent illustration of the diverse analytical methods that have been successfully applied to lignocellulosic materials. The practitioners of wood chemistry have rapidly assimilated and adapted modern instrumental chemistry to their specific problems. In contrast, the techniques of computational chemistry have not been widely used in such an environment. The current paper will attempt to describe the capabilities, opportunities, and limitations of such an approach, and discuss the results that have been reported for lignin-related compounds.

The term "computational chemistry" can refer in its broadest sense to a wide range of methods that have been developed to give insight into the fundamental behavior of chemical species. Such methods include, but are not necessarily limited to, those related to quantum mechanics (1), molecular mechanics (or force-field calculations) (2), perturbation theory (3), graph theory (4), or statistical thermodynamics (5). For the purposes of this chapter, comments will be restricted to force-field and quantum-based calculations, since these are the techniques that have been used in work on lignin. Furthermore, these methods have been reviewed in a very readable book by Clark (6).

A brief perusal of the previously cited references will reveal the highly involved nature of the mathematical formalism that has led to these techniques. This problem is offset, to some extent, by the ready availability of

0097–6156/89/0397–0262$06.00/0

well-developed software, and improvements in the speed, capacity, and costs associated with computer hardware. These advances enable users to concentrate on the interpretation of results, rather than the intricacies of the mathematics. A wealth of programs are available, for environments ranging from personal computers to supercomputers, at very reasonable cost from the Quantum Chemistry Program Exchange at Indiana University.

In spite of the minimal applications of computational chemistry to the chemistry of wood, the techniques have become highly developed and sophisticated in their ability to calculate chemical properties for a wide variety of compound classes. Methods based on quantum mechanics, commonly referred to as molecular orbital calculations, have been the topic of numerous books, reviews, and research papers (7,8,9,10). These techniques are concerned with the description of electronic motion, and the solution of the Schrödinger equation to determine the energy of molecular systems. Since the exact solution of the Schrödinger equation is only possible for two-particle systems, approximations must be invoked for even the simplest organic molecules.

Molecular Orbital Calculations. The most sophisticated and theoretically rigorous of the molecular orbital methods are *ab initio* calculations. These are performed with a particular mathematical function describing the shape of the atomic orbitals which combine to produce molecular orbitals. These functions, or basis sets, may be chosen based on a convenient mathematical form, or their ability to reproduce chemical properties. Commonly used basis sets are a compromise between these two extremes, but strict *ab initio* calculations use only these mathematical functions to describe electronic motion. Representative of *ab initio* methods is the series of GAUSSIAN programs from Carnegie-Mellon University (11). In general, these calculations are computationally quite intensive, and require a large amount of computer time even for relatively small molecules.

In order to perform calculations on larger molecules in a reasonable amount of time, approximations are made, which may involve the neglect of certain terms, or the inclusion of experimentally determined parameters. The best known and simplest example of this level of approximation are Hückel Molecular Orbital (HMO) calculations, which treat only pi-electrons, in conjugated hydrocarbons, with neglect of overlap (1). While obviously limited in use, HMO methods are still used in certain research applications.

A variety of more advanced, all-electron methods of this type are available, and are generally referred to as semi-empirical calculations. The acronyms used to name the individual methods are descriptive of the manner in which atomic overlap calculations are performed. Among the more widely used semi-empirical methods are those of complete neglect of differential overlap (CNDO/2) (12), modified intermediate neglect of differential overlap (MINDO/3) (13), and modified neglect of diatomic overlap (MNDO) (14).

The relative merits of a number of molecular orbital methods have been discussed in the literature (15,16,6). The methods cited differ fundamentally in that the CNDO/2 method was developed to mimic the results of *ab initio* calculations, while the latter two approaches are concerned with the reproduction of experimental results. It has been said that the aim of Dewar's group was to develop an "MO spectrometer" that can provide results with experimental accuracy in a reasonable time (6).

The results from molecular orbital calculations can be divided into two general categories, dealing with energy and electronic populations. The wealth of information that can be obtained is illustrated in Figure 1 (17). Such data are useful in the interpretation of experimental results, and the prediction of the course of novel reactions.

Molecular Mechanics Calculations. In sharp contrast to the molecular orbital calculations that are related to quantum mechanics, and attempt to describe the motion of electrons in a framework of fixed nuclei, molecular mechanics, or force field, calculations completely ignore the presence of electrons. These techniques, which have been reviewed by Osawa and Musso (18), Burkert and Allinger (2), and Allinger (19), treat molecular systems classically, representing the atoms as masses and the bonds which connect them as springs. In the simplest form, springs that obey Hooke's Law are used, while in more complicated systems, such as those involving hydrogen-bonded interactions, Morse potentials or other functions may be used (19, 2). Molecular mechanics calculations represent the energy of a compound as the sum of the energy of stretching, bending, vibration, torsion, non-bonded interactions, and terms that couple these motions. Initially, molecular mechanics methods were developed for the interpretation of vibrational spectra, but more recently they have been applied to a number of other phenomena and properties (19). A number of force-field methods have been described (20-24), and while conceptually similar, the potential functions and parameterizations that are used may be quite different (2).

Molecular mechanics calculations have been carried out on a wide range of chemical compounds including hydrocarbons, heteroatomic molecules, steroids, carbohydrates and proteins. Furthermore, a variety of information has been obtained such as heats of formation, rotational barriers, and rates of reaction (18, 2, 19). The major advantage of this method over other computational methods is that it is reasonably fast to perform in comparison to formal molecular orbital calculations.

Applications of Computational Methods to Lignin

With this introduction to the methods of computational chemistry, the attention of this paper will now turn to specific applications related to the chemistry of lignin. As promised at the beginning of this paper, the discussion will address not only the capabilities and opportunities that may accrue from this type of research, but will also consider the limitations of the techniques.

One of the basic assumptions is that all calculations are done on isolated gas-phase molecules. This is obviously a large shortcoming for lignin chemistry for a number of reasons. The reactions of importance in lignin

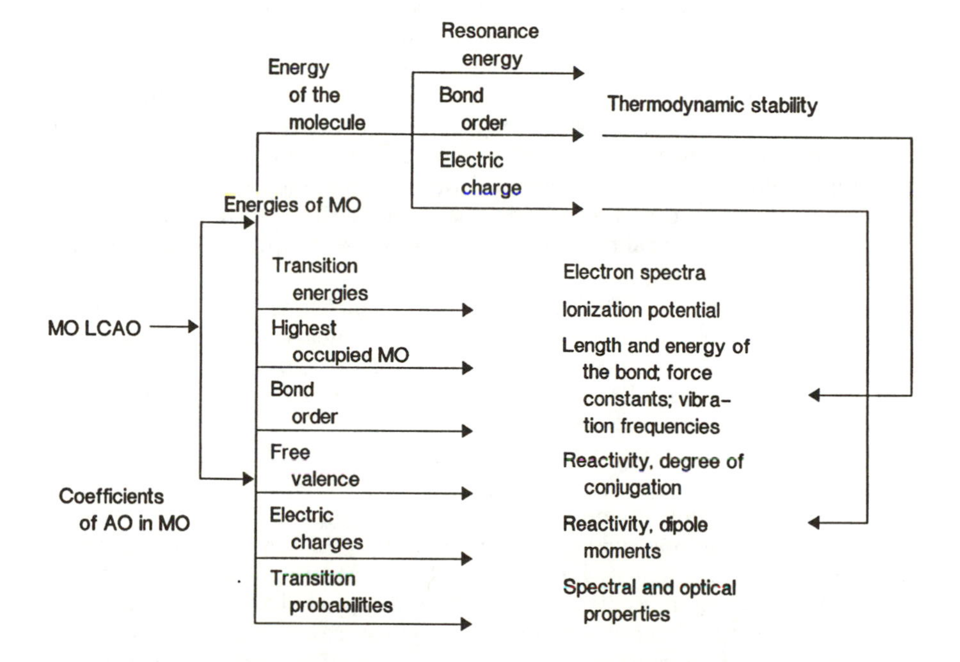

Figure 1: Electronic characters obtained by the molecular orbital method and their applications. (Reprinted with permission from ref. 17. Copyright 1977 Academic Press.)

work, specifically those related to pulping and bleaching, take place in aqueous systems, and lignin is a large heterogeneous polymer. The solvation questions may be overcome to some extent by additional calculations. For example, the molecular mechanics method described by Weiner and Kollman (24) has the ability to apply a given dielectric constant to the molecule, and corrections to gas-phase calculations have been reported (25,26).

The indeterminate and polymeric nature of lignin can, of course, be addressed by the utilization of judiciously chosen model compounds. This is usually the strategy in basic experimental studies on lignin, and the same logic should apply to computational methods.

In addition, the lack of basic thermodynamic data for compounds of interest is a limitation, since in most situations there are no experimental values to which computed results may be compared. The validation procedure to which the methods are subjected, however, includes a large range of compounds for which experimental data is documented. While this is not direct evidence that for specific compounds the results will be representative of experimental results, it is one of the assumptions that has been made in the work on lignin.

These difficulties notwithstanding, the methods of computational chemistry represent a unique approach to the questions of wood science, and rather than a summary dismissal, should be examined to determine their applicability. In spite of such difficulties, theoretical calculations have been successfully used in work related to materials science (27,28), and in numerous biochemical applications (29).

The application of computational methods to lignin chemistry are being researched mainly by individuals and groups in the United States, and Soviet Union, Czechoslovakia, and Finland. One of the earliest papers on this topic was concerned with the use of Hückel Molecular Orbital methods in the study of lignin reactivity (30). It was reported that the calculated localization energies were related to the reaction rates of phenols, and that calculations on free radical structures could be of importance in the elucidation of factors that influence the polymerization of lignin precursors. Furthermore, Glasser and co-workers (31-36) have used simulation techniques to model the polymerization and reactions of lignin.

In a study directly related to the coupling of phenols to form the lignin polymer, Martensson and Karlsson (37) used Pariser-Parr-Pople calculations to examine the π-electron spin densities of a variety of phenoxy radicals. The spin density, representative of unpaired electron character, was found to be consistently highest at the phenolic oxygen in each of the radical structures that were considered. This has been interpreted as theoretical evidence that is in accordance with experimental data indicating that the β-0-4 linkage is the predominant mode of combination in softwood lignin. The addition of methoxy, aroxy, and aryl substituents ortho to the phenolic oxygen was found to effectively dilute the spin densities throughout the radical structures reported. In particular, the spin density of the position para to the phenolic oxygen seems to be reduced at the expense of the carbon to which the substituent is attached. The relatively high spin density at this position as a result of substitution does not, however, represent a site of elevated reactivity, because of steric hindrance, or thermodynamic constraints (38).

Related research has been reported by Elder and Worley (39), in which MNDO was used to examine the structure of coniferyl alcohol, and its corresponding phenolate anion and free radical. This method represents an improvement over the PPP method, in that MNDO is an all-electron technique, and performs geometry optimizations. It was found that the calculated spin densities and charge values for the reactive sites did not correlate quantitatively with observed bond frequency, but it was observed that positions with partial negative charge and positive spin densities are the positions through which the polymerization has been found to occur.

The most extensive and systematic study of the chemistry of lignin with theoretical methods has been performed by Remko and co-workers. Their work has involved the nature of intramolecular (40-43) and intermolecular (44-48) hydrogen bonding of lignin model compounds, spectral transitions (49-52), and conformational analysis (53). The methods used have included CNDO/2 and PCILO (Perturbative Configuration Interaction using Localized Orbitals) (54).

Conformational studies of dimeric lignin model compounds have been reported by Remko and Sekerka (53), and Gravitis and Erins (55) from the Soviet Union. Gravitis and Erins (55) used the pairwise atom-atom potential function (AAPF) method, a force-field technique, to determine the conformation of β-0-4 linked dimers. It was found that the aromatic rings, in the most stable arrangement, exhibit a folded structure, with the rings in parallel planes. In a similar, but somewhat more extensive study of dimeric lignin model compounds, Remko and Sekerka (53), using PCILO, found that the lowest energy of the β-0-4 dimer corresponded to two conformations, one of which was a planar structure, while the other had a rotation of 120° between the aromatic rings. According to Remko and Sekerka (53), this discrepancy is due to the utilization of different computational methods. Furthermore, although the dimers in both papers were β-0-4 linked, they were not identical, and were not substituted with hydroxyl or methoxyl groups.

In conjunction with a study on the reactivity of dimeric quinone methides, Elder et al. (56) examined the physical and electronic structure of guaiacylglycerol-β-coniferyl ether, which is substituted in a manner representative of the lignin polymer. Calculations were performed using AMBER (Assisted Model Building with Energy Refinement) (24), which is a force-field method, and the energetic minimum was determined to be a folded structure similar to that reported by Gravitis and Erins (55).

Since the properties of any material are related to its structure, the conformational aspect of these studies cannot be neglected. The emphasis of the latter paper (56) was, however, primarily concerned with the electronic structure of the dimeric model compound, and how this information might be compared to experimental facts. Upon completion of the calculations, it began to appear that the results were seriously flawed. It was found, surprisingly, that the alpha carbon of the quinone methide has a slight negative charge. Since the principal reaction of kraft pulping is proposed to involve nucleophilic attack of this position by either sulfhydryl or hydroxide ions, how can these electronic data be reconciled?

Further examination of the results indicated that by invocation of Pearson's Hard-Soft Acid-Base (HSAB) theory (57), the results are consistent with experimental observation. According to Pearson's theory, which has been generalized to include nucleophiles (bases) and electrophiles (acids), interactions between hard reactants are proposed to be dependent on coulombic attraction. The combination of soft reactants, however, is thought to be due to overlap of the lowest unoccupied molecular orbital (LUMO) of the electrophile and the highest occupied molecular orbital (HOMO) of the nucleophile, the so-called frontier molecular orbitals. It was found that, compared to all other positions in the quinone methide, the alpha carbon had the greatest LUMO electron density. It appears, therefore, that the frontier molecular orbital interactions are overriding the unfavorable coulombic conditions. This interpretation also supports the preferential reaction of the sulfhydryl ion over the hydroxide ion in kraft pulping. In comparison to the hydroxide ion, the sulfhydryl is relatively soft, and in Pearson's theory, soft reactants will bond preferentially to soft reactants, while hard acids will favorably combine with hard bases. Since the alpha position is the softest in the entire molecule, as evidenced by the LUMO density, the softer sulfhydryl ion would be more likely to attack this position than the hydroxide.

The influence of frontier molecular orbitals has also been demonstrated in work on the chlorination of lignin model compounds (58). In this work, electronic parameters were compared to the difference in energy between ground state molecules and chlorinated intermediates. The results indicated that for coniferyl alcohol, the position exhibiting the greatest negative charge is the methoxyl oxygen, while the position with the greatest electron density in highest occupied molecular orbital area is para to the phenolic hydroxyl. The HOMO is the frontier molecular orbital of interest in this reaction, since the coniferyl alcohol is the nucleophile and the chloronium ion is the electrophile.

Favorable reaction indices notwithstanding, these positions have the two highest energy barriers to chlorination. Similarly, the β-carbon in the aliphatic sidechain has the lowest energy barrier, but does not have correspondingly great negative charge or HOMO electron density. It can be seen that steric considerations may be strongly influencing the reactivity of these sites, with the carbon being largely unhindered, while the others with greater reaction barriers are strongly hindered.

If these three positions are neglected, leaving aromatic positions and the α-carbon of the sidechain, it is found that the size of the negative charge does not reflect the order of the energy differences. In contrast, the HOMO electron density at the various positions increases in the same sequence as the energy barrier toward chlorination. It may be concluded, therefore, that the chlorination reaction is more sensitive toward orbital control, and steric arguments, than simple coulombic attraction.

The utilization of Pearson's hard-soft acid-base theory to interpret the reactions of lignin has also been described in work from the Soviet Union by Zarubin and Kirysun (59). Beyond this specifically related paper, researchers in the Soviet Union have been quite active in the application of numerical methods to lignin-related problems (60,61).

In an attempt to relate calculated results to experimental findings for monomeric, lignin model compounds, preliminary work has compared theoretically determined electron densities and chemical shifts reported from carbon-13 nuclear magnetic resonance spectroscopy (62). Although chemical shifts are a function of numerous factors, of which electron density is only one, both theoretical and empirical relationships of this nature have been explored for a variety of compound classes, and are reviewed by Ebraheem and Webb (63), Martin et al. (64), Nelson and Williams (65), and Farnum (66).

Regression analyses indicated that a model of C-13 NMR shift vs. electron density, for all compounds and carbons, resulted in a poor fit of the data with a significance level of 0.0879 and an R-squared of 0.0347. Inclusion of a factor defining the position of each carbon yielded a significant relationship, but still gave low correlation coefficient of 0.2508. Sorting the data by each carbon indicates significant linear relationships for carbons 4 and 6 within the aromatic ring, and the α and β carbons in the side chain. These results may provide a more general picture for the reactivity of lignin.

From this brief introduction, it is obvious that the execution of the calculations and the interpretation of the results are still in their initial stages. It is the contention of the author that the insight into lignin chemistry that may be obtained by these methods deserves careful consideration.

Literature Cited

1. Lowe, John P. *Quantum Chemistry*; Academic Press, Inc.: New York, 1978.
2. Burkert, Ulrich; Allinger, N. L. *Molecular Mechanics*; ACS Monograph 177; American Chemical Society: Washington, DC, 1982.
3. Dewar, M. J. S.; Dougherty, R. C. *The Theory of Organic Chemistry*; Plenum Press: New York, 1975.
4. Trinajstic, N. *Chemical Graph Theory*; CRC Press: Boca Raton, FL, 1983.
5. McQuarrie, Donald A. *Statistical Thermodynamics*; Harper and Row, Inc.: New York, 1973.
6. Clark, T. *A Handbook of Computational Chemistry. A Practical Guide to Chemical Structure and Energy Calculations*; Wiley-Interscience: New York, 1985.
7. Dewar, M. J. S. *The Molecular Orbital Theory of Organic Chemistry*; McGraw-Hill: New York, 1969.
8. Flurry, Robert L., Jr. *Molecular Orbital Theories in Bonding of Organic Molecules*; Marcel Dekker, Inc.: New York, 1968.

9. Murrell, J. N.; Harget, A. *Semi-Empirical, Self-Consistent-Field Molecular Orbital Theory of Molecules*; Wiley-Interscience: New York, 1972.
10. Pople, J. A.; Beveridge, D. L. 1970. *Approximate Molecular Orbital Theory*; McGraw-Hill: New York, 1977.
11. Binkley, J. S.; Whiteside, R. A.; Hariharan, P. C.; Seeger, R.; DeFrees, D. J.; Schlegel, H. B.; Frisch, M. J.; Pople, J. A.; Kahn, L. R. Gaussian 82 Release A. Carnegie-Mellon Univ., Pittsburgh, PA, 1982.
12. Pople, J. A.; Segal, G. A. *J. Chem. Phys.* 1966, **44**, 3289.
13. Bingham, R. C.; Dewar, M. J. S.; Lo, D. H. *J. Am. Chem. Soc.* 1975, **97**, 1285.
14. Dewar, M. J. S.; Thiel, W. *J. Am. Chem. Soc.* 1977, **99**, 4899.
15. Halgren, Thomas A.; Kleier, D. A.; Hall, John H., Jr.; Brown, Leo D.; Lipscomb, W. N. *J. Am. Chem. Soc.* 1978, **100**, 6595.
16. Dewar, M. J. S.; Ford, G. P. *J. Am. Chem. Soc.* 1979, **101**, 5558.
17. Volkenstein, M. V. *Molecular Biophysics*; Academic Press: New York, 1977.
18. Osawa, Eiji; Musso, H. *Angewante Chemie-International Edition* in English 1983, **22**, 1.
19. Allinger, N. L. *Adv. Phys. Org. Chem.* 1976, **13**, 1.
20. Lifson, S.; Warshel, A. *J. Chem. Phys.* 1968, **49**, 5116.
21. Engler, E. M.; Andose, J. M., Scheleyer, P. von R. *J. Am. Chem. Soc.* 1973, **95**, 8005.
22. Fitzwater, S.; Bartel, L. S. *J. Am. Chem. Soc.* 1976, **98**, 5107.
23. Allinger, N. L. *J. Am. Chem. Soc.* 1977, **99**, 8127.
24. Weiner, Paul K.; Kollman, Peter A. *J. Comput. Chem.* 1981, **2**, 287.
25. Tvaroska, Igor; Kosar, T. *J. Am. Chem. Soc.* 1980, **102**, 6929.
26. Tvaroska, I. *Biopolymers* 1982, **21**, 1887.
27. Hayns, M. R. *Int. J. Quantum Chem.* 1982, **21**, 217.
28. Calais, Jean-Louis. *Int. J. Quantum Chem.* 1982, **21**, 231.
29. Lowdin, Per-Olov. *Proceedings of the International Symposium on Quantum Biology and Quantum Pharmacology*; John Wiley & Sons: New York, 1987.
30. Lindberg, J. Johann; Henriksson, A. *Fin. Kemist. Medd.* 1970, **79**, 30.
31. Glasser, Wolfgang G.; Glasser, H. *Macromolecules* 1974, **7**, 17.
32. Glasser, Wolfgang G.; Glasser, H. *Holzforschung* 1974, **28**, 5.
33. Glasser, Wolfgang G.; Glasser, H. *Cell. Chem. Tech.* 1976, **10**, 23.
34. Glasser, Wolfgang G.; Glasser, H. *Cell. Chem. Tech.* 1976, **10**, 39.
35. Glasser, Wolfgang G.; Glasser, H.; Nimz, H. *Macromolecules* 1976, **9**, 866.
36. Glasser, Wolfgang G.; Glasser, H.; Morohoshi, N. *Macromolecules* 1981, **14**, 253.
37. Martensson, O.; Karlsson, G. *Arkiv Kemi* 1968, **31**, 5.
38. Glasser, Wolfgang G. In *Pulp and Paper*, Third Edition; Casey, James P., Ed.; John Wiley & Sons: New York, 1980.
39. Elder, T. J.; Worley, S. D. *Wood Sci. Tech.* 1984, **18**, 307.
40. Remko, Milan; Polcin, J. *Chem. Zvesti.* 1976, **30**, 170.
41. Remko, Milan; Polcin, J. *Z. Phys. Chem. (Leipzig)* 1977, **258**, 219.

42. Remko, Milan; Polcin, J. *Z. Phys. Chem. (Neue Folge)* 1980, **120**, 1.
43. Remko, Milan. *Adv. Mol. Relaxation and Int. Proc.* 1979, **14**, 315.
44. Remko, Milan; Polcin, J. *Z. Phys. Chem.(Neue Folge)* 1977, **106**, 249.
45. Remko, Milan; Polcin, J. *Z. Phys. Chem. (Neue Folge)* 1981, **125**, 175.
46. Remko, Milan; Polcin, J. *Z. Phys. Chem. (Neue Folge)* 1981, **126**, 195.
47. Remko, Milan. *Adv. Mol. Relaxation and Int. Proc.* 1979, **14**, 37.
48. Remko, Milan. *Z. Phys. Chem. (Neue Folge)* 1983, **134**, 129.
49. Remko, Milan; Polcin, J. *Chem. Zvesti* 1977, **31**, 171.
50. Remko, Milan; Polcin, J. *Monatsh. Chem.* 1977, **108**, 1313.
51. Remko, Milan; Polcin, J. *Z. Naturforsch* 1977, **32a**, 59.
52. Remko, Milan; Polcin, J. *Collect. Czech. Chem. Commun.* 1980, **45**, 201.
53. Remko, Milan; Sekerka, I. *Z. Phys. Chem. (Neue Folge)* 1983, **134**, 135.
54. Diner, S.; Malrieu, J. P.; Jordan, F.; Gilbert, M. *Theor. Chim. Acta.* 1969, **15**, 100.
55. Gravitis, J.; Erins, P. *J. Appl. Polym. Sci.: Appl. Polym. Symp.* 1983, **37**, 421.
56. Elder, T. J.; McKee, M. L.; Worley, S. D. *Holzforschung* 1988, **42**, 233.
57. Fleming, Ian. *Frontier Orbitals and Organic Chemical Reactions*; John Wiley & Sons: London, 1976.
58. Elder, T. J.; Worley, S. D. *Holzforschung* 1985, **39**, 173.
59. Zarubin, M. Ya.; Kirysun, M. F. *Fourth International Symposium on Wood and Pulping Chemistry*, 1987, p. 407.
60. Jakobsons, J., Gravitis, J. *Khim. Drev.* 1985, **1985**, 110 (Chemical Abstracts 103 : 87243).
61. Gravitis, J.; Kokorevics, A.; Ozols-Kalnins, V. *Khim. Drev.* 1986, **1986**, 107 (Chemical Abstracts 104 : 131736).
62. Kringstad, K. P.; Mörck, R. *Holzforschung* 1983, **37**, 237.
63. Ebraheem, K. A. K.; Webb, G. A. *Prog. NMR Spect.* 1977, **11**, 149.
64. Martin, G. J.; Martin, M. L.; Odiot, S. *Org. Mag. Res.* 1975, **7**, 2.
65. Nelson, G. L.; Williams, E. A. *Prog. Phys. Org. Chem.* 1976, **12**, 229.
66. Farnum, D. G. *Adv. Phys. Org. Chem.* 1975, **11**, 123.

RECEIVED May 29, 1989

WATER-SOLUBLE POLYMERS

Chapter 20

Differential Scanning Calorimetry and NMR Studies on the Water–Sodium Lignosulfonate System

Hyoe Hatakeyama[1], Shigeo Hirose[1], and Tatsuko Hatakeyama[2]

[1]Industrial Products Research Institute, 1–1–4 Higashi, Tsukuba, Ibaraki 305, Japan
[2]Research Institute for Polymers and Textiles, 1–1–4 Higashi, Tsukuba, Ibaraki 305, Japan

The phase transition and nuclear magnetic relaxation of the water-sodium lignosulfonate (NaLS) system with various water contents ranging from 0 to ca. 2.3 (grams of water per gram of sodium lignosulfonate) were evaluated by differential scanning calorimetry and nuclear magnetic resonance spectroscopy. It was found that at least two different kinds of water existed in the system: i.e., freezing water and non-freezing bound water. The longitudinal and transverse relaxation times of the water-NaLS system were measured as functions of water content and temperature. A minimum value for the ^{1}H longitudinal relaxation time (T_1) was observed at a temperature around −25°C. A sudden decrease in ^{1}H transverse relaxation time (T_2) was also observed at a similar temperature. The change of T_1 and T_2 of ^{23}Na in the system corresponded well with the motion of ^{1}H in water.

It is generally known that polyelectrolytes are highly soluble in water through hydration of ionic groups. However, attention has not been paid to the physico-chemical properties of highly concentrated aqueous solutions of polyelectrolytes. We have already reported that there are two kinds of water, i.e. freezing water and non-freezing water, around the ionic groups in the water-sodium poly(styrenesulfonate) (NaPSS), the water-sodium cellulose sulfate (NaCS), and the water-cation salts of carboxymethylcellulose (MeCMC) systems (1-3). Judging from the experimental results obtained by differential scanning calorimetry (DSC) and by nuclear magnetic resonance (NMR) spectroscopy, water molecules are strongly bonded to the ionic groups and form a hydration shell.

In this study, we chose the water-sodium lignosulfonate (NaLS) system, with various water contents ranging from 0 to ca. 2.3 grams of water per

0097–6156/89/0397–0274$06.00/0

gram of NaLS. The thermal properties and ^{1}H and ^{23}Na nuclear magnetic relaxation times of the system were investigated. The phase transition temperatures of the water-NaLS system were measured as a function of water content. At the same time, the bound water content was determined from the enthalpy of transition. Furthermore, a detailed study of the longitudinal and transverse relaxation times (T_1 and T_2) was carried out, and the molecular correlation times (τ_c) and activation energies (E_a) of water in the system were estimated.

Experimental

Sample Preparation. As the NaLS sample, WAFEX, which was supplied by Holmen AB, Sweden, was used after purification according to the method reported by Leopold (4). The amount of sulfonic acid groups was determined by titrating an aqueous solution of purified lignosulfonic acid with 1/5 N sodium hydroxide. The lignosulfonic acid was obtained from NaLS by passing it through an amberlite IR-120B (H^+) column. The sulfonate group content of the NaLS was 1.6 meq/g. The water content (W_c) of each sample was defined by

$$W_c(g/g) = W_w(g)/W_s(g) \tag{1}$$

where W_w is the weight of added water and W_s is the dry weight of each sample.

Differential Scanning Calorimetry (DSC). The phase transition of the water-polyelectrolyte systems was measured using a Perkin-Elmer differential scanning calorimeter, DSC-II, and a Du Pont model 910 differential scanning calorimeter. The scanning rate for heating and cooling experiments was 10°C/min. DSC curves were obtained in the temperature range from 50 to −120°C. Aluminum sample pans for volatile samples were pretreated by heating with water to 120°C in an autoclave in order to avoid any reaction between aluminum and water during heating and cooling runs. After the DSC measurement, the sample was weighed again to confirm that no weight loss had taken place. The temperature and enthalpy of crystallization of sorbed water were calibrated using pure water as the standard. The bound water content was calculated by the method reported previously (1).

Nuclear Magnetic Resonance (NMR) Spectroscopy. Longitudinal and transverse relaxation times (T_1 and T_2) of ^{1}H and ^{23}Na in the water-polyelectrolytes systems were measured using a Nicolet FT-NMR, model NT-200WB. T_2 was measured by the Meiboom-Gill variant of the Carr-Purcell method (5). However, in the case of very rapid relaxation, the free induction decay (FID) method was applied. The sample temperature was changed from 30 to −70°C with the assistance of the 1180 system. The accuracy of the temperature control was ±0.5°C.

Results and Discussion

DSC. Figure 1 shows DSC cooling curves of the water-NaLS system having various W_c's from 0.46 to 2.31 g/g. When a sample with water is cooled from 50°C at the rate of 10°C/min, a broad peak (P^*) is observed initially, followed by a sharp peak representing crystallization of water (P_I). However, as shown in Figure 1, only P^* is observed if W_c is lower than ca. 0.5 g/g.

The temperature for P_I increases with increasing W_c, while, on the other hand, the temperature for P^* decreases with increasing W_c. This feature of P^* is different from that of the broad peak representing freezing bound water, P_{II}, which is observed when water is bound to the hydroxyl groups of polymers such as lignin, cellulose and poly(hydroxystyrene) derivatives having no ionic groups (6-9). P_{II} appeared at a temperature lower than that of the crystallization peak of water (P_I). On the other hand, P^* appears at a temperature higher than that of P_I.

An exothermic peak similar to P^* was also observed in the case of highly concentrated aqueous solutions of NaCS and NaPSS (1,2). Therefore, it is supposed that P^* is observed when the molecules in the water-polyelectrolyte systems rearrange to a stable state having lower energy than the normal state of water. This suggests that the molecules in the water-NaLS system assume a loosely ordered state in the range between the temperatures where P^* and P_I appear.

The texture of the sample was observed using a polarized light microscope equipped with a temperature controller. Although the W_c of the sample could not be kept entirely constant owing to the structure of the sample cell, a texture showing a nematic type of molecular arrangement was observed at a temperature corresponding to that between P^* and P_I. However, the texture observed in the water-NaLS system was not as clear as those observed in the water-NaPSS and the water-NaCS systems. This unclear texture observed in the water-NaLS system may be attributed to the inhomogeneous chemical structure of NaLS.

Figure 2 shows the phase diagram of the water-NaLS system. This diagram was obtained by cooling the system at a rate of 10°C/min, indicating that the system changes from the isotropic liquid phase through the mesomorphic phase to the crystalline phase. The temperature of crystallization (T_c) increases with increasing W_c and levels off at W_c ca. 1.5 g/g. The temperature (T^*) corresponding to the appearance of P^* decreases with increasing W_c and becomes difficult to observe if W_c increases above ca. 2.3 g/g.

The amount of freezing water in the water-NaLS system can be calculated from the enthalpy, assuming the crystallization enthalpy of water to be 333 J/g (7). Thus,

$$W_f = (\Delta H_f/333)/W_s \qquad (2)$$

where W_f is the amount of freezing water and ΔH_f is the crystallization enthalpy of the system.

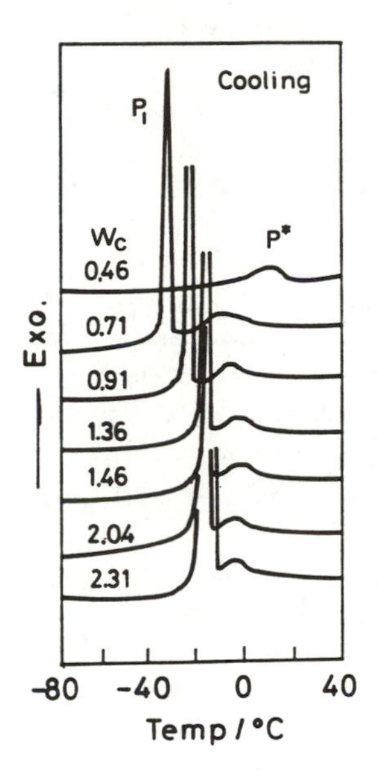

Figure 1. DSC cooling curves of the water-NaLS system with various water contents, $W_c(g/g)$.

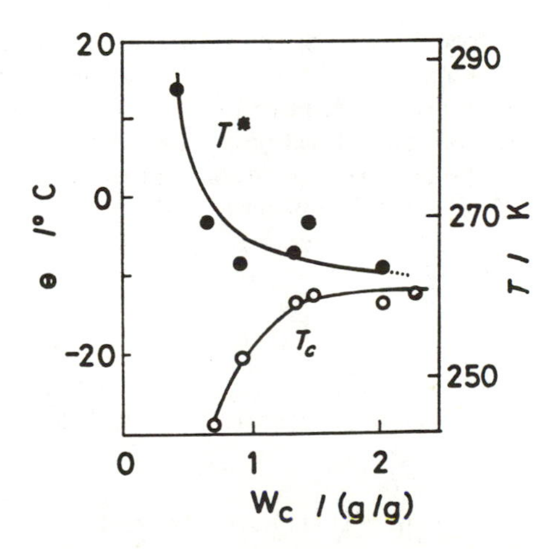

Figure 2. Phase diagram of the water-NaLS system.

However, the value of W_f is less than the total amount of water in the system, since water which is tightly bound to the hydrophilic groups of a polyelectrolyte cannot be frozen (1,2). Therefore, the total amount of water in the system is given by

$$W_c = W_f + W_{nf} \tag{3}$$

where W_f is the amount of freezing water and W_{nf} is that of non-freezing water.

Figure 3 shows the relationship between W_c, W_f and W_{nf}. The value of W_f increases in proportion to W_c at water contents higher than the specific critical amount which is obtained by the extrapolation of the linear W_f vs. W_c plots to the horizontal axis. The critical value obtained is 0.56 g/g. This amount corresponds to an amount less than the point at which the water in the water-NaLS system can no longer be crystallized. As shown in Figure 3, W_{nf} does not change appreciably with increasing W_c.

^1H NMR. Figure 4 shows the change of T_1 values for ^1H of water in the water-NaLS system with the inverse absolute temperature (K^{-1}). The T_1 value decreases with decreasing W_c and with decreasing temperature in the temperature range above −25°C, where a minimum value for T_1 is observed. The T_1 minimum occurs at a temperature lower than that at which water molecules in the system become rigid. In the temperature range below −25°C, a steep increase in the T_1 value is observed with decreasing temperature.

Figure 5 shows the change of T_2 values for ^1H of water with the inverse absolute temperature. A sudden decrease in T_2 values is seen at almost the same temperatures at which the minimum in T_1 is observed. The T_2 values give an average representation of the motion of water molecules in the system. Therefore, if we consider the molecular motion of water in the water-NaLS system having a certain W_c, the T_2 values at higher temperatures are characteristic of more mobile water, while the T_2 values at lower temperatures are characteristic of more restricted water.

It is usually not possible to distinguish various perturbed sites in an NMR experiment. Therefore, it is customary to limit the number of sites to two representing the free (P_f) and bound (P_b) fractions with $P_f + P_b = 1$. If we assume $R_i = 1/T_i$, ($i = 1, 2$), then

$$R_i = P_f R_{if} + P_b R_{ib} \tag{4}$$

In the case of free water, the extreme narrowing condition is fulfilled and thus $R_{1f} = R_{2f} = R_f$. Thus, from Equation 4, we can write

$$(R_1 - P_f R_f)/(R_2 - P_f R_f) = R_{1b}/R_{2b} \tag{5}$$

According to Woessner *et al.* (10,11), the relaxation rates of the bound water can be expressed as

$$\frac{1}{T_{1b}} = G \sum_i k_i \left(\frac{\tau_{ci}}{1 + \omega_0^2 \tau_{ci}^2} + \frac{4\tau_{ci}}{1 + 4\omega_0^2 \tau_{ci}^2} \right) \tag{6}$$

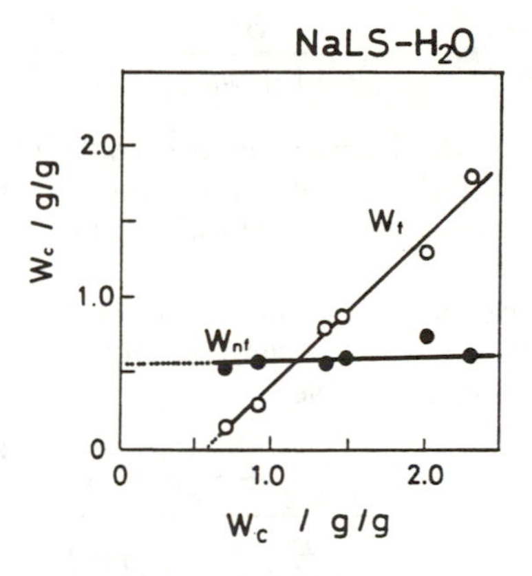

Figure 3. Relationship between water content (W_c), freezing water content (W_f) and non-freezing water content (W_{nf}).

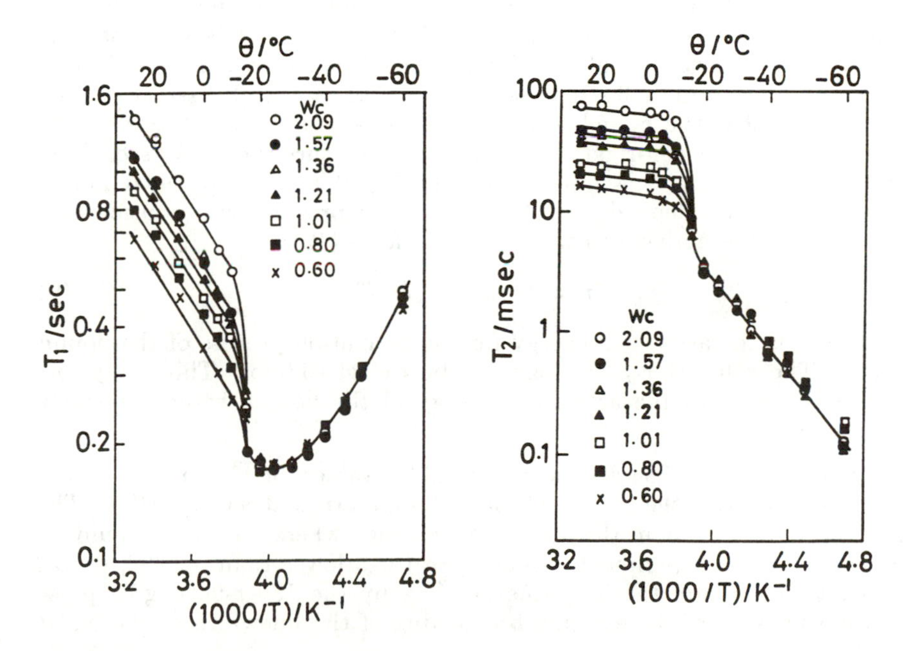

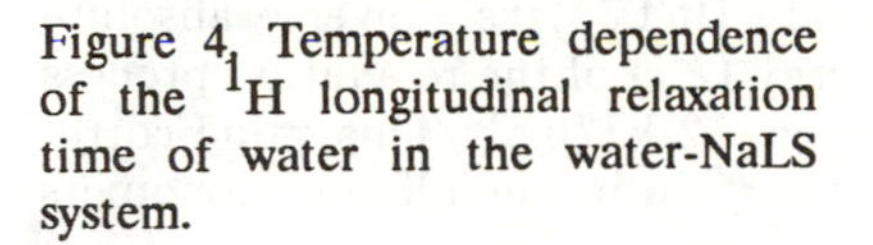

Figure 4. Temperature dependence of the ^{1}H longitudinal relaxation time of water in the water-NaLS system.

Figure 5. Temperature dependence of the ^{1}H transverse relaxation time of water in the water-NaLS system.

$$\frac{1}{T_{2b}} = \frac{1}{2} G \sum_i k_i \left(3\tau_{ci} + \frac{5\tau_{ci}}{1+\omega_0^2\tau_{ci}^2} + \frac{2\tau_{ci}}{1+4\omega_0^2\tau_{ci}^2} \right) \tag{7}$$

where $\sum_i k_i = 1$, and G is the interaction constant determining the magnitude of the relaxation, ω_0 the angular resonance frequency, and τ_{ci} the correlation time. The above expressions allow for multiple correlation times in the systems.

However, if we assume that the relaxation of bound water in the water-NaLS system is determined by one average correlation time τ_c, we can combine Equations 5, 6, and 7, to give

$$\frac{T_{2b}}{T_{1b}} = \frac{2\left(\frac{1}{1+\omega_0^2\tau_c^2} + \frac{4}{1+4\omega_0^2\tau_c^2}\right)}{\left(3 + \frac{5}{1+\omega_0^2\tau_c^2} + \frac{2}{1+4\omega_0^2\tau_c^2}\right)} \tag{8}$$

which is a function only of $\omega_0\tau_c$. It might be an oversimplification to only consider one type of average bound water. However, for practical purposes, it seems reasonable to use this calculation method as a starting point for the evaluation of bound water in the water-NaLS system.

Figure 6 shows the calculated τ_c values plotted vs. inverse absolute temperature. In the temperature range below −15°C, it is seen that the τ_c values are not dependent on W_c but dependent on temperature. The τ_c value increases from ca. 3×10^{-8} sec at −15°C to ca. 3×10^{-7} sec at −60°C. This shows that the bound water in the system is in the state between viscous liquid and non-rigid solid in this temperature range. As seen from the figure, the ln τ_c vs. temperature^{-1} (K^{-1}) plots are apparently linear. The temperature dependence of τ_c may be expressed with considerable accuracy by the Arrhenius equation in the form (12)

$$\tau_c = \tau_0 \exp\,(E_a/RT) \tag{9}$$

where E_a is the activation energy for the relaxation process of the bound water. The value of E_a was found to be ca. 24 kJ/mol. This value corresponds well with the activation energy of the bound water previously reported (2).

^{23}*Na NMR.* Figure 7 shows the change of T_1 values for ^{23}Na in the water-NaLS system with the inverse absolute temperature at various W_c's. The lnT_1 plots are linear in the temperature range where T_1 values could be observed in this experiment. At temperatures lower than −20°C, it was difficult to measure the T_1 value of ^{23}Na by the 180-τ-90 degree pulse method because of the extreme broadening of the linewidth of the NMR peaks.

From the slopes of the relaxation rate (lnT_1^{-1}) vs. inverse absolute temperature, the apparent activation energy (E_a) of the relaxation process was calculated. The value obtained was ca. 12 kJ/mol. This value corresponds well with the activation energy for ^{23}Na in persulfonate ionomers with water (13).

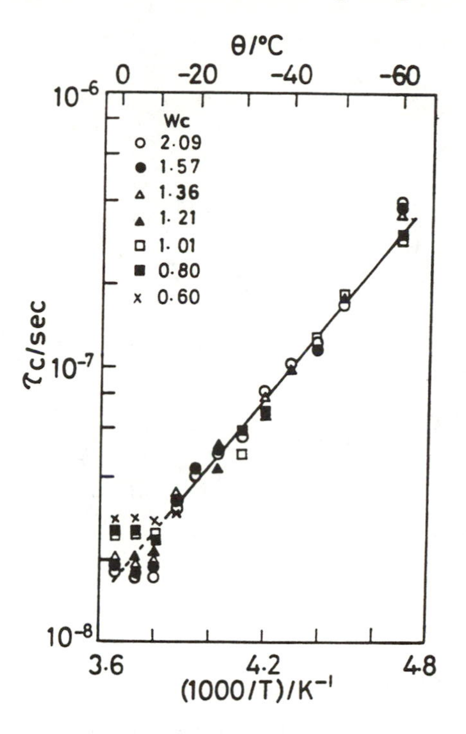

Figure 6. Temperature dependence of the average correlation time of water in the water-NaLS system.

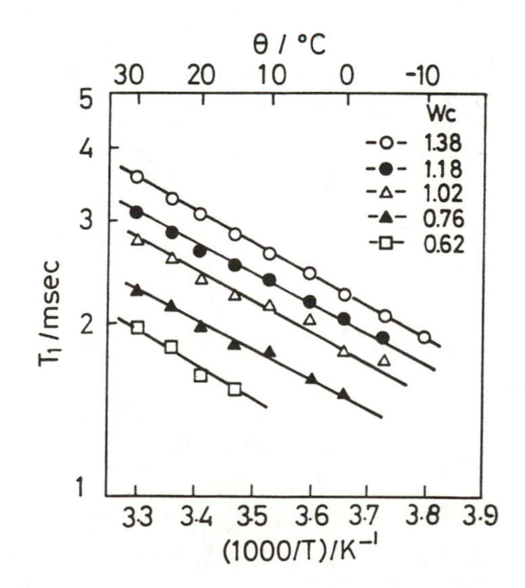

Figure 7. Temperature dependence of the ^{23}Na longitudinal relaxation time in the water-NaLS system.

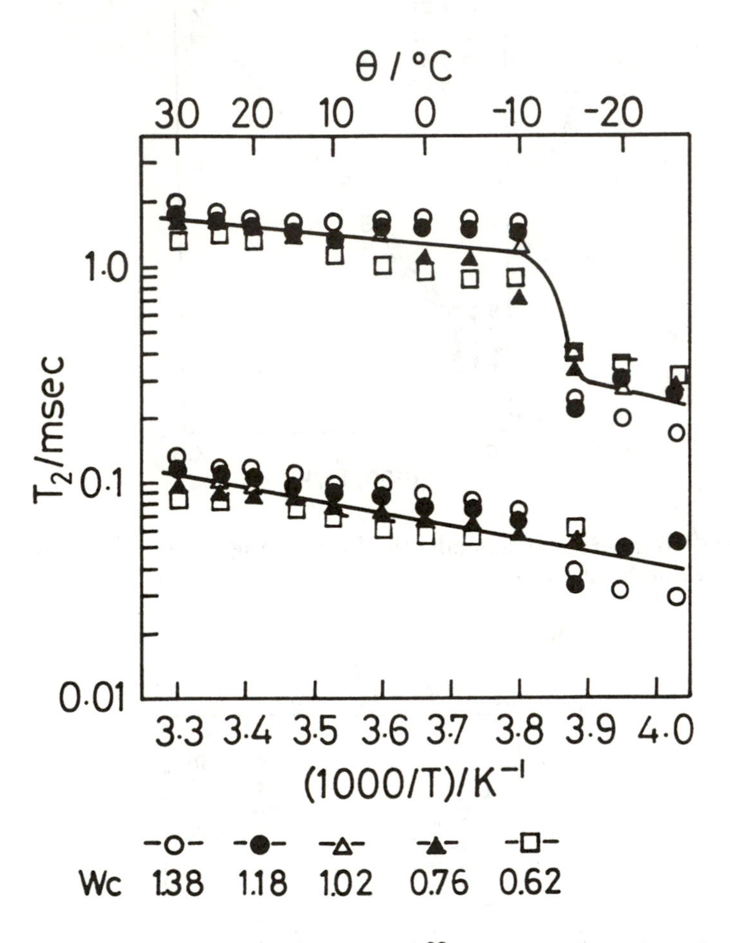

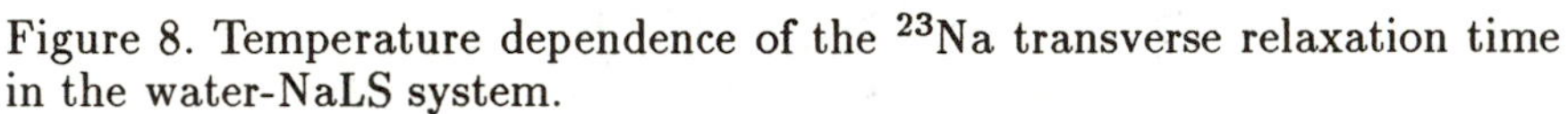

Figure 8. Temperature dependence of the ^{23}Na transverse relaxation time in the water-NaLS system.

Figure 8 shows the change of T_2 values of ^{23}Na with inverse absolute temperature. As seen from Figure 8, two types of transverse relaxations are observed. One is "slow" and the other a "fast" relaxation. This means that the transverse relaxation decays in a non-exponential manner. Non-exponential relaxation due to quadrupole relaxation was characterized by Hubbard (14). According to his calculation, the transverse relaxations produced by a quadrupole interaction are the sum of two or more decaying exponentials. As shown in Figure 8, the longer transverse relaxation time (T_{2s}) decreases with decreasing temperature, although the shorter transverse relaxation time (T_{2f}) does not change much with temperature. The sudden decrease of the T_{2s} value at around −15°C may reflect the influence of the crystallization of water in the water-NaLS system. This result is quite reasonably explained if we assume that the sodium ion is surrounded by the non-freezing water and that this non-freezing water is surrounded by the free water. Therefore, the motion of the free water in the system indirectly affects that of the sodium ion.

Literature Cited

1. Hatakeyama, T.; Nakamura, K.; Yoshida, H.; Hatakeyama, H. *Thermochmica Acta* 1985, **88**, 223.
2. Hatakeyama, H.; Iwata, H.; Hatakeyama, T. In *Wood and Cellulosics*; Kennedy, J. F., et al., Eds.; Ellis Horwood: Chichester, 1987; Ch. 4.
3. Nakamura, K.; Hatakeyama, T.; Hatakeyama, H. In *Wood and Cellulosics*; Kennedy, J. F., *et al.*, Eds.; Ellis Horwood: Chichester, 1987; Ch. 10.
4. Leopold, B. *Acta Chem. Scand.* 1952, **6**, 64.
5. Meiboom, S.; Gill, D. *Rev. Sci. Inst.* 1958, **29**, 688.
6. Hatakeyama, T.; Hirose, S.; Hatakeyama, H. *Makromol. Chem.* 1983, **184**, 1265.
7. Nakamura, K.; Hatakeyama, T.; Hatakeyama, H. *Textile Res. J.* 1981, **51**, 607.
8. Hatakeyama, T.; Ikeda, Y.; Hatakeyama, H. *Makromol. Chem.* 1987, **188**, 1875.
9. Nakamura, K.; Hatakeyama, T.; Hatakeyama, H. *Polymer* 1983, **24**, 871.
10. Woessner, D. E.; Zimmerman, J. R. *J. Chem. Phys.* 1963, **67**, 1590.
11. Woessner, D. E.; Snowden, B. S. *J. Colloid and Interface Sci.* 1970, **34**, 290.
12. Farrar, T. C.; Becker, E. D. *Pulse and Fourier Transform NMR*; Academic: New York, 1971; p. 57.
13. Komoroski, R. A. In *Ions in Polymers*; Eisenberg, A., Ed.; American Chemical Society: Washington, DC, 1980; Ch. 10.
14. Hubbard, P. S. *J. Chem. Phys.* 1970, **53**, 985.

RECEIVED March 17, 1989

Chapter 21

Preparation and Testing of Cationic Flocculants from Kraft Lignin

Erkki Pulkkinen, Airi Mäkelä, and Hannu Mikkonen

Department of Chemistry, University of Oulu, SF–90570 Oulu, Finland

Kraft lignin has been reacted to a quaternary ether derivative over several hours at temperatures ranging from room temperature to 65°C. Based on the nitrogen content (2.7-3.7%) of the purified product, 42-61% of the hydroxyl groups of lignin were etherified. The yield with respect to reagent was usually over 50%. In laboratory tests the lignin-based cationic ether effectively precipitated inorganic colloids from wastewaters.

It is well known that the acetylation, for example, of aromatic and aliphatic hydroxyl groups in kraft lignin can drastically change the solubility of lignin. The same occurs when hydroxyl groups of kraft lignin are reacted with epoxy reagents having quaternary ammonium groups in the molecule.

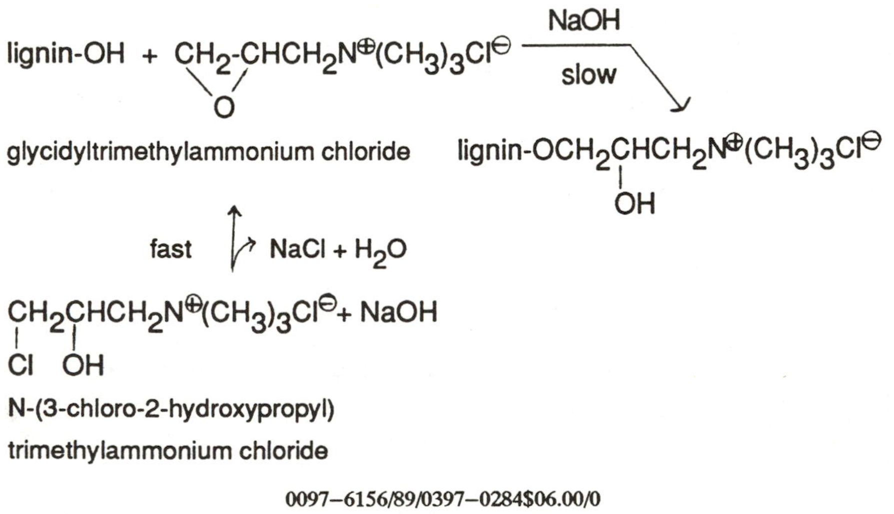

0097–6156/89/0397–0284$06.00/0

Even though 3.0-4.4 mmols (from a total of 7.2 mmols) of hydroxyl groups per gram of lignin were cationized by these reagents, the products were completely water soluble and effective flocculants. In this paper we discuss the preparation and testing these cationic lignin derivatives.

Experimental

Materials. Commercial kraft lignin, Indulin AT (Westvaco Corp.), has a combined phenolic and aliphatic hydroxyl group content of 7.2 mmol/g as determined by quantitative ^{29}Si NMR spectroscopy (1). The pine kraft lignin from the Nuottasaari mill of Veitsiluoto Corp., Finland, has a corresponding hydroxyl content of 7.7 mmol/g. The glycidyltrimethylammonium chloride reagent, containing 63.3% epoxy functionality, was a product of Raisio Corp., Finland. N-(3-chloro-2-hydroxypropyl)trimethylammonium chloride was prepared in the laboratory (2,3). Praestol 411K (Stockhausen Corp.) was used as a reference cationic flocculant.

Flocculation Tests (4). A conventional jar test procedure with a six-place multiple stirrer system (Phipps & Bird, Inc.), and crystalline silica particles (Min-U-Sil 5, Pennsylvania Glass and Sand Corp.), or alternatively material dredged from the bottom of a waterway, were used as a colloidal reference. After stirring the main batch of the silica (600 or 1800 mg Min-U-Sil 5 in 4000 mL of deionized water) for an hour, 2 mL aqueous HCl was added and the pH adjusted with solid $NaHCO_3$ to 4. The dredging effluent (3700 mg/liter, pH 5.1) was used as such as a settling reference. To perform the jar test, 600 mL of the batch dispersion was introduced into each 800 mL decanter flask and the dispersions were stirred at 100 rpm when the flocculant dosages were added. Thereafter stirring was continued at 100 rpm for 20 min., at 30 rpm for 20 min., and then allowed to settle for 30 min. undisturbed. Residual turbidities were measured by a Hach turbidimeter on 25 mL aliquots of the supernatant liquids taken 2 cm below the surface and plotted as formazin turbidity percent units (% FTU) against flocculation dosage in ppm solids.

Stability of Floc. After settling for 30 minutes and measuring the residual turbidity, a dispersion (600 mL) was poured into another flask and back to the original flask. The pouring back and forth was then repeated twice. After settling for 30 min., the residual turbidities were measured as before.

Gelling of Lignin with Formaldehyde. A 5.12 g sample of a sodium salt of lignin (40% nonvolatiles (N.V.), 15% NaOH:lignin) was reacted with 37% CH_2O (0.71 mole/100 g lignin) at 90°C in a test tube (height 10 cm, width 2 cm) until gelling occurred after 130 min. The gel time was measured with a Tecan Gelation Timer, GT3 (disc 14 mm).

Reaction of Lignin and Gelled Lignin with Glycidyltrimethylammonium Chloride. Glycidyltrimethylammonium chloride was added to a pre-prepared sodium salt of lignin (pre-stirred in a steam bath for 2 h) and the reaction mixture was stirred at 60-70°C for 3 h (Table I). Cationizing the gelled lignin (reacted with CH_2O for 130 min.) was essentially carried out in the same way except that the solids content of the reaction mixture was 29% instead of 44%. In the reaction of lignin with N-(3-chloro-2-hydroxypropyl)trimethylammonium chloride the molar ratios of reagent to NaOH was 1:1 and hydroxyl groups to reagent 1:1. The reaction was carried out at room temperature at 52% N.V.

Table I. Reaction of Kraft Lignin with Glycidyltrimethylammonium Chloride[a]

	Weight					
Components	grams	grams	mmoles	Molar Ratio		
Indulin AT	2.09	2.0				
Total OH			17.2			1
hydroxyl			14.4	1		
carboxyl			2.8			
Reagent (63.2%)	2.64	1.67	11.0	0.76	1	
NaOH	0.3	0.3	7.5		0.68	0.43
H_2O	4.0					
	9.03	3.97	Nonvolatiles		(N.V.):	44%

[a] After reacting at 60-70°C for 3 h, the isolated and dried product weighed 4.5 g and, after purification with both ethanol and ultrafiltration, provided 2.34 g product containing 2.7 and 2.9% N, respectively. The lignin recoveries in purification were 80.3 and 82.8%.

Purification of Cationized Lignin. The purification was carried out either by ultrafiltration of the neutralized product mixture through a Diaflo UM2 membrane (Amicon Corp., exclusion limit 1000 MW) or by slurrying a finely ground 1 g sample of dry crude cationic ether at room temperature in 160 mL 94% ethanol. The filtered powder was washed with 50 mL 94% ethanol and then with 10 mL diethyl ether and brought to a constant weight in a vacuum desiccator. The ultrafiltration was preferred when the nitrogen content of the purified cationic ether exceeded 3%.

Results

As is presented in Figure 1, the flocculation test with silica as a colloidal reference can be used to optimize the reaction time. By letting the reaction with lignin and N-(3-chloro-2-hydroxypropyl)trimethylammonium chloride advance at room temperature for 100 h instead of 50 h, the dosage requirement of the reaction mixture was reduced to about half. The residual turbidity already was sufficiently reduced after reacting for 50 h. Figure 2 indicates that the flocculation performance of the cationic lignin ether purified by ultrafiltration (2.9% N) or by washing with ethanol (2.7% N) was equal.

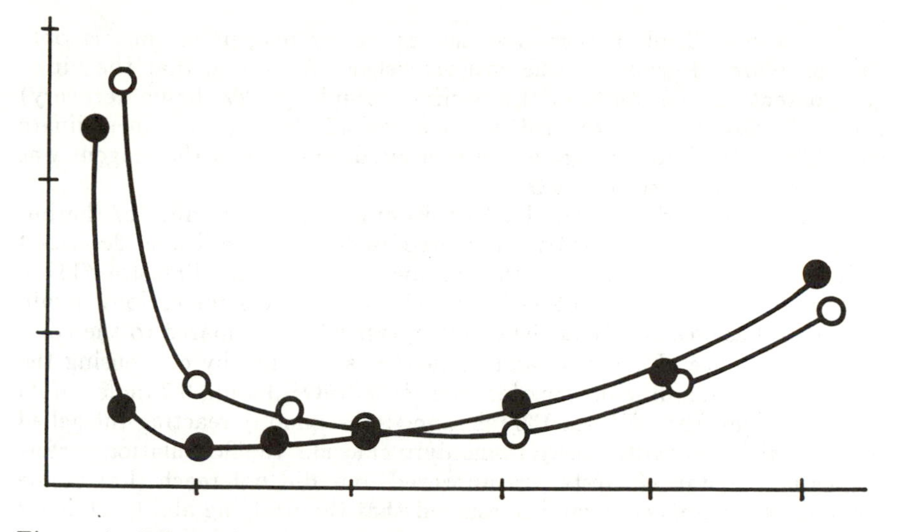

Figure 1. Effect of reaction time on flocculation performance: (●) 100 h; and (○) 50 h at room temperature.

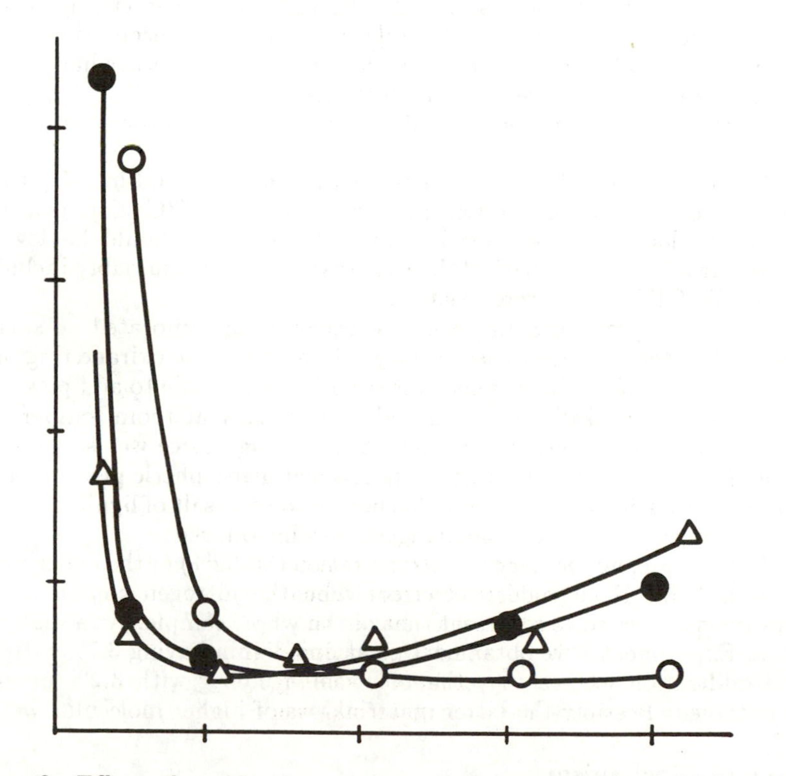

Figure 2. Effect of purification on flocculation performance: (○) unpurified; (●) purified by leaching with ethanol; (△) purified by ultrafiltration.

As seen in Table I, there was enough epoxy reagent to convert only 76% of hydroxyl groups to the cationic ether. Assuming that the nitrogen content (2.9%) found in the purified sample (80.2% lignin recovery) is equally distributed through the whole cationic lignin, one can estimate that 42% of the hydroxyl groups were reacted and 55% of the reagent was consumed in the derivatization.

Figures 3a and 3b show that the flocculation performance of the purified cationic lignin derivative, prepared under the conditions described in Table I, is equal to that of the commercial reference (Praestol 411K). Figures 4a and 4b, on the other hand, reveal that the same cationic lignin derivative behaves poorly in the stability test when compared to the commercial reference. Figures 4a and 4b further show that by condensing the sodium salt of lignin with formaldehyde (15% NaOH:lignin, 0.7 mole CH_2O per 100 g lignin) at 90°C for 130 min., and subsequently reacting the gelled product with glycidyltrimethylammonium chloride, the flocculation performance in the stability tests was improved, but did not reach that of the commercial reference. It can be assumed that the bridging ability of linear cationic polyacrylamides contributes positively in the stability test.

An improvement in the flocculation ability of the purified cationic lignin ether due to an increase in the molecular weight is clearly seen in Figure 5. The same figure reveals a moderate gain in the flocculation performance relative to Praestol 411K when the cationic lignin ether had 3.4% N instead of 2.9% N as in the jar test shown in Figure 3a.

A series of reactions was carried out in order to optimize the reaction conditions.

As seen in Table II, at best 3.7% of nitrogen was found in the purified cationic lignin, when the reaction time was 5 or 8 h at 55°C (Experiments 1 and 3). Perhaps the reaction time and temperature should be lowered further. At 3.7% N some 62% of the hydroxyl groups, presumably including all phenolic OH groups, were reacted.

Since the procedure in these experiments incorporated a smaller amount of water to avoid a base-catalyzed opening of the oxirane ring, high solids contents (52-64%) were encountered. It is advisable to add powdered lignin first to the alkaline solution and mix initially at room temperature and then at the reaction temperature. As a mixing device we used a round bottom flask attached to a rotary evaporator at atmospheric pressure aided with magnetic stirring in the flask. When the sodium salt of lignin appeared to be homogeneous, the cationic reagent was introduced.

The incomplete recovery by ultrafiltration (56-92%) in the experiments shown in Table II engenders an error when the nitrogen content of the retentate is projected to represent that of the whole sample. As a matter of fact, in Experiment 1 we obtained 1 g cationic lignin having 3.7% nitrogen by ultrafiltration whereas by ethanol washing 0.65 g with 3.2% nitrogen was obtained. Possibly the latter material was of higher molecular weight.

Reaction Mechanism

Under the experimental conditions there will be competition between a

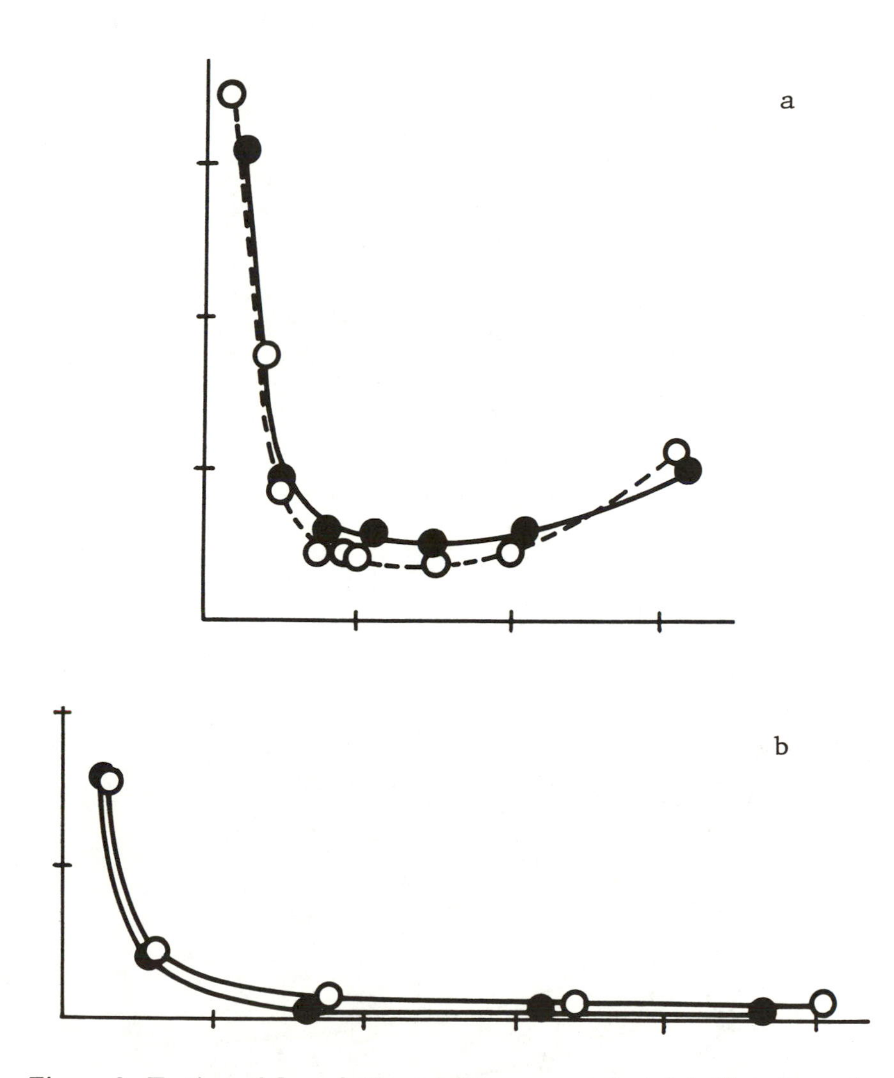

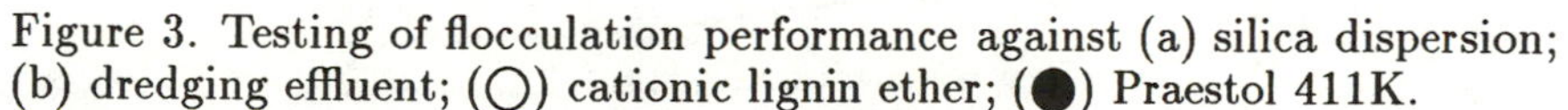

Figure 3. Testing of flocculation performance against (a) silica dispersion; (b) dredging effluent; (○) cationic lignin ether; (●) Praestol 411K.

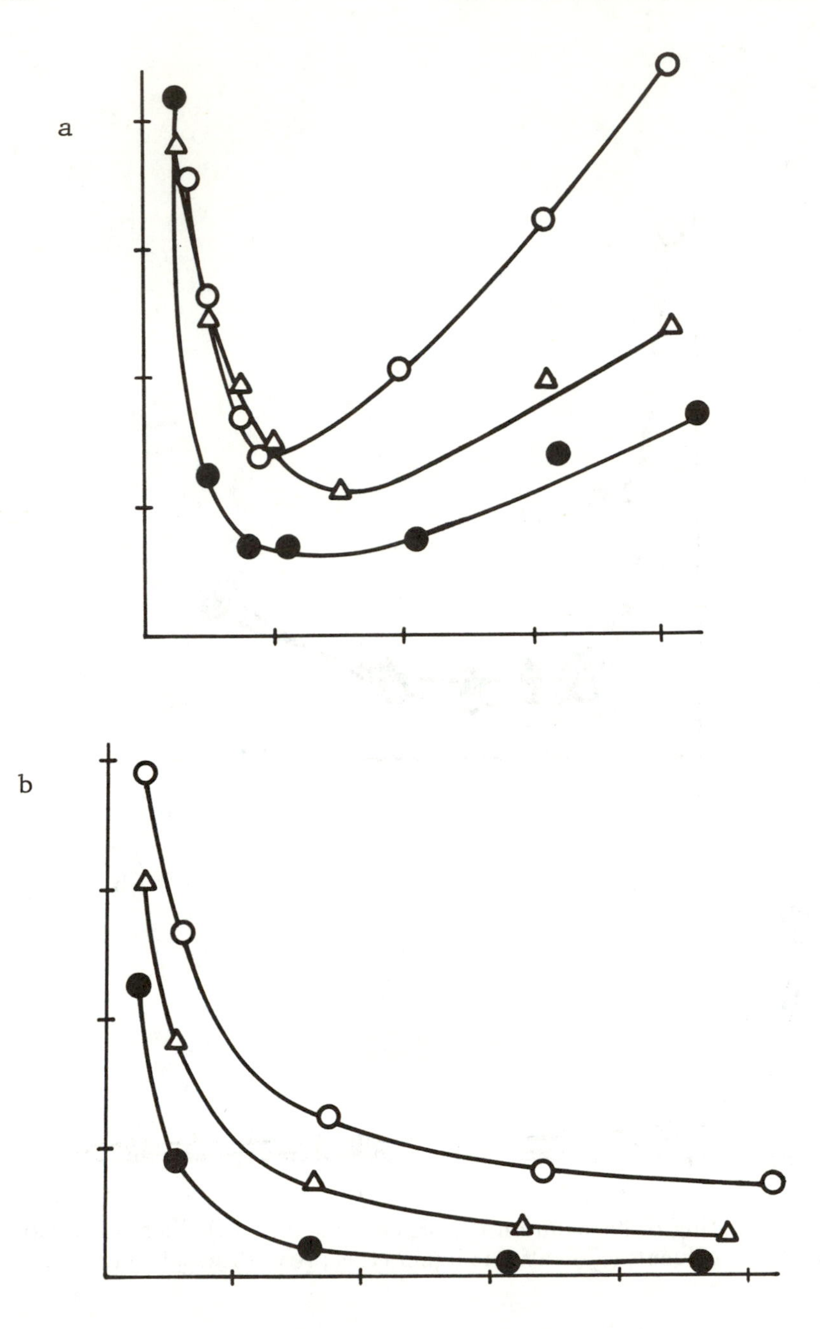

Figure 4. Comparison of flocculation performance in stability test (a) against silica dispersion; (b) against dredging effluent; (○) cationic lignin ether; (△) CH_2O-condensed cationic lignin ether; (●) Praestol 411K.

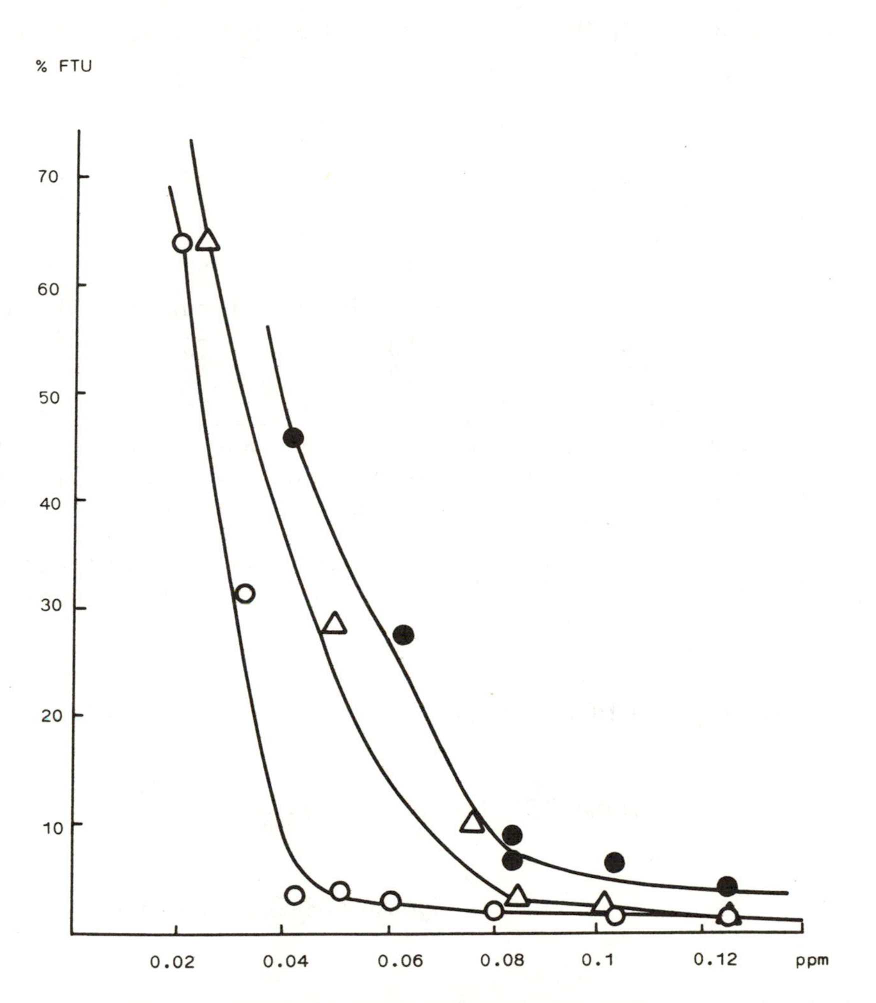

Figure 5. Effect of molecular weight on flocculation performance of cationic lignin derivative: (○) Indulin AT, $\overline{M}_w$ 23000, N: 3.3%; (△) Indulin AT, $\overline{M}_w$ 6300, N: 3.7%; (●) Praestol 411K, Min-U-Sil 5 450 mg/liter.

Table II. Conditions for Cationizing Lignin [a]

Exp. No.	Molar Ratios OH:Reagent:NaOH	Reaction Temp. °C	Reaction Time h	Cationic Lignin % N	Cationic Lignin % Recovery	Reagent % Yield
Reaction with Glycidyltrimethylammonium Chloride						
1	1:0.9:0.2	52	5	3.7	56	74
2	1:0.9:0.2	52	8	3.6	65	65
3	1:0.9:0.2	52	8	3.7	63	69
4	1:1.0:0.2	65	5	3.3	68	51
5	1:1.0:0.2	65	8	3.4	61	58
6	1.1.0:0.2	65	20	2.7	92	42
Reaction with N-(3-chloro-2-hydroxypropyl)trimethylammonium chloride						
7	1:0.9:0.9	r.t.[b]	50	2.9	70	50
8	1:1.0:1.0	65	10	3.2	65	51
9	1:0.6:0.6	65	12	2.1	74	44

[a] Experiments 1-6 were carried out in a 1 g sample of Indulin AT at 56% N.V. In experiment 7 a 2 g sample of Indulin AT was reacted at 52% N.V. Nuottasaari lignin (2 and 5 g) was reacted in experiments 8 and 9 at 64 and 58% N.V., respectively. All the samples were purified by ultrafiltration and the lignin % recovery was determined from the % nitrogen and the yield of the retentate. Reagent yields are corrected for ultrafiltration losses.

[b] Room temperature.

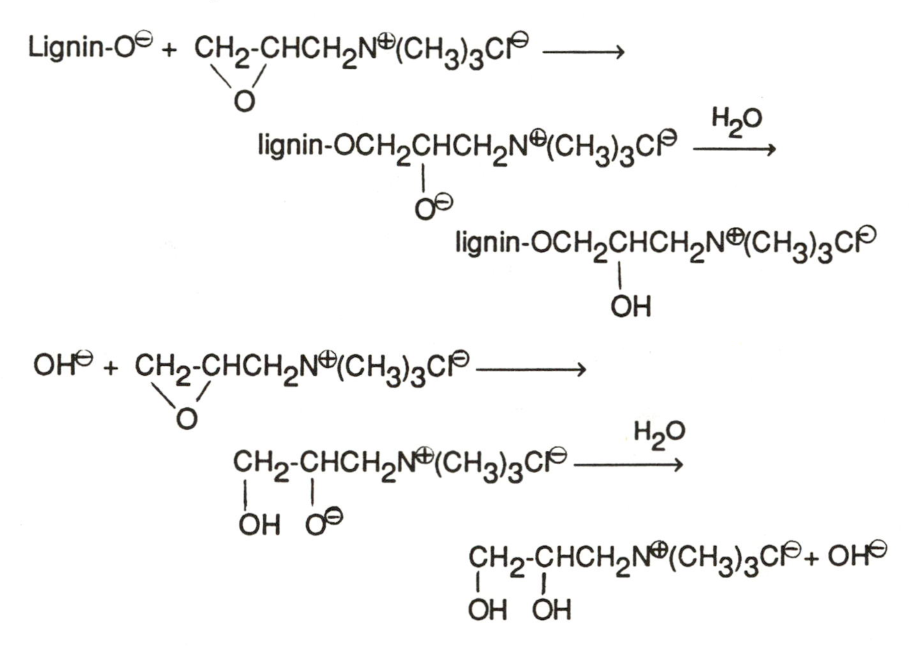

number of nucleophiles for opening the oxirane ring. Since the molar amount of sodium hydroxide, which is first introduced into the reaction mixture, nearly reaches the corresponding contents of phenolic and carboxylic acid groups (Table I), lignin phenoxide and carboxylate moieties must be the dominant nucleophilic species at the beginning. The main reactions can be expressed by the following equations.

Whether or not the reagent will react with the secondary hydroxyl group generated in each ring opening is at present unknown.

The chlorohydrin reaction with sodium hydroxide is assumed to be a two-step process in which a rate-determining intramolecular displacement of chloride ion by negatively charged oxygen follows a prior equilibrium between the hydroxide ion and the hydroxyl group of the halohydrin (5).

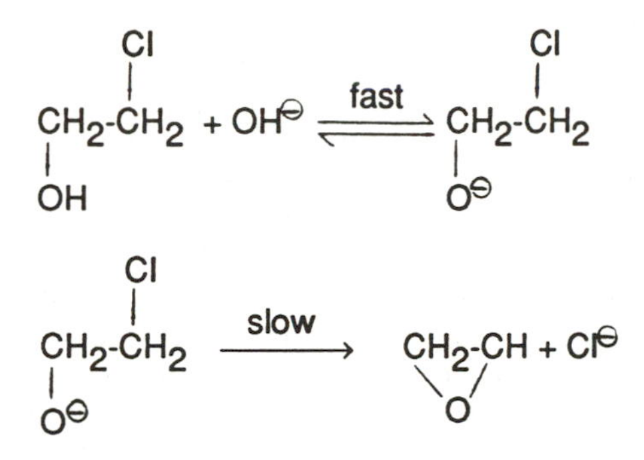

Indeed, we have found that N-(3-chloro-2-hydroxypropyl)trimethylammonium chloride is readily converted with sodium hydroxide to glycidyltrimethylammonium chloride, which obviously further reacts at a slower rate with lignin nucleophiles to form cationic lignin derivatives.

Conclusion

Even a partial conversion of the hydroxyl groups of kraft lignin into a cationic ether derivative containing quaternary ammonium groups imparts water solubility to the derivative which is a promising candidate as a potential flocculant in liquid/solid separations in industrial and municipal water treatment.

Literature Cited

1. Nieminen, M.; Pulkkinen, E.; Rahkamaa, E. *Holzforschung*, in press.
2. Paschall, E. F. U.S. Patent 2 876 217, 1959.
3. Paschall, E. F.; Minkema, W. H. U.S. Patent 2 995 513, 1961.
4. Hudson, R. E., Jr.; Wagner, E. G. *J. Am. Water Works Assoc.* 1981, **40**, 218-23.
5. Frost, A. A.; Pearson, R. G. *Kinetics and Mechanism*; John Wiley & Sons: New York, 1962; Ch. 12, p. 288.

RECEIVED March 17, 1989

Chapter 22

Synthesis and Characterization of Water-Soluble Nonionic and Anionic Lignin Graft Copolymers

John J. Meister, Damodar R. Patil, Cesar Augustin, and James Z. Lai

Department of Chemistry, University of Detroit, Detroit, MI 48221–9987

A general method of grafting lignin has been developed which allows solvent extracted lignin, steam exploded lignin, and kraft lignin to be converted to complex polymers. The lignins grafted have been obtained from aspen, poplar, and pine. The lignins are research samples, pilot plant products and commercial products from paper production. The types of materials made to date will be illustrated with a series of polymers made as industrial process chemicals. Nonionic copolymers of virtually any composition and molecular weight can be made from lignin and 2-propenamide. A graft terpolymer of lignin has been made by free radical reaction of 2-propenamide and 2, 2-dimethyl-3-amino-4-oxohex-5-ene-1-sulfonic acid in the presence of kraft pine lignin. The water soluble product is a thickening agent and has limiting viscosity number in water at 30°C which increases as the fraction of sulfonated repeat units in the molecule increases. The grafting reaction is rapid and yields of 80 weight % or more can be obtained in as little as 30 minutes from reactions run in 1, 4-dioxacyclohexane or dimethylsulfoxide. The reaction is initiated by a hydroperoxide, chloride ion, and lignin. Hydroxide radicals produced with iron (2+) do not appear to produce grafting. Adding 50 mole % sulfonated monomer to the reaction mixture produces graft copolymers with 12 to 24 times larger limiting viscosity numbers when compared to nonionic poly(lignin-g-1-amidoethylene). Adding 20 mole % sulfonated monomer to the reaction mixture increases product limiting viscosity number by a factor of 2 to 5.

0097–6156/89/0397–0294$06.00/0

Every year the U.S paper industry produces over 33 million metric tons of kraft lignin (1). Most of this biomass is burned as fuel but small amounts are used as binders, asphalt additives, or cement additives. Larger fractions of this waste would be used in other industrial or commercial processes if an economical way existed to convert lignin into a marketable product with sufficient profit margin to compensate for the loss of the fuel.

A way to make such a conversion has now been produced (2,3). Moreover, we have developed a chemistry for lignin that is apparently general. It is general in the lignins used in that a whole series of lignins withdrawn from wood by different techniques have been grafted by this method, as shown by the data of Table I. Since we recognize the potential for lignin utilization illustrated by the breadth of the chemistry we have developed, we have undertaken a broad and detailed study of thermoplastic and thermoset copolymers of lignin. These derivatives of lignin are being prepared and examined for their potential as process chemicals or commercial and engineering materials.

All of the samples received were used "as is" and were laboratory, pilot plant, or commercially-produced lignins.

The reaction converts lignin to a water-soluble copolymer or plastic by graft polymerization. The graft copolymer is formed by conducting a free-radical polymerization of an appropriate monomer on any of the lignins described in Table I. This report will describe preparation and testing of water-soluble, graft copolymers made with 2-propenamide and 2,2-dimethyl-3-amino-4-oxohex-5-ene-1-sulfonic acid in nitrogen-saturated, organic or aqueous/organic solvent containing lignin, calcium chloride, and a hydroperoxide. While the copolymerization can be run by a number of common methods, we have used solution polymerization to prepare laboratory or pilot plant scale samples of copolymer. We have shown grafting can be done using any of the liquids in Table II, which are now known to be effective in solution polymerization of graft copolymers. This polymerization process gives us easy heat control and rapid production of products for testing.

This reaction produces graft copolymers that possess side chains containing repeated polar units or multiple, ionic bonds which dissociate in polar solvents. The product can be an anionic or nonionic water-soluble copolymer with a limiting viscosity number in the range of 0.2 to 11 dL/g. The products increase the viscosity of aqueous solution, act as flocculating/deflocculating agents, thinning agents, dispersing agents, and sequester calcium ions. In the following sections, the synthetic procedure, purification procedures, characterization results, proof of grafting, tests of the role of iron in the initiation of grafting, and determination of extent of reaction as a function of time will be described.

Experimental

Synthesis. The polymerization can be run in any one of several solvents, listed in Table II. Dimethylsulfoxide has been used as the solvent for all reactions reported here. In other solvents, the product often precipitates as

Table I. Lignins Grafted with Hydroperoxy/chloride Chemistry

Source		
Pine[a]	Aspen[b]	Yellow Poplar[c]
Extraction Method		
Kraft	Solvent Extracted	Steam Exploded

[a]Pine lignins from the Westvaco Corporation of Charleston, SC.
[b]Aspen lignin from the Solar Energy Research Institute, of Golden, CO.
[c]Yellow poplar lignins from BioRegional Energy Associates, of Floyd, VA.

Table II. Liquids Useful in Solution Polymerization of Graft Copolymers

Dimethyl Sulfoxide[a] (DMSO)	1-Methyl-2-pyrrolidinone
1,4-Dioxacyclohexane[a]	Dimethylacetamide
Water[a]	Pyridine
Dimethylformamide	

[a]Most frequently used liquids.

the reaction proceeds. This reaction can be successfully run with concentrations or mole ratios of the reactants in the following ranges: (1) 25 weight % or less reactable solids content; (2) hydroperoxide to calcium chloride: 0.25 to 32; (3) hydroperoxide to lignin (M_n): 21 to 113; and (4) 0.01 to 0.95 weight fraction of monomer in reactable solids.

To a dry Erlenmeyer flask of appropriate size, add one half of the reaction solvent. All reactants, including the dry mass of hydroperoxide, should not constitute more than 23 weight % of the reaction mixture or an insoluble product may be produced. Add dry lignin and dry calcium chloride to the reaction vessel and cap with a septum or rubber stopper. In a separate vessel, dissolve 2-propenamide in about one quarter to one half of the solvent and, if appropriate, in a third vessel dissolve a second monomer in the final one quarter of the solvent. Saturate both monomer solutions with N_2 by bubbling with the gas for 10 minutes. Saturate the lignin solution with N_2 for 10 minutes. Add the hydroperoxide to the mixture, bubble with N_2 for 5 minutes, cap, and stir for 10 minutes. While stirring the lignin reaction solution, further saturate the monomer solutions with N_2. Add the 2-propenamide solution to the lignin solution with stirring and under an N_2 blanket. Wait 1 minute. Add the second monomer solution to the reaction vessel in the same way. Stir and cap the reaction under an N_2 blanket. Place the reaction vessel in a 30°C bath for 48 hours.

The reaction is terminated with a small volume of aqueous, 1% hydroquinone solution and a volume of water equal to 1/3 of the reaction solution volume is added to the product. This solution is added to 10 times its volume of 2-propanone and the polymer is recovered by filtration. The solids are redissolved in water. To remove calcium ion from the product, an amount of $Na_2C_2O_4$ equal to the moles of $CaCl_2$ added to the reaction is placed in the solution. The CaC_2O_4 precipitate is removed by filtration. The filtrate is dialyzed against distilled water for 3 to 5 days using #6 Spectropore dialysis tubing. The dilute aqueous solution is then freeze dried to recover the product.

Assays. Analytical procedures for determining oxidizing equivalents by iodine/thiosulfate titration, lignin content by UV assay, 1-amidoethylene content by Kjeldahl assay, limiting viscosity number, and elemental composition are given in ref. 4. The elemental assay for sulfur was done by ASTM method D 3177-82 with the correction that in step 7.3, the solution is brought to pH$\sim$ 3.8 with 6 *M* HCl rather than 2.5 *M* NaOH as incorrectly specified in the procedure. Size exclusion chromatography was done with a mobile phase of pH= 13, 0.1 *M* NaCl. All solutions were filtered through an 8 μm Nucleopore filter before use. The flow rate was 0.5 mL/min. and the mobile phase was not degassed. The injected sample size was 10 μL and the columns were maintained at 43°C during all separations. Spectra were taken from 200 to 420 nm and absorbance at 220 nm was plotted versus time for elution profile. Total permeation volume of the columns was 44 mL and total permeation time was 88 min.

Elemental analyses were also performed on most samples and these data were used to calculate the repeat unit content of the products using the following relationships:

$$WP_s = 7.1499 \times S$$

$$WP_N = 5.0745 \times N - 2.217 \times S$$

$$R_{N/S} = (2.2889 \times N - S)(S^{-1})$$

where S = weight percent sulfur in sample; N = weight percent nitrogen in sample; WP_s = weight percent of sample as sodium 1-(2-methylprop-2N-yl-sulfonate) amidoethylene repeat units; WP_N = weight percent of sample as 1-amidoethylene units; and $R_{N/S}$ is the molar ratio of amide repeat units to N-substituted, sulfonate containing repeat units.

Materials. Lignin, which makes up the backbone of the graft copolymers, is a crosslinked, oxyphenylpropyl polymer that acts as an intercellular glue in woody plants. Most lignin used in these studies is a commercial product. The material is a kraft pine lignin prepared in "free acid" form with a number-average molecular weight of 9600, a weight-average molecular weight of 22,000, and a polydispersity index of 2.29. The ash content of the lignin is 1.0 weight percent or less. The material was used as received. Elemental analysis is C 61.66, N 0.89, H 5.73, S 1.57, Ca 0.08, and Fe 0.014 weight percent. Other lignins used are described in the text with the results of the reaction or test.

2-Propenamide (common name acrylamide) used in all reactions was reagent grade monomer that was recrystallized from trichloromethane after hot filtration and dried under vacuum (P< 1.3 Pa) at room temperature for 24 h. The anionic monomer, 2,2-dimethyl-3-imino-4-oxohex-5-ene-1-sulfonic acid, was purified by heating a 15.2 weight percent solution in methanol to 65°C, filtering the hot solution, recovering the precipitated monomer from the cool solution, and vacuum drying the solid at room temperature for 24 hr. Dimethylsulfoxide was stabilized reagent grade material that was freshly vacuum distilled at 50°C before use. Calcium chloride and other salts used were reagent grade materials and were used as supplied. Gases used in the syntheses were standard commercial grade cylinder gases.

The dialysis membrane used was Spectrapor no.6, a 1000 upper-molecular-weight-cutoff cellulose, 45 mm diameter, membrane tubing made by Spectrum Medical Industries, Los Angeles, CA.

Equipment. Lignin spectra were run on a Perkin-Elmer Lambda 3, UV-visible spectrophotometer. Freeze drying was done on a FTS Systems Model FDX-1-84 lyophilizer. Weighings were done on a Mettler B6 balance. The viscometers used in fluid property measurements were Canon-Fenske capillary viscometers or a Brookfield LVT cone and plate viscometer. Size exclusion separations were performed with a Varian model 5000 high-performance liquid chromatograph equipped with a Rheodyne 10μL fixed-loop injector. The columns used in this work were TSK-Gel guard column; TSK-4000-pw column; and TSK-5000-pw column, plumbed in sequence. The detector was a Hewlett-Packard HP-1040A high speed spectrophotometric detector with its supporting computer, the HP-85, containing 16k bytes of additional memory. This detector can perform absorbance measurements at wavelengths from 190 to 600 nm and can detect and store an entire spectrum of the contents of the detector cell over the above wavelength range every second. This capacity allows spectra to be taken at numerous times during the elution of a chromatographic peak and is critical to proving the existence of graft copolymers by multivariate curve resolution (4,5).

Results and Discussion

Poly(lignin-g-(1-amidoethylene)). These nonionic molecules are small in size, readily adsorbed on silica surfaces, and prone to complex di- and trivalent metal ions from aqueous solution (2,3). Synthetic results for several samples of poly(lignin-g-(1-amidoethylene)) are given in Table III.

Note that these reactions show that copolymer can be produced with large weight fractions of lignin in the molecule. This chemistry can be used to place short sidechains on lignin. Other lignins can be reacted with this chemistry. Table IV shows synthetic data for the preparation of poly(lignin-g-(1-amidoethylene)) from several different lignins. Sample 1 is a kraft pine lignin grafted in a reaction coinitiated with sodium chloride. The lignin used in these studies is the commercial product described under *Materials*.

Sample 2 was run with a steam-exploded, solvent-extracted aspen lignin. This backbone, provide by the Solar Energy Research Institute, Golden, Colorado, as DJLX13 is an I-O-TECH process, wood extract. After steam decompression to disrupt the wood fiber, the wood was extracted with tetrachloromethane at approximately room temperature and reduced pressure. The wood was then extracted with methanol at 60°C and reduced pressure. The lignin sample used was recovered as the methanol extract. Samples 3 and 4 are results on a yellow poplar lignin. The material was produced by BioRegional Energy Associates of Floyd, Virginia. It is produced by steam exploding the wood, washing with water, extracting with alkali, and precipitating with mineral acid. The lignin has a high carboxylic acid content and a high level of phenolic hydroxyl groups. The molecular weight of the product was 1,000 to 1,200. Samples 5 and 6 are poly(lignin-g-(1-amidoethylene)) copolymers made with kraft pine lignin and run as controls at the same time as the reactions were run on poplar lignin. Yield and product properties are comparable for samples 4, 5, and 6 but the yield of sample 3 is low. This low yield may be due to loss of product during dialysis.

Data for a series of graft copolymers made using 2-propenamide are given in Table V for copolymers synthesized in dimethylsulfoxide. These data show that maximum yield is obtained when the chloride ion to lignin mole ratio is 492. The reaction on lignin with 2-propenamide is general in the composition and product properties that may be acheived. Table V shows the spectrum of lignin contents, 2-propenamide contents, and reagent ratios that can be used to produce a functional copolymer of differing composition, structure, and physical properties.

Table III. Yield and Limiting Viscosity Number for Lignin-(2-propenamide) Reactions[a]

sample no.	anhyd $CaCl_2$ in reaction mixture, wt %	yield g/wt%	limiting[b] viscosity no., dL/g	wt % lignin	wt % polymerized 2-propenamide	wt % Ca after ashing
1	4.0	3.36 / 90.8	0.59	7.5	67.6	4.91
2	2.0	3.64 / 98.4	0.46	6.95	73.4	2.25
3[c]	2.2	1.5 /100.0	0.21	12.4	48.3	6.94

[a] All reaction mixtures contained 20.0 mL of oxygen-bubbled, irradiated dioxane, 0.5 g of lignin, and 0.15 mL of ceric sulfate solution. Reactions run in a Pyrex flask and contained 1,4-dioxane irradiated for 3 h and 0.045 mol (3.2 g) of 2-propenamide.
[b] Determined in distilled water at 30°C.
[c] Reaction run with 0.014 mol (1.0 g) of 2-propenamide.

Table IV. Poly(lignin-g-(1-amidoethylene)) formed from Various Lignins and Coinitiators

Sample	Composition of Reaction(g) Lignin	2-propene amide	Chloride Salt[a]	Hydroperoxide	Solvent	Yield (g/wt.%)	Lignin Type
1	0.50	3.21	0.68	0.482 mL	21.28	3.46/93.3	Kraft
2	0.50	3.20	0.62	0.482 mL	21.28	3.48/94.05	I-O-Tech
3	0.51	3.21	0.62	0.482 mL	21.30	2.48/66.67	Poplar
4	0.50	3.20	0.62	0.482 mL	21.33	3.50/86.48	Poplar
5	0.50	3.27	0.64	0.482 mL	21.39	3.20/84.88	Kraft
6	0.50	3.22	0.63	0.482 mL	21.29	3.26/87.63	Kraft

[a] The same number of moles of chloride ion is used in sample 1 and samples 2 to 4. Sample 1 received sodium chloride while samples 2 to 4 received calcium chloride.

Table V. Lignin-co-(1-amidoethylene) samples.[a] DMSO Data

Reactants						Yield[c]			Product Composition			
Sample Number	2-propenamide (g)	Lignin (g)	$CaCl_2$ (g)	RO_2H[b] (g)	$Ce(^+4)$ (mL, .05M)	g	wt%	[η] (dL/g)	N	1-amido	Lignin	Ca
1	3.20	0.50	0.50	0.40/.3532	0.15	4.71/2.799	75.68	0.322	15.21	76.73	6.50	2.789
2	3.20	0.50	0.10	0.40/3532	0.15	3.30*/2.565	69.32	0.56	13.34	67.38	5.38	0.73
3	3.20	0.50	0.0503	0.40/.3532	0.15	2.72*/2.14	57.83	0.69	13.45	67.91	4.94	0.415
4	3.20	0.50	0.0102	0.40/.3532	0.15	3.16/2.58	70.07	0.77	15.047	76.045	4.601	0.137
5	3.20	0.50	0.50	0.25/.2208	0.15	4.31/3.306	89.36	0.35	13.284	66.946	7.00	2.89
6	3.20	0.50	0.50	0.152/.1342	0.15	4.40/3.34	90.35	0.44	13.034	65.79	6.03	2.86
7@	3.20	0.50	0.50	0.80/.7064	0.15	4.73/3.283	88.73	0.306	12.37	62.31	7.21	2.48
@	3.20	0.50	0.50	0.416	0.15							
@	3.20	0.50	0.50	0.416	0.15							
@	3.20	0.50	0.50	0.416	0.15							
8	3.20	0.50	0.1	0.15/.1325	0.15	4.07/3.307	89.37	0.615	13.55	68.41	5.48	0.757
9	3.20	0.50	0.0515	0.15/.1325	0.15	3.90/3.128	84.53	0.666	13.641	68.89	4.94	0.395
10	3.20	0.50	0.010	0.15/.1325	0.15	3.46/2.863	77.38	0.801	13.734	69.42	4.16	0.0949
11	2.00	0.50	0.50	0.15/.1325	0.15	3.10/2.119	84.77	0.372	11.74	59.12	6.88	3.54
12	2.00	0.50	0.10	0.15/.1325	0.15	3.04/2.091	83.64	0.395	11.67	58.81	6.26	1.007
13	2.00	0.50	0.0516	0.15/.1325	0.15	2.68/2.109	84.36	0.478	13.197	66.43	8.13	0.558
14	2.00	0.50	0.0107	0.15/.1325	0.15	1.84/1.44	57.58	0.565	14.162	71.43	6.54	0.33
15	1.00	0.50	0.50	0.15/.1325	0.15	2.14/1.105	73.64	0.192	7.950	39.53	12.24	5.45
16	1.00	0.50	0.10	0.15/.1325	0.15	1.12/1.063	70.87	0.288	11.515	57.52	13.74	1.65
17	1.00	0.50	0.0512	0.15/.1325	0.15	1.49/.857	57.14	0.275	10.63	52.95	14.95	0.88
18	1.00	0.50	0.0113	0.15/.1325	0.15	1.25/.919	61.29	0.276	11.516	57.41	15.52	0.325
19	3.20	0.50	0.5	0.4/.3532	0.15	3.27/	88.38	0.523	15.802	79.43	11.41	0.2
20	3.20	0.50	0.1	0.15	0.15	4.08/2.896	78.26	0.56	14.062	70.98	5.62	0.679

[a]All reactions run in 20.0 mL of dimethylsulfoxide save for the samples marked (@) run in 1,4-dioxacyclohexane.
[b]Results given as weight ratios of crude to pure 2-hydroperoxy-1,4-dioxacyclohexane.
[c]The yields listed are ratios of crude product recovered to pure product recovered. Weight percent yield is based on pure product recovered.
[d]1-amido = 1-amidoethylene repeat units in the polymer.
* = Some product lost during recovery.

The above data, when analyzed for the effects of individual reactants, shows that chloride ion is a critical reactant in controlling the yield and limiting viscosity number of the product copolymer. To quantify this effect as a first step in optimizing this synthesis, a series of tests were run in the above solvent system with each reaction having a different level of choride ion content. The results of the reactions are given in Table VI. The composition, reaction conditions, and yield for these reactions are also listed in Table VI. The concentration of lignin and 2-propenamide were kept at around 1.9 and 11.8 to 12.5% by weight, respectively, while varying the calcium chloride content from 0.97 to 3.78 weight %, as shown in Table VII. This nonionic graft copolymerization produces a maximum yield when concentration of the calcium chloride is at 2.41% by weight of total reaction mass, as shown in Figure 1.

Poly(lignin-g-((1-amidoethylene)-co-(sodium 1-(2-methylprop-2N-yl-1-sulfonate) amidoethylene))). A strongly anionic polyelectrolyte can be made from lignin by conducting a graft polymerization in the presence of 2-propenamide and 2,2-dimethyl-3-imino-4-oxohex-5-ene-1-sulfonic acid or its salts. Data from 22 reactions are given in Table VIII. This compound will be called copolymer 2. All reactions, with the exception of number 10, contained 0.50 g of kraft pine lignin. This series of reactions was run to determine: (1) the dependence of yield on reactant concentrations; (2) an estimate of extent of reaction as a function of time; and (3) the effect of iron contamination on reaction yield. Proof of graft copolymerization must be provided for these reaction products. All too frequently, materials synthesized in the presence of a possible backbone are assumed to be graft copolymerized (7). In place of assumed synthetic success, we have used a previously developed size exclusion chromatography method to verify that sidechain and backbone are chemically bound (2-5). In this technique, the absorbance spectra from 200 to 600nm of the effluent from the chromatography column is taken in real time throughout the elution of a peak. These spectra allow identification of the material in the detector cell and show the presence of backbone or sidechain at any point in the elution profile.

Size exclusion chromatograms for pure lignin and sample 8, Table VIII, are shown in Figure 2. The copolymer sample produces an absorbance peak at 34 min from the injection of 10 μL of 1.46 g/dL sample 8 in mobile phase. The lignin chromatogram shows an absorbance maximum at 38.5 min after injection of 10 μL of 0.45 g/dL of lignin. Elution of copolymer 8 starts at 25 min, 7 min ahead of the 32 min start of elution of lignin from the columns under identical conditions. Since the earlier elution of copolymer 8 shows it is a larger-molecular-size material, these chromatograms provide strong support for the formation of graft copolymer. Final proof is provided by the ultraviolet spectra of the eluting polymers. Two spectra of sample 8 effluent taken at 29.29 and 31.96 min both show the characteristic absorption maxima at 220 nm and shoulder at 286 nm characteristic of lignin. The 31.96 min spectra of sample 8 effluent is shown in Figure 3. Since reaction product lignin is eluting from the column before any pure lignin is seen in the detector, the reaction must have enlarged the reacted

Table VI. Synthetic Data of Poly(lignin-g-2-propenamide)

	Reactant (weight in grams)					Reaction parameter				
Sample Number	Lignin	$CaCl_2$	A	DMSO	E (mL)	Cl mmole	Ca/g	Cl/L	Cl/H	Yield %
24-124-1	0.50	0.25	3.21	21.29	0.50	4.50	0.97	9.00	1.07	92.17
24-124-2	0.50	0.38	3.20	21.21	0.50	6.85	1.47	13.7	1.63	94.59
24-124-3	0.50	0.50	3.20	21.27	0.50	9.01	1.93	18.02	2.15	96.56
24-124-4	0.50	0.63	3.20	21.28	0.50	11.35	2.41	22.70	2.71	99.20
24-134-1	0.51	0.77	3.20	22.03	0.50	13.87	2.85	27.40	3.31	93.00
24-134-2	0.51	0.92	3.21	21.35	0.50	16.58	3.47	32.51	3.96	87.33
24-134-3	0.50	1.05	3.23	21.53	0.50	18.92	3.92	37.84	4.34	84.18

Note :

A: 2-propenamide.
E: 30 % hydrogen peroxide (equivalent weight : 8.383 meq/mL).
Ca/g: calcium chloride content (wt %).
Cl/L : chloride content per unit weight of lignin.
Cl/H : molar ratio of chloride to hydrogen peroxide.

Table VII. The Composition of Reaction Mixtures Used to Make Lignin Graft Copolymers

Sample number	Total mass	Lignin wt%	$CaCl_2$ wt%	Monomer wt%	Monomer mmole/g	Yield %
24-124-1	25.75	1.94	0.97	12.47	1.75	92.17
24-124-2	25.79	1.94	1.47	12.41	1.75	94.59
24-124-3	25.97	1.93	1.93	12.32	1.73	96.56
24-124-4	26.11	1.91	2.41	12.26	1.72	99.20
24-134-1	27.01	1.89	2.85	11.85	1.67	93.00
24-134-2	26.49	1.93	3.47	12.12	1.71	87.33
24-134-3	26.81	1.86	3.92	12.05	1.69	84.18

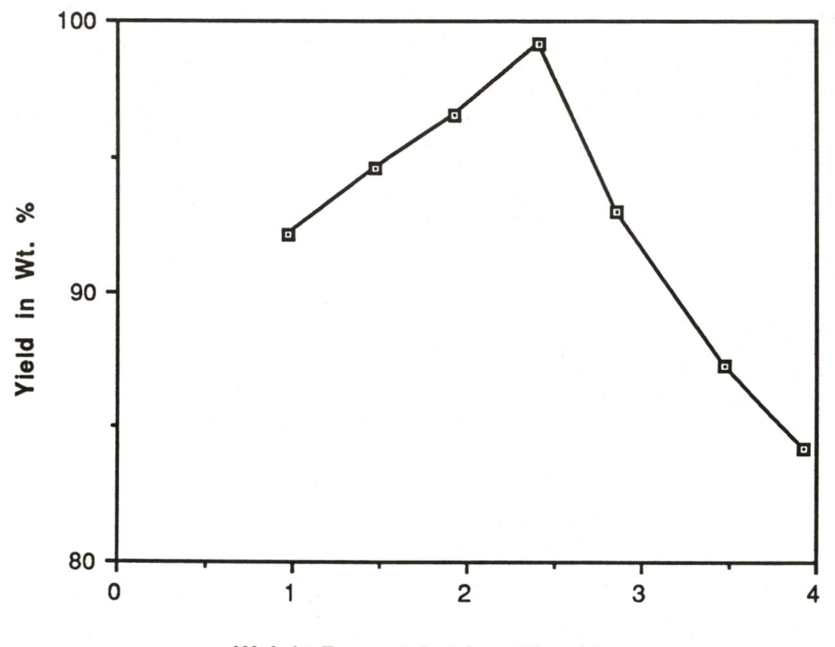

Figure 1. Yield vs. calcium chloride content per total mass in reactions of the nonionic copolymer.

Table VIII. Synthesis Data and Physical Characteristics of Graft Terpolymer

Reactants														
Sample Number	2-propen-amide (g)	A (g)	Ce(4+) solution (mL)	Dimethyl sulfoxide (mL)	$CaCl_2$ (g)	B (g)	Yield (wt %)	[η] (dL/g)	Assays (wt. %) C	Assays (wt. %) H	Assays (wt. %) N	Repeat Units (wt %) S	Repeat Units (wt %) 1-amido-ethylene	D
1	1.60	4.66	0.15	20	0.50	0.15	70.12	10.52	35.41	5.94	7.03	9.78	14.1	62.6
2	1.60	4.66	0.15	50	0.50	0.15	86.98	11.40	36.77	6.17	8.39	9.54	21.51	61.1
3	1.60	4.66	0.15	50	0.50	0.25	78.40	7.40	36.88	6.20	8.49	9.19	22.79	58.8
4	1.60	4.66	0.15	40	0.50	0.40	69.82	9.30	37.74	6.39	8.47	9.23	22.60	59.1
5	1.60	5.16	0.15	30	0.50	0.15	78.79	12.59	39.85	6.79	9.39	8.61	28.63	55.1
6	1.60	5.16	0.15	30	0.50	0.15	77.27	6.81	38.03	6.40	8.73	9.95	22.33	63.7
7	1.60	4.66	0.15	30	0.50	0.15	87.28	10.46	36.51	6.33	7.74	10.29	16.56	65.9
8	2.58	1.87	0.15	30	0.50	0.39	67.89	.953	43.31	6.65	10.92	5.65	42.89	40.4
9	2.56	1.86	0.15	30	0.53	0.39	79.49	2.46	42.47	6.29	10.76	5.66	42.1	36.2
10	21.98	15.99	1.28	219	4.35	3.35	91.02	1.97	42.25	6.49	11.34	6.21	43.78	44.4
11	2.56	1.86	0.15	20	0.50	0.36	80.69	-	44.38	6.83	10.84	4.79	44.39	34.2
12	2.56	1.86	0.15	20	0.50	0.34	315.64	-	44.73	6.46	9.62	4.70	38.40	33.6
13	2.57	1.87	0.15	20	0.50	0.34	69.89	-	43.88	6.58	10.82	4.32	45.33	30.9
14	2.56	1.87	0.15	20	0.50	0.34	66.79	-	44.90	6.82	10.73	4.69	44.05	33.5
15	2.57	1.76	0.20	15	0.50	0.34	18.50	-	42.80	6.74	11.41	4.75	47.37	34.0
16	2.57	1.86	0.15	20	0.50	0.34	45.06	-						
												$FeCl_2 \cdot 4H_2O$ g	$FeCl_2 \cdot 4H_2O$ moles x 10^5	
17	2.57	1.86	0.14	30	0.50	0.34	70.28		42.72	6.79	10.60	0	0	
18	2.56	1.87	0.15	20	0.50	0.335	51.63#		42.55	7.40	11.32	$1.5x10^{-3}$	.754	
19	2.56	1.86	0.15	20	0.50	0.338	72.72		44.36	6.89	9.91	$1.87x10^{-2}$	9.41	
20	2.56	1.87	0.15	20	0.51	0.34	18.26		42.26	6.50	9.25	.184	9.26	
21	2.56	1.87	0.15	20	0.50	0.34	68.29		41.77	6.67	10.78	0	0	
22	2.57	1.86	0.26	20	0.50	0.34	71.27		42.84	6.50	10.76	$1.55x10^{-3}$	.780	

= some product lost during purification.
a) All reactions, save #10, contained 0.50 g of lignin. Reaction #10 contained 4.39 g of lignin.

A) = 2,2-dimethyl-3-imino-4-oxohex-5-ene-1-sulfonic acid.
B) = Hydroperoxide. Samples 1 to 7: Values are weight of 1,4-dioxa-2-hydroperoxycycloexane in g. Samples 8 to 22: Values are amount of aqueous solution of 1,2-dioxy-3,3-dimethylbutane in mL. Equivalent/mL = $7.23x10^{-3}$.
D) = N-substituted 1-amidoethylene.

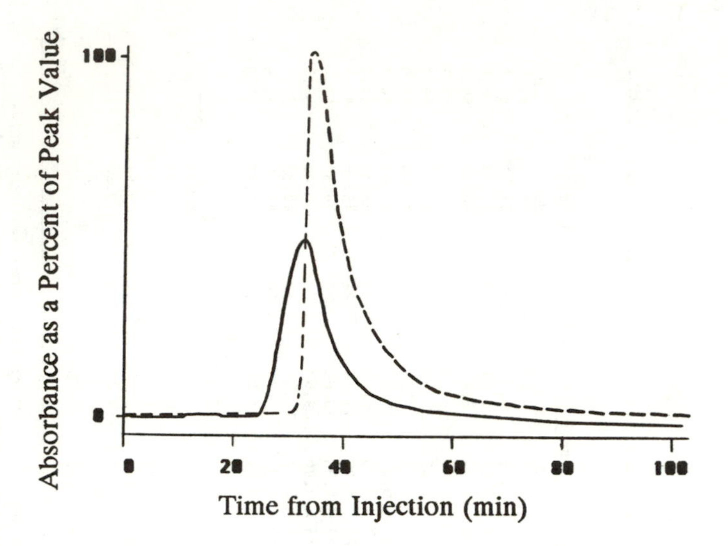

Figure 2. Size exclusion chromatogram of lignin and the anionic copolymer as monitored by absorbance at 220 nm.

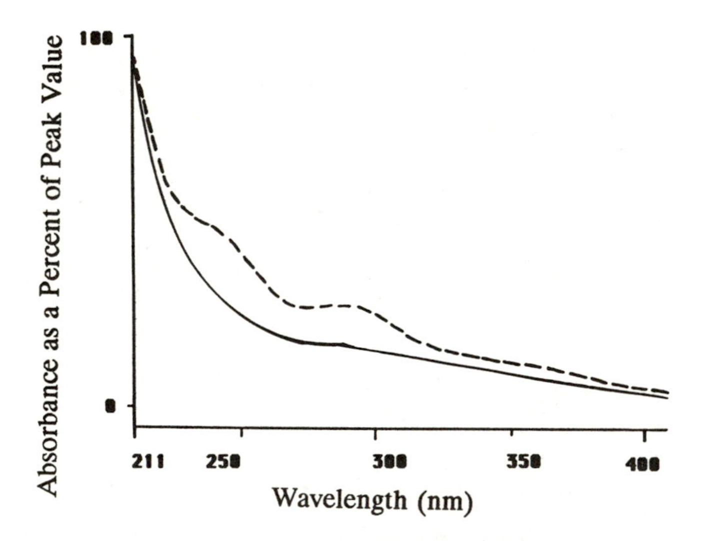

Figure 3. An ultraviolet absorbance spectrum of the anionic copolymer showing the characteristic absorbance pattern of lignin.

lignin molecule by forming a graft copolymer. A spectrum of pure lignin effluent at 36.96 min is also given in Figure 3. These data confirm that graft copolymer has been made.

Mechanistic Studies. Reactions 1 to 7 were run with different mole ratios of lignin to hydroperoxide and chloride ion to hydroperoxide. A maximum yield of 87 wt. % of polymerizable mass in the reaction is obtained when the lignin to hydroperoxide mole ratio is 4.17×10^{-2} and the chloride ion to hydroperoxide mole ratio is 7.21. These ratios occur in sample 2. Previous studies (2) have shown that the mole ratios between lignin, chloride ion, and hydroperoxide control yield in the graft copolymerization of poly(lignin-g-(1-amidoethylene)). These same ratios control yield in the formation of poly(lignin-g-(1-(2-methylprop-2N-yl)sulfonic acid)amidoethylene)-co-(1-amidoethylene))). These results imply that the grafting mechanism involves lignin, chloride ion, and hydroperoxide. An initiation reaction which is compatible with all findings to date is the attack on lignin by a hydroperoxide-(chloride ion) complex to create a site for free radical propagation. Such complexes have been seen in ESR studies of hydroperoxides (8).

Samples 8 and 9 of Table VIII show that this grafting reaction can also be run with an alternative hydroperoxide, 1,2-dioxy-3,3-dimethylbutane. This commercially available hydroperoxide can be used in place of 1,4-dioxacyclohexane-2-hydroperoxide. Sample 10 shows that amounts of copolymer as large as 40g can be made in single pot reactions. Reactions 11 to 16 of Table VIII were identical composition tests run for different amounts of time. Sample 11 was terminated after 31 min, sample 12 after 1 hr, samples 15 and 16 after 3 hr, sample 14 after 24 hr, and sample 13 after 48 hr. These data were gathered to determine the minimum duration of the reaction. The results showed that high yields (samples 11 and 16) can be obtained in reaction times as short as 30 min. Several samples (12 and 15) show low yields after reaction times as long as 3 hr but these results were obtained from a contaminated reaction or a reaction containing less than the appropriate amount of solvent, respectively. Previous reactions run in 1,4-dioxacyclohexane gave high yields in short reaction times (2,9). These data support a free radical polymerization mechanism for ethene monomers adding to lignin (2,9). Free radical reactions have rates which are insensitive to change of solvent.

Samples 17 to 22 were identical, 48 hr reactions run with different amounts of iron(2+) in the reaction mixture. Samples 17 and 21 contain only the iron added as a 104 ppm contaminant of lignin. The reaction mixture, in these cases, contains 9.28×10^{-7} moles of Fe^{2+}. The mole ratio between hydroperoxide and iron is 1,340. Since iron in neutral or acidic solution is not reduced in the presence of hydroperoxide, if the hydroperoxide-Fe^{2+} were acting as a Fenton's initiator for free radical polymerization, only 2% of the lignin in the reaction mixture could be grafted and the lignin content of the reaction product would be 0.2 wt. %. Reaction products usually contained between 6 and 10 wt. % lignin, a lignin concentration which could not occur unless extremely large amounts of chain transfer took place. Since lignin forms less reactive, quinoid-type structures when proton abstracted, high propagation rates could not be achieved by chain transfer mechanisms. To confirm that Fe^{2+} was not a reagent active in the initiation of this reaction, several more reactions were run with larger concentrations of Fe^{2+}.

Reaction 22 has an RO_2H/Fe^{2+} mole ratio of 160 but has about the same yield as an uncontaminated reaction, #17 and 21. Reaction 19 has an RO_2H/Fe^{2+} mole ratio of 13.3 but again shows no change in yield from that of an uncontaminated reaction. Sample 20 has a 1.35 mole ratio of RO_2H to Fe^{2+} and shows a sharp decrease in reaction yield and grafting. Here, the approximately 1:1 mole ratio of peroxide to iron should produce a high concentration of hydroxide radicals and extensive polymerization if these radicals are part of the polymerization process. Instead of a high yield, however, the reaction yield was less than one third of that obtained in the absence of iron. Therefore, a Fenton's initiation mechanism for this reaction is inconsistent with the data and probably does not occur. Elemental analysis data of Table VIII showed that product composition is proximate to, but not equal to, reaction mixture composition.

Properties. The limiting viscosity number values of these graft copolymers show that the molecules are sharply expanded by the addition of an ionic monomer to the chain. Reactions (3) run with the same number of moles of nonionic monomer (0.045) and producing approximately the same yield of product gave limiting viscosity numbers of 0.50 dL/g. In reactions 1 to 7 where 50 mole % of the monomer is now sulfonated, ionic monomer, the limiting viscosity number is 12 to 24 times higher. For products 8 to 10, the reaction mixture contained 20 mole % sulfonated monomer. The limiting viscosity numbers for samples 8 to 10 are 2 to 5 times higher than those of the nonionic copolymer, poly(lignin-g-(1-amidoethylene)) (3). The addition of sulfonated repeat units makes the polymer a better thickening agent and may make it a better complexing or flocculating/deflocculating agent.

Conclusions

A general method of grafting lignin has been developed which allows solvent extracted lignin, steam exploded lignin, and kraft lignin to be converted to complex polymers. The lignins grafted have been obtained from aspen, poplar, and pine. The lignins are research samples, pilot plant products, and commercial products from paper production. The types of materials made to date are industrial process chemicals and are nonionic and anionic graft copolymers of lignin. Extensive studies on the nonionic copolymer, poly(lignin-g-(1-amidoethylene)), show that the material can be made with a broad spectrum of lignins, that product properties can be controlled by control of reaction chloride ion content and monomer content, and that virtually any lignin content, molecular weight, and sidechain content can be achieved by control of synthesis variables.

A graft terpolymer of lignin has been made by free radical reaction of 2-propenamide and 2,2-dimethyl-3-amino-4-oxohex-5-ene-1-sulfonic acid in the presence of kraft pine lignin. The water soluble product is a thickening agent and has a limiting viscosity number in water at 30°C which increases as the fraction of sulfonated repeat units in the molecule increases. The grafting reaction is rapid and yields of 80 wt. % or more can be obtained in as little as 30 min from reactions run in 1,4-dioxacyclohexane or dimethylsulfoxide. The reaction appears to be initiated by a hydroperoxide, chloride ion, and lignin though the exact steps of the initiation are not known. Since the addition of Fe^{2+} to the reaction reduces yield at hydroperoxide to Fe^{2+} mole ratios of about 1, hydroxide radicals produced with Fe^{2+} do not appear to produce grafting. Adding 50 mole % sulfonated monomer to the reaction mixture produces graft copolymers with 12 to 24 times larger limiting viscosity numbers when compared to nonionic poly(lignin-g-1-amidoethylene). Adding 20 mole % sulfonated monomer to the reaction mixture increases product limiting viscosity number by a factor of 2 to 5.

For 2-propenamide, the reaction produced a maximum yield when the calcium chloride content is at 2.41% by weight of total reaction mass. In these nonionic copolymers, the limiting viscosity number is decreased with an increase of calcium chloride content. The elution volume of copolymers during size exclusion chromatography in basic aqueous mobile phase is smaller than that of lignin. Ultraviolet spectroscopy and size exclusion chromatography verify the formation of graft copolymer. These graft copolymers are highly water soluble, will increase the viscosity of aqueous solutions, and can be used as thickening agents and dispersing agents.

Acknowledgments

This work was partially supported by the National Science Foundation under award number CPE-8260766 and under National Science Foundation grant CBT-8417876. Support of the copolymer testing program by A and R Pipeline Company is gratefully acknowledged.

Keven Anderle, George Merriman, James Z. Lai, Damodar R. Patil, Mu Lan Sha, Nancy Chew, Chin Tia Li, Thomas Buchers, Cesar Augustin, Harvey Channell, and others completed a sizable portion of this work and their aid and effort is greatly appreciated and acknowledged.

Literature Cited

1. Goheen, D. W.; Hoyt, C. H. In *Kirk-Othmer Encycl. Chem. Technol.*, 3rd Ed., 1981, 295.
2. Meister, J. J.; Patil, D. R.; Field, L. R.; Nicholson, J. C. *J. Polym. Sci., Polym. Chem. Ed.* 1984, **22**, 1963-1980.
3. Meister, J. J.; Patil, D. R.; Channell, H. *J. Appl. Polym. Sci.* 1984, **29**, 3457-3477.
4. Meister, J. J.; Patil, D. R.; Field, L. R. *Macromolecules* 1986, **19**, 803.
5. Nicholson, J. C.; Meister, J. J.; Patil, D. R.; Field, L. R. *Anal. Chem.* 1984, **56**, 2447-2451.
6. Meister, J. J. In *Water-Soluble Polymers in Enhanced Oil Recovery*; Stahl, G. A.; Schulz, D. N., Eds.; Plenum Publ. Co.: New York, 1988; Ch. 2.

7. Meister, J. J. In *Renewable-Resource Materials: New Polymer Sources*; Carraher, C. E., Jr.; Sperling, L. H., Eds; Plenum Press: New York, 1985; pp. 305-322.
8. Dixon, W. T.; Norman R. O. C. *J. Chem. Soc.* 1963, **5**, 3119-3124.
9. Meister, J. J.; Patil, D. R. *Macromolecules* 1985, **18**, 1559-1564.

RECEIVED May 29, 1989

PHENOLIC COMPOUNDS

Chapter 23

Characteristics and Potential Applications of Lignin Produced by an Organosolv Pulping Process

J. H. Lora[1], C. F. Wu[1], E. K. Pye[1], and J. J. Balatinecz[1,2]

[1]Repap Technologies Inc., 2650 Eisenhower Avenue, P.O. Box 766, Valley Forge, PA 19482
[2]Faculty of Forestry, University of Toronto, Toronto, Ontario M5S 1A4, Canada

The ALCELL process, a proprietary organosolv pulping process, produces a novel lignin by-product. This lignin has many physical and chemical properties which distinguish it from lignins produced by the kraft and sulfite processes. It has a low molecular weight ($\overline{M}_n \sim 1000$), is highly hydrophobic and insoluble in neutral or acidic aqueous media, but soluble in moderate to strong alkaline solutions and certain organic solvents. It has a T_g in the range of 130°C. This material, soon to be produced in tonnage quantities from a commercial-scale demonstration plant in New Brunswick, Canada, has many potential applications. It shows strong promise as a partial replacement on an equal weight basis for PF resins in waferboard, OSB and other wood composites. The results of tests of boards made from this material having different characteristics will be presented. Other applications for this novel lignin will be discussed.

Organosolv pulping has been of considerable interest as a laboratory process for most of this century, but until recently has not received serious attention for development as a commercial process. Within the last decade, increasing concern about the capital cost and scale of economically viable new kraft pulp mills, together with the problems of environmental impact of conventional chemical pulp production, have caused several companies to seriously consider commercial development of alternatives. Among the advantages to organosolv pulping processes when compared to kraft are lower capital costs, smaller scale for economically attractive operation (which allows the use of smaller wood resource areas), significantly lower environmental impact, and the production of by-products of potentially significant commercial interest.

0097–6156/89/0397–0312$06.00/0

Montreal-based Repap Enterprises Corporation Inc., is embarked on a commercialization program for an organosolv pulping process, known as the ALCELL process. This process, developed by Repap's subsidiary, Repap Technologies Inc., of Valley Forge, Pennsylvania, from earlier work of C.P. Associates of Montreal, is an aqueous ethanol-based organosolv process which, in pilot plant studies, has produced hardwood pulps that after bleaching have comparable quality to bleached kraft pulps. In addition, it produces a lignin by-product having interesting and unique properties. This lignin recovered from pilot plant cooks of various hardwoods has been examined for its chemical and physical properties and has been tested in numerous applications with significant success and interest. In early 1989 the material will be available in tonnage quantities as a byproduct of a 33-ton of pulp/day commercial demonstration plant. It is anticipated that large amounts of ALCELL lignin will be available once the process is widely used industrially. This paper deals with the characteristics of ALCELL lignin as produced from different hardwoods and hardwood mixtures and discusses preliminary data and opportunities for commercial application.

Production

In the ALCELL process, conventional hardwood chips are cooked in batch extractors with an aqueous ethanol liquor at appropriate temperatures, pH, and time. In the process lignin, hemicellulose and other various components of wood are extracted from the chips into the aqueous ethanol forming a black liquor.

This black liquor is flashed and then the lignin is recovered by a patented precipitation technique followed by settling, centrifugation (or filtration) and drying. The result is a fine, brown, free-flowing powder. More details on how ALCELL lignin is obtained can be found in the literature (1,2).

Properties

ALCELL lignin has a low moisture content (less than 2% water). It has a bulk density of 0.57 kg/L. The material is soluble in some organic solvents and also in dilute aqueous alkali solutions. It is insoluble in water under neutral or acidic conditions.

ALCELL lignins that have been analyzed for molecular weight have a number average molecular weight lower than 1,000 and polydispersities between 2.4 and 6.3 (Glasser, W. G., personal communication, 1984, Virginia Polytechnic Institute and State University, Blacksburg, VA). More detailed studies are still required to correlate extraction and recovery conditions with molecular weight information.

ALCELL lignins normally have softening points in the 138-147°C range. Differential scanning calorimetry studies of aspen, birch and red oak ALCELL lignins (McGhie, A. R., personal communication, 1984, University of Pennsylvania, Philadelphia, PA) show an endothermic peak between 50 and 75°C (Figure 1). Thermogravimetric analysis suggests that this peak

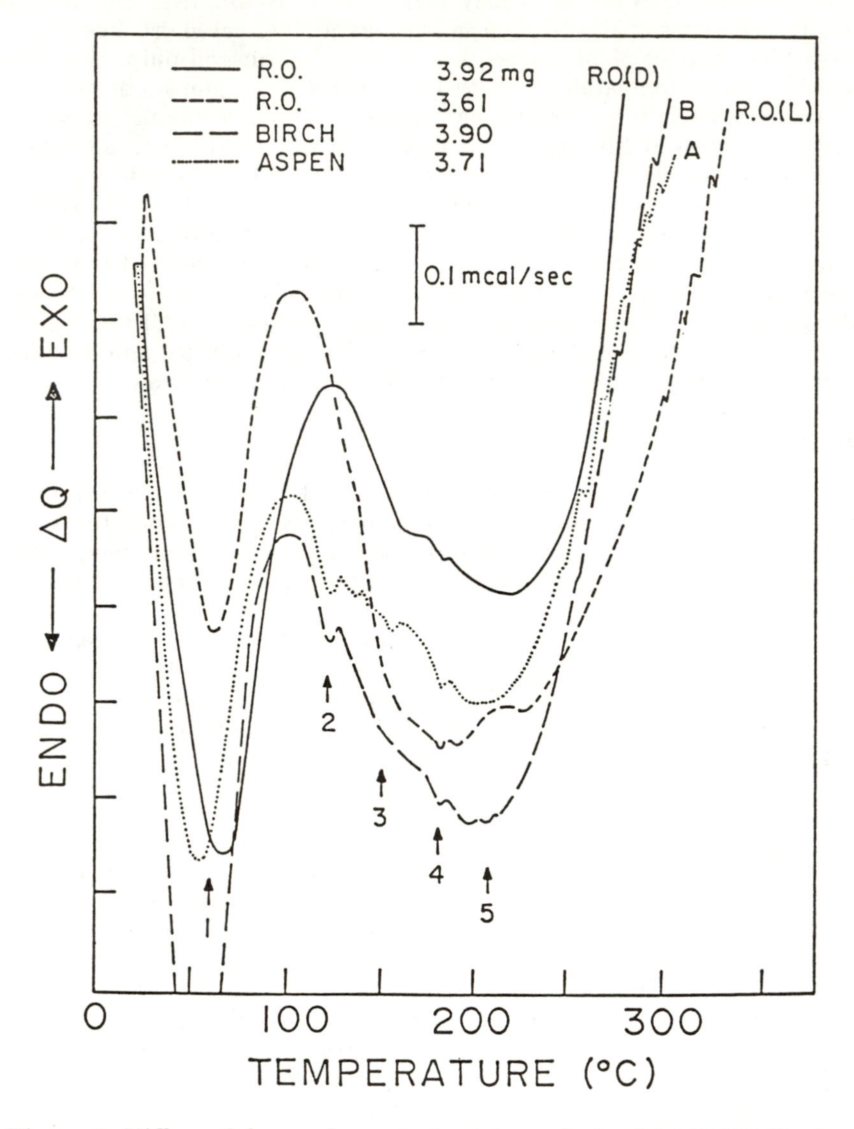

Figure 1. Differential scanning calorimetric analysis of ALCELL lignins.

probably corresponds to loss of absorbed water. Between 100 and 125°C a broad endothermic peak starts. This indicates the onset of chain segment motion and corresponds to the glass transition temperature. The broad endothermic peak has a maximum at about 200°C. At this temperature an exothermic reaction starts which is accompanied by increased weight loss. It must be noted that between 100 and 200°C there are three small peaks superimposed on the broad endothermic peak. They seem to occur at about the same temperatures for all samples analyzed. Further work is required to determine the significance of this observation.

In spite of its low degree of polymerization, ALCELL lignin to some extent resembles lignin in its native state. For instance, when reacted with phloroglucinol under acid conditions ALCELL lignin gives a purple coloration. This suggests that some of the cinnamaldehyde end units characteristic of lignin in its native state are still present in ALCELL lignin and/or that some have formed during the extraction process. Table I shows a typical elemental analysis for aspen ALCELL lignin. As a reference aspen milled wood lignin (MWL) has been included.

Table I. Elemental Analysis

	Aspen ALCELL Lignin (%)	Aspen MWL (3) (%)
C	64.04	59.96
H	6.20	6.19
O	29.65	33.85
Ash	0.11	
Methoxyl	17.09	20.42

C^9 formulae for the MWL and the ALCELL lignin are respectively $C_9H_{8.74}O_{3.02}\ (OCH_3)_{1.37}$ and $C_9H_{8.53}O_{2.45}\ (OCH_3)_{1.04}$. As observed, the ALCELL lignin has less oxygen and less methoxyl per C^9 unit than the MWL. These differences indicate that the ALCELL lignin has undergone some modifications which may include self-condensation, condensation with furfural generated in the ALCELL process, and/or incorporation of ethanol. The low carbohydrate and low ash content together with the high carbon content translate into a heating value of about 26,700 J/g (11,500 Btu/lb).

UV spectra of neutral solutions of ALCELL lignins exhibited maximum at 205-210 nm and at 275-281 nm which are characteristic of other lignin preparations. Alkali-neutral difference spectra exhibited three maxima at about 252-254 nm, 296-300 and 363-366 nm which indicate the presence of aromatic hydroxyl, α-conjugated hydroxyls, and conjugated carbonyl groups. The latter includes carbonyl groups in the α-position as well as those in cinnamaldehyde units mentioned above. The alkali-neutral difference spectrum of ALCELL lignins reduced with sodium borohydride shows an almost complete elimination of the peak at 360-366 nm and an increase in the 296-300 nm peak, thus confirming the presence of conjugated carbonyl groups. The alkali-neutral difference spectra of reduced samples also show a shoulder at about 330 nm, which indicates the presence of some α, β ethylenic groups probably in the form of phenylcoumarone type structures. Phenylcoumarone structures have been found among the products of acid degradation of lignins (4,5).

In general, the IR spectra of ALCELL lignins is very similar to the IR spectra of milled wood lignins. Perhaps the most striking difference is the increase in absorption at 1700-1720 cm^{-1} which is attributed to the presence of β-unconjugated ketone groups. This band is common to other lignins generated under mild acid conditions, such as autohydrolysis or steam explosion lignins (3,6,7). Recently, FTIR has been used on red oak ALCELL lignins (8). By the use of deconvolution and second-derivative spectroscopy, resolution was enhanced dramatically, and very weak bands not visible in the original spectrum became apparent (Fig. 2). Some of the data suggests the presence of vinyl and trans-disubstituted olefins in the sample examined. The use of these enhancement techniques is expected to play a very important role in future structural studies.

The phenolic nature of the ALCELL lignin translates into reactivity with formaldehyde. Thus, when formaldehyde and ALCELL lignin (2.2 moles of formaldehyde/C^9 unit) are heated at 96°C and pH 10.8 for 75 minutes, about 1.6 moles HCHO are incorporated for each C^9 unit of ALCELL lignin. This figure is in the range of figures reported for sulfite lignins (1.6-2.1 mole HCHO/C^9 unit) and is higher than what has been reported for kraft lignins (0.1-0.5 mole HCHO/C^9 unit) (9).

Applications

For many years wood chemists have tried to find applications for lignins other than their use as fuel. The amount of proposed products and applications is considerable. For instance, a recent literature survey on lignin utilization includes more than 3700 references, mostly patents (10). In spite of this massive amount of work on products, processes and applications, less than 2% of lignin available in spent liquors from conventional pulping processes are recovered and marketed for non-fuel applications in the U.S.

ALCELL lignin could compete in many applications now being filled by lignosulfonates and kraft lignin, but it is believed that unique higher-value markets can be developed. These applications will be helped by the low marginal cost of producing ALCELL lignin and will take advantage of its phenolic nature and reactivity with formaldehyde as well as of the properties that differentiate it from other lignins. The latter include hydrophobicity, lack of inorganic contaminants, low molecular weight, narrow molecular weight distribution, meltability, biodegradability, environmental acceptability, etc.

Some of the applications for ALCELL lignin that have been explored or considered are shown in Table II. One application that has been reported

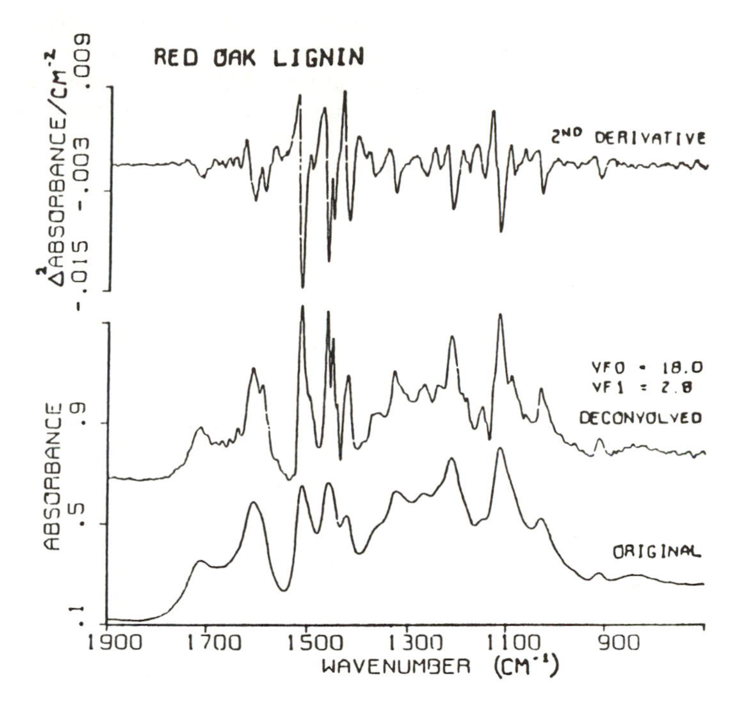

Figure 2. FTIR spectra of red oak ALCELL lignin. (Reprinted from ref. 8. Copyright 1987 American Chemical Society.)

Table II. ALCELL Lignin Applications

- Wood adhesives
- Molding compounds
- Flame retardants
- Diesel fuel additive
- Papermaking additives
- Slow release of agricultural, veterinary and pharmaceutical chemicals
- Insulation materials
- Friction materials
- Surfactants
- Asphalt extender
- Rubber reinforcement
- Medical applications
- Engineering plastics
- Antioxidants
- Lignin-derived chemicals

in the literature (11) is the use of this lignin after hydroxypropylation as a component in fire-resistant polyurethane foams.

Recent work has concentrated on the use of ALCELL lignin as a substitute for phenol-formaldehyde resins in wood adhesives, particularly waferboard. Some of the results obtained when a PF resin (Bakelite 9111) was replaced with different levels of hardwood ALCELL lignin in waferboard manufacture will be briefly discussed below. Table III shows the conditions used for waferboard manufacture.

Table III. Conditions Used for Waferboard Manufacture

Adhesive weight, % on wafers	2.6%
Wax emulsion (Paracol 802N)	1.7%
Pressure	3,800 KPA (550 psig)
Temperature	200°C
Time at temperature	5 min.

Initially three ALCELL lignin samples were evaluated: acid (pH 3.95), neutral (pH 7.51), and alkaline (pH 8.96). The moisture content of lignin-PF bonded boards ranged from 4.1 to 4.8%, which is well within the 8% allowable limit of commercial grade waferboard.

The thickness swelling of boards bonded with 50% or less lignin ranged from 14.7 to 17.1% after 24 hours immersion. It compared well with the control and was significantly better than the 25% maximum allowable in Canadian Standards Association (CSA) Standards. Similarly, water absorption values were within acceptable limits when lignin content was under 50%. Boards bonded with 100% ALCELL lignin did not give a waterproof bond and delaminated following 24 hours immersion.

As observed in Figure 3, boards bonded with 30% acid, neutral, or alkaline ALCELL lignin have almost identical internal bonding (IB) as the control. When lignin content increased, the IB became lower, especially for neutral ALCELL lignin. Boards bonded with 50% or less lignin met the minimum CSA requirement which is 50 psi. Boards bonded with alkaline lignin had the highest IB among all the lignin-containing boards.

Boards bonded with up to 50% ALCELL lignin had modulus of rupture (MOR) that surpassed the minimum CSA requirement by 40% or more. All of the boards containing lignin had a lower MOR than the control (100% = PF resin). As shown in Figure 4, the replacement of PF resin by ALCELL lignin under the conditions used in this study resulted in a decrease in MOR. Acid lignin showed the highest MOR among lignin-containing boards. At 100% substitution there was no significant difference in MOR among the lignins with different pH levels.

The MOR retention, which is the ratio of wet MOR to dry MOR, is shown in Figure 5. ALCELL lignin tends to preserve the MOR strength at

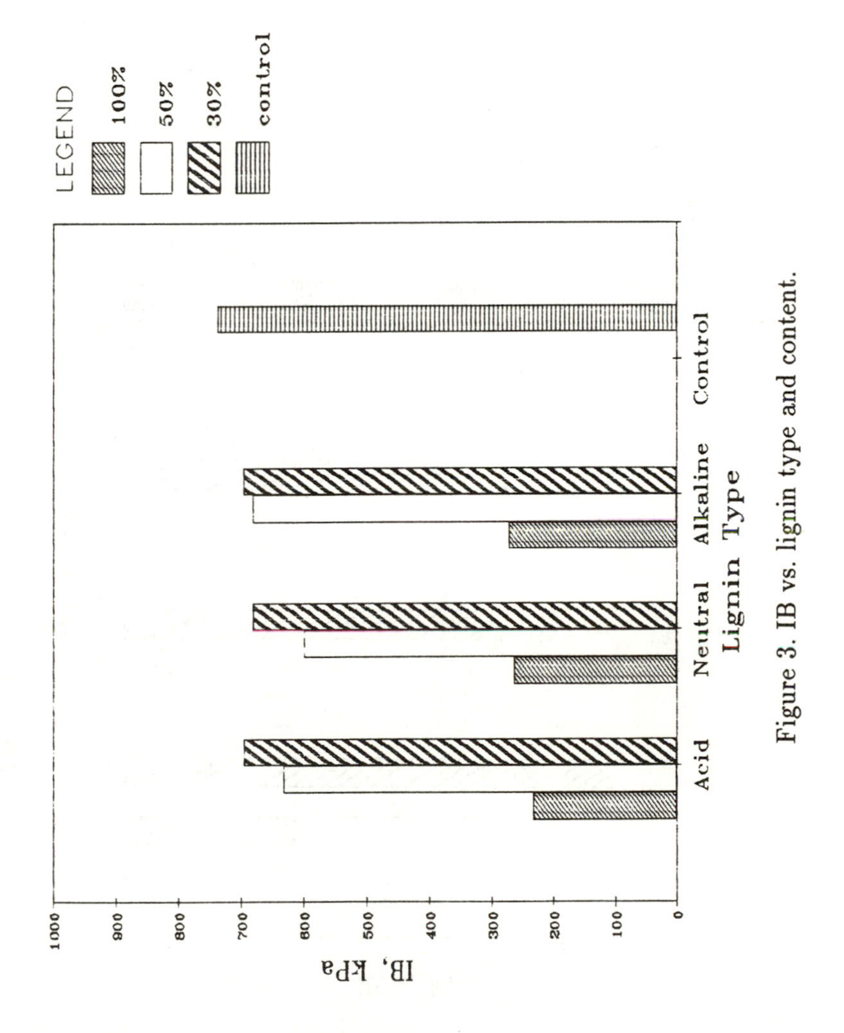

Figure 3. IB vs. lignin type and content.

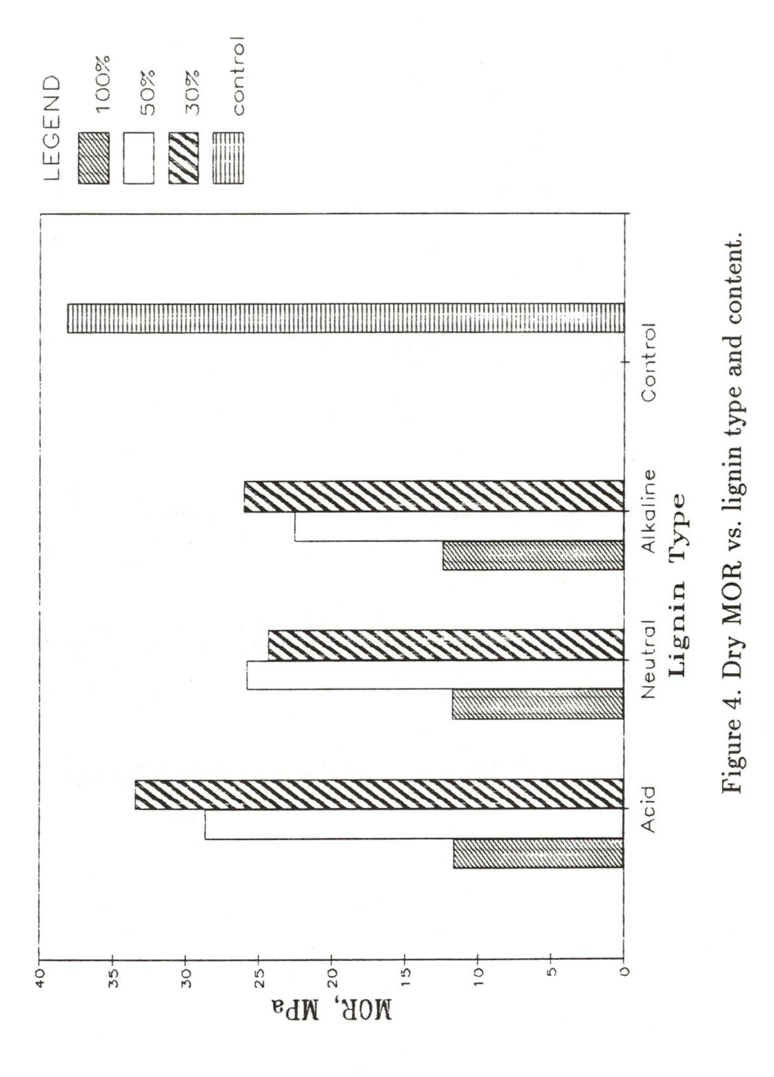

Figure 4. Dry MOR vs. lignin type and content.

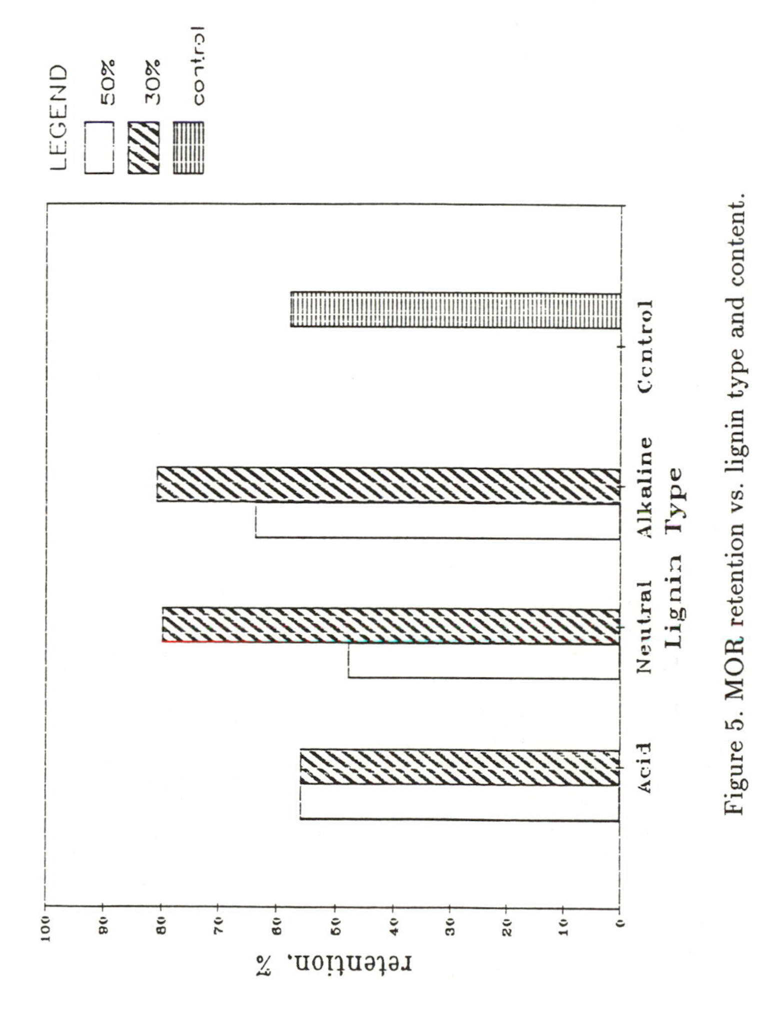

Figure 5. MOR retention vs. lignin type and content.

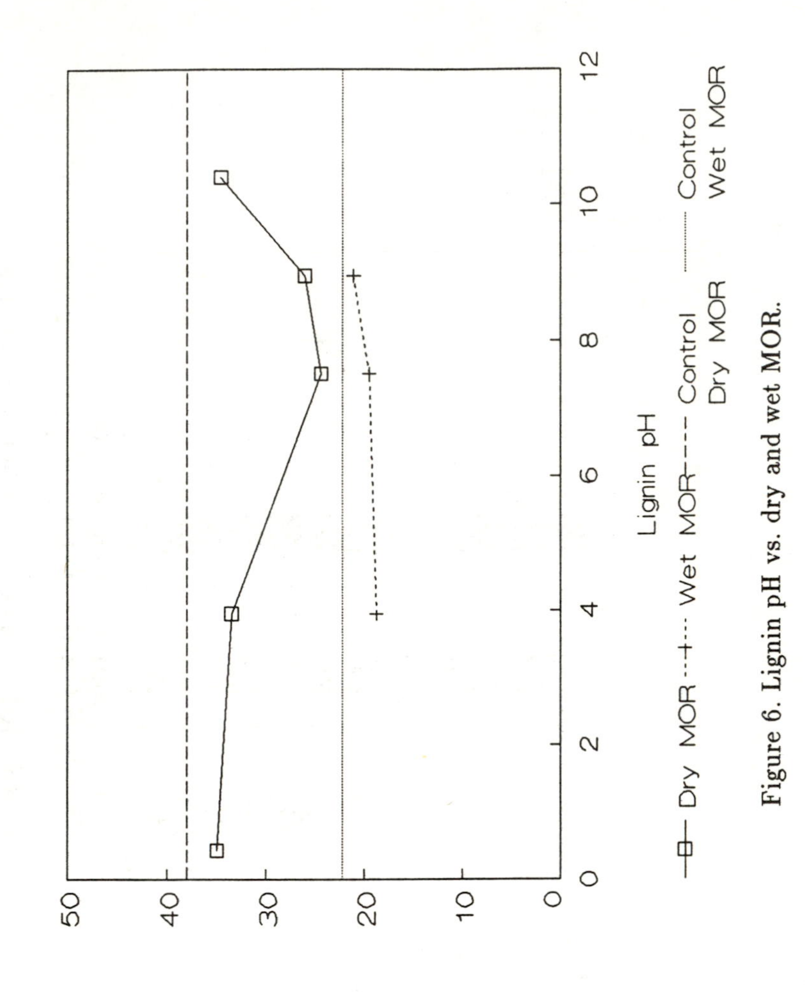

Figure 6. Lignin pH vs. dry and wet MOR.

least as well as PF resin alone when 50% or less lignin was used. Boards bonded with 30% alkaline and neutral lignin appeared to retain over 80% of their dry MOR strength. This indicates that boards bonded with certain lignin-PF mixtures may have a better weather resistance capability than the commercial PF resins.

Some of the most recent work has focused on evaluations of ALCELL lignins outside the pH range initially shown in Figure 6. When ALCELL lignin replaced 30% of the PF resin and its alkalinity or the acidity were increased, dry MOR values above 90% of the MOR of the control (100% PF resin) were obtained. This shows the importance of pH on performance. Future work will include a more detailed look at formulations, pressing conditions, and derivatization as a way of obtaining a better adhesive.

Conclusions

Several applications have been explored with significant success and interest using organosolv lignin available from a pilot plant. This lignin exhibits interesting physicochemical properties which distinguish it from lignins previously available on a commercial scale. Larger volumes are expected to be available by the end of the decade stimulating technical and market developments for this product.

Literature Cited

1. Lora, J. H.; Aziz, S. *Tappi J.* 1985, **68**(8), 94-97.
2. Williamson, P. N. *Pulp and Paper Canada* 1987, **88**(12), 47-49.
3. Lora, J. H.; Wayman, M. *Can. J. Chem.* 1980, **58**(7), 669-676.
4. Adler, E.; Lundquist, K. *Acta Chem. Scand.* 1963, **17**, 13-26.
5. Sarkanen, K. V.; Ludwig, C. H., Eds. *Lignins*; John Wiley and Sons: New York, 1971; 253.
6. Marchessault, R. H.; Coulombe, S.; Morikawa, H.; Robert, D. *Can. J. Chem.* 1982, **60**(18), 2372-82.
7. Chua, M. G. S.; Wayman, M. *Can. J. Chem.* 1979, **57**(19), 2603-11.
8. Byler, D. M.; Gerasimowicz, W. V.; Susi, H.; Cronlund, M.; Lora, J. H. *Polym. Prep.* (Am. Chem. Soc., Div. Polym. Chem.) 1987, **28**(2), 260-261.
9. Gillespie, R. H. FPRS Adhesives Symp. Proc. 1985.
10. Boye, F. *Utilization of Lignins and Lignin Derivatives*; Bib. Ser. No. 292; Inst. of Paper Chem., Appleton, WI.
11. Glasser, W. G.; Leitheiser, R. H. *Polym. Bull.* 1984, **12**, 1-5.

RECEIVED February 27, 1989

Chapter 24

Organosolv Lignin-Modified Phenolic Resins

Phillip M. Cook[1] and Terry Sellers, Jr.[2]

[1]Eastman Kodak Company, Eastman Chemicals Division, P.O. Box 1972, Kingsport, TN 37662
[2]Forest Products Utilization Laboratory, P.O. Drawer FP, Mississippi State, MS 39762

An adhesive system for structural wood panels is demonstrated in which at least 35% of the resin solids of a phenol-formaldehyde resin are successfully replaced with organosolv hardwood lignin. The bulk properties of the lignin were characterized before and after its purification with aqueous sodium bicarbonate solution. Phenolic resins are described that have been prepared from both purified and non-purified (crude) lignin. The nature of residues removed from the lignin during purification and their effects on resin adhesive properties are also briefly described. Evaluations of the lignin-modified phenolic resins were carried out on a small scale by measuring the lap-shear strength of parallel laminated maple blocks, and on a large scale by performing strength and dimensional tests on southern pine flake boards (waferboard or strandboard types). Flake board tests included internal bond, static bending (modulus of rupture, MOR), water absorption, and thickness swelling. The MOR was measured on both dry specimens and specimens subjected to accelerated aging wet-dry cycles. Both maple block and flakeboard evaluations included available commercial phenolic resins as controls.

Lignin makes up about one-quarter of the weight of dry wood and is second only to cellulose as the most abundant naturally occurring polymer. Pulping methods which use organic solvents are particularly well-suited to solubilize the lignins for lignin isolation and are readily coupled with solvent recovery (1). Despite these facts, there are no large volume organosolv pulping processes or commercial products based upon organosolv lignin (OSL) in the United States.

0097–6156/89/0397–0324$06.00/0

Lignins are phenolic-like polymers that from time to time have been considered for use as a phenol replacement in phenol-formaldehyde (PF) resins. Today, however, very little lignin is used in phenolic resins. Purified kraft lignins have been suggested as a phenol substitute, but the price of such lignins has been structured to be equal to that of phenol, and resin adhesive suppliers have resisted its use. Very little has been published on the use of OSL in phenolic adhesives despite the fact that these lignins are generally found to have a higher purity and reactivity than lignins from conventional pulping methods.

The most prominent wood adhesives used over the last quarter of a century have been aminoplast and polyphenolic types (2). In the United States, polyphenolic adhesives continue to be predominantly used for production of weather-resistant wood products, such as structural plywoods and flake boards (3). Phenolic resin prices have increased over the past decade, generally paralleling phenol prices. This increase has occurred in part due to a continuing erosion of United States phenol manufacturing capacity and the corresponding increase in availability of phenol from other countries. Any significant increase in the price of oil (the source of phenol) itself or interruption in supply will only compound the problem and raise phenol prices even higher.

The objective of this work was to demonstrate the utility of organosolv red oak lignin (a projected cheaper polyphenol than phenol) in phenolic adhesives for wood composites. This work involved three stages:

1. Analytical characterization of red oak OSL.
2. Resin synthesis of OSL-modified phenol-formaldehyde resins.
3. Evaluation of lignin-modified resins used to bond maple blocks and southern pine flake boards.

Results and Discussion

Analytical Characterization. As expected, red oak OSL contains much less sulfur and ash but more carbohydrates than kraft lignin (Table I) (kraft lignin included for comparison). The molecular weight and polydispersity (Table II) are less for red oak OSL than for kraft lignin (Indulin AT, Westvaco, Charleston, SC), and OSL has good solubility in organic solvents (Table I). Solubility in refluxing methylene chloride was surprisingly high. Purification of the OSL by aqueous reslurry in 10% sodium bicarbonate solution increases the softening temperature (Table II).

Thermogravimetric weight loss for the OSL is similar to that of kraft lignin. Except for a slightly elevated methoxyl content, the NMR and degradation results in Table II are within the expected range for a hardwood lignin.

OSL Impurities. It was also of interest to determine the impurities removed from crude OSL by reslurry in aqueous sodium bicarbonate, which was done to improve its usefulness in phenolic resins (Cook, P. M., Eastman Kodak at Kingsport, TN, personal communications, 1987). Extraction and acetylation procedures, involving methylene chloride, acetic anhydride pyridine

Table I. Lignin characterization (bulk and solubility properties)

Analytical Text	Basis	Amount		
		Crude OSL	Purified OSL	Kraft Lignin[1]
Non-volatiles	%	97	97	97
Ash, 775°C	%	1.94	1.13	3.48
Carbohydrates:				
(Extraction-HPLC)	%	3.0	1.1	0.7
Combustion:				
C	%	59.85	60.94	61.99
H	%	5.87	5.63	5.65
N	%	0.27	Trace	0.52
S	%	0.24	0.04	1.68
Extraction (weight loss):				
Ether, reflux	%	6	–	3
Heptane, reflux	%	< 1	–	< 1
Methylene chloride, reflux	%	50	–	11
Water, 25°C	%	5	–	2
Water, 60°C	%	12	–	6
Water, 100°C	%	5*	–	3*
*Sample agglomerates severely.				
Solubility (25°C):				
Methanol	g/L	105	–	–
Ethanol	g/L	53	–	–
n-Propanol	g/L	35	–	–
2-Propanol	g/L	15	–	–
Acetic Acid	g/L	108	–	–
Acetone	g/L	104	–	–
2-Butanone	g/L	96	–	–
Acetonitrile	g/L	86	–	–
Ethyl Acetate	g/L	59	–	–

[1] Indulin AT by Westvaco, Charleston, SC; a purified pine lignin.

and water, were followed which produced three residues from the impurities and each was submitted for GC/MS and ^{1}H NMR analyses. Residue 1 proved to be nearly all lignin-related products. Residues 2 and 3 were found principally to contain fully acetylated monomeric C_5 and C_6 sugars which were also C_1 alcoholic glycosides. Results of GC/MS investigations of the residues indicated that the low-molecular-weight impurities (e.g., mono-functional lignin-related species, waxes, and carbohydrates) should be removed to improve the OSL reactivity. The extent of impurity removal can be approximated by measuring gelation time (i.e., the higher the impurity content, the longer the gel time). For example, the gelation time (using

Table II. Physical and chemical characterization of lignin

Analytical Test	Basis	Amount		
		Crude OSL	Purified OSL	Kraft Lignin[1]
$T_{softening-initial}$:				
(TMA)	°C	89	124	124
$T_{softening-final}$:				
(TMA)	°C	111	138	169
TGA (weight loss):				
50°C	%	1	0	1
100°C	%	3	1	3
200°C	%	5	2	6
300°C	%	18	10	12
400°C	%	44	38	26
500°C	%	55	49	51
GPC molecular properties:				
(CH_2Cl_2/HFIP,				
Styragel Col.)	M_w	2227	2030	2479
	M_n	906	925	523
	M_w/M_n	2.46	2.19	4.32
NMR analysis:				
OCH_3	%	22.04	–	13.70
	(bonds/C_9 unit)	(1.55)	–	(0.81)
Syringyl/guaiacyl	ratio	1.33	–	< 1
Bonds:				
OH (Phenolic)	per C_9 unit	0.60-0.65	–	0.57-0.62
OH (Aliphatic)	per C_9 unit	0.55-0.60	–	0.70-0.72
H (Aromatic)	per C_9 unit	1.90-1.95	–	2.50-2.55
H (Aliphatic)	per C_9 unit	3.60-3.70	–	4.25-4.30
H (Hydroxyl)	per C_9 unit	1.15-1.25	–	1.35-1.40

Source: Glasser, W. G. Report to P. M. Cook. Located at Eastman Chemical Division, Research Laboratory, P. O. Box 1972, Kingsport, TN 27662 (1984).

[1] Indulin AT by Westvaco, Charleston, SC; a purified pine lignin.

a Sunshine Gel Tester) at 100°C (212°F) was 21 minutes for a PF resin solution, 26 minutes for a purified OSL/PF resin blend, and 53 minutes for an unpurified OSL/PF resin blend. The procedure involves blending dry lignin solids with a PF resin solution to a 23% lignin concentration and adjusting with water and 50% sodium hydroxide to the pH (e.g., 11.1±) and viscosity (e.g., 500 ± 50 cP) of the PF resin.

Block Lap-Shear Results. For laminated maple wood, this work indicated a maximum of 40% of the PF resin solids can be replaced with OSL without detrimentally affecting adhesive properties (Table III) (Cook, P. M., Eastman Kodak at Kingsport, TN, personal communications, 1987). Organosolv lignin-PF resins were at least equivalent to the commercial (control) resin and out-performed kraft lignin-based PF resins at the 35% and 40% solids replacement levels (Table III). Purified OSL generally yielded better results than unpurified. A distracting quality of kraft lignin resins is the sulfur-like odor produced during resin preparation and panel hot pressing. Use of OSL alleviates this problem. The reasons why OSL out-performed kraft lignin in this work are not clear, but are likely related to the molecular weight characteristics of the OSL and its ease of solubilization. Based upon the proposed structures of hardwood and softwood lignins, the contrary would have been predicted (i.e., the higher syringyl/guaiacyl ratios of hardwood would be expected to be more detrimental to bonding).

Table III. Selected maple block test results

Resin Type	Lignin Replacement (%)	Dry Bond		4-Hour Boil		Strength Retention (Dry/Wet) (%)
		Shear Strength (MPa)[1]	Wood Failure (%)	Shear Strength (MPa)	Wood Failure (%)	
Control PF	0	5.18	80	3.08	0	60
OSL-PF (crude lignin)[2]	24	4.99	100	2.54	30	51
OSL-PF (CH_2Cl_2 extracted)	34	5.35	100	3.02	40	57
Kraft L-PF (Indulin AT)	35	4.25	90	1.83	0	43
OSL-PF ($NaHCO_3$ washed)	40	5.60	60	3.37	30	60
Kraft L-PF (Indulin AT)	40	4.14	40	1.90	10	46
OSL-PF (crude lignin)[2]	45	2.68	20	1.00	0	37

[1] Multiply MPa by a factor of 145 to convert to psi.
[2] This resin also contained 1% melamine.

Flake Board Test Results. Results from initial screening tests looked promising for the use of kraft pine lignin and organosolv hardwood lignin at 25% substitution for phenol in PF resins used to bond flake boards. Therefore, this study was designed to concentrate on improving the OSL-PF cook procedure, increase the phenol substitution to 35%, and measure these effects by expanding the board test criteria (4). In general, the purified OSL-PF resins as well as the unpurified OSL-PF resins performed equal to, or better than, the control commercial resins and the control resins containing comparable substitution amounts of pecan shell flour (*vs.* OSL) in all test panel properties examined (Table IV).

Table IV. Selected test results of flake boards

Physical Property	Units	Sampling of Four USA Mills (1986)	Laboratory Test Results: Control Resin B		Laboratory Test Results: 35% OSL	
			100% PF	35% PS[1]	Purified	Crude
Density	avg. kg/m^3	650-700	(average of boards was 753)			
Internal bond	avg. kPa	340-460	510	386	717	538
Modulus of rupture	avg. MPa	22-38	33	33	34	34
Accelerated aging	% of dry MOR	50-70	76	75	77	70
Water absorption	%	25-34	33	35	25	19
Thickness swell	%	12-15	16	25	15	11
Resin solids applied	%	3-5	5	3.25	5	5

[1] These panels were bonded with a resin solution containing 65% PF and 35% pecan shell (PS) flour, delivering 3.25% resin solids to the furnish.

Note: To convert metric to English units—kg/m^3 to lb/ft^3, multiply by a factor of 0.0624; kPa to psi, multiply by a factor of 0.145; MPa to psi, multiply by a factor of 145.

Use of powdered lignins as an extender at 25% to 35% replacement of PF solids in liquid commercial resins is impractical because of problems of dispersion, viscosity, and stability which hinder subsequent uniform resin spray application. However, substitution of lignin for phenol at levels of 35 weight percent, or higher, in cooked lignin-phenol-formaldehyde copolymers is practical. While impure (crude) OSL cooked into a PF resin and used to bond flake boards yielded similar board properties to cooked purified OSL in PF resins, the impure lignin was troublesome during the cook procedures, forming gels upon long methylol-lignin condensation. Impure lignin was more gritty than purified lignin and some extraneous material may have settled rather than dissolve or suspend in the cooked resins. Impure lignin required an adjustment in the water in both steps of the cook to yield a satisfactory non-volatile solids content. If any coarse lignin settled in the cooked resin, this may partially explain the lower percent non-volatile solids phenomenon with resins incorporating impure lignin.

Conclusions

It has been demonstrated that red oak OSL could be used to replace 35% to 40% of the phenol (or phenolic resin solids) in phenol-formaldehyde resins used to laminate maple wood and to bond southern pine flake boards (waferboard and/or strandboard) without adversely affecting the physical bond properties. While this pulping process and by-product lignin do not commercially exist at this time in the United States, lignins from such processes are projected to cost 40% to 50% less than phenol as a polymer raw material.

It is recommended that a reslurry of crude OSL in an organic solvent or 10% aqueous salt (e.g., $NaHCO_3$) solution be performed to remove low-molecular-weight (mono-functional) species, waxes, and carbohydrates. This purification leads to an improvement in OSL reactivity and contributes to the usefulness of OSL as a PF resin extender or PF copolymer raw material. It is presumed that extraneous removed materials in the crude lignin react with formaldehyde but do not lead to productive cross-linking polymer formation.

The success obtained in this study on bonding flake board and maple blocks supports the possibility of using OSL in resins for bonding crosswise laminates (i.e., plywood). Subsequent research conducted at the Mississippi Forest Products Utilization Laboratory on OSL-PF resins for bonding plywood has been equally successful.

Experimental

Wood Pulping and Lignin Purification. The wood pulping involved red oak wood chips, aqueous organic solvent, and an acid catalyst, which were processed at 120 to 140°C. The resultant pulp was washed with more aqueous solvent, water-slurried and cooled. The lignin was isolated by removing the solvent under vacuum, replacement of solvent with water, lignin precipitation, filter cake water washing, and drying. The lignin was purified by slurring one part of dry OSL in five parts of a 10% aqueous sodium bicarbonate solution at 60°C for one hour. The slurry was cooled to room temperature and filtered. The filter cake was washed with water until the filtrate was less than 7.5 pH. The filter cake was then dried in a forced-air oven at 50 to 55°C, with a weight yield of 90% to 95%. The resultant dry and loose lignin was used without further processing such as grinding.

Analytical Characterization. The lignins were characterized analytically by the following methods: ^1H NMR spectra, gel permeation chromatography (5), gas chromatography (6), thermal measurements, elemental analysis, sugar content, extractions, solubility, and combustion properties.

Resin Preparation. Two approaches to resin preparation were used with regard to the initial stages of condensation, depending on whether the resin was intended for laminating maple blocks or for bonding southern pine flake boards. For maple block resins, the steps involved were as follows: addition of water, sodium hydroxide (optional) and lignin, which were heated and held until a homogeneous mixture was obtained; addition of either 100% of the phenol or 10% to 20% of the formaldehyde required in the cook and heated to 75 to 90°C for 30 to 90 minutes; addition of the remainder of the phenol (unless all was added initially), and formaldehyde, holding at 60 to 75°C for 30 to 90 minutes. For maple block bonding, a series of cooks were made in which 24% to 45% of the resin solids were replaced with OSL or kraft lignin. For flake board resins the steps included the following: a prepolymer synthesis of methylol-lignin condensation for 30 to 90 minutes, then a sequential methylol-lignin, phenol-formaldehyde condensation (resin synthesis) step, which varied from 3 to 5 hours cook time (4). For flake board resins, cooks were made with 25% and 35% of the phenol replaced with purified and unpurified OSL. Various properties of the resins were determined by standard methods, including non-volatile solids, viscosity, pH, free formaldehyde, free phenol, alkalinity, and gel time. In both studies, commercial resins were obtained and used as controls. Table V provides typical stoichiometric data of resin reactants investigated for both resin types.

Table V. Typical stoichiometric properties of resin reactants

Reactant	Amount by Resin Type	
	Blocks (mol)	Flake Board (mol)
Water (sufficient for desired non-volatile solids)		
Phenol	1.00	1.00
Formaldehyde	1.8-3.0	2.98
Organosolv lignin[1]	0.25-0.80	0.25
Sodium hydroxide	0.40-0.65	0.64
Urea	–	0.09

[1] Assume a mole equivalent unit of 200 for organosolv lignin, based on a typical analyzed C_9 lignin unit.

Maple Block Screening Method. A series of experimental procedures were performed on bonding maple block wood (Cook, P. M., Eastman Kodak at Kingsport, TN, personal communications, 1987). The procedure adopted was the ASTM D 905 standard, modified as follows: Sugar maple (*Acer saccharum*) wood, 76 by 25 by 5.7 mm in size (3 inches long, 1 inch wide, and 0.25 inch thick), with 6% moisture content was planed to obtain fresh surfaces for bonding. The desired amount of resin (with no mix additives) was weighed (58.6 g/m^2, 12 $lb/1000\ ft^2$, resin solids basis) and applied to one block surface and then a second clean block was overlapped so that 25 square mm (1 square inch) surface area common to each block was coated. The resin coated blocks were placed directly in the hot press (no clamp time). The blocks were hot pressed at 177°C (350°F) for 4 to 6 minutes at 3.44 MPa (500 psi). All bonded blocks were allowed to stand (post cure) at ambient temperatures for 24 hours prior to testing. Ten bonded blocks were tested dry and ten were tested after an accelerated aging regimen (4 h boil in water, cool, and test wet) for each resin. The lap-tension-shear strength of the test specimens was measured using an Instron machine, and subjective estimates of the percent wood failure (or bond failure) were observed and recorded. The data were subjected to statistical analysis.

Flake Board Screening Method. This phase of the study has been previously reported in more detail by Sellers *et al.* (4) but can be summarized as follows. Disc-cut southern pine flakes (wafers and/or strands) were obtained from commercial flake board plants in the Southern United States. The flakes were approximately 76 mm (3 in.) long, 19 to 38 mm (0.75 to 1.5 in.) wide, and 0.5 to 0.8 mm (0.020 to 0.030 in.) thick and adjusted to 2.5% moisture content at the time of use. Each resin type was applied at 5% and 7% resin solids rates. The mat configuration was homogeneous and hand felted in sufficient size to obtain a trimmed panel measuring 560 by 610 by 12.5 mm (22 by 24 by 0.5 in.). The mats were hot pressed at 205°C (400°F) for 6 minutes with a dual-pressure regimen (5.51/2.75 MPa, 800/400 psi). The initial high pressure was dropped after one minute into the cycle and the panels were pressed to 12.7-mm (0.5-in.) metal stops in the hot press. The target density was 745 kg/m^3 (46 lb/ft^3). Three panels per resin type and application rate were made. Resin types included the control commercial resins (no lignin content), lignin-modified PF resins, and control PF resins containing an inert filler (pecan shell flour) at the same loads as the lignin substituted resins. Screening tests for panel performance included internal bond (IB), dry static bending (modulus of rupture, MOR), accelerated aging MOR (strength retention after an American Plywood Association performance 6-cycle test), and dimensional tests [water absorption (WA) and thickness swell (TS)]. The statistical analysis was a two-factor experiment (resin type-resin application level), using an analysis of variance and T tests (LSD) for grouping for the various physical properties tested.

Product Disclaimer

The use of trade, firm, or corporation names in this publication is for the information and convenience of the reader. Such use does not constitute an official endorsement or approval by the Mississippi Forest Products Utilization Laboratory (MFPUL), Mississippi State University, or the Eastman Kodak Company of any product or service to the exclusion of others which may be suitable.

Acknowledgments

Appreciation is expressed to the following MFPUL employees: Dr. A. L. Wooten (retired) for resin synthesis; Gary Stovall and George Miller for resin and flake board sample preparation and testing; and Lynn Prewitt for GFC work. Materials were donated in support of this project by Borden

Incorporated, Chembond Corporation, Eastman Kodak Company, Georgia Pacific Corporation, Louisiana Pacific Corporation, and Southeastern Reduction Company. This work was supported by two research grants from Eastman Kodak Company.

Literature Cited

1. *Chemical Week* 1984, **134**(1), 26, 28.
2. Conner, A. H.; Lorenz, L. F. *J. Wood Chem. Technol.* 1986, **6**(4), 591-613.
3. Sellers, T., Jr. *Plywood and Adhesive Technology*; Marcel Dekker: New York, 1985; pp. 349-373.
4. Sellers, T., Jr.; Wooten, A. L.; Cook, P. M. *Proc. Structural Wood Composites: New Technologies for Expanding Markets*; Forest Products Research Society, Madison, WI, 1988, Proc. No. 47359, 43-50.
5. Glasser, W. G.; Glasser, H. R.; Morohoshi, N. *Macromol.* 1981, **14**(2), 252-262.
6. Glasser, W. G.; Barnett, C. A.; Sano, Y. *J. Appl. Polym. Sci., Appl. Polym. Symp.* 1983, **37**, 441-460.

RECEIVED February 27, 1989

Chapter 25

Wood Adhesives from Phenolysis Lignin

A Way To Use Lignin from Steam-Explosion Process

Hiro-Kuni Ono and Kenichi Sudo

Forestry and Forest Products Research Institute, Ministry of Agriculture, Forestry, and Fisheries, P.O. Box 16, Tsukuba Norin, Ibaraki 305, Japan

The lignin extracted from steam exploded pulp was phenolated in the presence of sulfuric acid. The degree of phenolation was calculated to be in excess of one mole/lignin (C_9) unit on the basis of ^{13}C NMR measurements. The phenolated lignin was methylolated in order to prepare adhesive resins. The cure behavior of the adhesive resins was examined by Torsional Braid Analysis (TBA). Results revealed that the phenolated steam explosion lignin-based resins had intrinsic retardation in cure as compared to a commercial phenolic resin. This defect, however, was partly overcome by increasing their pH values. The adhesives from these resins generally provide excellent bond strength comparable to phenolic resin.

The steam explosion process is a recent development in wood processing (1, 2). Much attention has been paid to this process from the viewpoint of total wood utilization. Cellulose and hemicellulose from this process can be converted into sugars of commercial value by enzymatic methods (3). However, the conversion of lignin from this process (steam explosion lignin) into useful materials continues to present difficulties. Preparation of adhesives from it is considered to be a feasible way to solve this problem.

Steam explosion lignin is reported to have more reactive functional groups (4), and to contain no sulfur, as compared to kraft (thio) lignin and lignin sulfonates. The absence of blocked reactive functional groups in the lignin must be an advantage to various chemical modifications by which the lignin would be utilized as an adhesive of commercial use.

The addition of phenol-formaldehyde precondensate to lignin or methylolated lignin has been known as a way to introduce phenolic reactivity into lignin, and this is usually applied to the preparation of lignin-based

0097–6156/89/0397–0334$06.00/0

adhesives (5-10). In this case, all the lignin does not combine with the precondensate. Some part of it is considered to function as filler. The amount of lignin added is usually limited by some strength requirement.

In order to make lignin molecules contribute to bond strength, lignin should be incorporated into the phenolic main chain structure by a phenolysis reaction, for instance. There have been two kinds of lignin phenolysis reactions reported. Muller and Glasser have prepared phenolated lignins by a two-step reaction (11). Kraft, acid hydrolysis, and steam explosion lignins were allowed to react with formaldehyde before the methylolated lignins were combined with phenol. The adhesion quality of the phenolated lignin/phenol-formaldehyde resins was investigated, and the phenolated steam explosion lignin/phenol-formaldehyde resin was found to have slightly lower bond strength than the kraft lignin/phenol-formaldehyde and a neat phenolic resin. It was suggested that a considerable amount of syringyl propane units in the steam explosion aspen lignin was partly responsible for the lower bond strength as compared to the neat phenolic resin (12).

The other phenolysis reaction is the direct reaction of phenol with the substituted propyl side chain of lignin. Wacek et al. elucidated the chemistry of this reaction by oxidation (13). They concluded that condensation occurred between the o- or p-position of phenol and lignin's α-position substituted by OH, O-R_1, =O or =C-R_2 (R_1 and R_2: lignin residue). This mechanism has been confirmed by Kratzl et al. by using radioactive ^{14}C labeled lignin (14). Kobayashi et al. have prepared phenolated lignin by applying this phenolysis method in order to obtain thiolignin-based molding compounds (15). This method is suitable for hardwood lignins so far as they have sufficient functional groups at their α-positions of the propyl side chain. Since the steam explosion process has usually been applied to hardwoods, the latter phenolysis method is of interest from the viewpoint of adhesive preparation as compared to the former phenolysis.

Experimental

Materials. Lignin (SEL) was extracted with dilute alkali from steam exploded white birch (*Betula platyphylla*) pulp. The pulp was prepared at 200°C for 10 minutes. The ^{13}C NMR measurement of its acetylated product provided a methoxy:aryl ratio of 1.48, and this is in good agreement with Obst and Landucci's value of 1.44 (16). The molecular weight of C_6-C_3 units of SEL was calculated to be about 180. Chemical reagents were the first grade in the Japanese industrial standard. They were used as received. A commercial phenolic resin (D-17 of Ohshika Co.) was used in order to compare with bond strength of lignin-based adhesives. A commercial additive (Hot P of Ohshika Co.) and an extender (wheat flour: Akahana of Nisshin Co.) were used for the adhesive formulation.

Phenolysis of SEL. Phenol (120 g) was charged into a 300 ml round bottom flask equipped with thermometer, stirrer and cooler. SEL (60 g) was added slowly so that it could dissolve thoroughly. Sulfuric acid (0.5 ml) was then added as a catalyst. The typical sulfuric acid:SEL charge ratio was 1.5 mMole/g. The reaction schedule consisted of 1 hr of warm-up period and 3 hrs of phenolysis at 170°C, followed by removal of unreacted phenol under reduced pressure. The phenolated SEL was a black substance which dissolved more easily in organic solvents, especially in alcohols, than did SEL. The softening point of the phenolated SEL was about 140°C. Other phenolysis lignins, with different charge ratios of sulfuric acid, were prepared to compare with the amount of bound phenol.

Preparation of Phenolated SEL-Formaldehyde Resins. Phenolated SEL-formaldehyde resins (LP's) were prepared in a manner similar to resole-type phenol-formaldehyde resin by using an alkaline catalyst. In a 100 ml flask equipped with thermometer, stirrer and cooler, an appropriate amount of formalin was combined with the equivalent amount of methanol as a dissolving aid, and 50% aqueous NaOH solution (3.2 g). The powder of the phenolated SEL (20 g) was then slowly added to the mixture, and it was forced to dissolve thoroughly by agitation. The reaction schedule consisted of 15 minutes of warm-up period, 60 minutes of methylolation at 60°C and 10 minutes of warm-up period to 80°C, followed by condensation at 80°C. After the reaction was complete, methanol and some water were removed under reduced pressure, resulting in a black viscous resin. Three formaldehyde charge levels of 1.5, 3 and 5 moles were employed based on functionality of the phenolated SEL for the resin preparation. Solid content of the three resins was adjusted with water to around 46%. Preparation of the three resins was carefully controlled in order to adjust their viscosities to around 0.3 Pa·s, which is supposed to be in a desirable range for adhesive application. LP's from formaldehyde charge ratios of 1.5, 3 and 5 were named as LP-A, LP-B, and LP-C, respectively. Methylolation of SEL itself was also carried out in the same manner as LP preparation in order to compare their bond strength with those of LP's.

Adhesive Formulation. Adhesives for testing consisted of 100 parts resin, 10 parts wheat flour as extender, 4 parts commercial additive, and 2 parts water. This formulation is recommended by Japanese board manufacturers. Adhesives without wheat flour were also formulated to examine the bond strength of the neat lignin based resins.

Specimen for Tensile Shear Adhesion Test. Single lap specimens for shear strength tests were made from birch test panels (length: 80 mm, width: 25 mm, depth: 3 mm). Two of the panels were glued with the adhesives in a spreading rate of 1.5 g/100 cm^2 for single glue line. Bonding area of the specimens was 25 × 13 mm. The specimens were first pressed in an ambient temperature for an hour under a pressure of 0.78 MPa and then hot-pressed at 140°C under a pressure of 0.98 MPa for 6 minutes.

^{13}C NMR Spectroscopy. ^{13}C NMR measurements were carried out using a JEOL JMN-GSX 400 spectrometer for quantitative analysis in order to examine the amount of bound phenol in the phenolated SEL's. The analysis was conducted in DMSO-d_6 by using gated decoupling technique.

Gel Permeation Chromatography (GPC). Molecular weight distributions of phenolysis lignis during the phenolysis were examined at 60°C in THF as mobile phase by using a Toyo Soda HPLC-802 UR gel permeation chromatograph equipped with two 60 cm polystyrene gel columns in series (TSK-GEL H G2500 and G1000). The chromatograms were monitored by refractometer.

Torsional Braid Analysis. LP-A, LP-B, and LP-C were coated on glass braids and their cure behaviors were examined by using a RHESCA RD-1100A torsional braid analyzer. Their relative rigidity changes along with cure temperature (heating rate was 1°C/min) were monitored in order to examine the cure speed of the LP resins. The relative rigidity changes during cure at 140°C were also measured to examine the cure speed dependence on resin pH.

Tensile Shear Strength Measurement. The tensile shear bond strength was determined in accordance with Japanese standard JIS K 6851 (normal and repeated boiling test) by using a Toyo Seiki Strograph W tensile tester. In the normal test, the test specimens were conditioned for at least 48 hrs at 20 ± 5°C under a relative humidity of 65 ± 20% before tensile shear strength was measured. In the repeated boiling test, the specimens were immersed in boiling water for 4 hrs, dried at 60 ± 3°C for 20 hrs and immersed again in boiling water for 4 hrs, followed by immersion in water at room temperature until cooled. The specimens were tested in wet state. Six specimens were tested for each resin.

Results and Discussion

Optimum Reaction Period between Phenol and Lignin. The end point of the phenolysis reaction can be estimated by the period when the consumption of phenol reaches equilibrium. Phenolysis products were monitored by GPC during the reaction. The change in molecular weight distribution during the phenolysis is shown in Figure 1. The increase of the peak at 19 min clearly demonstrates that polymerization of SEL occurs as the reaction proceeds. The ratio of peak area of phenolated lignin (16 to 28.7 min in elution time) to that of unreacted phenol (27.9 min) vs. reaction time is shown in Figure 2. Unreacted phenol diminishes rapidly at the early stage of the reaction before reaching equilibrium within 3 hrs. This was defined as optimum phenolysis reaction period.

Functionality Measurement of Phenolated Lignin. It is important to have knowledge of the functionality of the phenolated lignin from the point of view of further chemical modification. The amount of bound phenol in the phenolysis reaction has been measured by titrating the phenol extracted from the reaction mixture (15). This indirect method measures the unreacted phenol and determines bound phenol as the difference between the initial charge and the titrated phenol. This is sometimes misleading. ^{1}H NMR spectroscopy is another candidate for the determination of the amount of bound phenol. However, this calculation is difficult since the number of protons before and after the phenolysis reaction is unknown.

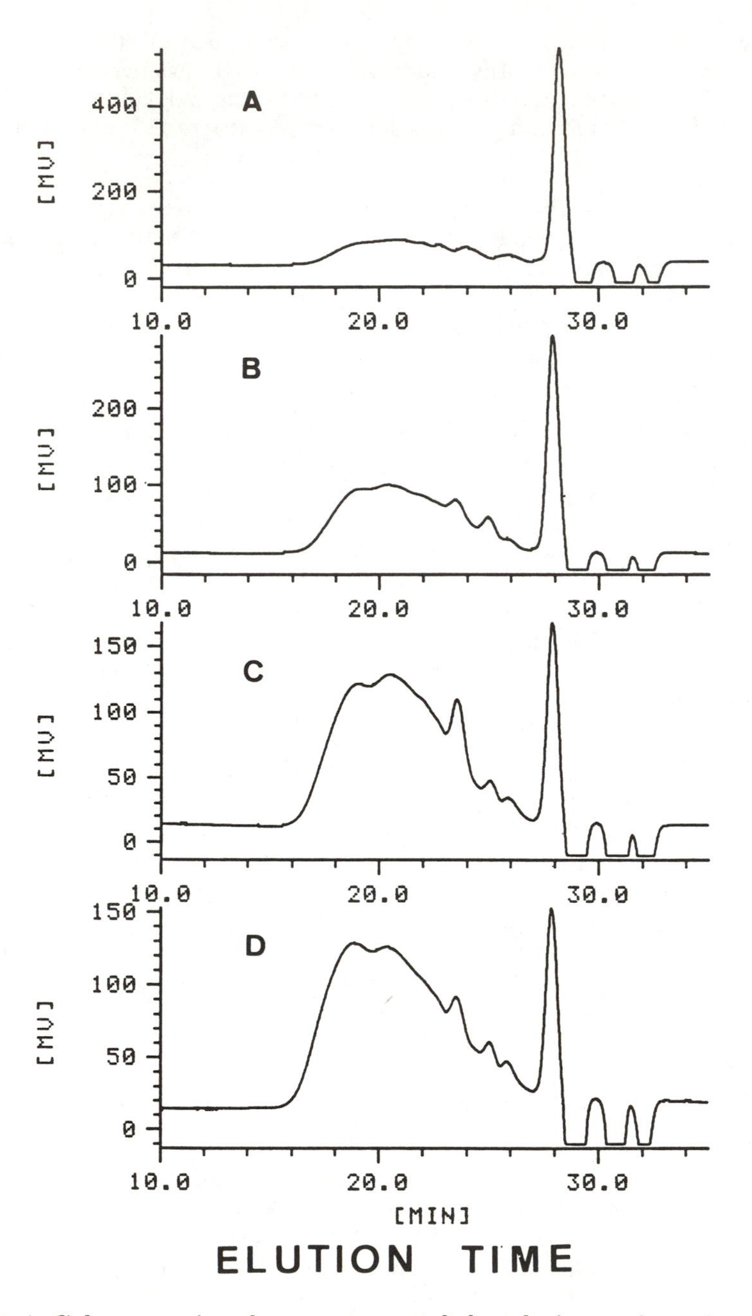

Figure 1. Gel permeation chromatograms of phenolysis reaction mixtures—Phenol mixtures reacted for (A) 10 min, (B) 60 min, (C) 120 min, and (D) 240 min.

Recently, quantitative ^{13}C NMR analysis of lignins has been developed (17) and improved (18). This method would provide more accurate values as compared to titration, if the NMR signals specific to bound phenol can be distinguished from the others. Partial ^{13}C NMR spectra of SEL and of phenolated SEL (sulfuric acid 1.5 mMole/g SEL, reaction period 3 hrs) are illustrated in Figure 3. Phenolysis gives rise to a new signal at 155.3 ppm which can be assigned to the carbon adjacent to the OH group in bound phenol. This new signal is distinguished from the signals of unreacted phenol (157.7 ppm) and of aromatic ring carbons of guaiacyl and syringyl units (152.2 ppm to 103.5 ppm) (19). Assuming that the steam explosion lignin is composed of guaiacyl and syringyl units, the amount of phenol bound to the lignin can be quantitatively determined by using the gated decoupling technique (18). Denoting the peak area at 155.3 ppm as A, and the peak group area ranging from 152.2 ppm to 103.5 ppm as B, the ratio of B/A can be expressed as follows:

$$B/A = (5P + 6R)/P \tag{1}$$

where P is the molar ratio of phenol bound to lignin, and R is the molar ratio of aromatic nuclei in the lignin. Therefore, P/R is the molar ratio of bound phenol to lignin unit and it is derived from Equation 1 as:

$$P/R = 6/(B/A - 5) \tag{2}$$

The spin-lattice relaxation times of lignins and phenolic precondensates have been reported to be 0.1 to 3 seconds (17) and 2.7 seconds (20), respectively. Landucci has found that more than 6 second relaxation delay time is required for a 45° tip angle (18). The tip angle used was 45°, and the relaxation delay time employed was 10 seconds. The results are summarized in Table I. As the addition of sulfuric acid increases, and the reaction period is extended, the amount of bound phenol increases. Although it has been reported that thiolignin has reacted with 0.34 moles of phenol (15), the results show that SEL has an ability of reacting with more than one mole of phenol. This is probably due to fewer blocked reactive functional groups in SEL as compared to thiolignin. The functionality of the phenolated lignin can be calculated to be about 2.9, assuming that o- and/or p-positions of the bound phenol and the 5-position of the lignin are reactive. The molecular weight of the phenolysis lignin unit is calculated to be about 290 by adding the amount of bound phenol to the molecular weight of the C_6-C_3 unit of SEL.

Cure Rate of the Phenolated SEL Resins. ^{13}C NMR spectra of the phenolated SEL formaldehyde-treated resins revealed the formation of methylol groups. A similar cure reaction to resole type phenolic resins is expected to occur with the phenolated lignin-based resins. Since cure rate normally determines production capacity of a board mill, it is important that new types of adhesives have at least the same cure rate as the conventional phenolic adhesives. Cure analysis of resins has usually been examined by

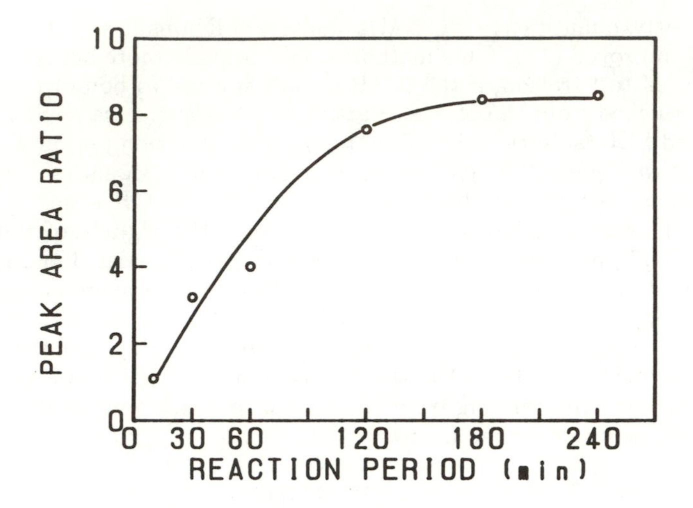

Figure 2. Ratio of peak areas of the phenolated lignin to unreacted phenol as an index of phenol consumption during phenolysis.

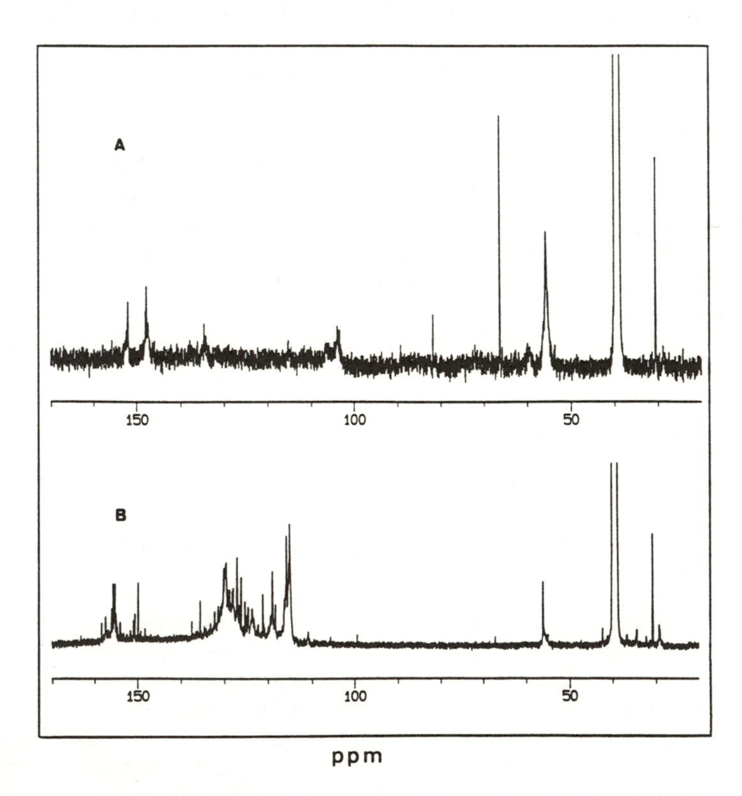

Figure 3. Partial ^{13}C NMR spectra of (A) the steam explosion lignin and (B) its phenolated product.

Table I. Amount of Phenol Bound to Lignin under Various Reaction Conditions

Amount of Sulfuric Acid (mMole/g lignin)	Reaction Period at 170°C (Hour)	Bound Phenol (Mole/lignin unit)
0.5	12	1.10
1.5	1	0.95
	3	1.19
	6	1.18
2.5	3	1.28
4	3	1.31

means of calorimetric methods like differential scanning calorimetry (12), or by mechanical analysis like TBA (21), which is considered to be sensitive to the development of adhesive strength. The latter was employed in this study in order to examine cure rate of the adhesive resins.

Relative rigidity vs. temperature curves of LP's are shown in Figure 4 in comparison with a commercial phenolic resin. The pH of these resins was previously adjusted to around 10.8. The phenolic resin is fully cured at around 75°C. By contrast, the curves of the three lignin-based resins exhibit slower cure as compared to the phenolic resin. The retardation increases as the charge ratio of formaldehyde increases. Some retardation had already been found, but neglected, for the phenolated lignin/phenol-formaldehyde resins (12). In this study, the neat phenolated SEL was used for resin preparation. It can be concluded that phenolated steam explosion lignin-based resins have an intrinsically retarded cure behavior as compared to phenolic resin at the same pH.

Cure Rate Dependence on pH for the Phenolysis Lignin Resins. Cure behavior at different pH's of the resins was measured at 140°C, which is the usual hot-pressing temperature of phenolic resins. Relative rigidity change curves of LP-B at different pH's are illustrated in Figure 5. Cure advances faster as the pH of the resin increases. When the pH is 11.9, LP-B provides faster cure than the phenolic resin. A similar tendency has been found for LP-C. These findings clearly demonstrate that increasing pH of the resins improves cure rate.

Evaluation of Adhesive Bond Strength. Since the cured product of LP-A was brittle, probably due to the low level of formaldehyde charged, LP-B and LP-C were selected as adhesive resins. In order to increase their cure rate, the pH's of LP-B and LP-C resins were adjusted to around 11.2 and 12.0, respectively, prior to adhesive formulation. For the purpose of comparison, the bond strength of the methylolated SEL resin adhesive was examined as well.

The results are summarized in Table II. The bond strength of the adhesives from the methylolated SEL expectedly exhibit poor adhesion as compared to the phenolated SEL adhesives, especially after repeated boil.

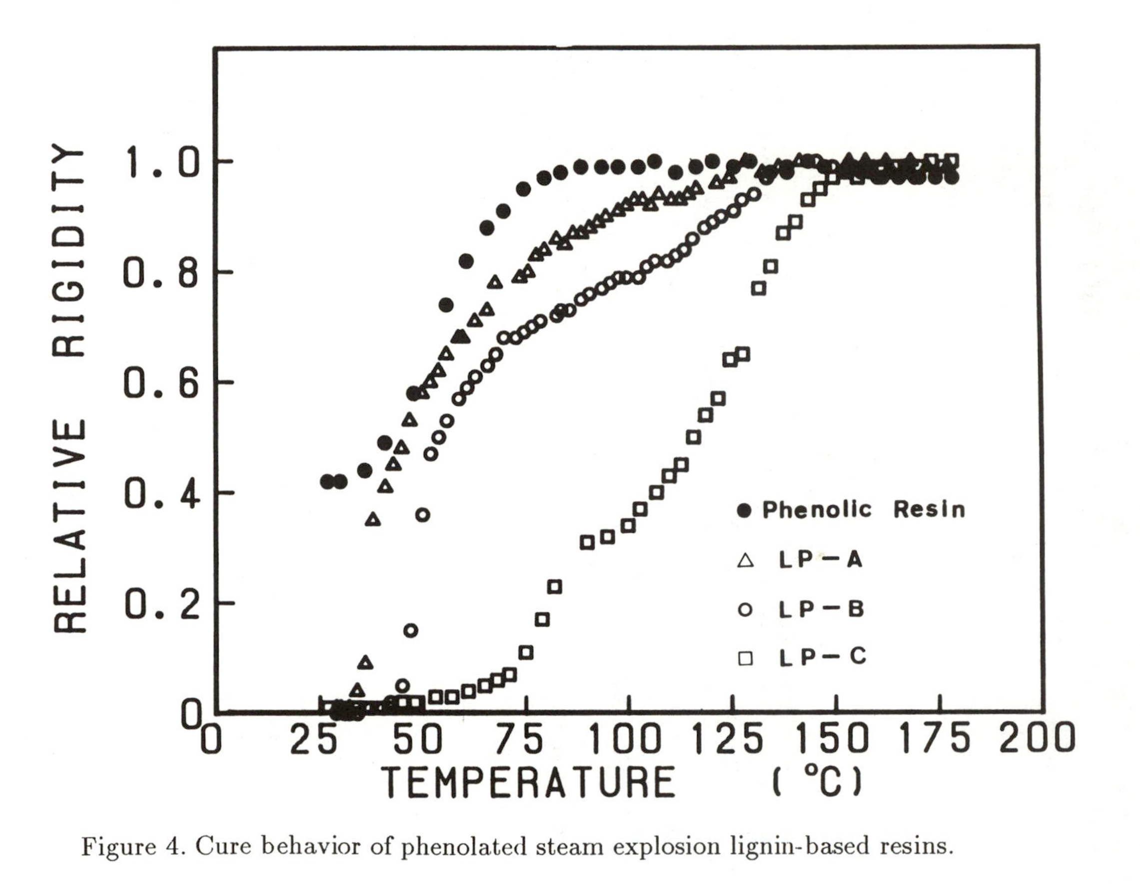

Figure 4. Cure behavior of phenolated steam explosion lignin-based resins.

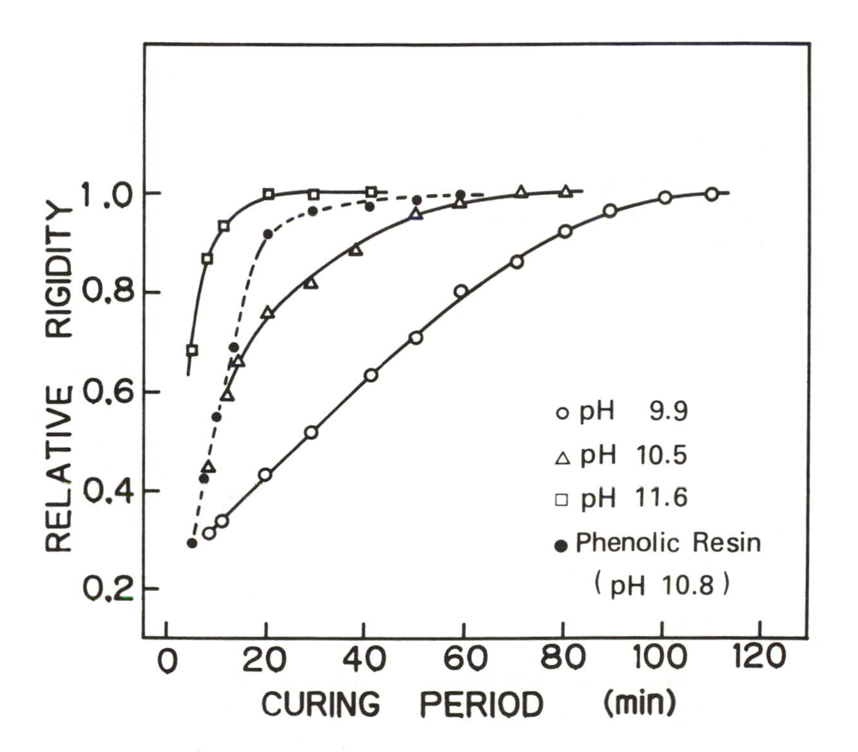

Figure 5. Cure rate dependence on pH for the phenolated steam explosion lignin-based resin LP-B.

This might be explained with failure to crosslink sufficiently. Sano et al. have also reported that phenolysis of hardwood lignin sulfonates enhances adhesion properties (22). It is thus clearly demonstrated that the introduction of phenol into lignin improves adhesion properties.

Table II. Tensile Shear Bond Strength of Adhesives from Phenolysis Lignin in Normal Test and after Repeated Boil Treatment

Adhesives	Bond Strength (10^5 Pa) (Standard Deviation)	
	Normal	Repeated Boil
Adhesives from LP-B		
with extender	47.2 (8.8)	52.9 (6.1)
without extender	65.3 (9.6)	43.5 (3.3)
Adhesives from LP-C		
with extender	41.5 (9.6)	55.6 (7.9)
without extender	68.4 (7.8)	51.4 (6.7
Adhesives from methylolated SEL		
with extender	45.9 (9.2)	0
without extender	32.0 (10.1)	0
Adhesive from commercial phenolic resin		
with extender	62.6 (6.6)	47.3 (11.5)

The two phenolated SEL adhesives with wheat flour do not provide good bond strength in the normal test, but provide excellent strength after repeated boil. The bond strength increment after repeated boil indicates the occurrence of post-cure in the phenolated SEL adhesives. The acidity of the wheat flour might retard resin cure by reducing the pH of the adhesives (23). LP-C provided better bond quality as compared to LP-B. Since the difference between LP-B and LP-C lies in the formaldehyde ratio charged, this finding suggests that a stoichiometric charge might not be adequate for the formation of methylol groups in LP's. There might be a problem with the reaction of formaldehyde with the phenolated SEL as compared to phenol. Generally speaking, the phenolated SEL adhesives without wheat flour provide bond strengths comparable to the phenolic resin in normal and repeated boil tests. It has been reported that the resin from the reaction of formaldehyde with the mixture of the two-step phenolated steam explosion lignin and phenol has also provided comparable bond strength (12). It is noteworthy that the phenolated SEL lignin was directly methylolated without any addition of phenol.

Although there are still some problems, such as the ineffective methylolation of the phenolated SEL and the selection of a suitable extender, it can generally be concluded that phenolysis is a promising method to develop steam explosion lignin into attractive adhesives comparable to commercial phenolic resin.

Conclusions

Quantitative analysis of phenolysis lignin by ^{13}C NMR has indicated that the amount of phenol bound to the steam exploded lignin is unexpectedly large as compared to thiolignin. The steam explosion lignin itself could react with formaldehyde, resulting in thermosetting resins. This resin, however, displayed poor adhesion properties, especially in the repeated boil test. The phenolated lignin-formaldehyde resins provided excellent adhesives comparable to a commercial phenolic resin. Phenolysis appears to be a promising method to utilize steam explosion lignins as adhesives.

Acknowledgments

This research was funded by the Biomass Transfer Program of the Japanese Ministry of Agriculture, Forestry and Fisheries. The authors would like to express their thanks to Mr. H. Ohara for extracting lignin and to Ms. Yao Xin of the Forest Products Research Institute of Heilongiang Province in China for preparing the phenolated lignin during her ten months leave at their Institute.

Literature Cited

1. Dietrichs, H. H.; Sinner, M.; Puls, J. *Holzforschung* 1978, **32**, 193.
2. Marchessault, R. H.; St.-Pierre, J. In *Future Sources of Organic Raw Materials–CHEMRAWN I*; St.-Pierre, L. E.; Brown, G. R., Eds.; Pergamon: New York, 1980; p. 613.
3. Shimizu, K.; Sudo, K.; Nagasawa, S.; Ishihara, M. *Mokuzai Gakkaishi* 1983, **29**, 428.
4. Sudo, K.; Shimizu, K.; Sakurai, K. *Holzforschung* 1985, **39**, 281.
5. Johansson, I. J. Ger. Offen. 2 624 673, 1976.
6. Forss, K. J.; Fuhrmann, A. G. M. Ger. Offen. 2 601 600, 1976.
7. Adams, J. W.; Schoenherr, M. W. U.S. Pat. 4 306 999, 1981.
8. Enkvist, T. U. E. U.S. Pat. 3 864 291, 1975.
9. Sakakibara, A. Japanese Pat. 51-22497, 1976.
10. Clarke, M. R.; Dolenko, A. J. U.S. Pat. 4 113 675, 1978.
11. Muller, P. C.; Glasser, W. G. *J. Adhesion* 1984, **17**, 159.
12. Muller, P. C.; Kelley, S. S.; Glasser, W. G. *J. Adhesion* 1984, **17**, 185-206.
13. von Wacek, A.; Daubner-Rettenbacher, H. *Monatsch. Chem.* 1950, **81**, 266.
14. Kratzl, K.; Buchtela, K.; Gratzl, J.; Zauner, J.; Ettingshausen, O. *Tappi* 1962, **45**, 113.
15. Kobayashi, A.; Haga, T.; Sato, K. *Mokuzai Gakkaishi* 1966, **12**, 305.
16. Obst, J. R.; Landucci, L. L. *Holzforschung* 1986, **40**, 87.
17. Robert, D. R.; Bardet, M.; Gagnaire, D. *Proc. Inter. Symp. Wood and Pulping Chem.*, Tsukuba Science City, Japan, 1983, Supp. Vol., p. 1.
18. Landucci, L. L. *Holzforschung* 1985, **39**, 355.
19. Lapierre, C.; Monties, B. *Holzforschung* 1984, **38**, 333.
20. Fujita, N.; Ogasawara, K. *Netsukoukasei Jyushi* 1982, **3**, 31.
21. Steiner, P. R.; Warren, S. R. *Holzforschung* 1981, **35**, 273.
22. Sano, Y.; Ichikawa, A. *Mokuzai Gakkaishi* 1987, **33**, 47.
23. Sellers, T., Jr. *Plywood and Adhesive Technology*; Marcell Dekker: New York, 1985; Ch. 19.

RECEIVED February 27, 1989

Chapter 26

Modification of Lignin at the 2- and 6-Positions of the Phenylpropanoid Nuclei

Gerrit H. van der Klashorst

Renewable Resources Chemicals Programme, Division of Processing and Chemical Manufacturing Technology, CSIR, P.O. Box 395, Pretoria, South Africa

Lignin model compounds were reacted with formaldehyde in acid medium at positions *meta* to the aromatic hydroxy groups. Both phenolic and etherified phenolic lignin model compounds were shown to give fast polymerization or hydroxymethylation of the *meta* positions depending on the conditions used. This reaction differs from the reaction of formaldehyde with phenolic lignin model compounds under alkaline conditions, where the reaction with formaldehyde always occurs at positions *ortho/para* to the aromatic hydroxy group. The conditions developed for the polymerization and hydroxymethylation were also evaluated on industrial lignin. The reaction of formaldehyde with the *meta* positions of lignin clearly have considerable potential for the use of lignin, particularly heavily condensed alkali lignin, in polymeric applications.

Lignin, present as a waste material in the spent liquors of the pulping industry, constitutes a potentially useful raw material for the production of various polymeric products. This has repeatedly been demonstrated in the past by its application in products ranging from wood adhesives to plastics (1). In these applications the lignin is crosslinked to increase its molecular mass or to form a rigid three-dimensionally crosslinked structure.

The features on lignin that were utilized for these polymerization or modification reactions are as follows (compare Fig. 1):

1. phenolic hydroxy group;
2. aliphatic hydroxy group;
3. unsubstituted 3- or 5-positions on C_9 units; and
4. structures that can form quinonemethide intermediates.

0097–6156/89/0397–0346$06.00/0

Examples where the phenolic hydroxy groups were utilized include the preparation of lignin epoxies by reaction with epichlorohydrin (2), esterifications with bis-acid chlorides (3) and cyanuric chloride (4), and polymerization with aziridines (5).

The aliphatic hydroxy groups have been used as the polyol component of urethane resins (6). In many of these applications phenolic hydroxy groups have been alkoxylated to afford more aliphatic alcohols.

The unsubstituted 3- and 5-positions on the C_9 units of alkali lignin have been used in two types of polymerization reactions—electrophilic displacement reactions and free radical oxidative coupling reactions. The electrophilic displacement reaction approaches included the use of lignin in phenol formaldehyde resins (7), urea formaldehyde resins (8), and the crosslinking of lignin with diazonium salts (9). Oxidative coupling reactions via free radical mechanisms yielded lignin based adhesives of high strength (8).

In this paper an additional approach that can be used for the modification or polymerization of especially alkali lignin is discussed. That is:

The Modification of Lignin at the 2- and 6-Positions of the Phenylpropane Units

It was shown by Sarkanen *et al.* that hard and softwood model compounds undergo protodedeuteration in acidic medium preferably on positions 2 and 6 (9). They suggested that the observed 2, 6-protodedeuteration reactions should also hold for other reagent species.

Electrophiles other than protons were indeed shown to react with lignin model compounds at the 2- and 6-positions in acidic media: Kratzl and Wagner investigated the reaction of paraphenolic benzyl alcohols in alkaline and acidic solutions (10). When 4-hydroxy-3-methoxybenzyl alcohol 4 was reacted with the 4-alkyl substituted phenol 5 under alkaline conditions, the expected *ortho* linked product (6) was isolated. However, under acidic conditions the methylene linkage formed *meta* to the phenolic hydroxy group (Fig. 2).

Yasuda and Terashima (11) recently showed that lignin under Klason lignin determination conditions afforded *meta*-linked products. Vanillyl alcohol (4) yielded after reflux in 5% sulfuric acid, *meta*-linked product (8) (Fig. 3).

These examples thus clearly show 4-hydroxy-3-methoxyphenyl and 4-hydroxy-3, 5-dimethoxyphenylalkanes (typical of the lignin structure) to be reactive at positions 2 and 6 towards electrophilic substitution reactions under acidic conditions. The subject matter of this paper deals with the development of procedures whereby the 2- and 6- (i.e., *meta*) positions are utilized for the polymerization and modification of lignin.

Polymerization of Lignin Model Compounds. The controlled polymerization of several lignin model compounds at the "*meta*" positions was subsequently attempted (13):

The hardwood lignin model 13 was reacted in acidic aqueous dioxane to afford dimers, trimers, tetramers or higher oligomers (Fig. 4A).

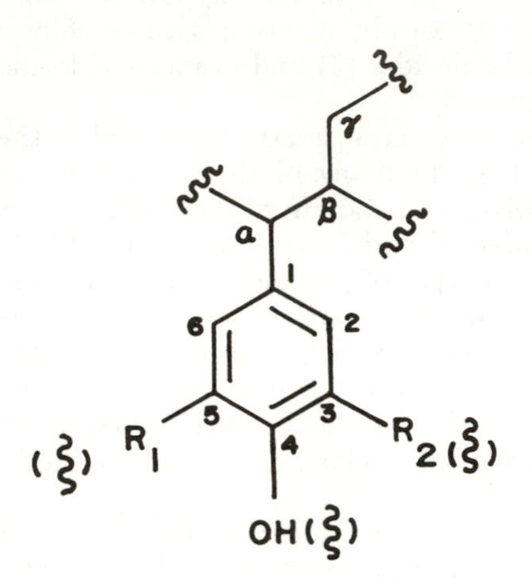

Figure 1. Phenylpropane building units of lignin. 1 = R_1, = R_2 = H. 2 = R_1 = OMe. 3 = R_1 = R_2 = OMe.

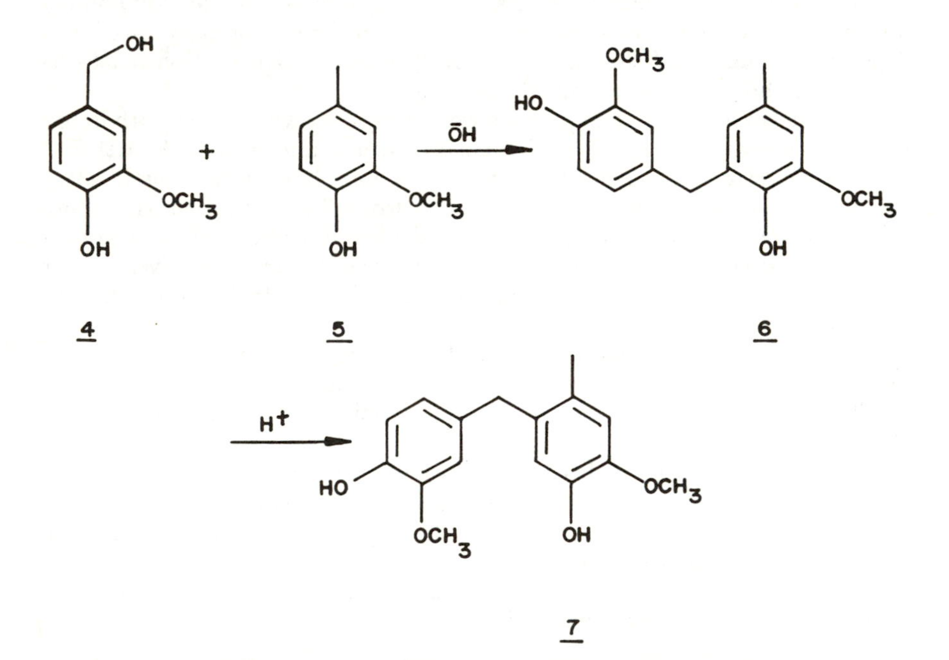

Figure 2. The reaction of vanillyl alcohol (4) with a reactive phenol (5).

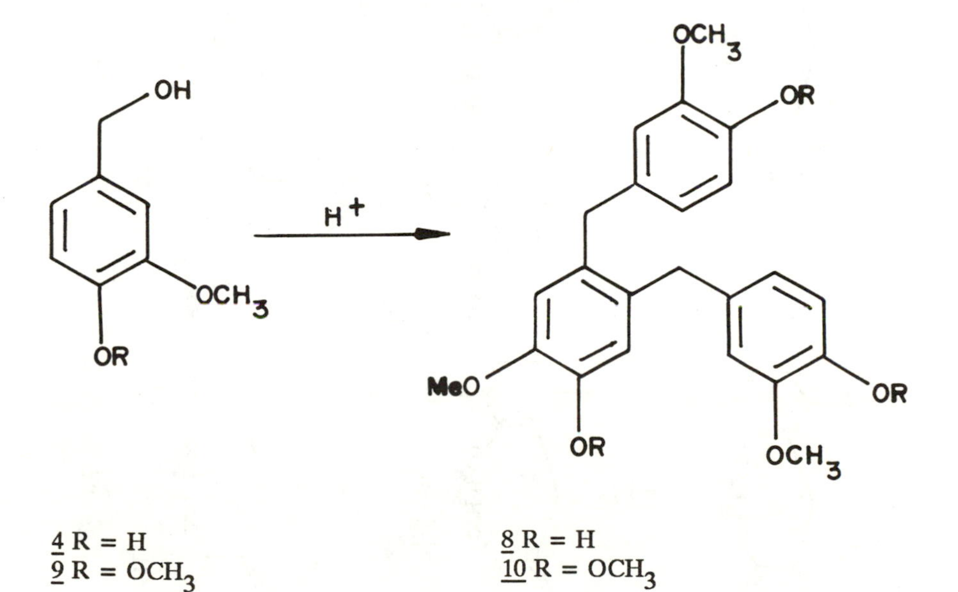

4 R = H
9 R = OCH_3

8 R = H
10 R = OCH_3

Figure 3. Self-condensation of vanillyl alcohol (4) and 3,4-dimethoxybenzyl alcohol (8) in acid (11).

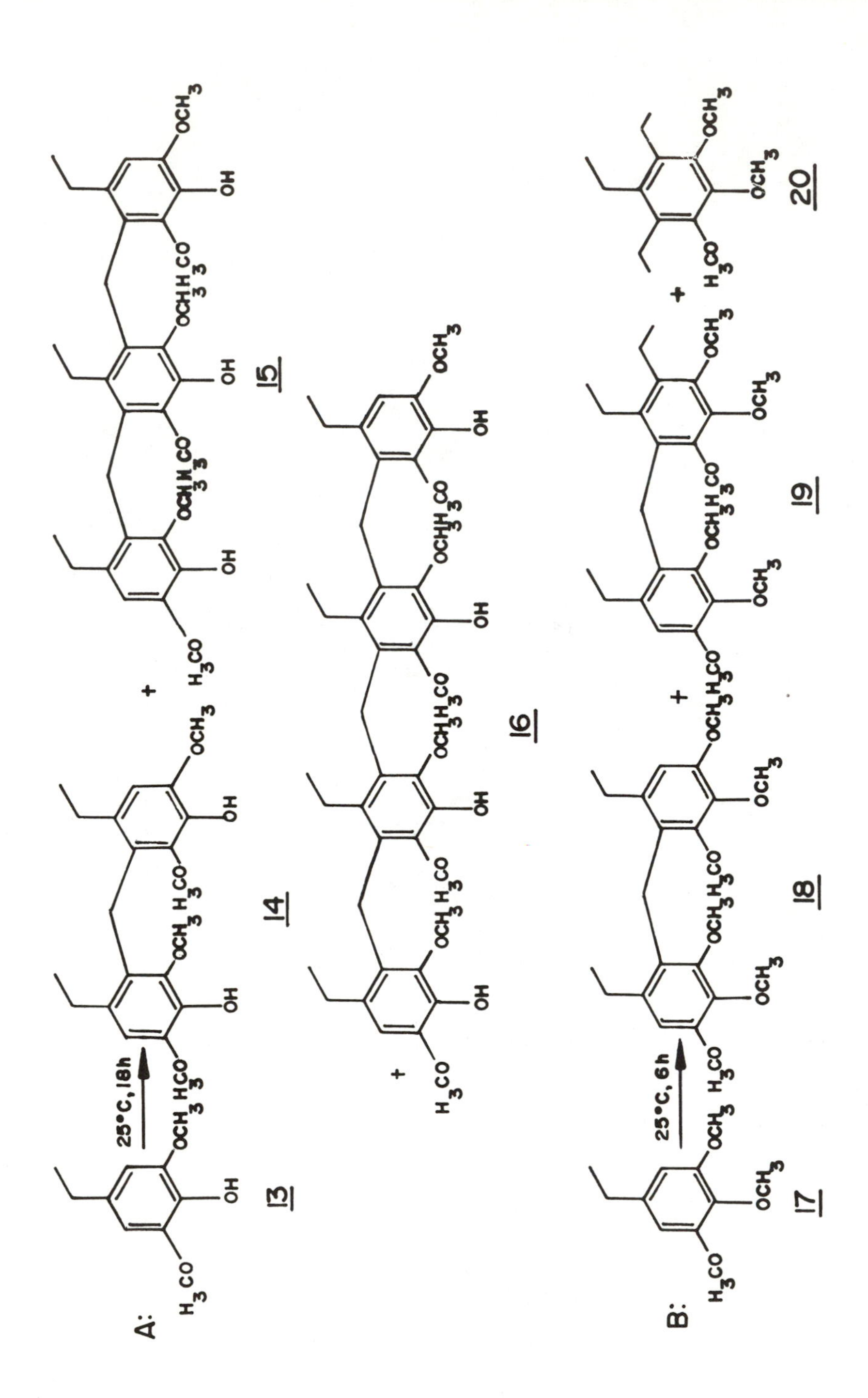
A:
25°C, 18h
13
14
15
16
B:
25°C, 6h
17
18
19
20

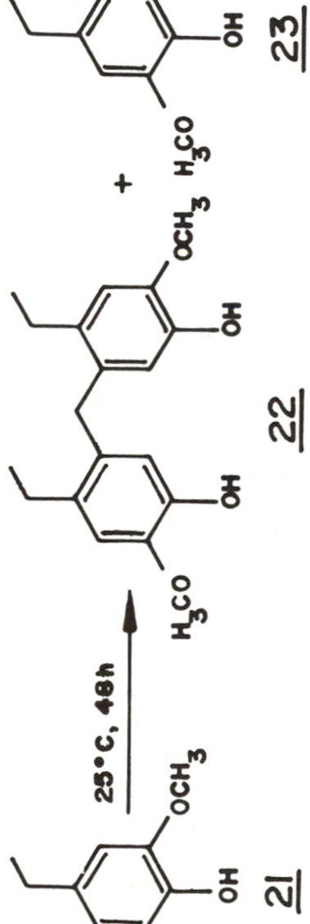

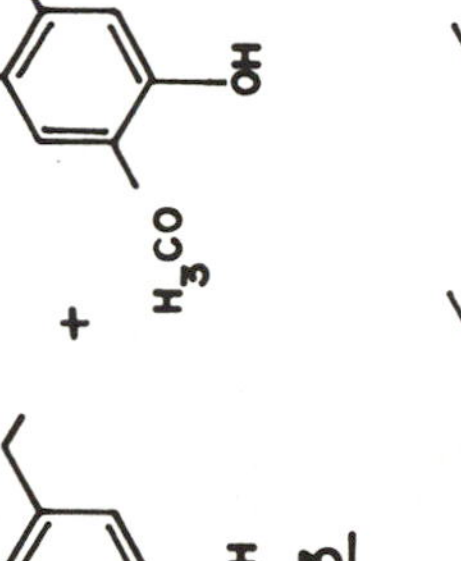

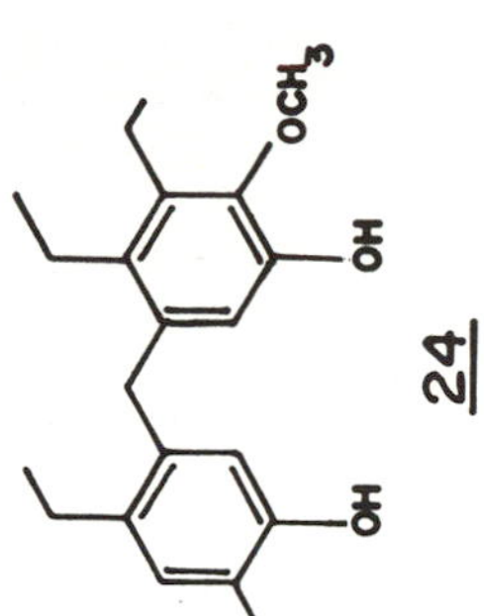

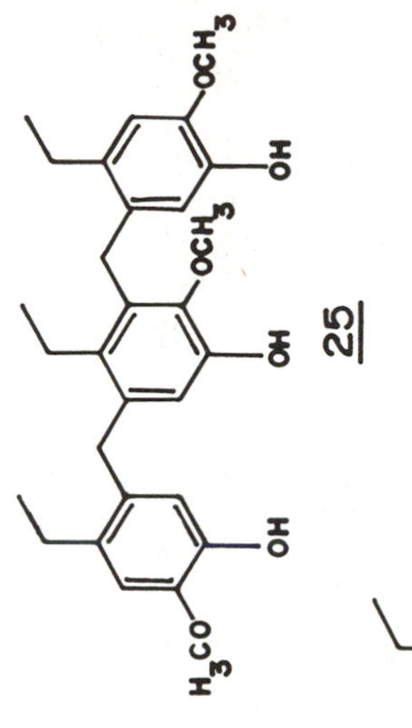

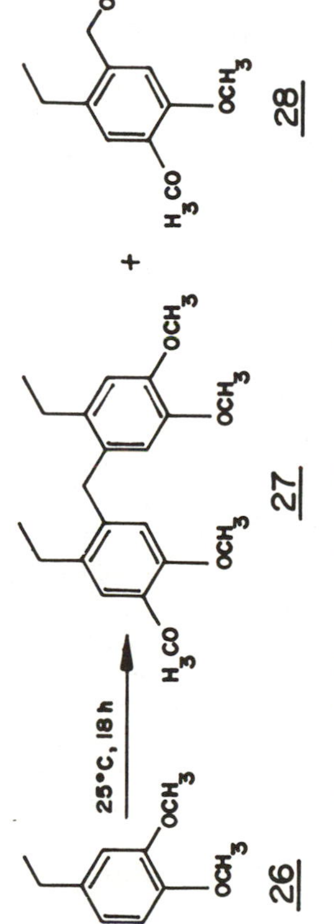

Figure 4. The reaction of lignin model compound with a twenty-fold excess formaldehyde in 0.82 mol/dm^3 hydrochloric acid in aqueous dioxane (13).

The etherified hardwood lignin model 17 reacted at a similar rate as the phenolic model indicating the etherification of the phenolic group has a small effect on the reaction rate. When this reaction was repeated at 55°C with an excess formaldehyde, some *meta*-hydroxymethylated products were obtained (Fig. 4B).

Phenolic and etherified softwood lignin model compounds were also successfully crosslinked at the *meta* positions. Special care was taken to assign the structure of the dimer (22) (Fig. 4C). With NMR-SPI (selective population inversion) techniques it was unevocably proved that the methylene linkage was situated at the 6-positions (13).

Mechanism

The mechanism for the reactions on the 2- and 6-positions is based on two factors (Fig. 5): induction by the alkyl group; and resonance effects of the methoxy group.

Alkyl groups without any oxy-function on the α-carbon are electron donating by induction and will enhance the electron density on the 2-, 4-, and 6-positions of the lignin alkyl rings. During alkaline pulping a major portion of α-oxy-functions are lost due to condensation reactions, i.e., methylene linkage formation with the α-carbon as bridge. Alkali lignin, and especially highly condensed alkali lignin, can therefore be expected to have a high number of non-α-oxy-substituted side chains. Alkali lignin can therefore be expected to have a high reactivity towards modification at positions 2 and 6.

During the investigation on hard- and softwood model compounds it was observed that the dimethoxy model compounds (hardwood) were about five times more reactive than the mono methoxy model compounds. This illustrates the beneficial effect of the presence of the second methoxy group (by resonance effects).

From the results on the model compounds it was clear that:

- 1-alkyl substituted syringyl and guaiacyl model compounds react with formaldehyde in acidic medium at positions 2 and 6.
- The formed benzylic alcohols were unstable in acid and reacted fast with a second model compound to form methylene linked polymers.
- Etherified phenolic models reacted at the same positions and similar rates as did the phenolic models.
- The dimethoxy, i.e., hardwood, model compounds reacted faster than did the monomethoxy model compounds.

Meta Hydroxymethylation

In the above paragraph it was shown that the lignin C_9 units can be polymerized at the 2- and 6-positions with formaldehyde in acidic aqueous dioxane. The polymerization reaction proceeds presumably via a hydroxymethylated intermediate which was isolated in low yields only when large excesses of formaldehyde were employed. If the hydroxymethylation of the *meta* position can be achieved in high yield it would clearly afford a very reactive intermediate. The hydroxymethylation of the *meta* position was subsequently attempted on model compounds (14).

The hydroxymethylation of the hardwood model (13) was first attempted by using a 20 molar excess formaldehyde in 1, 1N HCl in 50% aqueous dioxane (14). Since 13 has two reactive sites (2 and 6), this means a ten-fold excess of formaldehyde per reactive site. The ratio of hydroxymethylation to methylene linkage formation was determined by ^{1}H-NMR analysis.

Samples were taken at different intervals, neutralized, extracted and acetylated for ^{1}H NMR analyses. The coefficients of the integrals of the aliphatic (δ2.0-2.2) and the aromatic (δ2.3-2.5) acetoxy groups of the proton NMR spectra, were taken as the yields of hydroxymethylation.

The degree of dimer formation via a methylene linkage (crosslinking) was estimated by the coefficient of the integrals of the proton NMR spectra of methoxy groups situated adjacent to the newly formed bond—of which the resonances are shifted upfield (δ3.4-3.7)—and that of the total methoxy resonances (δ3.4-4.0).

The large excess of formaldehyde resulted in the hydroxymethylation of 13 in yields of about 50% hydroxymethylation. This yield was obtained within 15 minutes and remained constant for over 6 hours. The number of unsubstituted aromatic positions decreased from 0.6/model compound after 15 minutes to 0.2 after 1 hour reaction time. However, after only 15 minutes about one methylene linkage per model compound occurred, indicating substantial methylene linkage formation.

The *meta*-hydroxymethylation yield was very dependent on large excess formaldehyde used and insensitive to acid concentration and temperature. After extensive exploration of the reaction condition variables it was discovered that a decrease in water content of the solvent results in the increase of hydroxymethylation yields. Eventually conditions were devised to achieve close to 100% *meta*-hydroxymethylation with no methylene linkage formation. Both softwood and hardwood model compounds gave only mono-*meta*-hydroxymethylation (Fig. 6). This is probably due to the deactivation by the electron pulling effect of the introduced hydroxymethyl group.

The reactions on the model are summarized in Table I.

The suitability of the different conditions found for the *meta*-hydroxymethylation and crosslinking of model compounds were subsequently evaluated on three alkali lignins obtained from different industrial spent liquors (15).

The alkali lignins used were selected from different industrial origins in order to have a wide variation in their properties. The lignins were a kraft softwood, soda/AQ hardwood and soda bagasse lignin (Table II).

The kraft lignin is constituted mainly of guaiacylpropanoid units (2) of which about half are condensed, i.e., substituted at the 5-position (16). The number of aromatic sites reactive towards base catalyzed electrophilic substitution reactions, therefore, is fairly low at about 0.3 per C_9 unit (Table II).

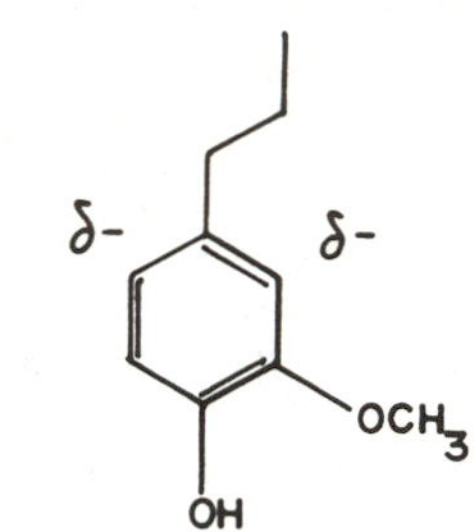

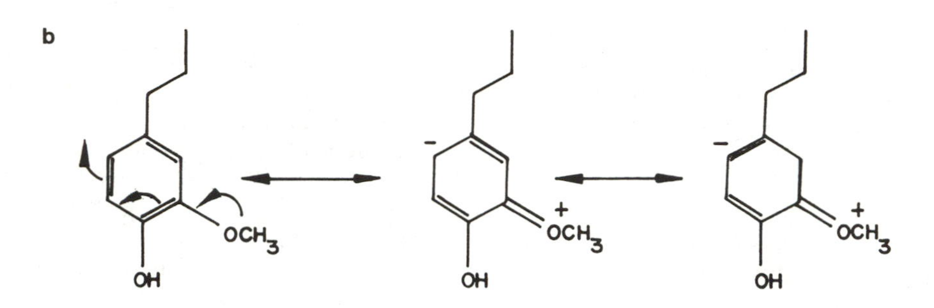

Figure 5. Mechanisms responsible for the higher electron density on positions 2 and 6 on lignin phenylpropane units in acidic medium. (a) the induction effect of the alkyl group at position 1; (b) the resonance effects of the electron pairs on the methoxy oxygen.

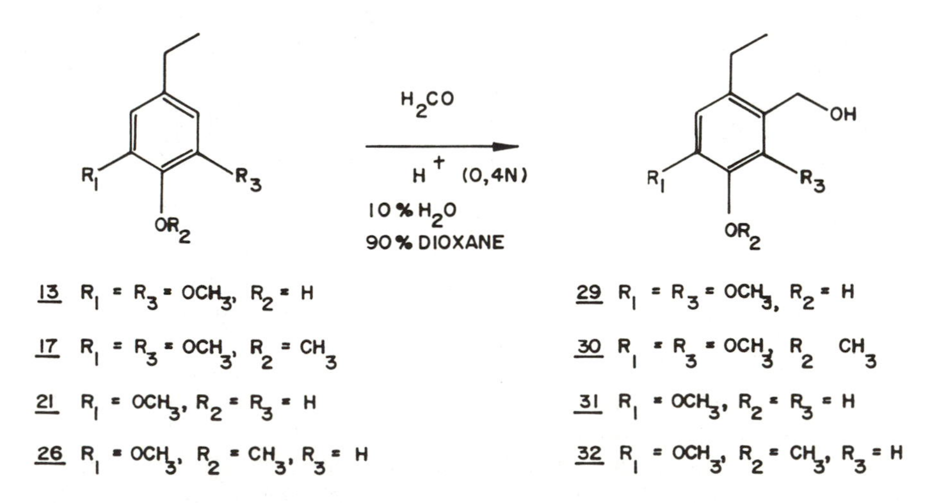

Figure 6. Model compounds hydroxymethylated at the 6 position in 10% aqueous dioxane and 0.4N HCl (1).

Table I. The reaction of 4-hydroxy-3,5-dimethoxyphenyl ethane (13) with formaldehyde in acidic aqueous dioxane at 80°C (14)

Mole H_2CO/ Mole 13	Solvent	$[H^+]$	Temp. (°C)	Time (h)	Product
1:1	50% aqueous dioxane	2.2N HCl	80	4	Mainly methylene linkages. No hydroxy-methylated compounds were isolated.
1:2	23% aqueous	1.0N HCl	80	4	100% monohydroxy-metela-dioxanetion + 20% dioxane methylene linkages
1:2	10% aqueous dioxane	0.4N HCl	80	4	100% monohydroxy-methylation

Table II. Properties of three industrial lignins (16-18)

Lignin	Type	Plant Material	Calculated Number of Unsubstituted 3/5 Positions on Aromatic Rings (/C_9)	Average Molecular (/C_9)
Kraft	Softwood	Pine	0.3	176
Soda/AQ	Hardwood	Eucalyptus	0.1	184
Soda	Grass	Bagasse	0.7	175

The soda/AQ hardwood lignin is constituted of both guaiacyl (2) and syringylpropanoid (3) units, of which the former have been condensed extensively during pulping (17). This is indicated by the low number of unsubstituted 5-positions (Table II).

The soda bagasse lignin, on the other hand, is a very reactive industrial lignin as is indicated by its high number of unsubstituted 3- and 5-positions of 0.7/C_9 on phenolic propanoid units. This lignin is only mildly condensed and contains 1.05 unsubstituted 3- or 5-positions.

The industrial lignins were firstly crosslinked with formaldehyde. Each lignin was reacted with an excess of 1.35 mole H_2CO/mol C_9 lignin unit. The reactions were performed in 2.2N hydrochloric acid in 50% aqueous dioxane, the conditions previously optimized for the polymerization of lignin model compounds (vide infra). Gelling of the lignin masses occurred only after a small proportion of the formaldehyde was consumed (Fig. 7) indicating that extensive methylane linkage formation (crosslinking) occurred.

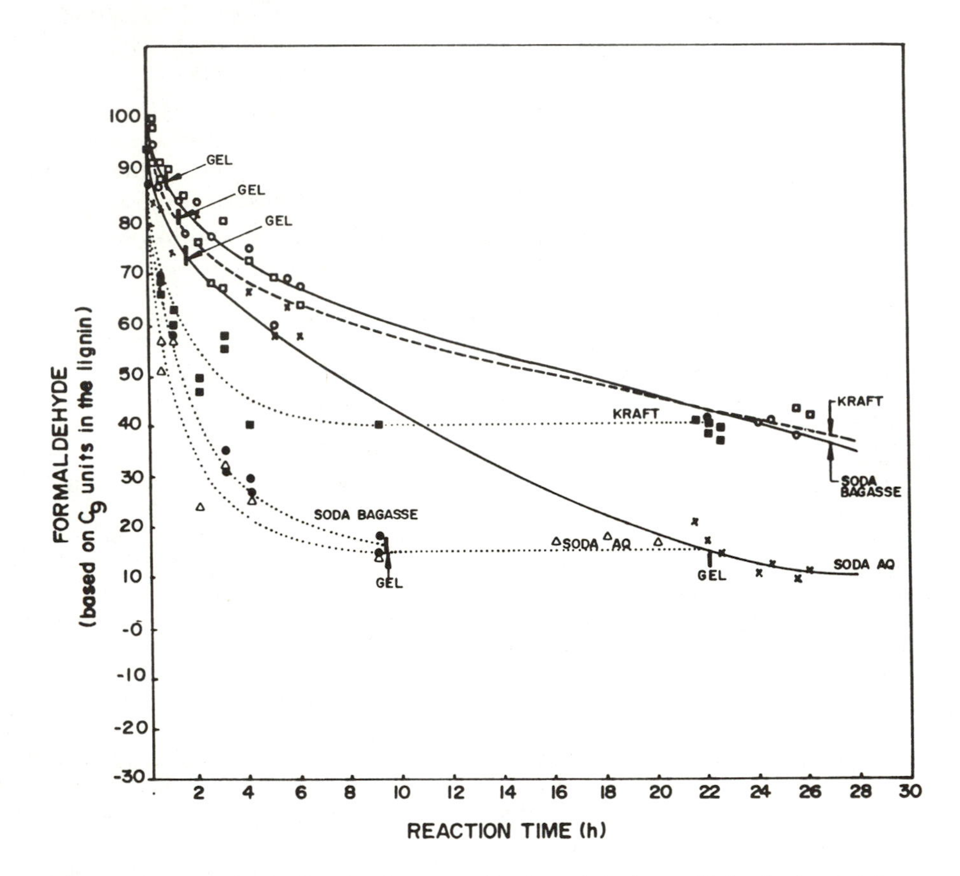

Figure 7. Reaction of soda bagasse lignin with formaldehyde in 1.0 hydrochloric acid in 23% aqueous dioxane at 80°C (—, – – – –) and in 0.4N hydrochloric acid in 10% aqueous dioxane at 80°C (......).

The three industrial lignins were subsequently reacted with formaldehyde under conditions where mainly mono-*meta*-hydroxymethylation is expected. In 10% aqueous dioxane containing 0,4N hydrochloric acid (Fig. 7), the lignin reacted very fast and consumed about half of the theoretical amount of formaldehyde in one hour.

The rate of formaldehyde consumption levelled off after about 5 hours. The total quantity consumed for both the soda/AQ and soda bagasse lignin is approximately 85% of the calculated theoretical amount. The formaldehyde consumed by the kraft lignin levelled off after 60% of the stoichiometric calculated quantity had been consumed. Some crosslinking of the lignin, however, still occurred since the soda bagasse lignin gelled after ca. 9 hours and the soda/AQ lignin after 20 hours. The kraft lignin showed no gelling after 28 hours reaction time. The results obtained nevertheless show that lignin can be extensively *meta*-hydroxymethylated without crosslinking. Prolonged reaction times should, however, be excluded.

The high formaldehyde consumption by the bagasse lignin can probably be attributed to mono-hydroxymethylation of syringyl and guaiacyl phenylpropanoid units at the 2- and 6-positions. The *para*-hydroxyphenyl units of the lignin were probably hydroxymethylated at the *ortho* (3- or 5-) positions. The formaldehyde consumption of the soda/AQ lignin can only be attributed to *meta* hydroxymethylation, owing to the low availability of unsubstituted 5-positions. The extensive consumption of formaldehyde by this lignin and its unpolymerized state after 6 hours thus clearly indicates that *meta* hydroxymethylation of the lignin was achieved in high yield.

The reactions applied above can be generalized as follows:

- Formaldehyde reacts with lignin at the 2- and 6-positions of the phenylpropane medium in 10 to 20% aqueous dioxane containing 0.4 to 1N hydrochloric acid, to afford *meta* hydroxymethylated intermediates. This results in a reactive *meta*-hydroxymethylated lignin intermediate capable of being modified with various reagents or by reacting with itself.
- In 50% aqueous dioxane and acidic conditions, lignin reacts with formaldehyde at positions 2 and 6 to give methylene crosslinks which result in the formation of a gelled or polymerized products (Fig. 8).

The usefulness of *meta*-hydroxymethylated lignin was illustrated by the preparation of a cold curing resin (15). Kraft lignin was hydroxymethylated (10% H_2O) in dioxane and 0.4N HCl) at the *meta* positions. After removal of excess formaldehyde the reactive kraft lignin intermediate was reacted with resorcinol in an aqueous solution to afford a resorcinol grafted lignin adduct (Fig. 9). The lignin-resorcinol adduct was alkalified to pH = 10 and paraformaldehyde added resulting in gelling of the product in about 6 minutes. The polymerized product was hard and resisted solvation in sodium hydroxide solutions. This result clearly indicates at least some of the potential of the *meta* hydroxymethylation of lignin to result in the production of, for example, cold set wood adhesives.

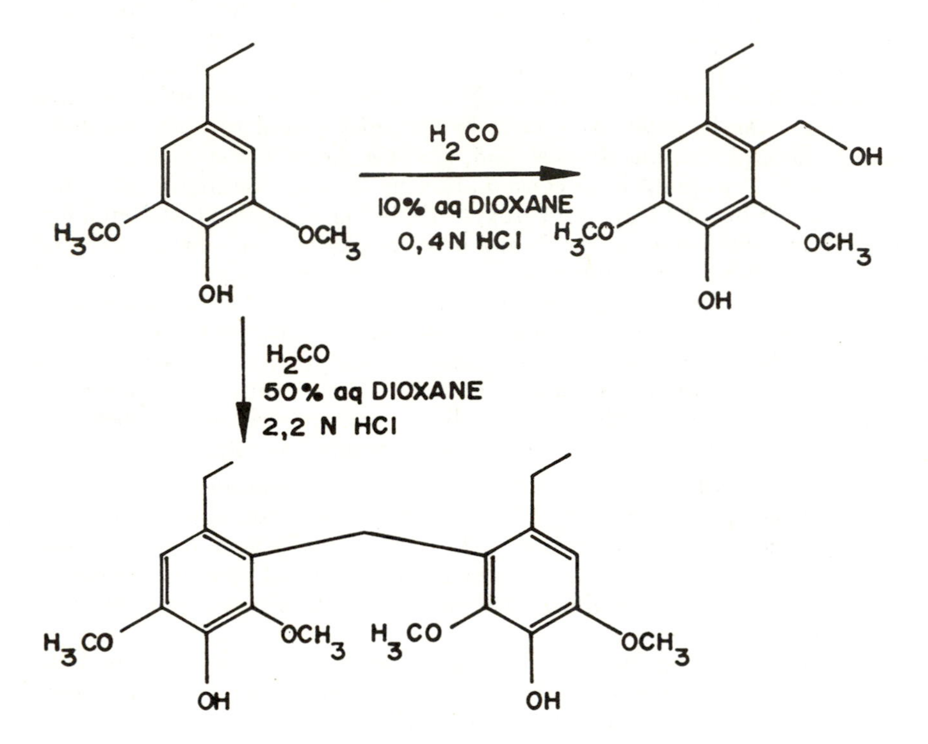

Figure 8. Generalized reaction of formaldehyde with lignin at the *meta* positions in acidic aqueous dioxane.

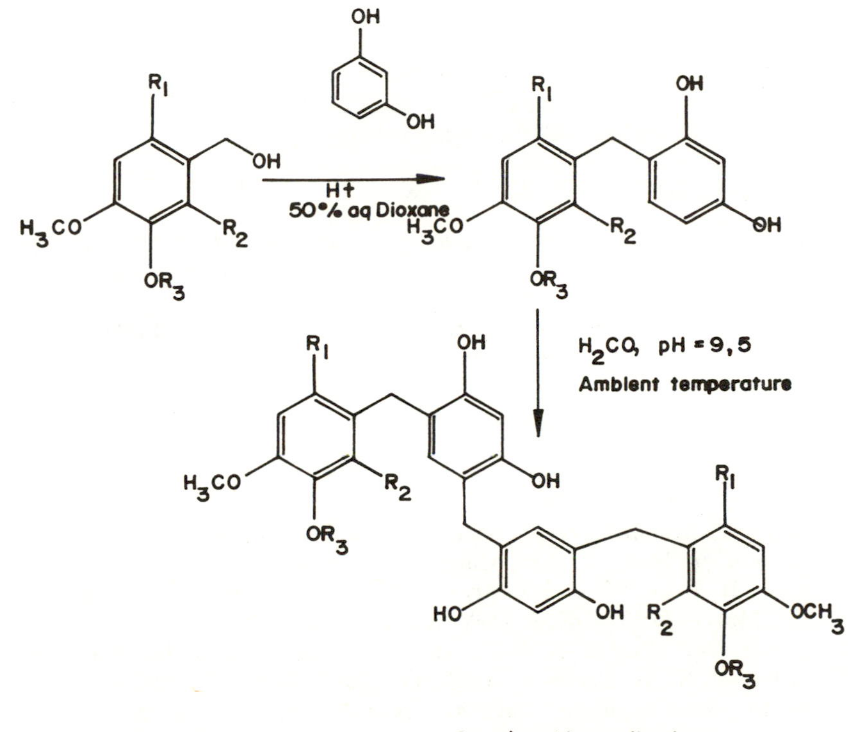

Figure 9. Diagrammatic illustration of the utilization of the *meta*-hydroxymethylation of lignin for the production of cold set wood adhesives.

Conclusions

From the results presented it is evident that the 2- and 6-positions of the phenylpropanoid nuclei can be used for the modification of alkali lignin. The positions can be used for the controlled polymerization of alkali lignin. Secondly, the positions were also used to introduce hydroxymethyl groups. The hydroxymethyl groups are reactive towards nucleophiles such as phenol and resorcinol. The modification of the 2- and 6-positions of the C_9 units holds tremendous potential for the utilization of alkali lignin in polymeric applications.

Acknowledgments

This research was partly sponsored by the CSIR Foundation for Research Development, which we gratefully acknowledge.

Literature Cited

1. Hoyt, C. H.; Goring, D. W. *Lignins: Occurrence, Structure and Reactions*; Sarkanen, K. V.; Ludwig, C. H., Eds.; Wiley-Interscience: New York, 1971; Ch. 20, p. 833.
2. Ito, H.; Shiraishi, N. *Mokuzai Gakkaishi* 1987, **33**(5), 393-399.
3. Van der Klashorst, G. H.; Forbes, C. P.; Psotta, K. *Holzforschung* 1983 **37**(6), 279-286.
4. Forbes, C. P.; Van der Klashorst, G. H.; Psotta, K. *Holzforschung* 1984, **38**(1), 42-46.
5. Doughty, J. B. *For. Prod. J.* 1963, **13**, 413.
6. Glasser, W. G.; Hsu, O. H. U.S. Patent 4 017 474, 1977.
7. Van der Klashorst, G. H.; Pizzi, A. In *Wood Adhesives Chemistry and Technology*; Pizzi, A., Ed.; Marcel Dekker: New York, Vol. 2, in press.
8. Nimz, H. H. In *Wood Adhesives Chemistry and Technology*; Pizzi, A., Ed.; Marcel Dekker: New York, 1983; Ch. 5, p. 248.
9. Forbes, C. P.; Psotta, K.; Nimz, H. *Holzforschung* 1983, **37**(3), 147-152.
10. Sarkanen, K. V.; Ericsson, B.; Suziki. *Adv. Chem. Ser.* 1966, 59.
11. Yasuda, S.; Terashima. *Mokuzai Gakkaishi* 1978, **24**(3), 56.
12. Yasuda, S.; Terashima. *Mokuzai Gakkaishi* 1982, **28**(6), 383.
13. Van der Klashorst, G. H.; Strauss, H. F. *J. Polym. Sci., Polym. Chem. Ed.* 1986, **24**, 2143-69.
14. Van der Klashorst, G. H. *J. Wood Chem. Technol.* 1988, **8**(2), 209-220.
15. Van der Klashorst, G. H. The Modification of Lignin at the 2- and 6-Positions of the Phenylpropanoid Nuclei. Part III. Hydroxymethylation of Industrial Alkali Lignin. *J. Wood Chem. Technol.*, in press.
16. Van der Klashorst, G. H. *Holzforschung* 1988, **42**(1), 65.
17. Van der Klashorst, G. H.; Strauss, H. F. *Holzforschung* 1987, **41**(3), 185.
18. Van der Klashorst, G. H.; Strauss, H. F. *Holzforschung* 1986, **40**(6), 375.

RECEIVED May 29, 1989

Chapter 27

Enzymatic Modification of Lignin for Technical Use

Strategies and Results

A. Hüttermann, O. Milstein, B. Nicklas, J. Trojanowski, A. Haars, and A. Kharazipour

Forstbotanisches Institut der Universität Göttingen, Büsgenweg 2, D–3400 Göttingen, Federal Republic of Germany

Extracellular phenoloxidases from white-rot fungi are promising tools for the biotechnological conversion of technical by-product lignins into valuable polymers. A short review of lignin-degrading enzymes of white-rot basidiomycetes is given as well as two examples for the potential application of laccases in lignin transforming processes. (1) In nature, enzymes react generally in aqueous solution. However, the bulk of technical by-product lignins (e.g., kraft lignin) is water insoluble, causing reaction rates of phenol-oxidases to be slow. This problem was addressed by the development of a process which uses immobilized phenoloxidases in an organic-aqueous system. (2) The same class of enzymes can efficiently be used for the biologically catalyzed bonding of particle boards. Adhesive cure is based on the oxidative polymerization of lignin using phenoloxidases as radical donors. This lignin-based "bio-adhesive" can be applied under conventional pressing conditions. The resulting particle boards meet German performance standards without emission of harmful vapors. An added advantage is the total utilization of lignin from spent pulp liquor.

The concept of using the natural polyphenol lignin as a feedstock for polymer production or as an adhesive in wood composites has encouraged numerous scientific endeavors (1), the basic rationale for which concerns replacing expensive petrochemical resins with this comparatively low-cost renewable raw material. Although lignin occurs as a polymer with several attractive structural features in the plant cell wall, having macromolecular architecture and many types of reactive functional groups, the lignin available from cellulose production is much less attractive for use in polymers. The harsh reaction conditions of the pulping process transform lignin

0097–6156/89/0397–0361$06.00/0

into a rather inert material. The content of phenolic groups, for instance, is drastically changed by all commercially applied pulping processes, and the number of alkyl aryl ether bonds declines. Thus, several important properties are unfavorably altered in the commercially available lignins as compared to the original plant material.

The properties of today's commercially available lignins can be summarized as follows (2): Lignins from the sulfite process (lignin sulfonates) are extremely polar owing to the presence of sulfonate groups, and they are highly water soluble. Kraft lignins, which are derived from the kraft (or sulfate) process, have low polarity and are water insoluble. These compounds are therefore rather unattractive as feedstocks for polymer production, and little progress has been made in this direction so far. In addition, the lignin coming from the organosolv process is water insoluble and its molecular weight is possibly too small to be suitable as a prepolymer for plastics production.

A strategy for the enzymatic modification of lignin for its technical use should thus concentrate on the following goal: to produce a homogeneous, pure lignin preparation of reasonably high molecular weight with high reactivity provided by reactive functional groups.

Extracellular Enzymes of Lignin-Transforming Fungi

The following enzymatic activities have been characterized which are able to change native lignin:

- *ligninase*, i.e., lignin peroxidase resp. veratryl peroxidase, an enzyme isolated and characterized by Kirk and his collaborators (3) (for an updated review see ref. 4-5). This enzyme is considered to be the main lignolytic system in white-rot fungi. It is determined via its ability to catalyze the oxidation of veratryl alcohol.
- *laccase*, i.e., polyphenol oxidase (E.C.1.10.3.2.). This enzyme is common in many microorganisms, and it is reported to have many different physiological functions (6). It has been known for a long time that laccase is able to polymerize phenols. Its role in lignin degradation has first been discussed by Ander and Eriksson (7).
- *poly-blue-oxidase*. This enzyme was first described by Glenn and Gold (8); it oxidizes the lignin model compound poly-blue which is a polymeric dye.

All three enzymes have the advantage of being subject to convenient spectroscopic assays.

The most important gross changes in the lignin molecule which enzymes can catalyze, and which are important for possible industrial uses, are the following:

- solubilization,
- demethylation,
- changes in phenolic and aliphatic hydroxyl contents, and
- changes in molecular weight distribution.

In our laboratory, we have screened various microorganisms for solubilization and demethylation activity in addition to testing for the presence of ligninase, laccase and poly-blue-oxidase. Data on changes in the molecular weight distribution have been given elsewhere (9,10). Although the biochemical work on lignin degradation has focused mainly on a single (although rather suitable) organism, the white rot fungus *Phanerochaete chrysosporium* (5), we decided to analyze the lignin transforming capacity of fungi coming from different niches in the ecosystem. These fungi represent a wide ecological range, and they take into consideration the many different conditions under which fungi degrade wood resp. lignin.

The results are summarized in Table I. The data indicate that, within this variety of different species, no correlation exists between the activities of the different enzymes and demethylation or solubilization. Surprisingly, no correlation, either, was observed between the ability to solubilize organosolv lignin and the activity of the lignin peroxidase. By contrast, a relation was found between lignin solubilization and poly-blue oxidase activity (Table I). Statistical analysis of the data given in Table I revealed a correlation coefficient of $r = 0.82$, which indicates a significant relation between these two parameters. We therefore consider this enzyme as a tool for getting lignin into aqueous solution. Part of our future work will concentrate on this specific enzymatic activity.

Laccase from White-Rot Fungi. Although there is no stringent correlation between laccase activity and lignin decomposition (e.g., Table I), a prominent role of this enzyme in lignin degradation has been discussed by several authors (4,7,11). The presence of laccase has also been shown to result in polymerization both *in vivo* (12,13) and *in vitro* (14,15). Laccase acts on phenolics via a non-specific oxidation which generates quinoid intermediates. This results in the formation of reactive intermediates which may subsequently polymerize (16-18). It is therefore reasonable to expect that the low molecular weight substances coming from enzymatically degraded lignin are readily repolymerized, and polymerization may be dominating over depolymerization processes during lignin transformation *in vitro* (19).

In addition to polymerization, which amounts to a crosslinking of lignin via oxidative coupling, this group of enzymes also catalyzes another important reaction in lignin: It hydroxylates phenolic substrates, thereby introducing phenolic hydroxyl groups which serve as new reactive sites on the molecule. These catalytic properties make the enzyme an interesting candidate for a variety of possible uses in lignin biotransformation.

Although the role of the enzyme during lignin degradation *in vivo* does not seem to have been elucidated so far, there are several indications that this activity is important for the fungi which generate the enzyme. We have studied this using the white-rot fungus *Heterobasidion annosum*, which causes root and butt rot in conifers. In these investigations we found the following:

- Although *H. annosum* has an extremely high variation in the isoenzyme pattern of other enzymes, it produced only a single band in the isoelectric focusing experiment of the laccase preparations isolated from more than 60 different strains (20).
- Although *H. annosum* is readily mutagenized by a variety of agents, it

Table I. Peroxidase and Oxidase Production by Selected Wood-Inhabiting Fungi, and their Capacity to Demethylate and Solubilize Lignin

	Units x 1000/1 mg Mycelium				
Fungus	Lignin-peroxidase[1]	Poly-Blue Oxidase	Laccase[3]	Demethylation[4]	Solubilization[5]
Trametes versicolor	1.6	20.8	2.9	29.5	52.1
Polyporus pinsitus	1.7	33.3	5.6	28.0	65.1
Phallus impudicus	2.5	14.0	0.0	27.1	13.3
Oudemansiella radicata	2.2	5.1	1.4	16.0	20.3
Bjerkandera adusta	7.6	8.6	0.0	16.8	19.5
Pleurotus florida F6	1.7	26.5	1.7	14.6	35.3
Pleurotus florida PP	1.7	20.6	3.9	14.0	45.3
Polyporus platensis	1.6	24.0	0.9	14.0	60.0
Ustulina deusta	1.0	5.6	0.8	14.0	19.2
Polyporus varius	3.2	7.1	0.0	12.0	16.9
Xylaria polymorpha	1.7	0.9	0.3	12.0	15.8
Phlebia radiata	3.9	25.0	4.7	11.0	56.8
Polyporus brumalis	1.5	23.4	1.8	10.6	56.1
Merulius tremellosus	2.8	27.0	1.2	8.0	58.9
Daedaleopsis confragosa	3.4	13.4	0.0	5.0	31.2

[1] 1 Unit = increase of $\underline{A}_{310}$/min/1 ml medium using veratryl alcohol and H_2O_2 at pH 3.

[2] 1 Unit = decrease of the adsorbance ratio $\underline{A}_{595}/\underline{A}_{482}$/ml medium after 24 h incubation using 0.01% Poly Blue.

[3] 1 Unit = change of $\underline{A}_{645}$/min/ml medium using 0.01% TMB.

[4] Figures indicate accumulative release of $^{14}CO_2$ using $^{14}CH_3O$-organosolv-lignin.

[5] Figures indicate ^{14}C-water solubles in culture in % of initial ^{14}C-activity using ^{14}C-β-organosolv lignin.

was impossible to detect laccase-free mutants in the many thousands of mutagenized clones which were inspected (21).

- The sophisticated regulation of the synthesis and secretion of the enzyme after induction with a suitable phenol as revealed by density labelling studies (6) indicates that this enzyme is under stringent metabolic control of the organism.

Laccase Activity and Properties of the Immobilized Enzyme in Organic Solvents. It has recently been shown that a number of enzymes, when immobilized, can express their activity in a reaction medium in which the bulk of water has been replaced by organic solvents (22-26). Furthermore, Klibanov and co-workers (27) have shown that lignin could be depolymerized with horseradish peroxidase in organic media, an observation however which is still surrounded by controversy (28). Because of these findings, and in view of the fact that the bulk of the commercially available lignins are water-insoluble, we have recently concentrated on how to use the enzyme in solvents in which organosolv and kraft lignins would be soluble, too.

For our studies we used laccase from the basidiomycete *Trametes versicolor*, which was purified using DEAE-Sephadex A-50 chromatography. When a lyophilized powder of the purified laccase was added to a solution of dimethoxyphenol (DMP) or syringaldazine in water-saturated ethylacetate, the color of the substrates turned rapidly yellow or red-violet, indicating that the usually observed laccase reaction takes place indeed under those conditions. However, the above described enzyme reaction system has several disadvantages and inconveniences, similar to those described for the horseradish peroxidase in organic solvents (26): Since the enzyme does not dissolve in the organic solvent, it forms particles which clump together and stick to the walls of the reaction vessel. Moreover, the overall reaction is slow and irreproducible, since only the active sites at the surface of the clumps are available for catalysis and the size of the individual clumps varies considerably. No correlation could therefore be established between the amount of enzyme and the extent of its reaction.

These problems were overcome by immobilizing the enzyme. Since the usual methods for immobilization of laccase did not work, we adopted a new method, details of which will be described elsewhere (29). Reasonable measurements were possible with this technique. Typical patterns of laccase activity could be monitored via the changes of absorbance of 2,6 DMP and syringaldazine. When the reaction took place in organic solvents, the absorption spectra of the products were similar to those obtained for the same reaction in buffer. Furthermore, the catalytic action of the *T. versicolor* laccase followed Michaelis-Menten-kinetics in most of the organic solvents which were tested (see Table II for specific examples).

Expression of the catalytic capacity of the immobilized laccase was also observed in more than a dozen different solvents, provided that they were either saturated with water or, in the case of solvents miscible with water, small amounts of water had been added (Table III). No enzymatic reaction was observed when the solvents tested were free of water. No correlation was found between the activity of the immobilized laccase and the hydrophobicity of the solvent in which the reaction took place. The rate of laccase reaction in ethylacetate was only twice that in toluene, despite the fact that water-saturated ethylacetate contains 50 times more water than

Table II. Kinetic Parameters of Oxidation of 2,6 DMP and Syringaldazine with Laccase in Organic Solvents

Substrate	Incubation Media	Km mM	Reaction Rate
2,6 DMP	Ethyl acetate [a]	0.52	11.6Δ468 nm · min^{-1}· mg E^{-1}
	Acetonitrile[b]	1.10	12.6Δ468 nm · min^{-1}· mg E^{-1}
	Acetone[b]	0.65	21.3Δ468 nm · min^{-1}· mg E^{-1}
	Tetrahydrofuran	1.06	3.4Δ468 nm · min^{-1}· mg E^{-1}
Syringaldazine	Ethyl acetate[a]	0.15	216 μmol · min^{-1}· mg E^{-1}
	Acetonitrile[b]	0.08	282 μmol · min^{-1}· mg E^{-1}
	Acetone[b]	0.08	206 μmol · min^{-1}· mg E^{-1}
	Tetrahydrofuran[b]	0.31	36 μmol · min^{-1}· mg E^{-1}

[a] Ethylacetate was presaturated with water.
[b] Containing 7% (v/v) water.

water-saturated toluene (30). Toluene, on the other hand, reacted ten times as fast as methanol. The reaction did not take place in either 3.5% aqueous dimethylsulfoxide or 3.5% aqueous dimethylformamide. Thus, immobilized laccase seems to be tolerant of a much wider range of solvents in which it can function than was reported for peroxidase (26). This may be due to the fact that different methods of immobilization have been employed.

Addition of water always enhanced the efficiency of immobilized laccase. The optimal amounts of water were surprisingly low: rates of syringaldazine oxidation comparable to those observed in buffer were observed in 65% aqueous acetonitrile, 50% aqueous acetone, 50% aqueous dioxane, and others. Immobilized laccase is surprisingly stable when stored in organic solvents. Stored in n-hexane at 30°C, it was stable for more than eight days with less than 10% loss of initial activity, whereas at the same conditions the system lost about 90% of its activity when stored in buffer. The addition of water to the pure solvents had a detrimental effect on stability.

In summary, the data indicate that immobilized laccase retains much of its activity in organic solvents. This may have important technological implications.

Application of Laccase as a Radical Donor in Adhesives for Particle Boards. The properties of laccase with regard to lignin render it a candidate for application in technical processes. A two-component adhesive was formulated with lignin as the phenolic component and laccase as radical donor. The process is described in more detail elsewhere (1).

It is evident that the application of a biological catalyst, the enzyme, in a technical process (in this case particle board production) demands as *conditiones sine qua non* the following parameters:

i. The enzyme has to be produced on inexpensive substrates in large quantities;
ii. It must be applicable in crude form without prior purification;
iii. It has to be reasonably stable at room temperature;

Table III. The Rates of Laccase-Catalyzed Oxidation of Syringaldazine in Different Organic Solvents

Solvent	Reaction Rate μmole/min/mg enzyme
Ethyl acetate [a]	216
Toluene[a]	104
Benzene[a]	81
Ether[a]	75
Isooctane[a]	68
n-Hexane[a]	62
Cyclohexane[a]	48
Chloroform[a]	15
Dichlorethane[a]	8
Acetonitrile[b]	116
Acetone[b]	80
Ethanol[b]	28
1.4-Dioxane[b]	16
Tetrahydrofuran[b]	11
Methanol[b]	10
Dimethyl sulfoxide[b]	0
N,N-Dimethylformamide[b]	0

[a] No reaction was observed in the anhydrous solvents; the solvents were presaturated with distilled water at room temperature.
[b] Reaction took place in solvent following addition of 3.5% (v/v) of distilled water.

iv. Since particle boards are pressed at high temperatures, the enzyme must be heat tolerant;
v. Board pressing has to follow current state-of-art, with press times as short as possible; and
vi. The price of the final product has to be competitive with conventional petrochemical resins.

We have succeeded with the development of an enzyme-catalyzed adhesive system on lignin basis that meets all the requirements stated above. This system is based on a basidiomycete (*Trametes versicolor*), grown in a fermenter on spent sulfite liquor and aminophenol as additional enzyme inducing agent. Among many other basidiomycetes, this fungus was chosen because of its high extracellular laccase activity which produced best adhesion results. The culture broth obtained after 4 days of cultivation is concentrated by evaporation or ultrafiltration, and the resulting concentrated, crude enzyme solution is applied directly in the binding system. A sterile storage of the enzyme concentrate at room temperature is possible for at least one month.

Compared to the chemically cured adhesive systems in common use today, this enzymatic binder has certain advantages:

1. It is based on the total utilization of waste lignin, first as binding

component in the adhesive and second as nutrient source for enzyme production.
2. Because of the high catalytic activity of the enzyme, the binding process can be performed under mild conditions without application of large amounts of harmful chemicals.
3. In contrast to particle boards bonded with synthetic resins, the "biobonded" boards do not emit any harmful vapors as do, for example, formaldehyde-bonded boards.

The production of the "biobonded" particle boards has been described recently in detail (1, 31). Analytical results by gel permeation chromatography and ultracentrifugation have been presented in support of an *in vivo* and *in vitro* polymerization of lignin sulfonates. Besides polymerization, the lignin molecule was found to be modified by carboxylation, and this was shown by difference spectroscopy (32). The following is a short summary of pertinent results.

Milled wood lignin was mixed with the crude enzyme solution of *Trametes versicolor* extracellular phenoloxidases produced on spent sulfite liquor in a ratio of approximately 2:1. This comprised the main part of the two-component "bio-adhesive." Industrial particles were bonded with 15% bio-adhesive under conventional pressing conditions to have 19 mm particle boards (40 × 50 cm) of the properties described in Table IV. The bonding reaction (crosslinking) took place in *aqueous* solution at room temperature. If conventional pressing technology is applied, the temperature should be elevated in order to maintain water evaporation within a reasonable press time.

Table IV. Technological and Mechanical Properties of "Biobonded" Particle Boards: Effect of Lignin Type, Enzyme Activity and Pressing Time

Lignin Type	Press Time (min	Density (kg/m³)	Versal Tensile Strength (N/mm²)	Thickness Swelling** After	
				2h	24h
1 + enzyme	5	774	0.47	5.4%	24.0%
2 + enzyme	5	775	0.42	5.4%	19.9%
2 + enzyme	3	745	0.44	6.7%	29.3%
Controls:					
1 – enzyme*	5	734	0.18	40.1%	<60.0%
Urea formaldehyde (66%)	3	700	0.37	10.0%	30.0%

Note: The lignins were technical by-product lignins of different polarity, 1 = lignin sulfonates, 2 = milled wood lignin.
* The enzyme was inactivated by autoclaving.
** Performed by soaking in cold water.

Table IV shows that the enzyme has a significant effect on both the versal tensile strength and the thickness swelling. The lignin type, by contrast, affected mostly the swelling properties of the board: the use of the less polar milled wood lignin resulted in better water resistance than the corresponding lignin sulfonate products. Press times of 3 min in combination with enzyme and milled wood lignin meets the German standard requirements for V20 particle boards.

Acknowledgments

The work described was supported by Grant 86 NR 0063 from the Bundesministerium für Ernährung, Landwirtschaft und Forsten, Grant PBE 18938 A from the Bundesministerium für Forschung und Technologie, and funds from G. A. Pfleiderer, Neumarkt/Opf.

Literature Cited

1. Haars, A.; Trojanowski, J.; Hüttermann, A. In *Bioenvironmental Systems*; Wise, D. L., Ed.; Boca Raton, FL, 1987; Vol. I, pp. 89-129.
2. Glasser, W. G. *For. Prod. J.* 1981, **31**, 24-29.
3. Tien, M.; Kirk, T. K. *Science* 1983, **221**, 661-63.
4. Kirk, T. K.; Farrell, R. L. *Ann. Rev. Microbiol.* 1987, **41**, 465-505.
5. Odier, E., Ed. In *Lignin Enzymic and Microbial Degradation*; INRA Symposia, Vol. 40, Paris 1987; INRA Publications.
6. Haars, A.; Hüttermann, A. *Arch. Microbiol.* 1983, **134**, 309-13.
7. Ander, P.; Eriksson, K. E. *Arch. Microbiol.* 1976, **109**, 1-8.
8. Glenn, J. K.; Gold, M. H. *Appl. Environ. Microbiol.* 1983, **45**, 1741-47.
9. Haars, A.; Majcherczyk, A.; Trojanowski, J.; Hüttermann, A. In *Energy from Biomass*; Palz, W.; Coombs, J.; Hall, D. O., Eds.; Elsevier: London and New York, 1985; pp. 973-77.
10. Trojanowski, J.; Milstein, O.; Majcherczyk, A.; Haars, A.; Hüttermann, A. In *Lignin Enzymic and Microbial Degradation*; Odier, E., Ed.; INRA Symposia, Vol. 40, Paris 1987; INRA Publications.
11. Ishihara, T. In *Lignin Biodegradation: Microbiology, Chemistry, Potential Application*; Kirk, T. K.; Higuchi, T.; Chang, H. M., Eds.; CRC Press: Boca Raton, FL, 1980; pp. 17-31.
12. Hüttermann, A.; Gebauer, M.; Volger, C.; Rösger, C. *Holzforschung* 1977, **31**, 83-89.
13. Hüttermann, A.; Herche, C.; Haars, A. *Holzforschung* 1980, **34**, 64-66.
14. Haars, A.; Hüttermann, A. *Naturwissenschaften* 1980, **67**, 39-40.
15. Leonowicz, A.; Szklarz, G.; Wojtas-Wasilewska, M. *Phytochemistry* 1985, **24**, 393-96.
16. Ishihara, T.; Miyazaki, M. *Mokuzai Gakkaishi* 1972, **18**, 415-19.
17. Liu, S. Y.; Minard, R. D.; Bollag, J. M. *Soil Sci. Soc. Am. J.* 1981, **45**, 1100-05.
18. Lundquist, K.; Kristersson, P. *Biochem. J.* 1985, **229**, 277-79.
19. Kaplan, D. L. *Phytochemistry* 1979, **18**, 1917-19.
20. Hüttermann, A.; Volger, C.; Schorn, R.; Ahnert, G.; Ganser, H. K. *Eur. J. For. Pathol.* 1979, **9**, 265-74.

21. Cwielong, P. Ph.D. Thesis, Georg-August University, Göttingen, 1986.
22. Martinik, K.; Levashov, A. V.; Klyachko, N. L.; Berezin, J. V. *Dokl. Akad. Nauk SSSR* 1977, **236**, 920-23.
23. Butler, L. G. *Enzyme Microb. Technol.* 1979, **1**, 253-59.
24. Antonini, E.; Carrea, G.; Cremonesi, P. *Enzyme Microb. Technol.* 1981, **3**, 291-96.
25. Zaks, A.; Klibanov, A. M. *Proc. Natl. Acad. Sci. USA* 1985, **82**, 3192-96.
26. Kazandjian, R. Z.; Dordick, J. S.; Klibanov, A. M. *Biotechnol. and Bioeng.* 1986, **28**, 417-21.
27. Dordick, J. S.; Marletta, M. A.; Klibanov, M. A. *Proc. Natl. Acad. Sci. USA* 1986, **83**, 6225-57.
28. Lewis, N. G.; Razal, R. A.; Yamamoto, E. *Proc. Natl. Acad. Sci. USA* 1987, **84**, 7924-27.
29. Milstein, O.; Nicklas, B.; Hüttermann, A. Patent Application submitted 1988.
30. Weast, R. C.; Selby, S. M.; Hodgman, C. D., Eds. In *CRC Handbook of Chemistry and Physics*; The Chemical Rubber Co.: Cleveland, OH, 1965; D1-D21.
31. Hüttermann, A. *Holzforschung* 1977, **31**, 45-50.
32. Haars, A.; Bauer, A.; Hüttermann, A. In *Aerobic Digestion and Carbohydrate Hydrolysis of Waste*; Ferrero, G. L.; Ferranti, M. P.; Naveau, H., Eds.; Elsevier: London, New York, 1984; pp. 467-69.

RECEIVED February 27, 1989

POLYOLS, POLYURETHANES, POLYBLENDS, AND GRAFTS

Chapter 28

Estimating the Adhesive Quality of Lignins for Internal Bond Strength

Natsuko Cyr and R. George S. Ritchie

Alberta Research Council, P.O. Box 8330, Station F, Edmonton, Alberta T6H 5X2, Canada

Quantitative C-13 NMR spectrometry and size exclusion chromatography afforded sufficient data on lignins from steam-exploded aspen wood, steam-exploded aspen bark waste, and a Kraft hardwood to determine the number, and kind, of reactive sites which undergo reaction with phenol-formaldehyde resin. A model was proposed to predict the quality of the resin produced, when such lignins are methylolated. Board performance (as IB) was taken as a measure of the binder quality. Preliminary testing of boards made from methylolated lignin:phenol-formaldehyde compositions with up to 50% lignin content not only passed the Canadian Standards for random-oriented waferboard but agreed well with the prediction model.

The cost of wood adhesives in reconstituted board industries can be a substantial portion of production costs. The fact that lignins are less expensive than PF resins alone makes them attractive to use in resins as extenders. However, the chemical structure of lignins isolated from various lignocellulosics depends on the plant species, the method of isolation, and age of the lignin fraction (1). Adhesives which are derived in part from lignin tend to have a high degree of variability in performance, due to such lignin variations. One approach to reduce these variations is to mask the lignin under a surface of grafted copolymer (2-5). An alternative is to develop a method whereby lignin suitability for a binder can be assessed, for the purpose of pre-screening lignin samples without resorting to extensive formulations and testing; such is the objective of this paper.

0097-6156/89/0397-0372$06.00/0

Experimental

Steam-exploded lignins were prepared from aspen by the Tigney process (6) and purified as described (1). Acetylations were accomplished in acetic anhydride-pyridine and the acetates isolated in the usual fashion. Methylolations were performed as described by Marton *et al.* (7). Quantitative C-13 NMR spectra were obtained using "inverse gated decoupling" on a Bruker WP-80 spectrometer at 20.1 MHz. Typically 10,000 scans were accumulated. Chromium acetoacetonate was added to the chloroform-d and DMSO-d_6 solutions to shorten T_1. Spectra in DMSO were obtained at 363 K to increase the chromium salt solubility and decrease viscosity. Size exclusion chromatography was performed on a Milton-Roy HPLC using three Waters columns (one 1000 A, two 10,000 A). Lignin acetates were analyzed as 1% solutions in dichloromethane, using a flow rate of 2 mL/min, and a UV detector (254 nm). Elution times were calibrated against polystyrene standards. Calculations on molecular weights were performed by the method of Yau *et al.* (8).

The lignins were separately mixed with PF resin (dry blending) at 30% and 50% levels prior to application as a binder. The formation and testing (9) of the waferboards (30 cm × 30 cm) were done by the Alberta Research Council Panel Testing Laboratory.

Parameters

Back (10) has indicated that superior board performance is achieved with covalent bonding of the adhesive to the wood. A binder, then, must have at least the minimum number of reactive sites per molecule. If there is one or fewer such sites, then the lignin should behave as a filler, which may or may not be chemically bound to the resin. In the case of two reactive sites, a "linear" macromolecule is possible, or the lignin may be considered to behave as an extender for a resin. When three or more sites are available, crosslinking can occur and the lignin could then become a full partner in the crosslinked binder. One may project how the lignin could behave, once the reactive sites on the lignin molecule have been mapped. For this chapter, the interactive sites will be alcohols and benzyl alcohols, to simulate the reaction of PF resins with the carbohydrates in the wood.

Unlike PF resins, lignins do not have methylol groups attached to the aromatic rings. However, methylols can be grafted onto lignin at p-hydroxybenzyl and guaiacyl sites C-3, C-5 and C-5, respectively, provided these C_9 units of the lignin have a free phenolic OH group at C-4. No reaction occurs with syringyl groups. In hardwoods, the proportion of p-hydroxybenzyl groups is very small compared to guaiacyl and syringyl units, so C-4 free guaiacyl groups measure the number of potential methylolation sites.

The methoxyl signal in C-13 NMR spectra of lignins (both free, in DMSO, and acetylated, in chloroform) is unambiguous and was used as a reference for the benzene ring. The integrated methoxyl signal will vary depending upon the amounts of syringyl, guaiacyl, and p-hydroxyphenyl groups present. Since the last is ignored here, the ratio of syringyl to guaiacyl or "s/g" ratio can be determined by NMR by a number of methods. From the lignin spectrum in DMSO define:

$$a = \frac{s}{g} = \frac{[\text{area in region } 100 - 109 \text{ ppm}]}{[\text{area in region } 109 - 123 \text{ ppm}]} * \frac{2}{3} \tag{1}$$

The number of methoxyl groups per average benzene ring:

$$N = \frac{1 + 2a}{1 + a} \tag{2}$$

From the same spectra, C-3 and C-5 in bound syringyl groups are assigned to the region 150-155 ppm, and the region 143-150 ppm contains C-3, C-5 of the free syringyl groups, C-3 and C-4 of both types of guaiacyl units, and the alpha carbon of any olefins present (11-14). From the Proton NMR data of the lignins studied, the amount of olefinic material was small, so this last term can be neglected. Then:

$$b = \frac{s_b}{s_f + g_b + g_f} = \frac{[\text{area under region } 150 - 155 \text{ ppm}]}{[\text{area under region } 143 - 150\text{ppm}]} \tag{3}$$

By expanding Equation 1 and restraining the parameters to fit a single average benzene ring:

$$a = \frac{s}{g} = \frac{s_b + s_f}{g_b + g_f} \tag{4}$$

$$s_b + s_f + g_b + g_f = 1 \tag{5}$$

From the spectra of acetylated lignins (15) the relative amounts of primary, secondary, and phenolic OH groups are determined for the average benzene ring:

$$x = \frac{[\text{area of peak at } 170.8 \text{ ppm}]}{[\text{area of OMe peak at } 56 \text{ ppm}]} * N \tag{6}$$

$$y = \frac{[\text{area of peak at } 170.0 \text{ ppm}]}{[\text{area of OMe peak at } 56 \text{ ppm}]} * N \tag{7}$$

$$z = \frac{[\text{area of peak at } 168.9 \text{ ppm}]}{[\text{area of OMe peak at } 56 \text{ ppm}]} * N \tag{8}$$

Since Equation 8 refers to the free phenolic groups, then:

$$z = s_f + g_f \tag{9}$$

Solving the set of four simultaneous equations (Equations 3, 4, 5, and 9) gives the number of methylolation sites:

$$g_f = z - \frac{a - b}{[1 + a][1 + b]} \tag{10}$$

From the profile of the molecular weights obtained by the use of GPC data on the acetylated derivatives, the average molecular weight was calculated. From x, y, z, and N the average molecular weight of the basic structural unit was determined, and finally the number of benzene rings per molecule was realized, and hence the number of reactive sites per molecule.

Results

Figure 1 shows the variation by C-13 NMR in the lignin samples: (a) EXWL; (b) EXBWL; and a kraft hardwood lignin (c) TomL. Partial C-13 spectra of lignin acetates showing the carbonyl region are illustrated in Figure 2. Molecular weight measurements were determined using the size exclusion chromatograms shown in Figure 3. The data are summarized in Table I.

Table I. Lignin parameters obtained from C-13 NMR and GPC

Parameter	EXBWL	EXWL	TomL	B-PF	R-PF
Syringyl/guaiacyl ratio	0.85	1.50	1.30	–	–
No. of OMe per avg. ring	1.46	1.60	1.57	–	–
No. of primary OH/ring	0.53	0.42	0.27	1.04	1.15
No. of secondary OH/ring	0.59	0.37	0.19	–	–
No. of phenolic OH/ring	0.55	0.53	0.66	1.	1.
No. avg. molecular weight	1490	2230	1730	1870	1570
Avg. molecular weight/unit	280	269	243	223	231
Dispersity	5.1	4.0	5.5	4.1	3.3
No. of sites/ring (CH_2O)	0.33	0.21	0.19	1.04	1.15

Predictions of IB Quality

Table II gives a number of derived parameters used to assess the lignin's suitability as a binder. For a full binder, lignins required a minimum of three sites per average molecule for formaldehyde grafting. None of the lignins studied approach this level. At least twice the number of sites found is required for steam exploded lignins and more for the kraft lignin. PF resins have an average of eight sites per molecule, a much higher density than projected, so PF resins are excellent wood binders.

Although the lignins do not perform adequately as full binders, they can be used as extenders. Now the resin supplies the benzylalcohol functionality and the lignins (and wood chips) need only supply alcohol groups to make ether linkages. The results for data on methylolated lignins are expressed in Table II as indices. The index is obtained by dividing the

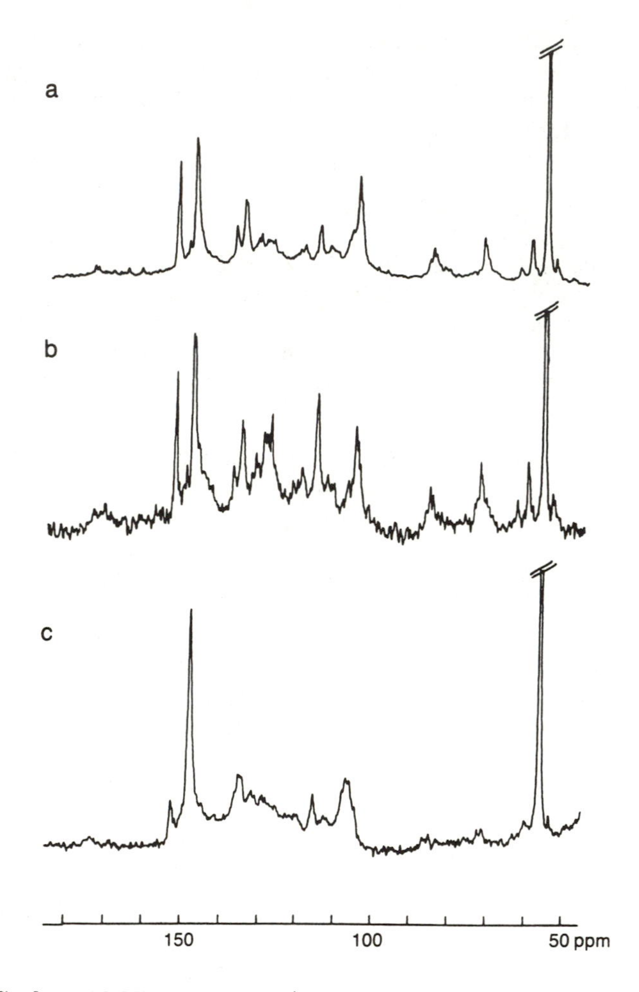

Figure 1. Carbon-13 NMR spectra (in DMSO-d_6) of steam-exploded aspen lignins: (a) EXWL; (b) EXBWL; and (c) TomL.

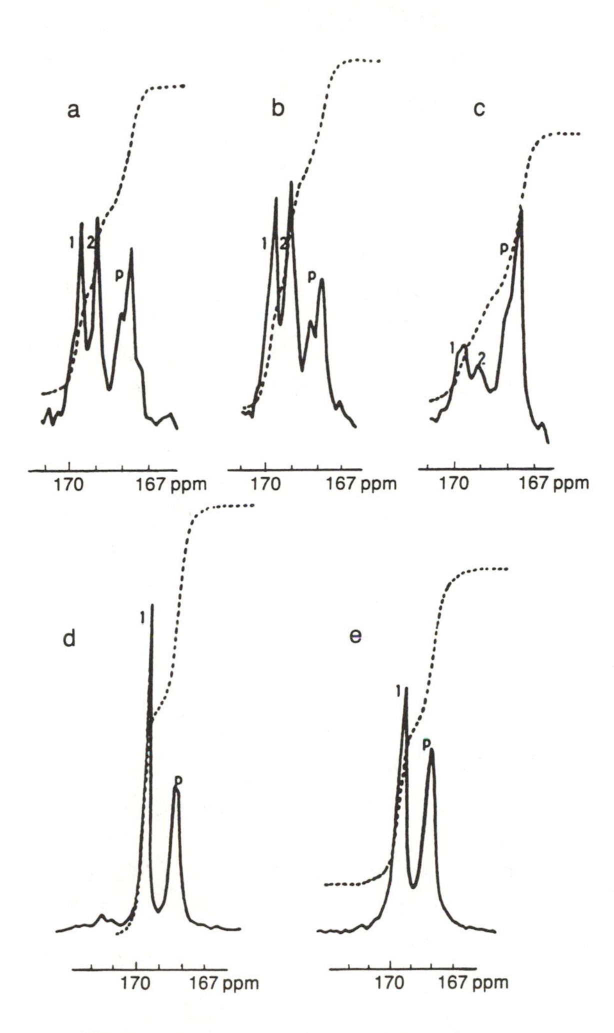

Figure 2. Partial ^{13}C-NMR spectra (in DMSO-d_6) of the acetylated materials: (a) EXWL; (b) EXBWL; (c) TomL; (d) Reichold PF resin; and (e) Bakelite PF resin.

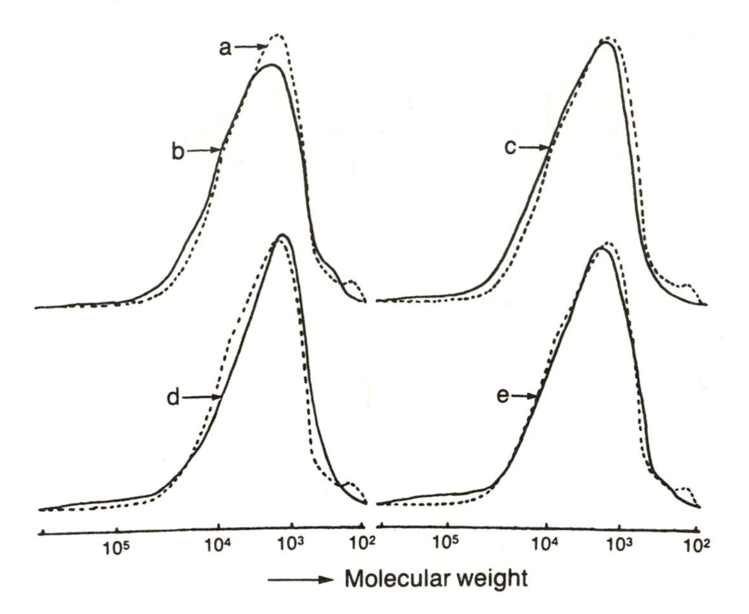

Figure 3. Gel permeation chromatograms of the acetylated materials: (a) Reichold PF resin; (b) Bakelite PF resin; (c) EXWL; (d) EXBWL; and (e) TomL.

Table II. Lignin parameters for assessing adhesive quality

Parameters	EXBWL	EXWL	TomL	B-PF	R-PF
Values based on one average molecule:					
No. of benzene rings	5	8	7	8	7
No. of aliphatic OH's	5.7	6.3	3.2	8.3	8.1
No. of phenolic OH's	2.8	4.2	4.6	8	7
Total no. of OH's	8.5	10.5	7.8	16.3	15.1
No. of CH_2O sites	1.7	1.7	1.3	8.3	8.1
Relative Indices:					
Hydroxymethyl index	21	15	15	85	100
Total aliphatic OH index	94	69	50	85	100
Total alcohol index	70	57	55	90	100
Phenolic OH index	42	42	60	96	100

parameter for one molecule by its average molecular weight, and comparing that ratio relative to the one obtained for Reichold PF resin. None of the indices for methylolated lignins exceeds that for R-PF resin, but the predicted, relative effectiveness of such lignins is high.

Board Testing Results

In order to evaluate the method, the three lignins were methylolated, formulated with R-PF resin and waferboards made. The boards were tested for dry MOR, MOE, wet MOR and IB, but only the last term was the parameter chosen for quality determinations by the authors. Limited number of testing results are given in Table III. The data are not corrected for board density. The "pass" level for Canadian requirements is IB = 0.345 MPa, so all of the boards prepared passed, even those containing binder with 50% of the PF resin replaced with methylolated lignin.

Table III. Board testing results (IB)

Binder		Density	IB	Predictions (Eqn. 11)		
PF-Resin	Lignin	kg m^3	Actual	Aliph.	Total	Phenol.
2.0%		679	0.448	–	–	–
1.4%	EXWL-F (0.6%)	667	0.448	0.41	0.39	0.37
1.0%	EXWL-F (1.0%)	650	0.356	0.39	0.35	0.32
1.4%	EXBWL-F (0.6%)	629	0.396	0.44	0.41	0.37
1.0%	EXBWL-F (1.0%)	641	0.384	0.43	0.38	0.32
1.4%	TomL-F (0.6%)	675	0.407	0.38	0.39	0.39
1.0%	TomL-F (1.0%)	739	0.408	0.34	0.35	0.36

The basis for prediction of board quality requires the quality of the board (as measured by the IB) be proportional to the effective adhesive content of the board. Therefore:

$$IB = 0.224 * (Wt_{PF} + (Wt_{ML} * R.I.)) \tag{11}$$

where R.I. is one of the relative indices in Table II. Clearly the R.I. term is not a constant at all PF/lignin ratios. For example, as the lignin approaches 100% binder content, the R.I. chosen must be the hydroxymethyl index, all values of which are dramatically lower than either aliphatic index or total alcohol index. Since the phenolic index, which is based on reactivity with the acidic phenol groups, consistently underestimated the actual results, this index was rejected.

Conclusions

If Equation 11 is valid, then the quality of the board will drop as the PF resin content is lowered and there is no interaction with the lignin (i.e., a filler). While there is a drop in board performance with increasing lignin content (at 2% binder), it is not as low as it would be for a filler (1.4%PF $\Rightarrow$ 0.31; 1.0%PF $\Rightarrow$ 0.22). Because the levels of IB obtained by testing are much higher than this, the lignin is undergoing chemical reaction with the PF resin and contributes to board strength, at levels up to 50% PF resin replacement. The degree to which board quality decreases is approximated well by an index based either upon the aliphatic OH content or total OH content, but not one based on phenolic OH's alone.

Exploded wood and Kraft methylolated lignins make poor quality binders at 2% loading, but should give better binders at 12%, which is not economical. The bark material is only marginally better.

Acknowledgments

The Canadian Forestry Service through the Canada-Alberta Forest Resource Development Agreement has funded this work.

Legend of Symbols

PF	=	phenol-formaldehyde
s	=	syringyl unit (3, 5-dimethoxy, 4-hydroxyphenyl-)
g	=	guaiacyl unit (3-methoxy-4-hydroxyphenyl-)
f	=	free, unit which is not etherified at C-4
b	=	bound, unit which is etherified at C-4
N	=	number of methoxyl groups per average benzene ring
x	=	number of primary OH groups per average benzene ring
y	=	number of secondary OH groups per average benzene ring
z	=	number of phenolic OH groups per average benzene ring
EXWL	=	steam exploded aspen heartwood
EXBWL	=	steam exploded aspen bark and wood rot (waste)
TomL	=	Tomlinite, a hardwood lignin product (Domtar)
R-PF	=	phenol-formaldehyde resin (Reichold RCI947)
B-PF	=	phenol-formaldehyde resin (Bakelite 9175)

IB	=	internal bond strength
F	=	suffix used to indicate methylolated sample
Wt_{PF}	=	weight of PF resin in a 100g board
Wt_{ML}	=	weight of methylolated lignin in a 100g board
R_f	=	total aliphatic OH index
R_a	=	total alcohol index

Literature Cited

1. Cyr, N.; Laidler, J.; Ritchie, R. G. S. *Holzforschung* 1987, **41**, 321-24.
2. Glasser, W. G.; Hsu, O. H.-H.; Reed, D. L.; Forte, R. C.; Wu, L. C.-F. In *Urethane Chemistry and Applications*; ACS Symp. 1980, **172**, 311-38.
3. Glasser, W. G.; Saraf, V. P.; Selin, J.-F. *Org. Coat. Plast. Chem.* 1981, **45**, 551-55.
4. Glasser, W. G.; Barnett, C. A.; Rials, T. G.; Saraf, V. P. *J. Appl. Polym. Sci.* 1984, **29**, 1815-30.
5. Muller, P. C.; Glasser, W. G. *J. Adhesion* 1984, **17**, 157-74.
6. DeLong, A. E. Canadian Patents 1 096 374 and 1 141 376.
7. Marton, J.; Marton, T.; Falkehag, S. I.; Adler, E. In *Adv. Chem. Ser.* 1966, **59**, 125-44.
8. Yau, W. W.; Kirkland, J. J.; Bly, D. D. In *Modern Size Exclusion Chromatography: Practice of Gel Permeation and Gel Filtration Chromatography*; Wiley: New York, 1979; pp. 315-21.
9. Canadian R1 Waferboard Standard: CAN3-0437.0-M85, 1985.
10. Back, E. L. *Holzforschung* 1987, **41**, 247-58.
11. Nimz, H. *Angew. Chem. Internat. Ed.* 1974, **13**, 313-21.
12. Luedemann, K. P.; Nimz, H. *Makromol. Chem.* 1974, **175**, 2393-407.
13. Gagnaire, D.; Robert, D. *Makromol. Chem.* 1977, **178**, 1477-95.
14. Kringstad, H.-D.; Mörck, R. *Holzforschung* 1983, **37**, 237-44.
15. Robert, D. R.; Brunlow, G. *Holzforschung* 1984, **38**, 85-90.

RECEIVED February 27, 1989

Chapter 29

Heat-Resistant Polyurethanes from Solvolysis Lignin

Shigeo Hirose[1], Shoichiro Yano[1], Tatsuko Hatakeyama[2], and Hyoe Hatakeyama[1]

[1]Industrial Products Research Institute, 1–1–4 Higashi, Tsukuba, Ibaraki 305, Japan
[2]Research Institute for Polymers and Textiles, 1–1–4 Higashi, Tsukuba, Ibaraki 305, Japan

Solvolysis lignin was obtained by a cooking using cresol-water. Polyurethanes (PU's) were prepared from solvolysis lignin, polyethylene glycol and 4,4'-diphenylmethane diisocyanate. Thermal degradation was investigated by thermogravimetry. It was found that lignin in PU's retarded the thermal degradation of PU's in air. The retardation was not observed in the degradation of PU's in nitrogen. These results indicate that the retardation was caused by the oxidative condensation of lignin. PU's containing polyol with phosphorous were also prepared. It was found that PU's became nonflammable in air with the addition of polyol with phosphorous.

In the past, researchers attempted to obtain many kinds of three-dimensional polymers from lignin, utilizing the characteristics of lignin as a polymer. It is generally recognized that polyurethane (PU) is one of the most useful three-dimensional polymers, because PU has unique features, various forms of materials can be obtained, and their properties can easily be controlled. Therefore, many attempts to obtain PU's using lignin as a raw material have been made. Recently, Saraf and Glasser *et al.* (1,2) studied the relationship between the structures and properties of PU's which were obtained from hydroxyalkylated lignins. They also studied adhesive capacity, adhesive suitability, etc., of PU's containing hydroxyalkylated lignins (3). Yoshida *et al.* (4) reported the mechanical properties of PU's containing fractionated kraft lignin which are obtained by successive extraction with organic solvents.

In the present study, PU's containing lignin were investigated with reference to thermal stability, which is one of the basic thermal properties of polymers. Thermal degradation of PU's which were obtained from solvolysis lignin was studied using thermogravimetry (TG). The inflammability of

0097–6156/89/0397–0382$06.00/0

PU's containing phosphorous was also studied in connection with thermal stability.

Experimental

Materials. Solvolysis Lignin (SL). A sample of crude lignin was obtained from Japan Pulp and Paper Research Institute, Inc. This crude lignin had been prepared in the following way. Beech (*Fagus crenata* Bl.) was cooked at 185°C for 210 min in a mixture of cresol and water (8:2, v/v) with the wood to liquor ratio of 1:5. Crude lignin was obtained as a precipitate when the cresol layer of cooking liquor was poured into benzene. The crude lignin was purified by reprecipitation using 90% acetic acid-water and a mixture of 1,2-dichloroethane and ethanol (2:1, v/v)-ethyl ether. The purified solvolysis lignin (SL) was used for the preparation of PU's.

Polyurethanes (PU's). SL and 4,4'-diphenylmethane diisocyanate (MDI) were dissolved in tetrahydrofuran (THF), and the solution was stirred for 1 hr at 60°C. A THF solution of polyethylene glycol (PEG 400) and diethyl bis(2-hydroxyethyl)aminomethylphosphonate (polyol containing phosphorous) was added to the reaction mixture, and the reaction time was extended for 1 hr. In all reactions, the molar ratio of the total amount of isocyanate groups to the total amount of hydroxyl groups (NCO/OH) was maintained at 1.2. The lignin content in PU was 20 wt%. Each solution was drawn on a glass plate, and allowed to dry for 48 hr. The residual solvent in a sample was removed under vacuum and curing of each PU film was carried out at 120°C for 3 hr under a pressure of 50 kg/cm^2.

Measurements. A Shimazu TG 30 thermogravimeter was used for thermogravimetry (TG). The measurements were carried out using ca. 5 mg of sample at a heating rate of 10°C/min and at a flow rate of 15 ml/min of either nitrogen or air. In order to determine activation energies, measurements were carried out at heating rates of 2, 5, 10, and 20°C/min. Inflammability of samples was tested using samples with a width of 5 mm and a thickness of 0.6 mm according to the procedure of Japan Industrial Standard (JIS) K7201, which is a testing method for inflammability of polymeric materials using the oxygen index method.

Results and Discussion

Thermal Stability of Polyurethanes (PU's). Figure 1 shows TG curves for PU's with lignin contents of 20 and 0 wt% in air. As can clearly be seen from Figure 1, the difference between the two curves is scarcely evident until ca. 7% of weight loss. However, in the region where weight loss exceeds 7%, the rate of degradation for PU with lignin is smaller than that for PU without lignin. For example, the difference in weight loss between the two PU's at 350°C is about 20%. This suggests that the incorporation of lignin into PU's is responsible for the retardation of the thermal degradation of PU's.

Figure 2 shows TG curves for PU's with a lignin content of 20 wt% in air and in nitrogen. No difference between the two curves can be observed

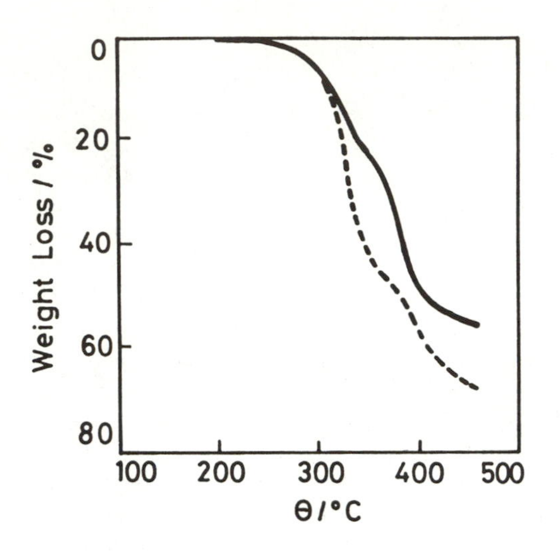

Figure 1. TG curves for polyurethanes (PU's) with 20 wt% of lignin and without lignin measured in air. Heating rate: 10°C/min. Flow rate: 15 ml/min. Solid line: PU with lignin and broken line: PU without lignin.

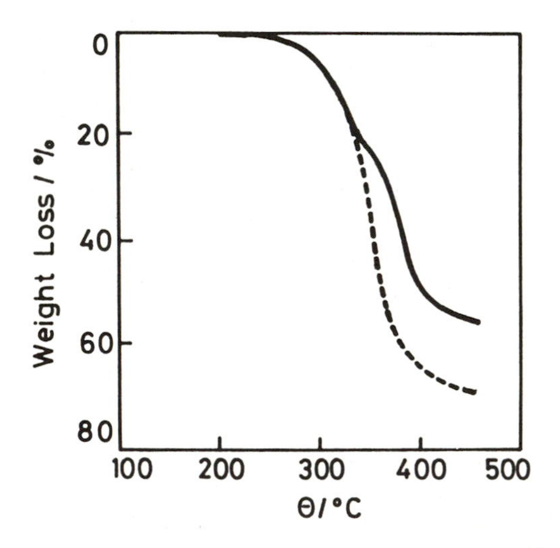

Figure 2. TG curves for polyurethane (PU) containing 20 wt% of lignin measured in air and in nitrogen. Heating rate: 10°C/min. Flow rate: 15 ml/min. Solid line: in air; broken line: in nitrogen.

until the weight loss reaches ca. 15%. However, the weight loss in nitrogen increases rapidly after the weight loss exceeds 15%. On the other hand, the TG curve obtained in air shows a shoulder which is not observed in that obtained in nitrogen. This suggests that there is a difference between the mechanism of thermal degradation in air and in nitrogen. For comparison, the TG curves of solvolysis lignin in air and in nitrogen are shown in Figure 3. It was found that lignin heated in air degraded less than that heated in nitrogen. It is assumed that the retardation of thermal degradation of lignin in air is caused by the condensation reactions under the participation of radicals which are formed by the reaction of lignin with oxygen in air. Therefore, it is considered that the retardation of degradation of PU in air is caused by lignin in PU.

In order to characterize thermal degradation of PU in air in detail, activation energies (E) for thermal degradation were calculated. An integral method (5), which is generally valid for analysis of thermal degradation of polymers using TG, was applied to the determination of E for PU and lignin.

Assuming that the degradation reaction is represented by the Arrhenius relationship, the equation finally obtained by the integral method when the weight fraction is constant is as follows:

$$-\log\phi_1 - 0.4567E/RT_1 = -\log\phi_2 - 0.4567E/RT_2 = ----$$

where E represents activation energies; R, gas constant; ϕ, heating rate; T, absolute temperature where weight fraction becomes a given value at each heating rate. Activation energy (E) can be calculated using the above relationship between $\log\phi$ and 1/T.

TG curves were obtained by measurements at heating rates of 2, 5, 10 and 20°C/min for the degradation of PU with a lignin content of 20 wt%. Figure 4 shows the plots of $\log\phi$ versus 1/T obtained from each TG curve. These plots give straight lines and E can be calculated from the slope of each line. However, the integral method can be applied only for a process having a single E. The process at weight losses below 10% can be treated as an apparent single process, because the slopes of the lines are almost the same at weight losses of 5 and 10% as shown in Figure 4. Therefore, the integral method is applicable to analysis of degradation at weight losses below 10%. The calculated value of E was 121 kJ/mol.

The measurements under the same conditions were carried out for PU without lignin and for lignin, and E's were calculated. The obtained value of E for each sample is tabulated in Table I. E for PU with a lignin content of 20 wt% is 17 kJ/mol higher than that for PU without lignin. This indicates that PU with lignin is thermally more stable than PU without lignin. Furthermore, E for lignin itself is 132 kJ/mol, which is the highest value among those for three samples. Therefore, it can be concluded that PU is thermally stabilized by the presence of lignin, which is more stable than PU itself.

Polyurethanes (PU's) Containing Phosphorous. Nonflammability as well as thermal stability is an important property of polymers. In the present

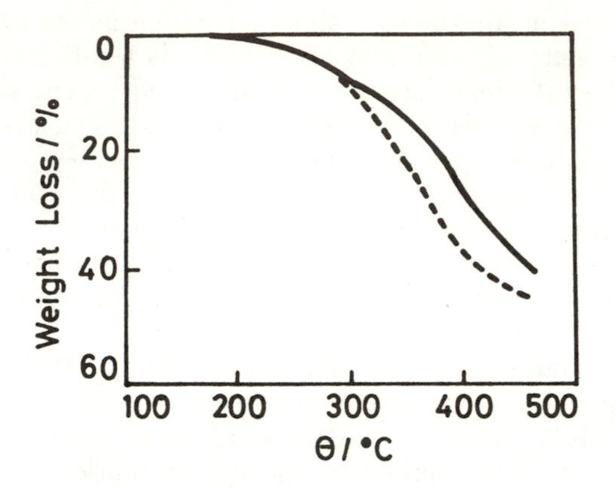

Figure 3. TG curves for solvolysis lignin measured in air and in nitrogen. Heating rate: 10°C/min. Flow rate: 15 ml/min. Solid line: in air; broken line: in nitrogen.

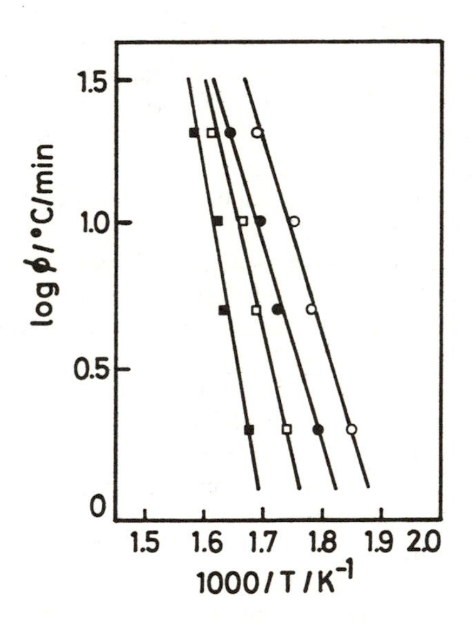

Figure 4. Relationship between logarithmic heating rate (logϕ) and reciprocal absolute temperature (1/T) for the degradation of polyurethane (PU) with 20 wt% of lignin in air. Weight loss, ○: 5%, ●: 10%, □ : 15% and ■ : 20%.

Table I. Calculated Activation Energies (E's) for Thermal Degradation of Polyurethanes (PU's) and Lignin

Sample	E (kJ/mol)
PU (lignin 20 wt%)	121
PU (lignin 0 wt%)	104
Lignin	132

study, PU's containing phosphorous were prepared using a polyol containing phosphorous (see Experimental), which is known as a reactive flame resistant reagent. These modified PU's containing phosphorous were investigated with reference to inflammability and thermal stability.

Table II shows the correlation of wt% of polyol containing phosphorous in PU with 20 wt% of lignin with oxygen indices which were obtained according to the procedure of JIS method. In this method, the inflammability of a sample is tested in a mixture of oxygen and nitrogen under a constant flow rate of the mixture. The oxygen index is the minimum concentration of oxygen in which a sample is inflammable. The oxygen index for PU without phosphorous and those for PU's with phosphorous are 19.3 and 24.6, respectively. That is, qualitatively, PU's with phosphorous were not inflammable in air and PU without phosphorous was inflammable. This indicates that PU becomes less inflammable by the addition of 3 wt% of polyol containing phosphorous.

Figure 5 shows TG curves for PU with 0-8% of polyol containing phosphorous. The starting temperature of degradation of PU decreases with increasing the amounts of added polyol with phosphorous. This indicates that the addition of polyol with phosphorous reduces thermal stability of PU. However, it can be said that the retardation of degradation is promoted by the addition of polyol containing phosphorous at temperatures over approximately 400°C.

Table II. Oxygen Indices of Polyurethanes (PU's)[a] Containing Polyol with Phosphorous (P-polyol)

P-polyol (wt%)	Oxygen Index[b]
0	19.3
3	24.6
8	"
16	"

[a] PU (lignin content: 20 wt%).
[b] Oxygen indices were obtained by a method according to JIS K7201.

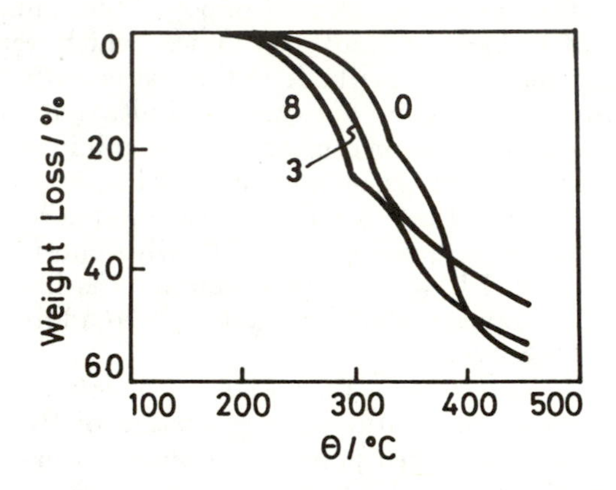

Figure 5. TG curves for polyurethanes (PU's) (containing 20 wt% of lignin) prepared with polyol containing phosphorous measured in air. Number indicated for each line shows the content of polyol with phosphorous expressed in wt%. Heating rate: 10°C/min. Flow rate: 15 ml/min.

Conclusions

From the results obtained in this study, it is concluded that lignin in PU's retards the thermal degradation of PU's in air and also that PU's containing lignin become nonflammable in air with the addition of a polyol containing phosphorous.

Literature Cited

1. Saraf, V. P.; Glasser, W. G. *J. Appl. Polym. Sci.* 1984, **29**, 1831.
2. Saraf, V. P.; Glasser, W. G.; Wilkes, G. L.; McGrath, J. E. *J. Appl. Polym. Sci.* 1985, **30**, 2207.
3. Newman, W. H.; Glasser, W. G. *Holzforschung* 1985, **39**, 345.
4. Yoshida, H.; Mörck, R.; Kringstad, K. P.; Hatakeyama, H. *J. Appl. Polym. Sci.* 1987, **34**, 1187.
5. Ozawa, T. *Bull. Chem. Soc., Japan* 1965, **38**, 1881.

RECEIVED February 27, 1989

Chapter 30

Elastomeric Polyurethanes from a Kraft Lignin–Polyethylene Glycol–Toluene Diisocyanate System

Roland Mörck, Anders Reimann, and Knut P. Kringstad

STFI, Box 5604, S–114 86 Stockholm, Sweden

Polyurethane (PU) films were produced by solution casting from a three-component PU system consisting of a non-derivatized, low molecular weight fraction of kraft lignin (KL), polyethylene glycol (PEG) of various molecular weights and toluene diisocyanate (TDI). At low contents of KL ($< 10\%$) and low molecular weights of PEG (300-600), weak but very flexible (ultimate strain $\sim$ 1000%) PU films were obtained whereas some of the films obtained with intermediate KL contents (15-25%) and low molecular weight PEG showed considerable toughness. Use of higher molecular weight PEG (4000) resulted in high values of ultimate strain (500-600%) also at high ($\sim$ 30%) KL content. The crosslink densities of the films were generally very low.

Many of the lignin-derived polyurethanes (PU's) so far reported in the literature have been extensively crosslinked due to the high functionality of lignins in the synthesis of PU's (1-6). The content of hard segments has also in most cases been very high (1-6). As a result, the lignin-derived PU's described have generally been hard, glassy, and sometimes brittle materials.

It is anticipated that the possibility of producing elastomeric lignin-containing PU's strongly depends on whether crosslink density can be held on a low level or not. It is also necessary to include soft segments, so that the glass transition temperature (T_g) of the PU falls below room temperature.

We observed in a previous investigation of PU's derived from a kraft lignin (KL)-polyether triol-diphenylmetane diisocyanate (MDI) system (Yoshida, H.; Mörck, R.; Kringstad, K. P.; Hatakeyama, H., submitted to *J. Appl. Polym. Sci.*) that KL fractions of low molecular weight yielded more flexible and less crosslinked PU's than the medium and high molecular weight KL fractions. This was attributed to the lower functionality of the low molecular weight KL fractions.

0097–6156/89/0397–0390$06.00/0

Preliminary experiments in our laboratory indicated very high values of ultimate strain for PU's derived from a 2,4-toluene diisocyanate (TDI)-polyethylene glycol (PEG)-KL system in which the KL component was a low molecular weight fraction of non-derivatized KL. This PU system was therefore subjected to a more comprehensive investigation in which the KL content, the soft/hard segment ratio and the molecular weight (chain length) of the soft segment (PEG) were varied systematically. The present paper reports the results of this investigation.

Experimental

Kraft Lignin. A softwood kraft lignin (KL) was isolated from a partly evaporated, industrial kraft black liquor by precipitation through the addition of dilute sulfuric acid as described elsewhere (7). The lignin was thereafter fractionated by successive extraction with organic solvents (7). The KL fraction used in the present investigation was the second of five fractions obtained (propanol soluble - methylene chloride insoluble).

Polyethylene glycol (PEG) and toluene diisocyanate (TDI). PEG of various molecular weights (300-4000) and TDI (80:20 mixture of 2,4- and 2,6-isomers) were obtained from MERCK, Darmstadt, West Germany.

Polyurethane synthesis. PU films were synthesized by pre-polymerization of all starting materials in tetrahydrofuran solution at room temperature followed by solution casting as described previously (6). Dibutyltin dilaurate was used as catalyst (2% of total sample weight). The PU films obtained were cured for 35 hours at 95 ± 1°C.

Differential scanning calorimetry (DSC). The DSC analyses were carried out using a Perkin-Elmer DSC-7 and a DuPont 910DSC. T_g was defined as the midpoint of the change in heat capacity occurring over the transition. The samples were first scanned to 95°C, thereafter cooled and recorded a second time. The T_g was determined from the second run. The measurements were carried out under an atmosphere of dry nitrogen at a heating rate of 10°C/min.

Swelling tests and determination of tensile properties. The procedure for estimating cross-link density from equilibrium swelling data is described in detail in a previous paper (6). The tensile properties of the PU films were carried out at 23°C and 60% relative humidity. The crosshead speed and distance were 10 mm/min and 30 mm, respectively. A more detailed description of the tensile tests is given elsewhere (6).

Infrared spectroscopy. Infrared spectra of thin PU films were obtained using a Perkin-Elmer 983 infrared spectrophotometer.

Results and Discussion

The composition of starting materials for the PU's studied in the present investigation are listed in Table I. Five series of PU films were synthesized in which PEG of different molecular weights were applied. The isocyanate/hydroxyl group (NCO/OH) ratio was kept constant (1.2) in all samples since variations in the NCO/OH ratio may also lead to variations in crosslink density (8).

The TDI used was an 80:20 mixture of the 2,4- and 2,6-isomers, which is commonly used in industrial applications (8).

The low molecular weight KL fraction used in this study amounted to 22% of the unfractionated lignin (7). The hydroxyl group content of this fraction was determined to 7.3 mmol/g of which a major part (5.0 mmol/g) was found to be phenolic (7). Its weight average molecular weight ($\overline{M}_w$) was determined to 1100 (relative value related to polystyrene) using high performance size exclusion chromatography (7). The molecular weight distribution was relatively narrow ($\overline{M}_w/\overline{M}_n = 1.4$). The functionality of this KL fraction in the PU synthesis was estimated to 6-8.

Crosslink density of PU films. The PU films were subjected to swelling tests in dimethylformamide. The crosslink densities of the films were thereafter estimated from equilibrium swelling data using a modified version of the Flory-Rehner equation. The swelling tests as well as the calculation of crosslink density are described in detail in a previous paper (6).

Figure 1 shows that the crosslink density increases, as expected, with increasing KL content. This is mainly due to the higher functionality of KL in comparison with PEG. The fact that the content of hydroxyl groups and isocyanate groups increases with rising KL content at each molecular weight of PEG studied (a consequence of the constant NCO/OH ratio) may also contribute, since this may increase the number of positions where crosslinking can occur.

The swelling behavior does, however, strongly indicate that the crosslink densities of the TDI-based PU's studied here are considerably lower than those previously found (10) for PU's from a corresponding KL-PEG-MDI system where the same KL fraction and the same NCO/OH ratio were used. This is at least partly due to the fact that the MDI used in the previous investigation was a mixture of di- and triisocyanates having a somewhat higher functionality than the TDI used here. It may also to some degree be due to a combined effect of the low reactivity of phenolic hydroxyl groups toward isocyanates (11-14) and of the fact that the *ortho*-isocyanate group in 2,4-TDI is far less reactive than the *para*-isocyanate group at room temperature (15-17). This effect, if present, may reduce the effective functionality of non-derivatized KL in the PU system studied here. It is furthermore possible that the reaction between the *ortho*-isocyanate group in 2,4-TDI and some of the hydroxyl groups in KL may be sterically hindered which would also reduce the effective functionality of non-derivatized KL in the KL-PEG-TDI system. No attempts were, however, made in this investigation to establish whether these effects were

Table I. Polyurethane Composition

Sample	KL %	PEG %	TDI %	NCO/OH
PEG 300				
1	0.0	59.0	41.0	1.2
2	5.0	53.8	41.2	1.2
3	10.0	48.6	41.4	1.2
4	16.9	41.4	41.7	1.2
PEG 400				
5	0.0	65.7	34.3	1.2
6	5.0	59.9	35.1	1.2
7	10.0	54.2	35.8	1.2
8	17.0	46.0	37.0	1.2
9*	24.9	37.3	37.8	1.2
10*	31.9	28.3	39.8	1.2
PEG 600				
11	0.0	74.1	25.9	1.2
12	5.0	67.6	27.4	1.2
13	10.0	61.1	28.9	1.2
14	17.0	51.7	31.3	1.2
15	24.9	41.5	33.6	1.2
16	31.9	32.4	35.7	1.2
PEG 1000				
17	5.0	75.4	19.6	1.2
18	10.0	68.0	22.0	1.2
19	17.0	57.8	25.2	1.2
20	25.0	46.3	28.7	1.2
21	31.9	36.1	32.0	1.2
PEG 4000				
22	17.0	66.1	16.9	1.2
23	25.0	53.0	22.0	1.2
24	31.9	41.5	26.6	1.2
25	34.9	36.4	28.7	1.2
26	38.0	31.3	30.7	1.2

*Too brittle for tensile testing.

present or not, even though the existence of unreacted isocyanate groups in the cured PU's was qualitatively confirmed for some selected samples by infrared spectroscopy via observation of the isocyanate stretching band at 2270 cm^{-1}.

Differential scanning calorimetry analysis. The KL-containing PU's from three of the PU series studied (PEG 400, 600, and 4000) were analyzed by DSC. The PU's were cured to stable properties at 95°C (35 h). The samples were, however, unstable at higher temperatures, probably due to additional curing. All samples analyzed by DSC were therefore initially scanned through T_g up to 95°C, where the first scan was interrupted. After cooling, the scan was repeated. This second scan was in most cases terminated at 120 – 135°C, where baseline instabilities and exotherms began to appear.

The results of the DSC analysis are summarized in Table II. Examples of representative DSC scans are shown in Figures 2, 4, and 5. Figure 2 shows DSC scans (second scan) for the PU's from the PEG 600 series (PU's 12-16). Each of these DSC scans display a clearly defined glass transition for the soft segment. The transition shifts to higher temperatures and broadens as the KL content (and thereby the weight fraction of hard segments) increases. The former effect is further illustrated by Figure 3, which shows that the T_g decreases, in each PU series, with increasing weight fraction of soft segments (or increases with increasing weight fraction of hard segments) over the total range studied. The T_g has been described as a sensitive index of phase mixing in PU's (18) and the behavior of T_g observed in the present study is usually interpreted as a strong indication of phase mixing (18-21). The PU's are not necessarily completely free of phase separation between the soft and hard segments, but the phases are extensively mixed.

No melting endotherm appeared in the first DSC scan for any of the KL-containing PU's from the PEG 400 and PEG 600 series, which indicates the absence of any soft segment crystallinity in these samples. This is further indicated by the fact that all KL-containing PU's from the PEG 400 and PEG 600 series were clear and transparent. The PU's in the PEG 4000 series were, however, turbid, which may be an indication of soft segment crystallization. The turbidity was most pronounced in PU's 22 and 23 and decreased rapidly in extent with increasing KL content. Soft segment crystallinity is further indicated by the presence of a melting endotherm in the first DSC scan for PU's 22 and 23 (Fig. 4). Although PU #24 displayed some turbidity, no melting endotherm was observed in the first DSC scan of this sample.

The first DSC scan for some samples having their glass transition located somewhat above room temperature (PU's 9, 10, 16, 25 and 26) showed an endothermic peak overlapping with the glass transition. On cooling and repeating the scan only the glass transition appeared. The DSC scans of PU #9 are shown as an example in Figure 5. Since these samples were glassy and transparent (except PU's 25 and 26, which were slightly turbid), this endotherm is most likely due to enthalpy relaxation. It is well known that the polymer chains are frozen in a non-equilibrium conformation in the glassy state. However, some motion occurs even in the glassy state (22), especially when the polymer is kept at (or annealed at) a temperature moderately below T_g. As a result of these restricted motions, a slow relaxation of the non-equilibrium state towards the equilibrium state takes place. This process, which is accompanied by a decrease in enthalpy and free volume, is known as physical aging or enthalpy relaxation.

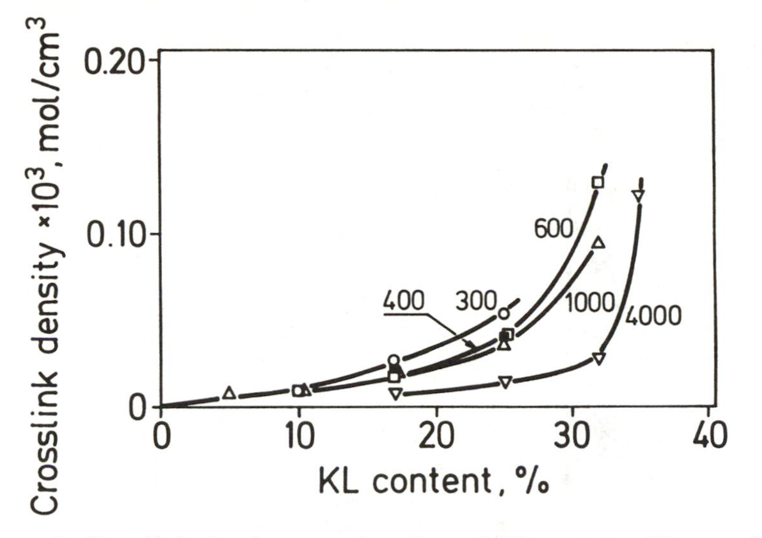

Figure 1. Crosslink density as a function of KL content. The numbers in the figure indicate different molecular weights of PEG.

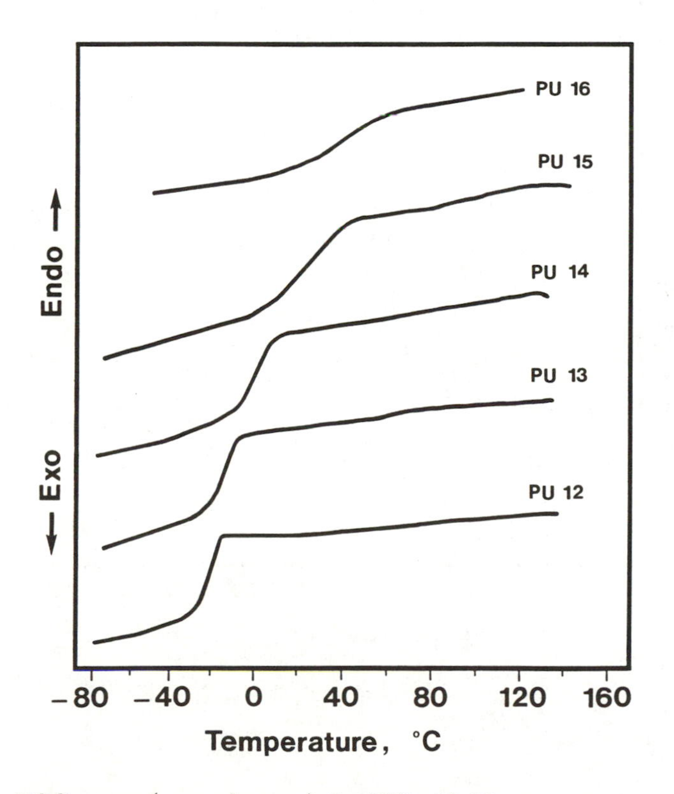

Figure 2. DSC scans (second scans) for PU's 12-16.

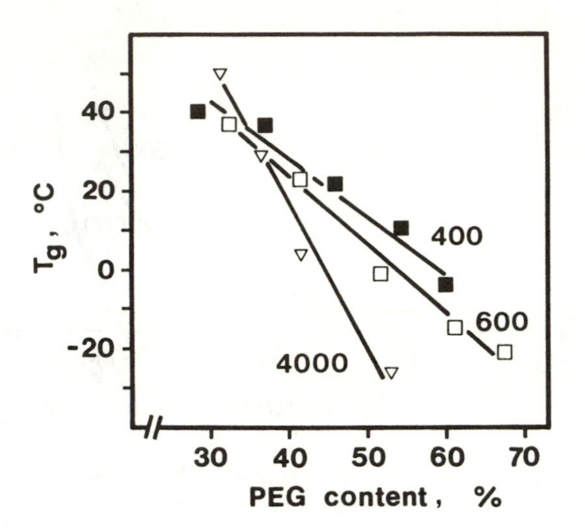

Figure 3. T_g as a function of PEG content. The numbers in the figure indicate different molecular weights of PEG.

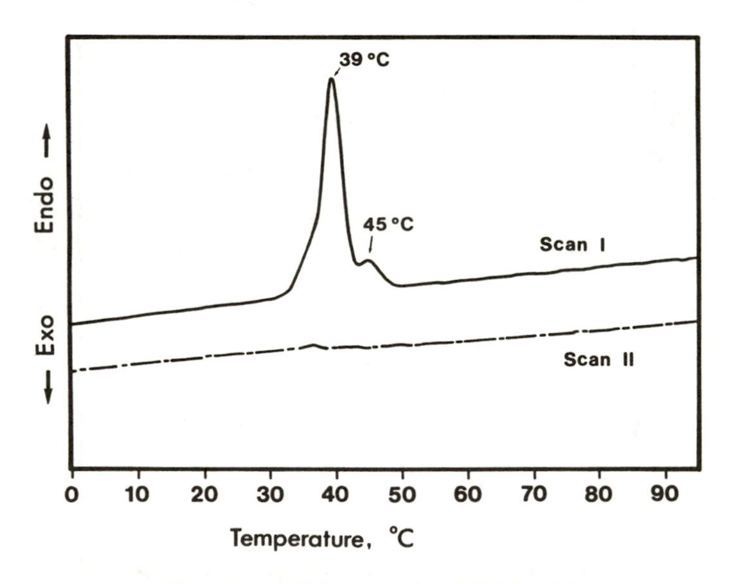

Figure 4. DSC scans for PU No. 23.

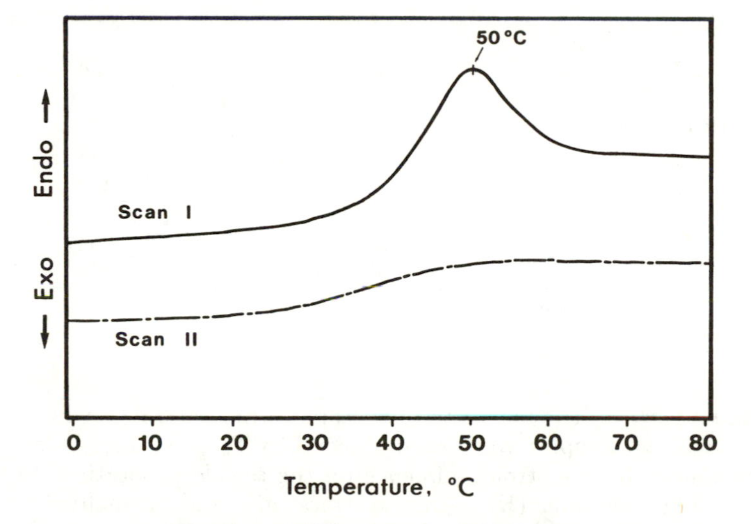

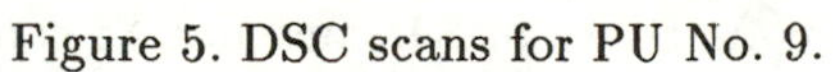

Figure 5. DSC scans for PU No. 9.

Table II. DSC Analysis

Sample	T_g, °C	Endothermic peaks (first scan), °C
PEG 400		
6	- 4	–
7	11	–
8	22	–
9	37	50
10	40	50
PEG 600		
12	-21	–
13	-15	–
14	- 1	–
15	23	–
16	37	54
PEG 4000		
22	n.d.*	48
23	-26	39,45
24	4	–
25	29	54
26	50	64

*n.d.= not determined.

Tensile properties. The tensile tests were performed at 23°C, which means that some of the samples were tested above their T_g and others below. This will, of course, have a strong influence on the tensile properties. Figure 6 shows Young's modulus (E), ultimate stress (σ_b), and ultimate strain (ϵ_b) of the various PU's as functions of KL content. As can be seen, the PU's are, at the lower molecular weights of PEG studied, (300, 400 and 600) initially weak but very flexible. The strength increases as the KL content increases, whereas the flexibility is gradually lost. Some of these PU's are quite tough, especially some of those prepared from low molecular weight PEG (400, 600) at intermediate KL contents (PU's 8 and 15). In the PEG 1000 and 4000 series, the most flexible PU's were obtained at intermediate KL contents (PEG 1000 series) or at intermediate to fairly high KL contents (PEG 4000 series).

The KL content at which steep increases in E (Fig. 6a) and σ_b (Fig. 6b) are initiated indicates, in each PU series, the transition from the rubbery state to the glassy state. The transition is shifted towards higher KL contents as the molecular weight of PEG increases, which is due to differences in composition between the five series of PU's studied. This is discussed somewhat more in detail below.

Some of the PU's show very high values of ϵ_b (Fig. 6c). These are

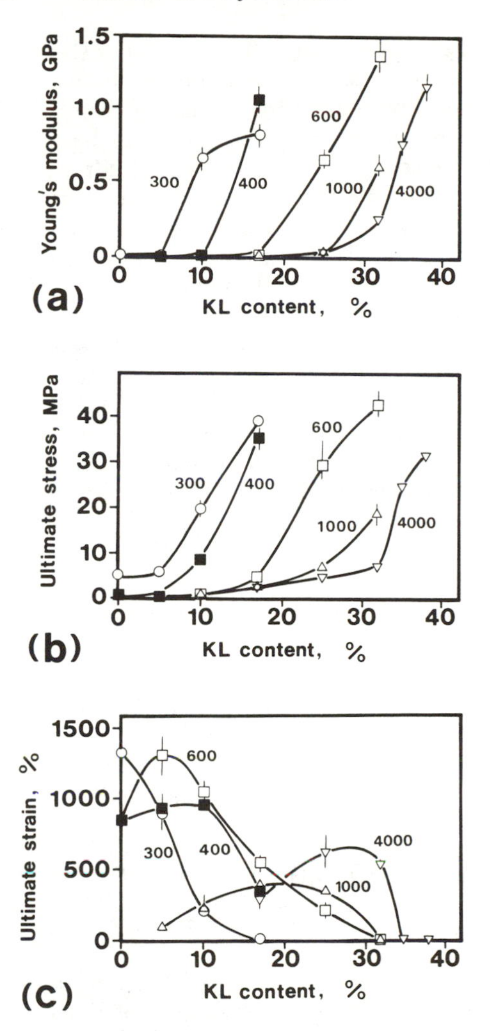

Figure 6. Young's modulus (a), ultimate stress (b), and ultimate strain (c) as functions of KL content. The numbers in the figure indicate different molecular weights of PEG.

generally much higher than the highest ϵ_b-values found (200-250%) for the previously studied MDI-based, KL-containing PU's (6, 10). This must be mainly a result of the markedly lower crosslink densities of the TDI-based PU's studied here. The elongation of TDI-based PU's has previously been shown to be sensitive to variations in crosslink density (23).

Previous studies of KL-polyether-derived PU's have shown that high contents of KL ($> 30 - 35\%$) result in hard and sometimes brittle PU's regardless of the NCO/OH ratio used (6) and regardless of the molecular weight of KL (Yoshida, H.; Mörck, R.; Kringstad, K. P.; Hatakeyama, H., submitted to *J. Appl. Polym. Sci.*). The results of the present investigation show that the upper limit for the KL content at which elastomeric PU's can be produced can be extended by increasing the chain length (molecular weight) of the soft polyether segment. This can be explained as an effect of the decrease in the content of PEG-derived hydroxyl groups that follows with increasing molecular weight of PEG. In order to fulfill the requirements of a constant NCO/OH ratio, an increase in the chain length of the PEG segments will, at any given PEG content, result in an increase in KL content and a decrease in the content of diisocyanate. In other words, the KL content increases whereas the soft/hard segment ratio remains unchanged. An interesting example of this effect in the present investigation is PU film #24, which was synthesized using PEG 4000 as soft segment. This PU film showed a very high value of ϵ_b (540%) at a KL content as high as 31.9%. The PEG content of this PU was 41.5%, which in the PU system studied here, approximately corresponds to the minimum level of PEG content required to produce PU's with high room-temperature values of ϵ_b (Fig. 7).

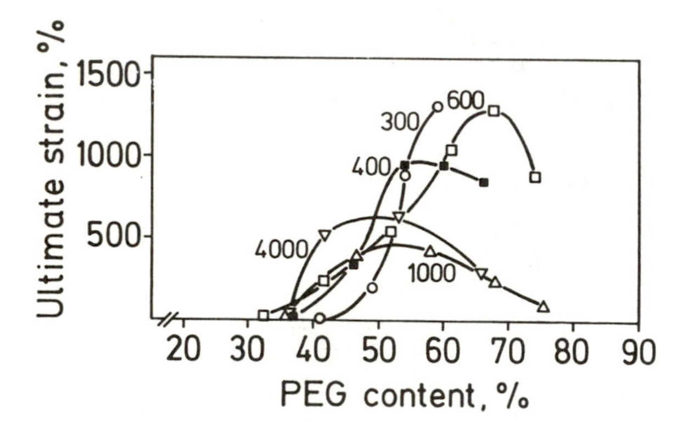

Figure 7. Ultimate strain as a function of PEG content. The numbers in the figure indicate different molecular weights of PEG.

Conclusion

Elastomeric, lignin-derived polyurethanes can be produced from a three-component system, in which a polyether diol (like PEG) is used as soft segment, provided that the crosslink density is held at a low level and that the soft segment content is high enough to keep T_g below room temperature.

Acknowledgments

The authors wish to thank Dr. K. Nakamura, Industrial Research Institute of Kanagawa Prefecture, Yokohama, Japan, for recording the DSC scans shown in Figure 2.

Financial support for this investigation from the Swedish National Board for Technical Development is gratefully appreciated.

Literature Cited

1. Saraf, V. P.; Glasser, W. G. *J. Appl. Polym. Sci.* 1984, **29**, 1831-1841.
2. Rials, T. G.; Glasser, W. G. *Holzforschung* 1984, **38**, 191-199.
3. Rials, T. G.; Glasser, W. G. *Holzforschung* 1984, **38**, 263-269.
4. Saraf, V. P.; Glasser, W. G.; Wilkes, G. L.; McGrath, J. E. *J. Appl. Polym. Sci.* 1985, **30**, 2207-2224.
5. Rials, T. G., Glasser, W. G. *Holzforschung* 1986, **40**, 353-360.
6. Yoshida, H.; Mörck, R.; Kringstad, K. P.; Hatakeyama, H. *J. Appl. Polym. Sci.* 1987, **34**, 1187-1198.
7. Mörck, R.; Yoshida, H.; Kringstad, K. P.; Hatakeyama, H. *Holzforschung* 1986, **40**, suppl. 51-60.
8. Saunders, J. H. *Rubber Chem. Technol.* 1960, **33**, 1259-1292.
9. Schauerte, K. In *Polyurethane Handbook*; Oertel, G., Ed.; Hanser Publishers, Munich 1985; p. 62.
10. Kringstad, K. P.; Mörck, R.; Reimann, A.; Yoshida, H. *Proc. Fourth International Symposium on Wood and Pulping Chemistry*, Paris, April 27-30, 1987, Vol. 1, pp. 67-69.
11. Baker, J. W.; Gaunt, J. *J. Chem. Soc.* 1949, 19-24.
12. Kratzl, K.; Buchtela, K.; Gratzl, J.; Zauner, J.; Ettingshausen, O. *Tappi* 1962, **45**, 113-119.
13. Ulrich, H.; Tucker, B.; Sayigh, A. R. *J. Org. Chem.* 1967, **32**, 3938-3941.
14. Kresta, J. E.; Garcia, A.; Frisch, K. C.; Linden, G. In *Urethane Chemistry and Applications*; Edwards, K. N. Ed.; ACS Symposium Series No. 172; American Chemical Society: Washington, DC, 1981; pp. 403-417.
15. Simons, D. M.; Arnold, R. G.; *J. Am. Chem. Soc.* 1956, **78**, 1658-1659.
16. Czerwinski, W.; Ciemniak, G. *Angew. Makromol. Chem.* 1985, **134**, 23-35.
17. Aranguren, M. I.; Williams, R. J. J. *Polymer* 1986, **27**, 425-430.
18. Schneider, N. S.; Paik Sung, C. S. *Polym. Eng. Sci.* 1977, **17**, 73-80.
19. Schneider, N. S.; Paik Sung, C. S.; Matton, R. W.; Illinger, J. L. *Macromolecules* 1975, **8**, 62-67.
20. Seefried, C. G.; Koleske, J. V.; Critchfield, F. E. *J. Appl. Polym. Sci.* 1975, **19**, 3185-3191.
21. Aitken, R. R.; Jeffs, G. M. F. *Polymer* 1977, **18**, 197-198.
22. Voigt-Martin, I.; Wendorff, J. In *Encycl. Polym. Sci. Eng.*; 2nd ed., John Wiley & Sons, 1985, Vol. 1, 789-842.
23. Smith, T. L.; Magnusson, A. B. *J. Polym. Sci.* 1960, **42**, 391-416.

RECEIVED February 27, 1989

Chapter 31

Effect of Soft-Segment Content on the Properties of Lignin-Based Polyurethanes

Stephen S. Kelley[1,4], Wolfgang G. Glasser[1,2], and Thomas C. Ward[1–3]

[1]Department of Wood Science and Forest Products, Virginia Polytechnic Institute and State University, Blacksburg, VA 24061
[2]Polymeric Materials and Interfaces Laboratory, Virginia Polytechnic Institute and State University, Blacksburg, VA 24061
[3]Department of Chemistry, Virginia Polytechnic Institute and State University, Blacksburg, VA 24061

Alternative methods of preparing lignin-based polyurethanes are compared in relation to thermal and mechanical properties. Generally the glass transition temperature, modulus and tensile strength all increase with increasing lignin content. The elongation at break is consistently low except for the networks prepared with chain-extended hydroxypropyl lignins, where the elongation decreases as the lignin content increases. Polyurethanes prepared with aromatic isocyanates generally show better mechanical properties than those prepared with hexamethylene diisocyanate. Polyurethanes from fractionated (unmodified) lignin with added polyethylene glycol, and those from hydroxypropyl lignin, both provide high strength, homogeneous networks. Only chain-extended hydroxypropyl lignins produce networks with high elongation.

Over 750,000 tons of lignin per year are currently used worldwide, mostly as lignin sulfonates for aqueous dispersions (1). While this is a substantial quantity, it is only a fraction of the lignin available at pulp and paper mills.

One goal of researchers has been to devise methods for incorporating lignin into solid materials. Lignin has been suggested as a filler or additive in several systems (3-6). However, for most of these uses lignin serves as a low value component rather than as an integral, active ingredient. Some of the features which have limited lignin to these low value uses include its heterogeneous and complex structure, high modulus, poorly defined thermal transitions, and poor solubility (7, 8). Thus, lignin is difficult to melt blend

[4]Current address: Eastman Chemicals Division, P.O. Box 1972, Kingsport, TN 37662

0097–6156/89/0397–0402$06.00/0

with other materials. The poor solubility of kraft lignin, the most abundant source, has limited commercial incorporation of lignin into urethane (9), phenol formaldehyde (10), or epoxy (11) systems.

In an effort to overcome these limitations researchers have looked at new sources of lignin or ways to modify lignins which are currently produced (12,13). The state of the art has progressed to the point that many of the traditional limitations of lignin have been addressed. The salient structural features of lignin may be identified and quantified (2,7). The thermal transitions can be measured, modified and related to the lignin structure (14). The solubility of lignin can be improved by new isolation processes, by fractionation or by derivatization. Thus, a more realistic view of lignin, based on current technology, is that of a multifunctional polyol which may be incorporated into several types of crosslinked polymers.

Although the use of lignin as an additive to polyurethanes is not new (15-20), even the most judicious selection of lignin isolation or modification schemes has not allowed researchers to overcome the incorporation limit of 25 to 40 weight percent of lignin as an active component in polyurethanes. Solvent fractionation allows for the isolation of lignin fractions with well defined solubilities and functionalities (21,22). Both of these features are critical for the practical inclusion of lignin into liquid polyol systems.

Fractionation can address the solubility considerations and reduce the structural and molecular heterogeneities (18,20). However, it is of limited usefulness in overcoming the problems of high modulus and high thermal transitions. Fractionation also does not address the consideration of differential reactivities of the various functional groups present in lignin. The presence of carboxyl groups in addition to phenolic as well as primary and secondary hydroxyls can lead to loss of control over gelation of polyurethane networks (23). Premature gelation can produce inhomogeneous materials and, in extreme cases, phase separation. To address these problems, various derivatization schemes have been used to modify the lignin structure.

Reaction of lignin with ethylene or propylene oxide produces a material which has essentially a single type of functionality (8). The derivatives also show improved solubility and lower thermal transitions compared to their unmodified counterparts. These lignin derivatives have been incorporated into homogeneous polyurethane networks (14,15). The properties of these lignin-based polyurethanes depend on the diisocyanate crosslinker and to a lesser extent on the source of the lignin. In general, the polyurethanes have been high modulus, brittle materials with very low elongation to break.

To improve the toughness and ultimate strain characteristics of these polyurethanes, oligomeric glycols were added to the polyols based on hydroxypropyl lignins prior to crosslinking. The addition of polyethylene glycols (PEG) of varying molecular weights increased the ultimate strain from under 10% to over 40%, in some cases (16). The strength and modulus remained high as long as the isocyanate to hydroxyl ratio remained above 2.0. Addition of butadiene glycol to the hydroxypropyl lignin produced films with ultimate strains over 150%, but the films were two-phase materials over most of the composition range (17). The strength and modulus of the polyurethanes decreased as the lignin content decreased. Depending on which phase was continuous, these multiphase materials were regarded as rubber-toughened or lignin-reinforced networks.

An alternative approach to modifying the ultimate strain of lignin-based polyurethanes involved the reaction of hydroxypropyl lignin with additional propylene oxide (24, 25). This reaction produced a material that was viewed as a multi-armed star (26). The center of the star was the high modulus, aromatic lignin, while the arms were flexible polyether segments with hydroxyl groups on their terminus. Polyurethane networks based on these "chain-extended" hydroxypropyl lignins were homogeneous across their entire composition range (25). They showed ultimate strain properties of over 100%, but their strength properties decreased as the elongation increased.

The purpose of the current study is to offer a critical review of these various approaches to producing lignin-based polyurethane networks. The thermal and mechanical properties of various lignin-based polyurethanes are compared after being normalized for their lignin contents. The relative advantages of the various approaches to incorporating lignin into polyurethanes are discussed.

Materials and Methods

Materials. The materials are described in detail in the studies summarized in Table I. Most of these results are based on kraft lignin. The NCO/OH ratio of all of these networks were high; greater than 1.5. All of the networks were prepared from homogeneous solutions of the lignin-based polyol, added (soft segment) polyol, and diisocyanate. Films were cast and cured under mild conditions with a controlled loss of solvent. The films were post-cured to insure complete reaction (25).

Methods. Thermal properties were determined by differential scanning calorimetry (DSC). Mechanical properties were all measured in tension on samples cut from the films.

Results

Thermal properties of the lignin-based networks are drawn from three published sources (15, 16, 25). The mechanical properties of the materials are taken from four publications (15, 16, 19, 25). All of these lignin-based polyurethanes are reported to be homogeneous materials. These sources can be roughly divided into two classes. The first class represents polyurethanes which contain only lignin as the polyol, although in both cases the lignin has been modified by reaction with propylene oxide. The second class contains both lignin polyol and added polyethylene glycol (PEG). In one case the molecular weight of the added polyol was varied, and this series is represented with two samples containing highest and lowest molecular weight soft segment.

In several studies the NCO/OH ratio was also varied. To facilitate the comparisons, only networks with NCO/OH ratios between 1.5 and 3.0

Table I. Lignin source and soft segment type for polyurethanes discussed in this study

Lignin Source	Soft Segment Type	Formulation Type	Reference
Kraft lignin Organosolv lignin	Propylene oxide oligomers covalently bound to lignin (EHPL)[1]	A	25
Kraft hydroxypropyl lignin (HPL) (MS = 1.0)	PEG with M_w between 400 and 4,000	B	16[3]
Middle molecular weight fraction of kraft lignin	Propylene oxide triol with a M_w of 600	C	19[4]
Kraft and steam explosion hydroxypropyl lignins (HPL)	None	D	15

[1] Hydroxypropyl lignin (HPL) typically has a degree of substitution (DS) or 1.2 and a molar substitution (MS) of 1.0; and HPL with an extended propylene oxide chain (CEHOL) has an MS > 1.0.
[2] Recovered in 45% yield of parent kraft lignin.
[3] Only the values for PEG 400 and 4,000 are included.
[4] Values for NCO/OH ratio of 1.5 and 2.1 were averaged to give a single point.

are included in this discussion. It should be noted that this NCO/OH is quite high, though several studies have shown that this is required for lignin-based polyurethanes. Even with these limits, the NCO/OH ratio influences the observed glass transition temperature (T_g) and ultimate strain of some of the networks (15). However, generally this influence is relatively small compared to the other differences which are discussed. The effect of the NCO/OH ratio is noted where it is considered important.

Glass Transition Temperature. The effect of lignin content on the T_g of films crosslinked with HDI and aromatic isocyanates is shown in Figures 1 and 2, respectively. There is a positive relationship between lignin content and T_g. The T_g of the PEG-extended HDI polyurethanes is consistently above that of the networks containing chain-extended hydroxypropyl lignin (CEHPL). The T_g of the networks which are based solely on hydroxypropyl lignin (HPL) are higher than any of the soft segment-containing materials.

The T_g of polyurethanes crosslinked with aromatic isocyanates shows no difference between the PEG-extended and the CEHPL networks. The T_g values for both types of networks follow the same relationship. Networks without soft segment content generally have a higher T_g than those with

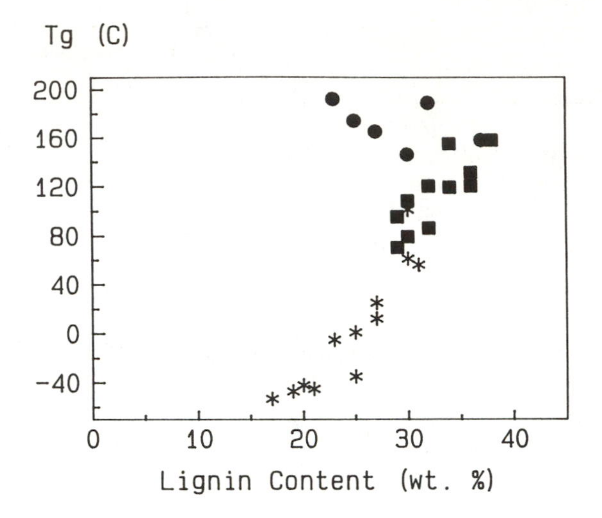

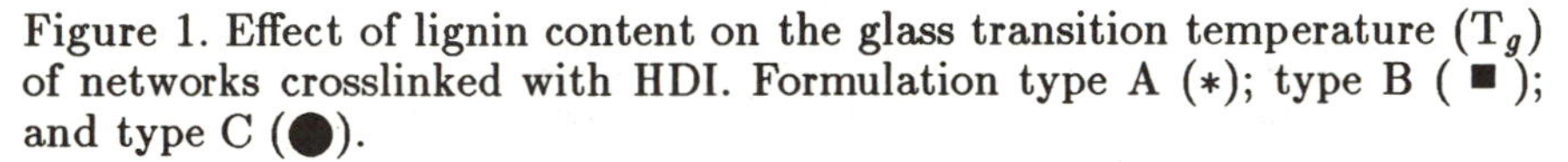
Figure 1. Effect of lignin content on the glass transition temperature (T_g) of networks crosslinked with HDI. Formulation type A (∗); type B (■); and type C (●).

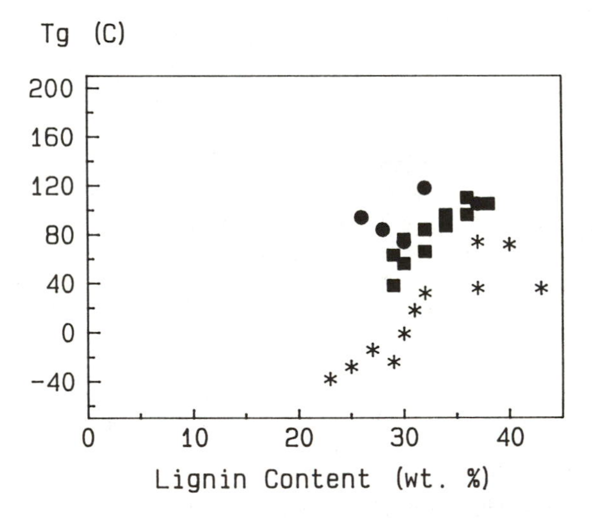

Figure 2. Effect of lignin content on the glass transition temperature (T_g) of networks crosslinked with aromatic isocyanate. Formulation type A (∗); type B (■); and type C (●).

added soft segment. The T_g values of non-extended networks are essentially insensitive to lignin content.

In most cases the T_g values are related directly to the lignin content of the network. Soft segment-free networks show more scatter than those containing alkylether components, and this may be attributed to the crosslink density of the materials. The method of soft segment incorporation, simple addition or covalently bonded, has a minimal impact on T_g. A comparison of the HDI and aromatic isocyanate-crosslinked networks, at a given lignin content, shows that the HDI-crosslinked polyurethanes generally have a lower T_g than the aromatic isocyanate-crosslinked networks.

Modulus of Elasticity. The effect of lignin content on modulus of elasticity (MOE) depends on the method used to add the soft segment (Figures 3 and 4). The moduli of networks without soft segment, or networks where the soft segment (PEG) has simply been added to the mixture, are relatively insensitive to lignin content. At high lignin contents, above 30 wt.%, the MOE is similar for all networks. The moduli of the aromatic isocyanate-crosslinked networks were generally higher than those crosslinked with HDI for the HPL-PEG networks (Fig. 4).

Only in the case of CEHPL can the MOE of the networks be substantially modified by raising the soft segment content. As the lignin content is reduced from 30 to about 20 weight percent, the MOE declines by more than two orders of magnitude. It is interesting to note the differences between the networks prepared with the fractionated lignin and added PEG (i.e., formulation type C, Table I), and the CEHPL polyol (i.e., formulation type A, Table I). The fractionated lignin produced networks that were relatively insensitive to lignin content. The moduli of these networks were slightly below those produced from the HPL with added PEG. In contrast, the CEHPL-based networks showed a predictable decrease in modulus with declining lignin content. In a fashion similar to T_g, the MOE values for the CEHPL-based networks are generally lower at a given lignin content when crosslinked with HDI compared to those networks crosslinked with aromatic isocyanates.

Ultimate Strength. In contrast to T_g and MOE, the ultimate (tensile) strength of lignin-based polyurethanes is very sensitive to the method of preparation (Figures 5 and 6). At a given lignin content, the networks crosslinked with aromatic isocyanates are consistently stronger than those employing HDI. Within an isocyanate type, and at a constant lignin content, network strength was found to progressively decrease from networks without soft segment to networks with added PEG to CEHPL-based networks. Within a given set of networks, there does appear to be a positive relationship between lignin content and ultimate strength, except for soft segment-free networks (i.e., type D).

For the polyurethanes crosslinked with HDI (Figure 5), networks with added PEG and CEHPL-based networks followed similar trends. The networks with added PEG show slightly higher strength values at a given lignin content. In contrast, the strength values of networks without soft segment

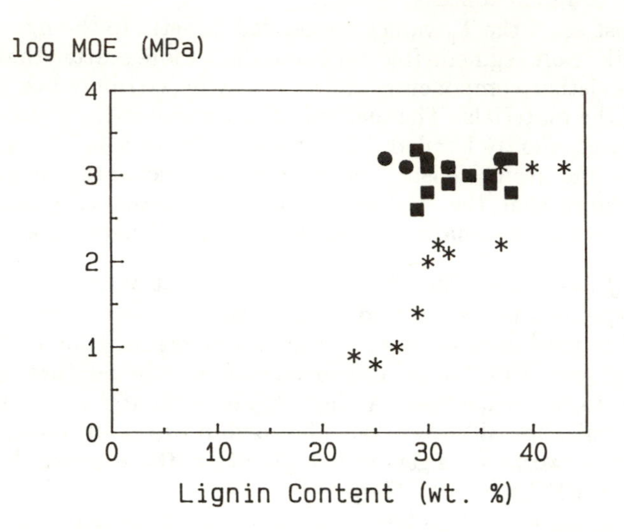

Figure 3. Effect of lignin content on the modulus of networks crosslinked with HDI. Formulation type A (∗); type B (■); and type C (●).

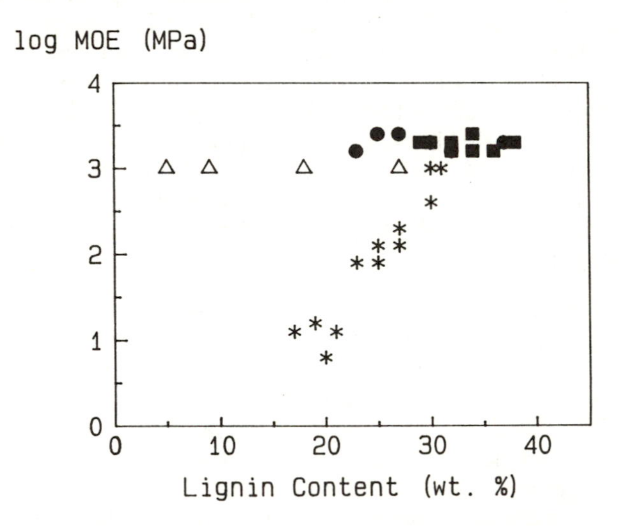

Figure 4. Effect of lignin content on the modulus of networks crosslinked with aromatic isocyanates. Formulation type A (∗); type B (■); type C (●); and type D (△).

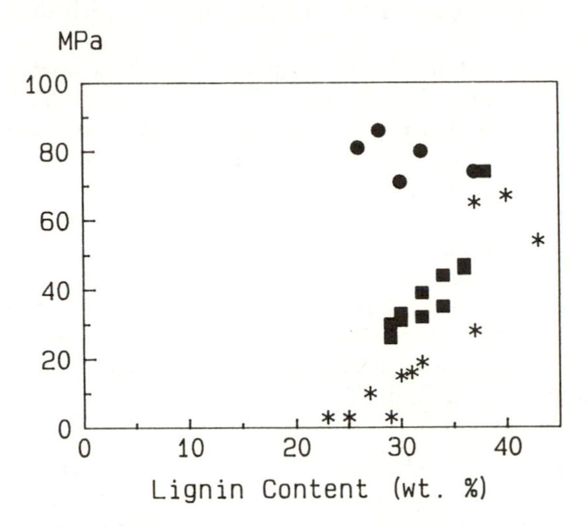

Figure 5. Effect of lignin content on the ultimate strength of networks crosslinked with HDI. Formulation type A (∗); type B (■); and type C (●).

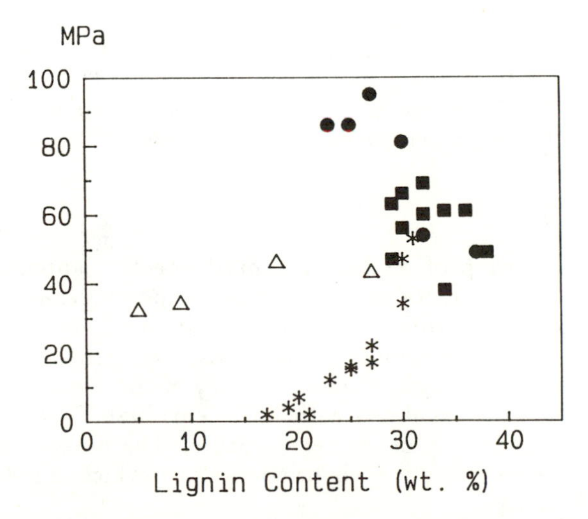

Figure 6. Effect of lignin content on the ultimate strength of networks crosslinked with aromatic isocyanates. Formulation type A (∗); type B (■); type C (●); and type D (△).

content actually increase with decreasing lignin content. This anomalous trend is due to an increase in the crosslink density with decreasing lignin content (25).

In the case of polyurethanes crosslinked with aromatic isocyanates, the trends are less clear than for the HDI-based materials. The CEHPL-based networks do show a predictable, monotonic decrease in strength as the lignin content declines (Fig. 6). For networks with added PEG, there is considerable scatter in the strength values. The data of Kringstad *et al.* (20), which covers a wide range of lignin contents, shows a slight decline in strength as the lignin content decreases. The strength values of Saraf *et al.* (15,16), that are based on HPL and that cover a limited range of lignin contents, show no consistent relationship between lignin content and strength. Again, networks prepared without any added soft segment show an increase in strength with decreasing lignin content. As with the HDI-crosslinked networks, this is due to an increase in the crosslink density with decreasing lignin content.

Ultimate Elongation. The ultimate elongations of polyurethanes crosslinked with HDI and aromatic isocyanates are shown in Figures 7 and 8, respectively. Generally, the ultimate elongation of the aromatic isocyanate-crosslinked networks is below that of the HDI-crosslinked materials. With two exceptions, the ultimate elongations of all networks are low. This is consistent with the relatively high MOE and strength values for the HPl-PEG and HPL-based materials.

The exceptions to the generally low elongation values are for networks prepared from fractionated lignin with added PEG, and for CEHPL-based networks. The polyurethanes prepared from fractionated lignin with added PEG show moderate ultimate elongations of 20-25%. The elongation properties of the CEHPL networks are even larger; ultimate elongations of greater than 50% are observed for both HDI and TDI-crosslinked materials.

Discussion

Considering this group of studies, the most effective approach to incorporating large weight fractions of lignin into a polyurethane network depends on the properties which are desired of the final product. At a given lignin content, the highest moduli are consistently provided by aromatic isocyanate-crosslinked polyurethanes. The T_g of these materials is consistently higher than that achieved with HDI. This high T_g should translate into a high heat deflection or use temperature. The limitation of the aromatic isocyanate materials is their brittle nature, which is reflected in low elongations at break.

If toughness is not the limiting consideration, then the simple addition of PEG soft segment appears to offer an attractive route to the production of polyurethanes with 25 to 30 wt.% lignin. It is important to note that both of the studies considered here have employed modified lignins. The relative advantage of hydroxypropylation (16) over fractionation (19) would depend on the relative cost of either procedure.

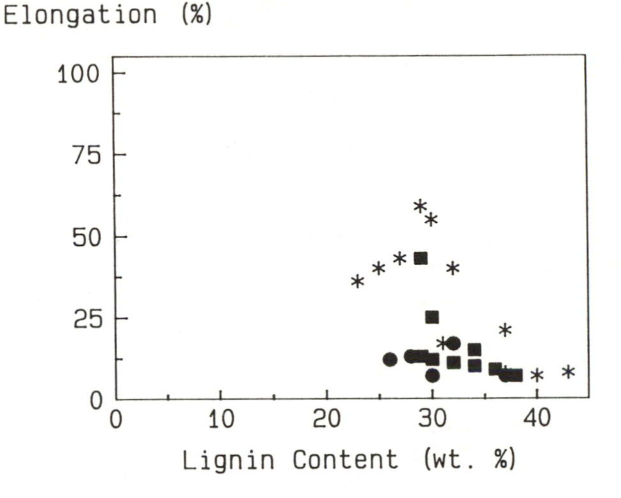

Figure 7. Effect of lignin content on the ultimate elongation of networks crosslinked with HDI. Formulation type A (∗); type B (■); and type C (●).

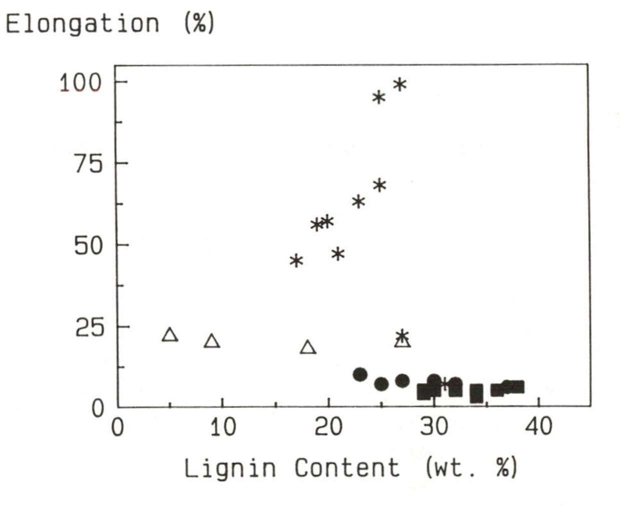

Figure 8. Effect of lignin content on the ultimate elongation of networks crosslinked with aromatic isocyanates. Formulation type A (∗); type B (■); type C (●); and type D (△).

Only one of the approaches considered here, chain-extension with propylene oxide of hydroxypropyl lignins, allowed for the preparation of networks with substantial elongation at break. Typical of most polyurethanes, an increase in the elongation at break resulted in a corresponding decrease in modulus and strength. This represents the most complex modification procedure of those discussed, although process development could probably simplify this modification for adoption on industrial scale.

It is worth noting that a recent study has shown that it is possible to prepare polyurethane networks with high ultimate strains from fractionated kraft lignin, PEG and toluene diisocyanate (27). The lignin content of these films ranged from 10 to 40 percent. The ultimate strength and modulus values were generally below those of the HPL-PEG networks. Phase separation was noted for some of the films, which could account for the large ultimate strains.

Conclusions

Several alternative schemes for the preparation of polyurethane networks with lignin have been discussed. For the studies reviewed, properties generally varied directly with lignin content. In general, T_g, MOE, and ultimate strength all increased as the lignin content of the networks increased. Ultimate strain decreased with increasing lignin content. Hexamethylene diisocyanate generally produced networks with inferior mechanical properties when compared to aromatic diisocyanates.

If a particular end use application does not require a high degree of toughness, then there exist two approaches for the preparation of high strength networks. Fractionation of unmodified lignin or modification of the whole lignin, through hydroxypropylation, both provide polyols for networks with high strength and modulus. Which of these approaches is most practical would depend on the relative cost of the fractionation or hydroxypropylation procedure. If a high ultimate strain is required for the network, then the preparation of chain-extended hydroxypropyl lignin becomes necessary.

Literature Cited

1. Lin, S. Y. In *Progress in Biomass Conversion*; Tilman, D. A.; Jahn, E. C., Eds.; Academic Press: New York, 1983; Vol. 4.
2. Glasser, W. G.; Kelley, S. S. In *Encyclopedia of Polymer Science and Engineering*; Mark, H. F.; Bikales, N. M.; Ovenberger, C. G.; Menges, G., Eds.; John Wiley: New York, 1987; Vol. 8, 2nd Ed., pp. 795-852.
3. Nichols, R. F. U.S. Patent 2 854 422, 1958.
4. Kratzl, K.; Buchtela, K.; Gratzl, J.; Zauner, J.; Ettinghausen, O. *Tappi* 1962, **45**, 113.
5. Glasser, W. G.; Hsu, O. H.-H. U.S. Patent 4 017 474, 1977.
6. Lambuth, A. L. U.S. Patent 4 279 788, 1981.
7. Falkehag, S. I. *Appl. Polym. Symp.* 1975, **28**, 247.
8. Glasser, W. G.; Barnett, C. A.; Rials, T. G.; Saraf, V. P. *J. Appl. Polym. Sci.* 1984, **29**, 1815.

9. Glasser, W. G.; Wu, C.-F.; Selin, J.-F. In *Wood and Agricultural Residues—Research, on Use for Fuels and Chemicals*; Soltes, E. J., Ed.; Academic Press: New York, 1983; pp. 149-166.
10. Forss, K.; Fuhrmann, A. *Paperi ja Puu* 1976, **11**, 817.
11. Ball, F. J.; Dougherty, W. K.; Moore, H. H. Can. Patent 654 728, 1962.
12. Hsu, O. H.-H.; Glasser, W. G. *Wood Sci.* 1976, **9**, 97.
13. Lange, W.; Faix, O.; Beinhoff, O. *Holzforschung* 1983, **37**, 63.
14. Rials, T. G.; Glasser, W. G. *J. Wood Chem. Tech.* 1984, **4**, 331.
15. Saraf, V. P.; Glasser, W. G. *J. Appl. Polym. Sci.* 1984, **29**, 1831.
16. Saraf, V. P.; Glasser, W. G.; Wilkes, G. L.; McGrath, J. E. *J. Appl. Polym. Sci.* 1985, **30**, 2207.
17. Saraf, V. P.; Glasser, W. G.; Wilkes, G. L. *J. Appl. Polym. Sci.* 1985, **30**, 3809.
18. Rials, T. G.; Glasser, W. G. *Holzforschung* 1984, **38**, 191.
19. Yoshida, H.; Mörck, R.; Kringstad, K. P.; Hatakeyama, H. J. *J. Appl. Polym. Sci.* 1987, **34**, 1187.
20. Kringstad, K. P.; Mörck, R.; Reimann, A.; Yoshida, H. *Proc. 4th Inter. Symp. Wood and Pulping Chem.*; Paris, 1987; Vol. 1, 67-69.
21. Mörck, R.; Yoshida, H.; Kringstad, K. P.; Hatakeyama, H. J. *Holzforschung* 1986, **40** (Suppl.), 51.
22. Kelley, S. S.; Ward, T. C.; Rials, T. G.; Glasser, W. G. *J. Appl. Polym. Sci.*, in press.
23. Pigoff, K. A. In *Encyclopedia of Polymer Science and Engineering*; Mark, H. F.; Gaylord, N. G.; Bikales, N. M., Eds.; John Wiley: New York, 1981; Vol. 11, pp. 507-561.
24. Kelley, S. S.; Glasser, W. G.; Ward, T. C. *J. Wood Chem. Tech.*
25. Kelley, S. S.; Glasser, W. G.; Ward, T. C. *J. Appl. Polym. Sci.* 1988, **36**, 759.
26. de Oliveira, W.; Glasser, W. G. *J. Appl. Polym. Sci.*, in press.
27. Reimann, A.; Mörck, R.; Kringstad, K. P. In *Lignin: Structure and Materials*; Glasser, W. G.; Sarkanen, S., Eds.; ACS Symp. Ser., in press.

RECEIVED February 27, 1989

Chapter 32

Starlike Macromers from Lignin

Willer de Oliveira[1,2] and Wolfgang G. Glasser[1,2]

[1]Department of Wood Science and Forest Products, Virginia Polytechnic Institute and State University, Blacksburg, VA 24061
[2]Polymeric Materials and Interfaces Laboratory, Virginia Polytechnic Institute and State University, Blacksburg, VA 24061

Multifunctional polymeric segments with controlled chemical and molecular characteristics (i.e., macromers) can be prepared from hydroxypropyl lignin (HPL) by either partial capping of OH groups followed by alkyl ether chain extension; or by grafting with mono functional linear chain segments. Star-like macromers may either serve as segmental components in block copolymers, or as thermoplastic elastomers with distinct crystallinity of one phase. Examples given include partially ethylated HPL which was chain-extended with propylene oxide; and star-block polymers of HPL with monofunctional cellulose tri-acetate of DP 12. Analytical results correspond to structural characteristics.

The Encyclopedia of Polymer Science and Engineering defines macromers as "polymers of molecular weight ranging from several hundred to tens of thousands, with a functional group at the chain end that can further polymerize" (1). Branched or spherical molecules may produce macromers with more than two terminal reactive functional groups, and these resemble star-like architecture. Macromers have gained usefulness in the synthesis of graft, block, and segmented polymers, often with multiphase morphology. The macromer must have a well-defined molecular weight and molecular weight distribution (1).

Lignins isolated from organosolv pulping and steam explosion have been shown to be of low molecular weight, in the range of 1000 or below ($\overline{M}_n$), and of narrow molecular weight distribution ($\overline{M}_w/\overline{M}_n$ ca. 3) (2-5). This reflects pentameric composition on the average, and this is within the molecular weight range that is typical for macromers. Recent advances in the art of fractionating lignins and lignin derivatives with regard to molecular weights (6-8) create the promise of isolating macromeric lignin fractions of narrow molecular weight distribution; and advances in the field of

0097–6156/89/0397–0414$06.00/0

chemical modification of lignin, which have produced lignins with uniform terminal functionality and good solubility (9-11), help meet the requirement for controlled reactivity. Hydroxypropyl lignin with extended propyl ether chains have recently been reported (12) that are viscous liquids with uniform terminal hydroxyl functionality on alkyl ether chains that radiate from a central lignin core. With an average hydroxyl functionality of about 1.2 per phenyl propane (C_9) repeat unit, a lignin macromer would possess a functionality of about 6 (13). Star-like macromers result from a partial reduction of functionality, such as by capping with a monofunctional ether or carbamate substituent, followed by chain extension with an alkylene oxide (13); or by the addition of preformed, monofunctional chain segments (14). Whereas degree of capping controls the number of arms per star in the former case, stoichiometry determines it in the latter.

This paper examines two types of hydroxypropyl lignin based macromers, and these are illustrated schematically in Figure 1. Macromers with propylene oxide (PO) are formed by reducing the number of available hydroxyl groups on HPL followed by chain extension with PO; and macromers with cellulose triacetate (CTA) are synthesized by attaching a monofunctional CTA chain to a limited number of terminal OH groups on HPL via a suitable grafting reaction.

Experimental

Materials. Hydroxypropyl lignin (HPL): Organosolv (aqueous ethanol) lignin from aspen, supplied by Repap Technologies Inc. (formerly the Biological Energy Corporation) of Valley Forge, PA, was hydroxypropylated by reaction with propylene oxide in the usual manner (9). This derivative was isolated and purified as described previously (9).

Cellulose triacetate segments: Cellulose triacetate, supplied by Eastman Kodak Company of Kingsport, TN, was depolymerized in accordance with a method by Steinmann (15), and this was NCO capped by reaction with toluene diisocyanate as reported previously (14). A CTA segment with an average degree of polymerization of 12 (corresponding to an $\overline{M}_n$ of 3600 $g\overline{M}^{-1}$) and an average of 0.9 equivalents of NCO/mole, was isolated (14).

Methods. Reaction with $DESO_4$: The hydroxyl functionality of HPL was reduced by reaction with diethyl sulphate in accordance with earlier work (13). The reaction product was isolated by liquid-liquid extraction and dialysis.

Chain extension with PO: Partially blocked hydroxypropyl lignin derivatives were reacted with propylene oxide in toluene, using KOH as catalyst, for the purpose of creating extended propyl ether chains. This has been reported elsewhere (13).

Grafting of CTA segments: Monofunctional CTA segments were attached to HPL by dissolving both components in ethylene chloride (ca. 20% concentration) and heating the mixture for 1 hour in the presence of T9 catalysts (Union Carbide). The reaction product was poured onto a

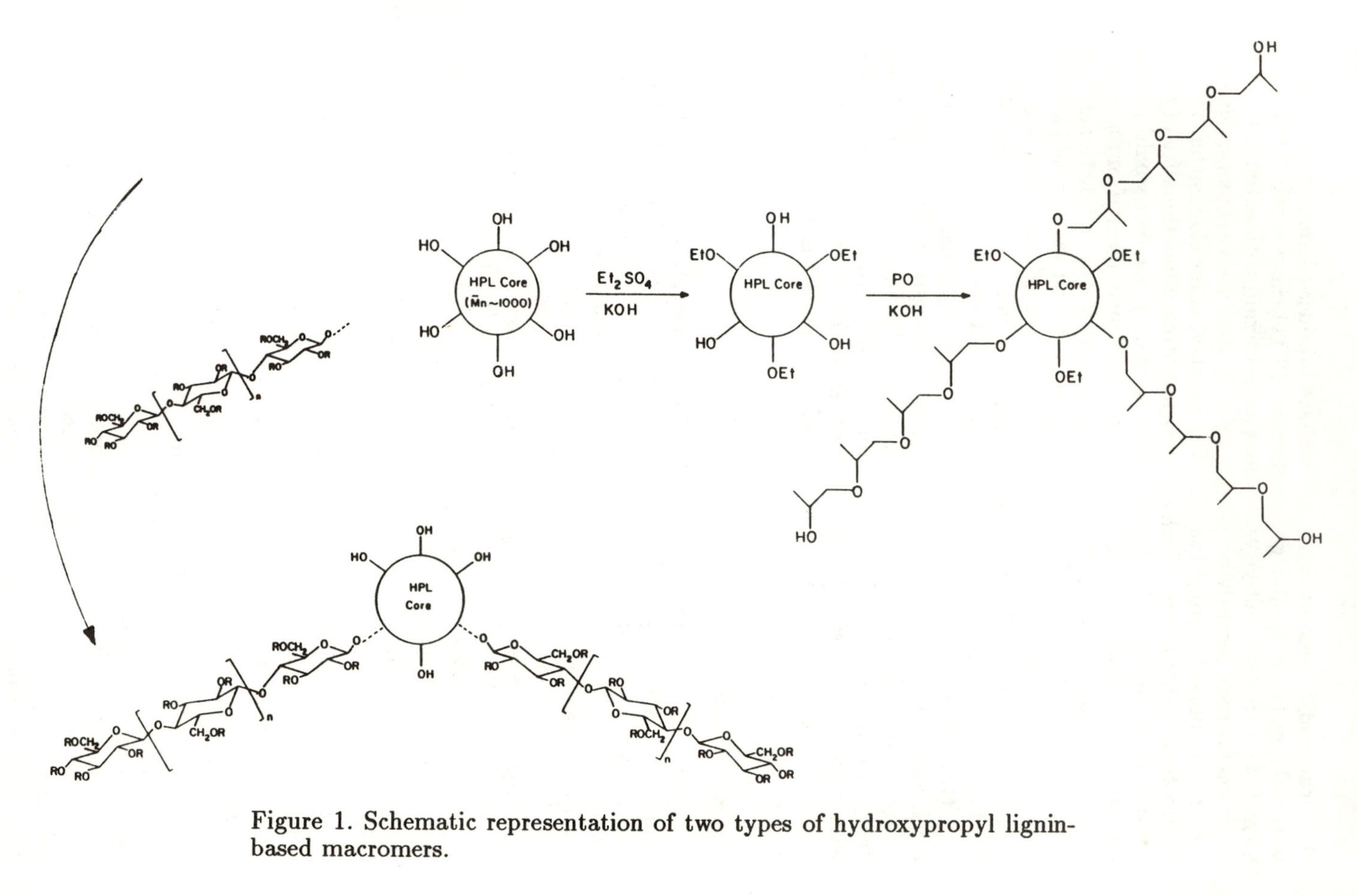

Figure 1. Schematic representation of two types of hydroxypropyl lignin-based macromers.

TEFLON mold; the solvent was allowed to evaporate; and the resulting film was kept in a desiccator for 1 week before further testing.

Analytical Methods: UV spectroscopy was performed on a Varian/Cary 219 Spectrometer. HI hydrolysis followed the method of Hodges, et al. (16) in conjunction with the separation of alkyl iodides by gas chromatography. Total hydroxyl content was determined as usual (17). Glass transition temperatures (T_g) were determined by preparing polymeric blends of lignin derivatives with commercial thermoplastics having T_g values at least 20° different from those of the derivatives. These blends were mixed in the melt followed by injection molding into a dog bone shaped mold. These molds were tested by dynamic mechanical thermal analysis and tan delta peak temperatures were taken as T_g (18). Swelling studies were performed in dimethyl formamide (DMF). X-ray scattering (WAXS) was performed on a Philips Defractometer using a Cu Ka source in the usual manner.

Results and Discussion

Macromers with propylene oxide. Hydroxypropyl (organosolv) lignin (HPL) is a low molecular weight (pre)polymer with a number average molecular weight of ca. 1200 $g\overline{M}^{-1}$ and approximately 6 equivalents of hydroxyl groups per $\overline{M}_n$ (13). Star-like macromer configuration with propylene oxide is determined by both degree of capping of hydroxyl groups and degree of chain extension. Although numerous alternatives exist regarding capping chemistry, convenience dictates the use of diethyl sulfate for OH reduction (13). A partially ethyl ether capped hydroxypropyl lignin derivative yields a mixture of methyl, ethyl, and isopropyl iodide when treated with HI. Their quantitative separation by gas chromatography (Fig. 2) yields information on lignin and propyl ether content, and on degree of capping (13). The degree of chain extension with propylene oxide can also be analyzed by H-NMR spectroscopy of acetylated derivatives (Fig. 3), and by UV spectroscopy (Fig. 4). Where H-NMR spectroscopy is complicated by the presence of overlapping ethoxy and propoxy signals, UV spectroscopy is limited to the determination of non-UV absorbing mass (i.e., ethoxy plus propoxy groups).

The relationship between target macromer functionality, degree of capping, and average macromer molecular weight is illustrated in Fig. 5. This relationship reveals that, for a target trifunctional star-like macromer of $\overline{M}_n$ 1500, approximately 60% of OH groups in the parent HPL need to be blocked. The degree of chain extension on the remaining hydroxy groups determines the polyether nature of the macromer.

Earlier work on chain extended HPLs has shown that these derivatives produce uniform (i.e., single phase) polymers with T_g varying in accordance with the Gordon-Taylor relationship (12). Polyurethanes from chain-extended HPLs were found to be rubber-like at room temperature with modulus declining as lignin content is reduced (8). Star-like structure determines functionality, T_g, viscosity, and several other properties that influence utility as polymer segment.

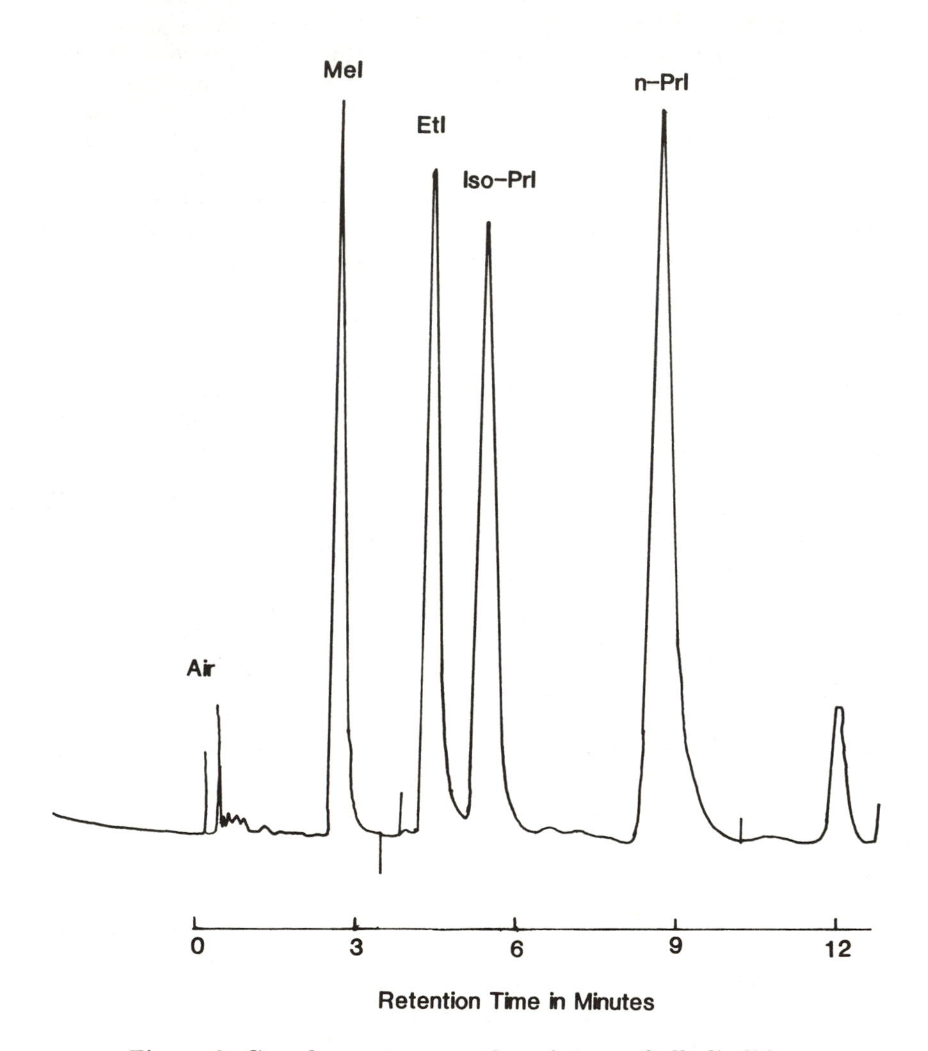

Figure 2. Gas chromatogram of a mixture of alkyliodides.

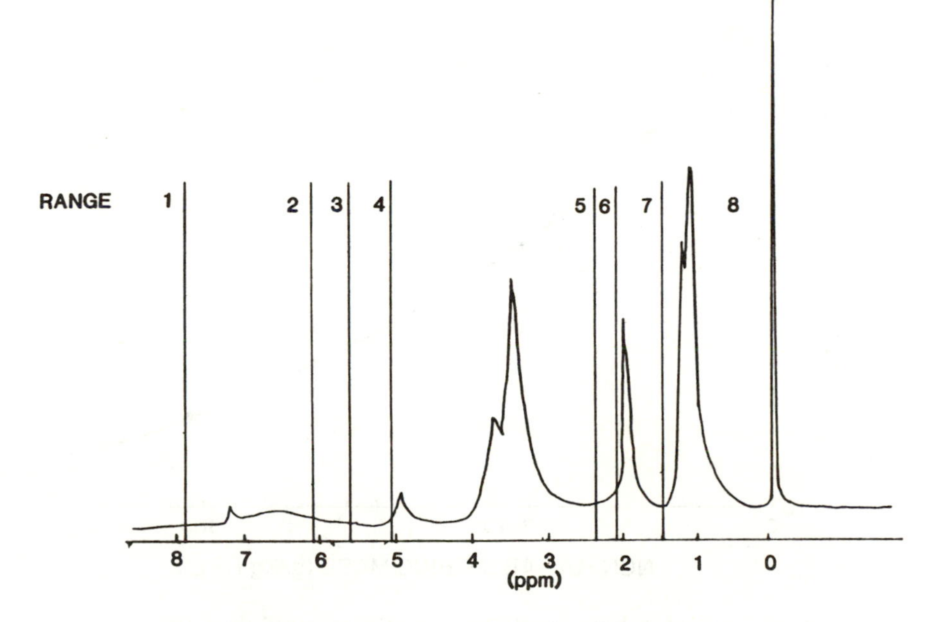

Figure 3. Typical H-NMR spectrum of acetylated HPL with assignment by range.

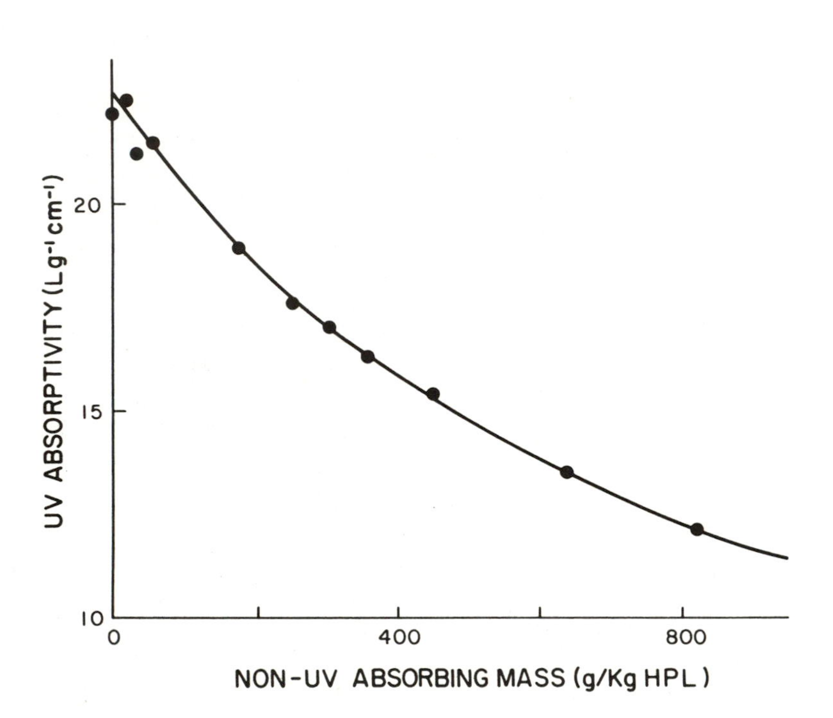

Figure 4. Relationship between UV-absorptivity coefficient (280 nm) and non-UV absorbing mass in chain extended HPL.

The engineering of lignin-based macromers with propylene oxide to target specifications appears as a useful technique for formulating components for block copolymers and segmented thermosets.

Macromers with cellulose triacetate. An alternative macromer synthesis route involves the grafting of monofunctional, preformed (linear) polymer chains onto a lignin prepolymer. Where a difunctional chain segment serves as crosslinking agent, monofunctional segments result in star-like structures. Thermoplastic elastomers are produced if the monofunctional, linear segment is glassy or crystallizable (19). The schematic illustration of Fig. 1 reveals that stoichiometry dictates the average number of radiating arms, and thus functionality per $\overline{M}_n$. Depending on the chemical and molecular characteristics of the arms being attached, two phase morphology results from phase demixing. This may be revealed analytically by differential scanning calorimetry (DSC), among other methods.

Cellulose triacetate (CTA) with monofunctionality (terminal NCO groups) was prepared and reacted with HPL by solvent casting (14). Some experimental results are summarized in Table I. Between 1 and 12 CTA segments per HPL macromer produce copolymers with between 60 and 95% CTA content. DSC reveals single glass transition temperatures which vary with composition, and melting points (T_m). The results suggest that T_g varies in accordance with the Fox relationship indicating a miscible amorphous phase. The melting point (T_m) also varies slightly in relation to composition, and it is found to be highest with the highest HPL content. Heat of fusion (H_m) declines from 1 to 0.5 cal g^{-1} when CTA content declines from 95 to 60%. Sol fraction is usually above 90%, unless CTA content increases to 90% or above. This suggests that the CTA component contained a small amount of derivative with higher functionality.

Table I. Properties of Macromers with CTA

CTA arms per $\overline{M}_n$	CTA Content (wt.%)	T_g(°C) Experimt.	T_g(°C) Theory	T_m (°C)	H_m ($calg^{-1}$) (%)	Sol Fraction (%)
1	60	113	112	252	0.5	93
2.5	80	123	124	250	0.5	93
6	90	125	130	239	0.8	89
12	95	126	133	246	1.0	82

The resulting star-like macromers with CTA were solid materials with distinct crystallinity, even at 20 and 40% HPL (Fig. 6). This suggests that even short CTA chains (i.e., DP 12) have the ability to organize into a crystalline lattice, thereby serving as pseudo-crosslinks. Depending on the chemical nature, and the molecular weight, multiphase materials result with variable mechanical properties. Above T_m of the chain segment,

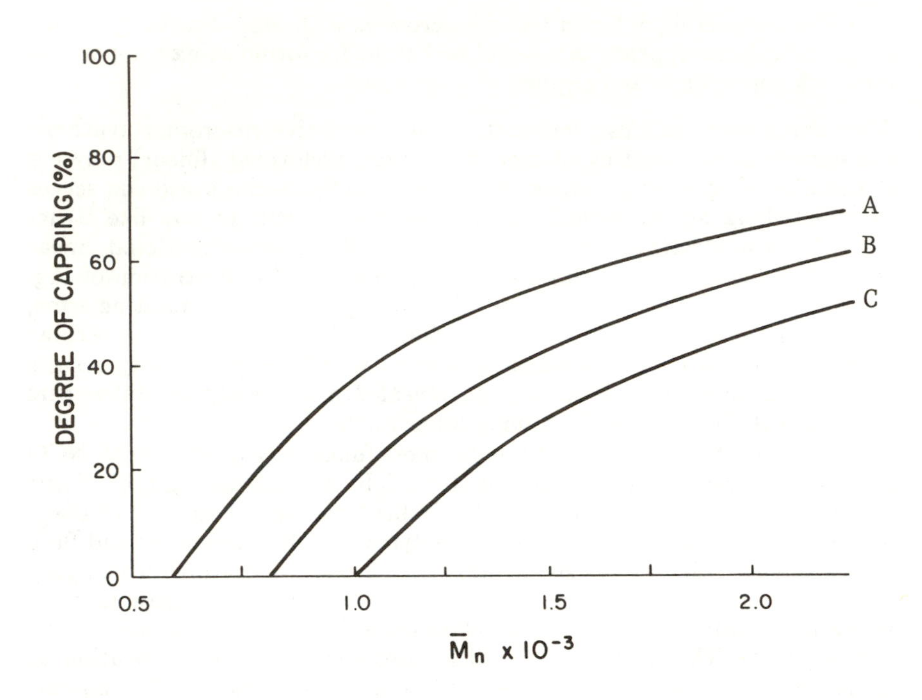

Figure 5. Theoretical relationship between required degree of capping and $\overline{M}_n$ for star-like macromers having (A) 3, (B) 4, and (C) 5 functional groups. (Note: For HPL having 1.3 OH groups C_9 on the average, and a molar substitution of 1.0.)

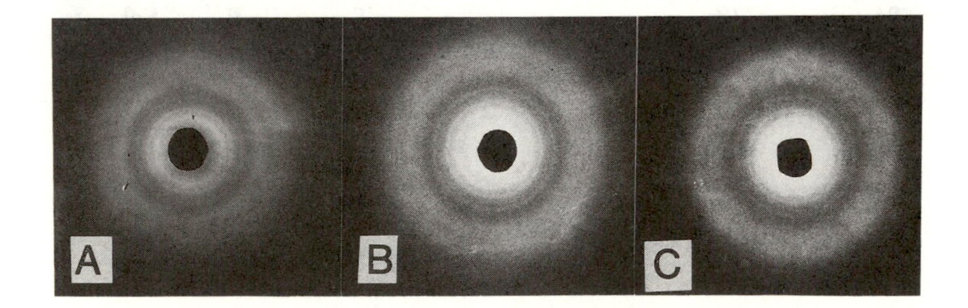

Figure 6. Wide angle X-ray scattering (WAXS) results of CTA-HPL copolymers. (A) CTA (control); (B) copolymer with 80% CTA content; and (C) copolymer with 60% CTA content.

these materials are fluids that may be processed with typical thermoplastic process technology; and below T_m they are tough structural materials.

Conclusions

The low molecular weight character of lignin and lignin derivatives (especially those derived from organosolv pulping and steam explosion), the superior solubility character of lignin derivatives, and the availability of molecular fractionation know-how, all invite the synthesis of macromers with controlled properties from lignin for use in graft, block, and segmented copolymers. This may be achieved by either capping hydroxyl functionality followed by chain extension with alkylene oxides; or by attaching preformed monofunctional chain segments. Both method of synthesis and chemistry of linear chain segment, determine processability and material properties.

Acknowledgment

Financial support for this study was provided, in part, by the government of Brazil, and by an industry-university cooperative with the Center for Innovative Technology of Virginia. Thanks also go to Mrs. Barnett for experimental assistance with several instrumental analyses.

Literature Cited

1. Kawakami, Y. In *Ency. Polym. Sci. Eng.*; Kroschwitz, J. I., Ed.; John Wiley & Sons: New York, 1987, Vol. 9, pp. 195-204.
2. Glasser, W. G.; Barnett, C. A.; Rials, T. G.; Saraf, V. P. *J. Appl. Polym. Sci.* **29**(5), 1815-30 (1984).
3. Glasser, W. G.; Barnett, C. A.; Muller, P. C.; Sarkanen, K. V. *J. Agric. Food Chem.* 1983, **31**(5), 921-930.
4. Lora, J. H.; Aziz, S. *Tappi J.* 1985, **68**(8), 94-97.
5. Marchessault, R. H.; Coulombe, S.; Morikawa, H.; Robert, D. *Can. J. Chem.* 1982, **60**(18), 2372–2382.
6. Faix, O.; Lange, W.; Beinhoff, O. *Holzforschung* 1980 **34**, 174-6.
7. Morck, R.; Yoshida, H.; Kringstad, K. P. *Holzforschung* 1986 **40** (Suppl.), 51-60.
8. Kelley, S. S.; Glasser, W. G.; Ward, T. C. *J. Appl. Polym. Sci.*, in press.
9. Wu, L. C.-F.; Glasser, W. G. *J. Appl. Polym. Sci.* 1984, **29**, 1111-23.
10. Christian, D. T.; Look, M.; Nobell, A; Armstrong, T. S. U.S. Patent 3,546,199, 1970.
11. Mozheiko, L. N.; Gromova, M. F.; Bakalo, L. A.; Sergeyeva, V. N. *Polym. Sci. USSR* 1981, **23**(1), 126-132.
12. Kelley, S. S.; Glasser, W. G.; Ward, T. C. *J. Wood Chem. Technol.* 1988, **8**(3), 341-359.
13. de Oliveira, W.; Glasser, W. G. *J. Appl. Polym. Sci.*, in press.
14. Demaret, V.; Glasser, W. G. *J. Appl. Polym. Sci.*, in press.
15. Steinmann, H. W. *Polym. Prep.* 1970, **11**(1), 285.

16. Hodges, K. L.; Kester, W. E.; Wiederrick, D. L.; Grover, T. A. *Anal. Chem.* 1979, **51**(13), 2172.
17. Siggia, S.; Hanna, J. G. *Quantitative Organic Analysis via Functional Groups*; J. Wiley & Sons, 1978, pp. 12-14.
18. Glasser, W. G.; Knudsen, J. S.; Chang, C.-S. *J. Wood Chem. Technol.* 1988, **8**(2), 221-234.
19. Legge, N. R.; Holden, G.; Schroeder, H. E., Eds. *Thermoplastic Elastomers*; Hauser Publishers, 1987; 575 pp.

RECEIVED May 29, 1989

Chapter 33

Hydroxypropyl Lignins and Model Compounds

Synthesis and Characterization by Electron-Impact Mass Spectrometry

John A. Hyatt

Research Laboratories, Eastman Chemicals Division, Eastman Kodak Company, Kingsport, TN 37662

An iterative method for the controlled rational synthesis of chain-extended hydroxypropyl derivatives of both aliphatic and aromatic hydroxyl compounds, based on oxymercuration of allyl ethers, is described. It is shown that the degree of chain extension in such compounds can be determined by electron impact mass spectrometry. Degradation of hydroxypropylated lignins by catalytic hydrogenolysis followed by capillary gas chromatography-mass spectrometry is demonstrated to be a method for establishing the presence and degree of chain extension in the hydroxypropyl lignin polymers.

One attractive approach, pioneered by Glasser and co-workers (1), to the conversion of lignin into a higher-value material comprises reaction of lignin with propylene oxide or ethylene oxide to produce an aliphatic polyol suitable for use in preparing network polymers such as polyurethane foams and plastics. Lignins are complex mixtures of moderate-to-high molecular weight phenylpropanoid polymers which bear phenolic and primary and secondary aliphatic hydroxyl groups (2), and conversion of this material from, in effect, a mixture of different types of alcohols to a substance homogeneous in terms of hydroxyl type and reactivity greatly improves its utility as a polyol.

Hydroxypropylation of lignins can be carried out under a variety of conditions, but the most successful approach has involved reaction of lignin with propylene oxide in the presence of basic catalysts; the lignin is suspended in toluene or dissolved in aqueous base (3, 4). The variation of polyol properties with preparative method (4) brings up a fundamental question of hydroxypropyl lignin structure: In the reaction of lignins with propylene oxide (Fig. 1), what is the product distribution and degree of chain extension of alkoxypropyl groups? Are there a significant number of

0097–6156/89/0397–0425$06.00/0

hydroxypropylated aliphatic hydroxyls in the product, or does most of the propylene oxide reside on the phenolic hydroxyls of the lignin nucleus?

Although techniques such as NMR spectroscopy can provide valuable information about the structure of hydroxypropyl lignins (4), we perceive a need for the development of synthetic methodology for model compounds which incorporate the structural features thought to be present in hydroxypropyl lignins. Such compounds can be used for studies of reaction kinetics as well as for confirming spectral assignments.

Synthesis and Mass Spectroscopy of Model Compounds

Although it is a simple matter to attach a single hydroxypropyl residue to a phenolic hydroxyl, the controlled synthesis of chain-extended hydroxypropyl ethers is much more difficult (Fig. 2) (5). The difficulty is that of selectively alkylating an aliphatic hydroxyl on the substrate in the presence of aliphatic hydroxyls on hydroxypropyl groups. Thus we required a method for the preparation of compounds of the type shown in Figure 2 where n is precisely known and controlled (rather than being a statistical range of values).

Our solution to this synthetic problem was the development of an iterative technique for preparing hydroxypropyl ethers from allyl ethers via oxymercuration-reduction. Figure 3 illustrates the process for the preparation of a series of three chain-extended hydroxypropyl derivatives of 2,6-dimethoxyphenol. Conversion of phenol 1 to the allyl ether 2 under phase-transfer conditions (6) was followed by oxymercuration (7) to give the intermediate organomercurial 3, which was reduced without isolation to give hydroxypropyl ether 4 in 64% overall yield. Ether 4 was then allylated to provide 5, which upon oxymercuration-reduction afforded hydroxypropyl derivative 6. One further iteration of the allylation-oxymercuration-reduction sequence yielded the hydroxypropyl compound 7.

We have found that this sequence can be generally applied to the synthesis of hydroxypropyl derivatives of alcohols and phenols; yields are uniformly acceptable and products of the desired degree of chain extension can be prepared completely free of lower and higher oligomers. The compounds shown in Table I were prepared using this chemistry.

In the course of characterizing the hydroxypropyl ethers prepared in this study, we noticed an intriguing pattern in the mass spectral fragmentation of these compounds. As shown in Table I, the degree of chain extension is reflected in the fragmentation pattern: Alcohols modified by a single hydroxypropyl ether unit have a strong M-58 peak, ethers of chain extension degree of 2 show a loss of m/e 116 and 58, and tripropylene glycol derivatives (degree of chain extension = 3) lose m/e 174, 116, and 58. Figure 4 shows the mass spectrum of compound 6. The loss of m/e 116 for this dipropylene glycol ether leads to the base peak at m/e 168, and the loss of m/e 58 to m/e 226 is the second most important fragmentation. In the case of all of the compounds in Table I, loss of the entire propylene glycol ether side chain provided either the base peak of the mass spectrum or the second strongest peak present. It is clear from this data that the length

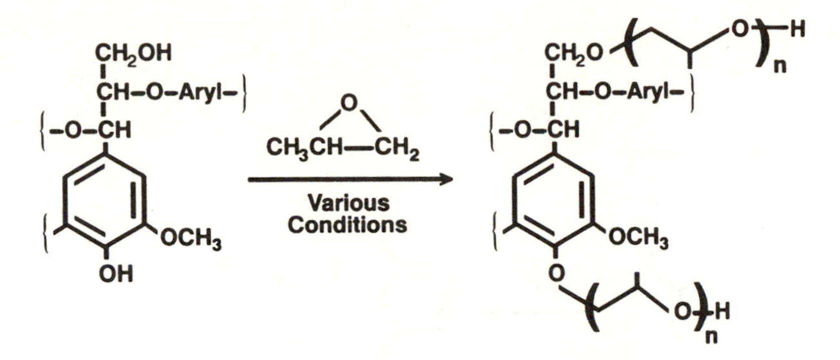

Figure 1. Hydroxypropylation of lignin.

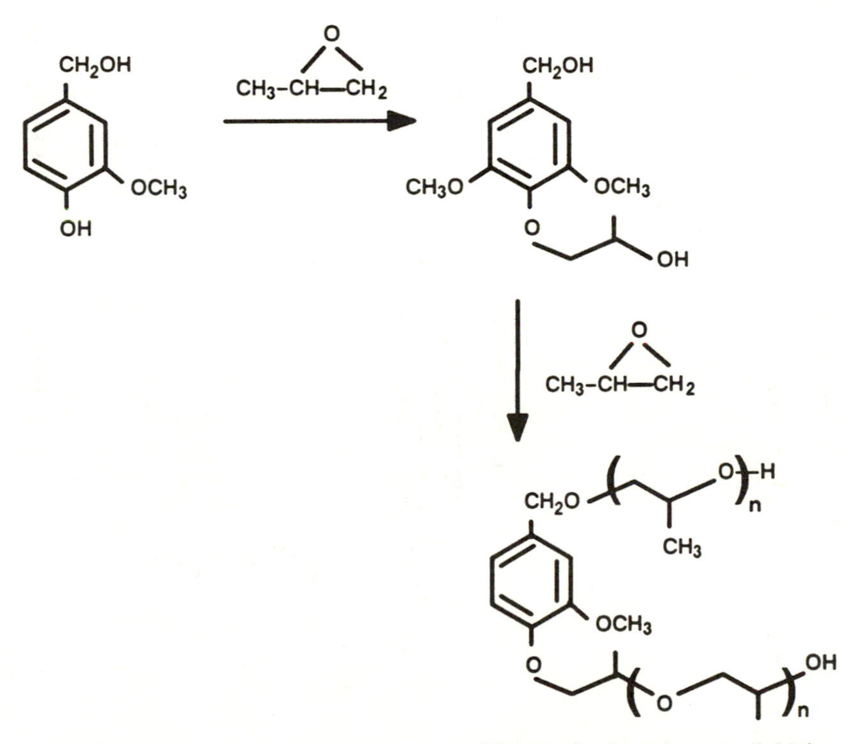

Figure 2. Hydroxypropylation of phenols and alcohols.

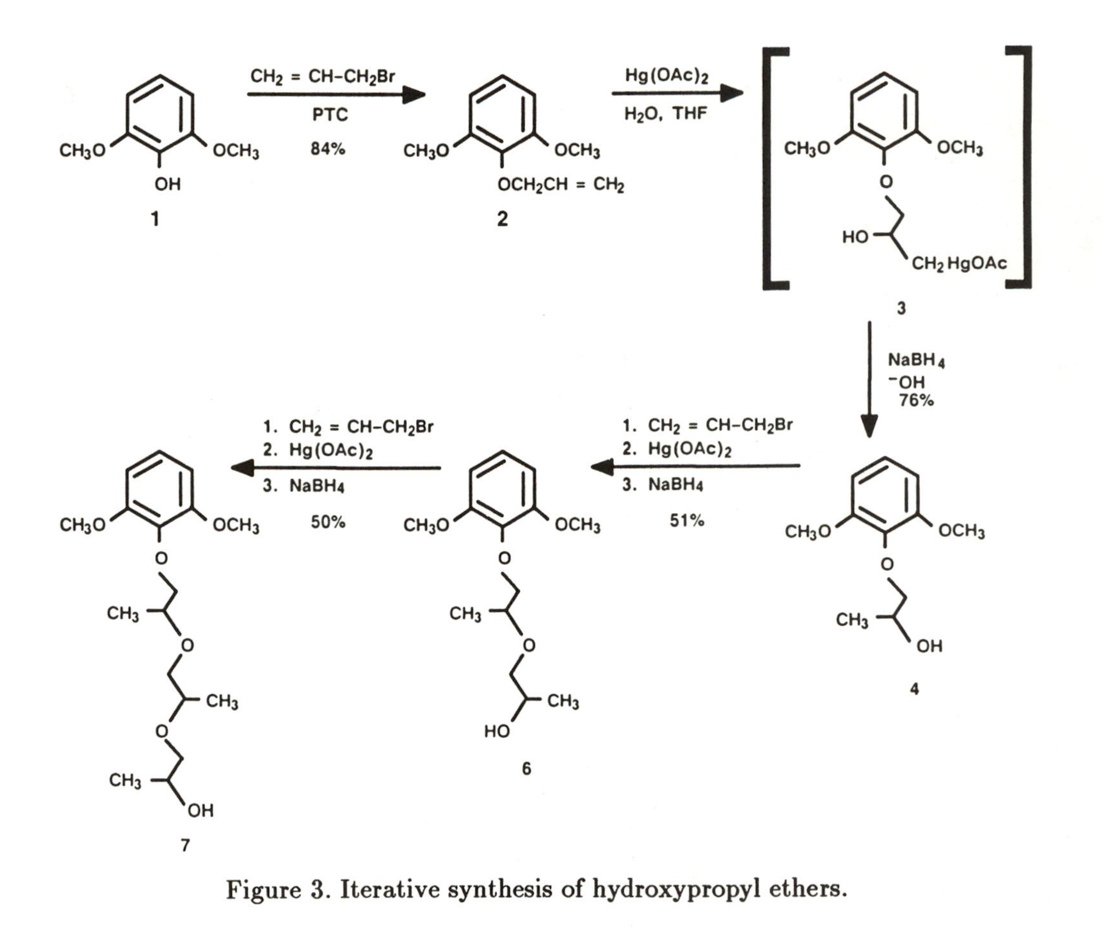

Figure 3. Iterative synthesis of hydroxypropyl ethers.

Table I. Electron Impact Mass Spectral Fragmentation of Hydroxypropyl Lignin Model Compounds

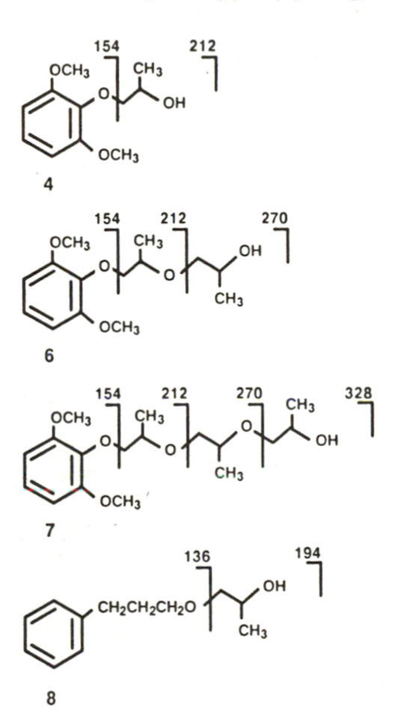

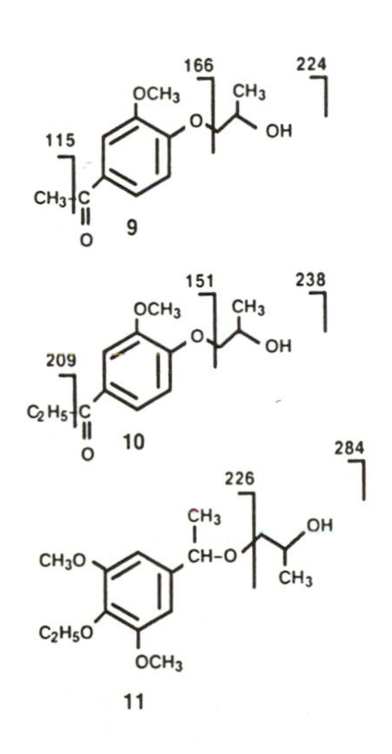

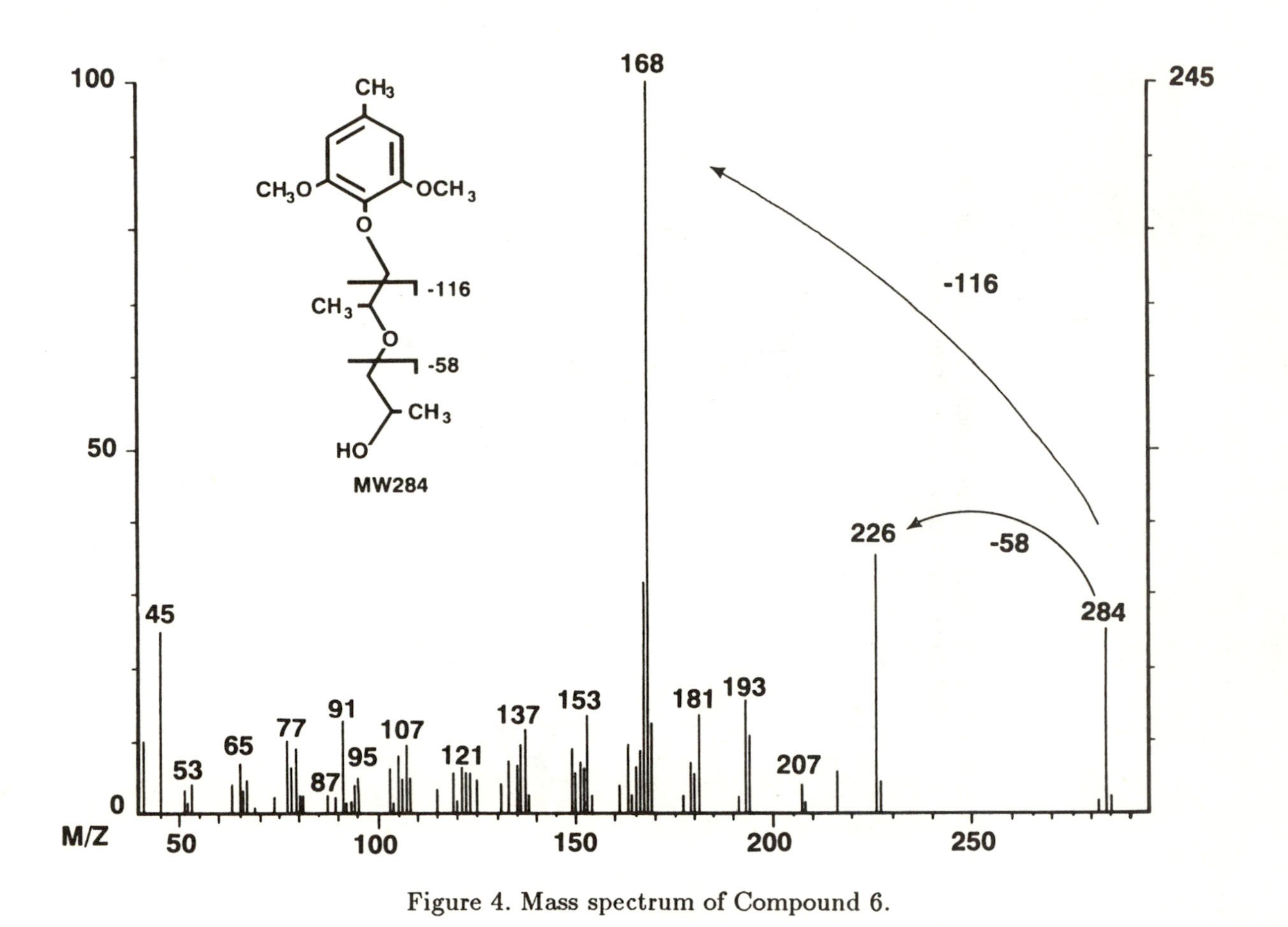

Figure 4. Mass spectrum of Compound 6.

of an oligo(propylene glycol) chain attached to an organic alcohol can be determined by examination of its mass spectrum.

Degradation and Analysis of Hydroxypropyl Lignin

The results described above indicated that, if a way could be found to degrade a hydroxypropyl lignin to volatile fragments suitable for analysis by capillary gas chromatography/mass spectrometry, much could be learned about the distribution and length of hydroxypropyl residues present on the lignin. The known ability to degrade lignins by hydrogenolysis (8), coupled with the reductive stability of simple alkyl and aromatic ethers, prompted us to examine the scheme shown in Figure 5.

Two control experiments were first conducted. In the first, model compound 7 was subjected to the hydrogenolysis conditions and then analyzed by VPC/MS. Despite hydrogenolysis at 285°C over 5% Pd/C catalyst under 5000 psi H_2, the model compound was recovered intact; no evidence of ring reduction or hydroxypropyl chain cleavage was seen in the mass spectrum. In the second control experiment, a sample of methanol organosolv lignin (hardwood) was hydrogenated under the same conditions and the volatile fractions recovered by high vacuum distillation. VPC/MS analysis of the volatile fractions (which amounted to 28-38% of the lignin charged) led to identification of the compounds listed in Table II. It is crucial to note that none of the compounds produced by hydrogenolysis of unmodified lignin displayed a mass spectral fragmentation of m/e 58 or a multiple thereof. Thus we have established that hydroxypropylated lignin moictics survive the hydrogenolysis conditions, and that the hydrogenolysis products of unmodified lignin are free of structures which would interfere with mass spectral identification of hydroxypropylated components.

When the hydrogenolysis/distillation/VPC/MS sequence was carried out on a sample of lignin which had been modified by reaction with propylene oxide in toluene in the presence of KOH/18-crown-6 catalyst, inspection of the resulting mass spectra for compounds having m/e 58 fragmentations (and multiples thereof) led to identification of the materials listed in Table III. It will be seen that several mono-hydroxypropylated materials and three di-hydroxypropyl (chain extension degree = 2) compounds were found. No products having hydroxypropyl chains longer than degree 2 were found; thus it appears that this particular modified lignin contains many short hydroxypropyl chains, rather than a few chains of greater degree of extension.

Similar analysis of a lignin modified by hydroxypropylation in aqueous base yielded only products bearing single hydroxypropyl residues. This apparent lack of chain extension is of course consistent with the fact that the hydroxypropylation was carried out under conditions favoring reaction only on phenolic hydroxyl groups.

Areas of uncertainty in this method of analyzing modified lignins include the problem that little more than one-third of the lignin is converted to volatile products by the hydrogenolysis reaction; the non-volatile fraction cannot be analyzed by VPC/MS. This will constitute a difficulty only if the

Hydroxypropyl Lignin

↓ H_2, Pd/c, >250 ,>3000 PSI H_2

Hydrogenated Hydroxypropyl Lignin

↓ High Vacuum Distillation

Volatile Fractions

↓ Capillary VPC - Mass Spectral Analysis

M-58, M-116, M-174,...etc.

Figure 5. Hydroxypropyl lignin degradation and analysis scheme.

Table II. Products identified in the VPC/MS Analysis of the Volatile Fraction from Unmodified Lignin

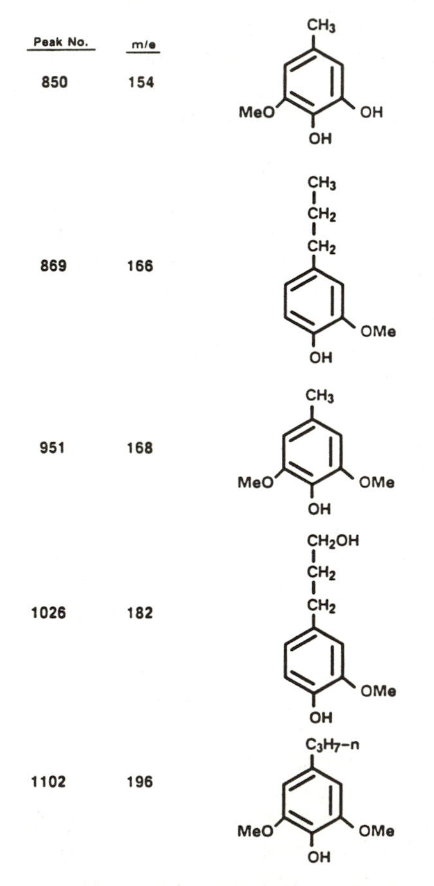

Peak No.	m/e
850	154
869	166
951	168
1026	182
1102	196

Note: None of these compounds showed m/e M-58, or multiples thereof.

Table III. Compounds Identified in the VPC/MS Analysis of a Hydrogenated Hydroxypropyl Lignin

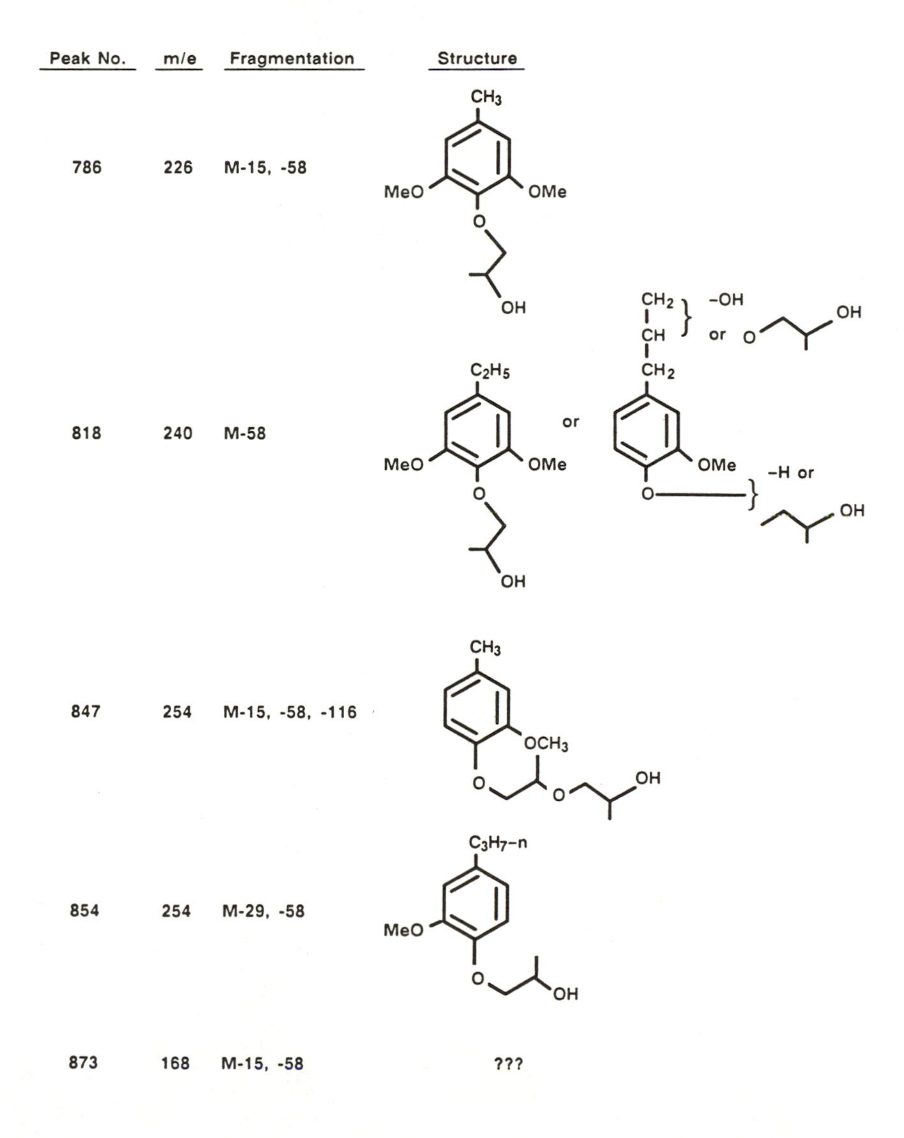

Peak No.	m/e	Fragmentation	Structure
786	226	M-15, -58	
818	240	M-58	
847	254	M-15, -58, -116	
854	254	M-29, -58	
873	168	M-15, -58	???

Continued on next page

Table III. Continued

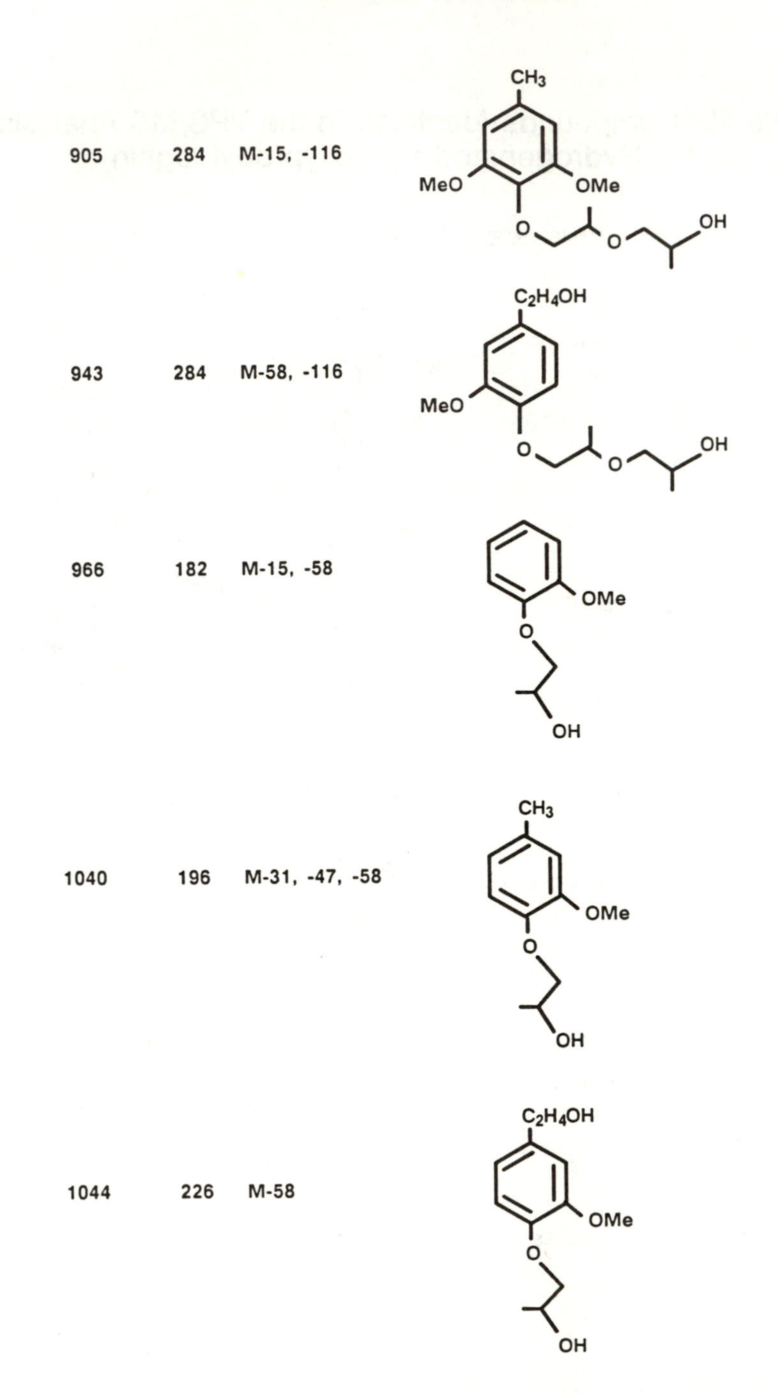

905	284	M-15, -116
943	284	M-58, -116
966	182	M-15, -58
1040	196	M-31, -47, -58
1044	226	M-58

distribution and degree of chain extension on the part of the lignin molecules which remain non-volatile differ from that on the part of the molecules which are converted to volatile fragments. While this seems an unlikely scenario, it is at present impossible to establish the point with any certainty. A second area of uncertainty arises from the fact that the method has been tested only with model compounds up to degree of chain extension = 3 (i.e., compound 7). We do not yet know whether longer hydroxypropyl residues would be degraded by these procedures; this point will be established in time by the synthesis of additional model compounds.

Summary

We have developed a simple, iterative synthetic method for the preparation of hydroxypropyl derivatives of phenolic and aliphatic alcohols which allows complete definition and control of the degree of chain extension in the products. This methodology has been applied to the preparation of a series of lignin model compounds having hydroxypropyl chain extension degrees of 1-3. We have shown that the degree of chain extension in such compounds can be unambiguously defined by analysis of their electron impact mass spectra. It has furthermore been demonstrated that such compounds are stable to a hydrogenolysis/distillation/VPC/MS sequence which, when applied to hydroxypropyl lignins, defines the site and degree of hydroxypropylation on the volatile fraction of the lignin residues. Topics for further investigation include determination of the degree of chain extension at which the method breaks down, and the nature of the lignin segments which are not rendered volatile by hydrogenolysis.

Acknowledgment

We thank Prof. W. Glasser for providing several hydroxypropylated phenols and lignin samples used in this study.

Literature Cited

1. Rials, T.; Glasser, W. *Holzforschung* 1986, **40**, 353, and references cited therein.
2. Sarkanen, K.; Ludwig, C. *Lignins. Occurrence, Structure, and Reactions*; Wiley-Interscience: New York, 1971.
3. Wu, L.; Glasser, W. *J. Appl. Polym. Sci.* 1984, **29**, 1111.
4. Glasser, W.; Barnett, C.; Rials, T.; Saraf, V. *J. Appl. Polym. Sci.* 1984, **29**, 1815.
5. Gee, G.; Higginson, W.; Levesley, P.; Taylor, K. *J. Chem. Soc.* 1959, 1338.
6. Starks, C.; Liotta, D. *Phase Transfer Catalysis*; Academic Press: New York, 1978.
7. House, H. *Modern Synthetic Reactions*; 2nd ed.; W. A. Benjamin: New York, 1972; pp. 387-396.
8. Hoffmann, P.; Schweers, W. *Holzforschung* 1975, **29**, 74.

RECEIVED March 17, 1989

Chapter 34

Bleaching of Hydroxypropyl Lignin with Hydrogen Peroxide

Charlotte A. Barnett[1,2] and Wolfgang G. Glasser[1,2]

[1]Department of Wood Science and Forest Products, Virginia Polytechnic Institute and State University, Blacksburg, VA 24061
[2]Polymeric Materials and Interfaces Laboratory, Virginia Polytechnic Institute and State University, Blacksburg, VA 24061

The bleaching of non-phenolic hydroxypropyl lignin (HPL) with hydrogen peroxide in homogeneous phase results in 50-90% loss of color depending on the definition of color. This is more realistically represented by a weighted absorptivity average in the visible range of the spectrum rather than by a fixed wavelength. The bleaching effect varies with the amount of hydrogen peroxide charged; with reaction time; with pH; and with other parameters. The oxidation reaction is best carried out in aqueous alcohol, at either pH 2 or above 12-13; and with 25-50% hydrogen peroxide on lignin derivative. At 80°C, the reaction is 90% complete after 1-2 hours with hardwood organosolv lignin, but it takes longer with other lignin types. Treatment with H_2O_2 introduces carboxylic acid functionality.

The typically dark brown color of isolated lignin products and their[1] derivatives constitutes a severe drawback to their utilization in such structural materials as engineering plastic (1). Although not of significant detriment to the physical and mechanical properties of lignin-derived materials, color is perceived as a significant handicap to marketing in commercial polymer markets (1). The light absorbing characteristics of isolated lignins exceed those of lignin in its native state (i.e., wood and other biomaterials), and this has been attributed to the formation of various chromophoric functional groups during isolation (2). Most prominently among them are conjugated quinonoid structures which are commonly removed from the lignin

0097-6156/89/0397-0436$06.00/0

of high yield pulps by the use of hydrogen peroxide (3). A vast literature is available on this type of "lignin-retaining" bleaching of pulps, which is contrasted by the "lignin-removing" bleaching technology normally involving chlorine and/or chlorine dioxide (4). Lignin-retaining bleaching is known to be transient since light absorbing structural features reappear upon exposure to the elements, especially oxygen and UV-rays. This has been attributed to the ease of quinone regeneration by phenolic functional groups (5).

A more permanent removal of color has been described by Lin (6), and by Dilling and Sarjeant (7), for lignin derivatives in which the phenolic functionality has been partially blocked. These largely non-phenolic (and sulphonated) lignin derivatives were bleached in homogeneous aqueous phase with hydrogen peroxide and chlorine dioxide. Reductions in color of 80-93% were reported for these water soluble derivatives (6, 7).

It was the intent of the present investigation to examine the one step bleachability of non-phenolic, water-insoluble hydroxypropyl lignin (HPL) derivatives with hydrogen peroxide in homogeneous phase (i.e., aqueous ethanol).

Experimental

Materials. Hydroxypropyl lignin (HPL) derivatives were available from the following lignin sources: organosolv (methanol) lignin from spruce, supplied by Organocell, Munich, FR Germany; organosolv (methanol) lignin from red oak, supplied by an undisclosed industrial source; organosolv (ethanol) lignin from aspen, supplied by Repap Technologies, Inc. (formerly the Biological Energy Corp.), Valley Forge, PA; kraft lignin from pine (Indulin AT), supplied by Westvaco Corp., N. Charleston, SC; and kraft lignin from hardwood, supplied by Westvaco Corp., N. Charleston, SC. All lignins were derivatized with propylene oxide by batch reaction as reported previously (8, 9), and the products were separated from homopolymer fractions of propylene oxide by liquid-liquid extraction.

Methods. A typical hydrogen peroxide bleaching operation consists of treatment of a 20% HPL solution in 60-75% aqueous ethanol with 25-50% hydrogen peroxide (by weight) on lignin derivative at 80°C. The reaction medium is adjusted to pH 2 by the addition of HCl, or to pH >12-13 by NaOH. Color loss is determined by UV/VIS absorbance measurements on suitably diluted samples.

Absorbance was measured at 25 nm intervals from 325 nm to 500 nm. These values were converted to absorptivity (L g^{-1} cm^{-1}) and the area under the curve was integrated from 350 nm to 500 nm using software by PolyMath Control Data (©1984). This value was compared to corresponding values obtained with the starting material (lignin or HPL), and the color loss was calculated from

$$\text{Color Loss}(\%) = \left[1 - \frac{\text{Area of bleached sample}}{\text{Area of starting material}}\right] \times 100$$

Similar calculations were done using seven individual wavelengths between 350 and 500 nm (i.e., at 25 nm intervals) and, using the sum of the absorptivities at the 7 wavelengths, color loss was computed from

$$\text{Color Loss}(\%) = \left[1 - \frac{a_1 + 2a_2 + 2a_3 \ldots + a_n(\text{bleached sample})}{a_1 + 2a_2 + 2a_3 \ldots + a_n(\text{starting material})}\right] \times 100$$

where a_x = absorptivity at wavelength x.

FTIR-spectroscopy was performed in a Nicolet 5 SXC instrument using KBr pellets. The spectra were baseline corrected and the 1125 cm^{-1} band was used as reference peak to compare bleached and unbleached spectra. The 1505 cm^{-1} peak was used as the reference peak to obtain the subtraction spectra.

Results and Discussion

Definition of color. The absorptivity coefficient of lignin and lignin derivatives in the visible range of the electromagnetic spectrum declines from nearly 15 L g^{-1} cm^{-1} at 325 nm wavelength to 1 to 2 L g^{-1} cm^{-1} at 500 nm (Fig. 1). Hydroxypropylation and H_2O_2-bleaching cause absorptivity values to decline throughout the visible spectrum. The extent to which the absorptivity declines varies, however, with wavelength. If expressed in percent of absorptivity at a given wavelength of the starting material, color loss is most dramatic at 500 nm, and much less significant at 325 nm (Fig. 1). Although color loss and bleaching have been defined using several individual wavelengths (6, 7, 10, 11), a more realistic expression of color loss employs a weighted average of absorptivity values in the range of 350-500 nm, i.e., the area under the absorptivity vs. wavelength curve in the region of 350-500 nm. This is illustrated by a relationship between the color loss values by the integration method *vs.* those by the individual wavelength method (Fig. 2). Whereas an absorptivity loss of 90% can be achieved at a wavelength of 500 nm, the resulting product is neither colorless (i.e., "white") nor has it lost more than 50-70% of its absorptivity (between 350 and 500 nm) as measured by the integration method. Thus, absorptivity averaging over a wider range of the visible spectrum provides for a more realistic color (or bleaching) determination method than does that based on any particular wavelength. This method has therefore been adopted in the current study in which color loss is expressed as the change of area under the absorptivity vs. wavelength curve between the starting material and the bleached product.

Hydrogen peroxide charge. The removal of color from non-phenolic lignin derivatives, and thus bleaching effect, varies with the amount of H_2O_2 charged to the reaction mixture. In the case where charge is varied from 20 to 150% w/w H_2O_2, at constant time (30 min.), pH (2) and temperature (80°C), color loss to lignin is 53-69%, and that to HPL is 22-48% (Fig. 3), depending on lignin source. Maximum bleaching requires maximum H_2O_2 (e.g., 150% on HPL), but H_2O_2 consumption declines to $<40\%$

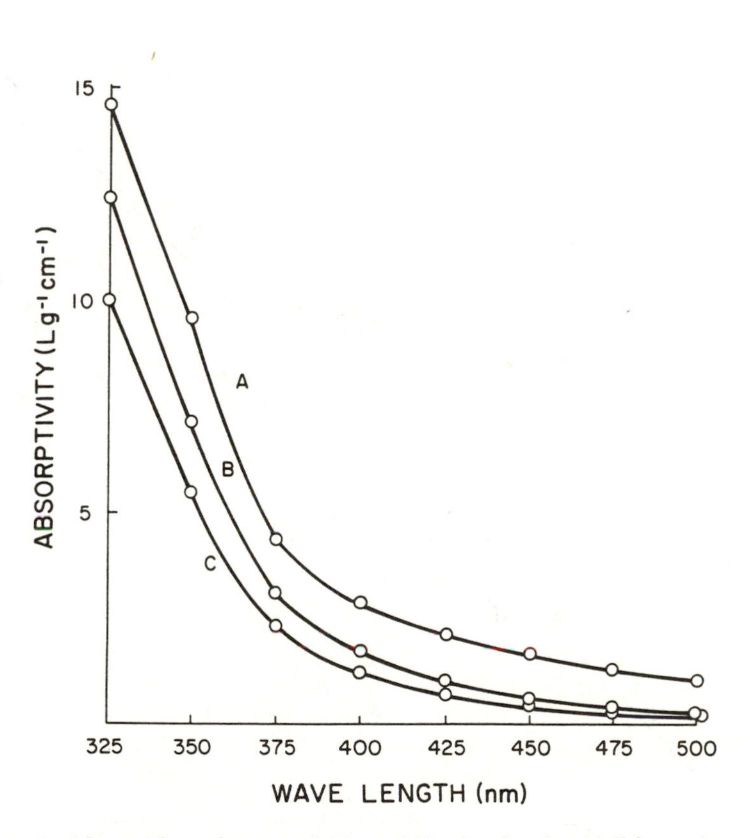

Figure 1. Absorption characteristics of lignins in the visible region of the electromagnetic spectrum; hardwood kraft lignin (A), HPL (B), and a typical bleached HPL (C).

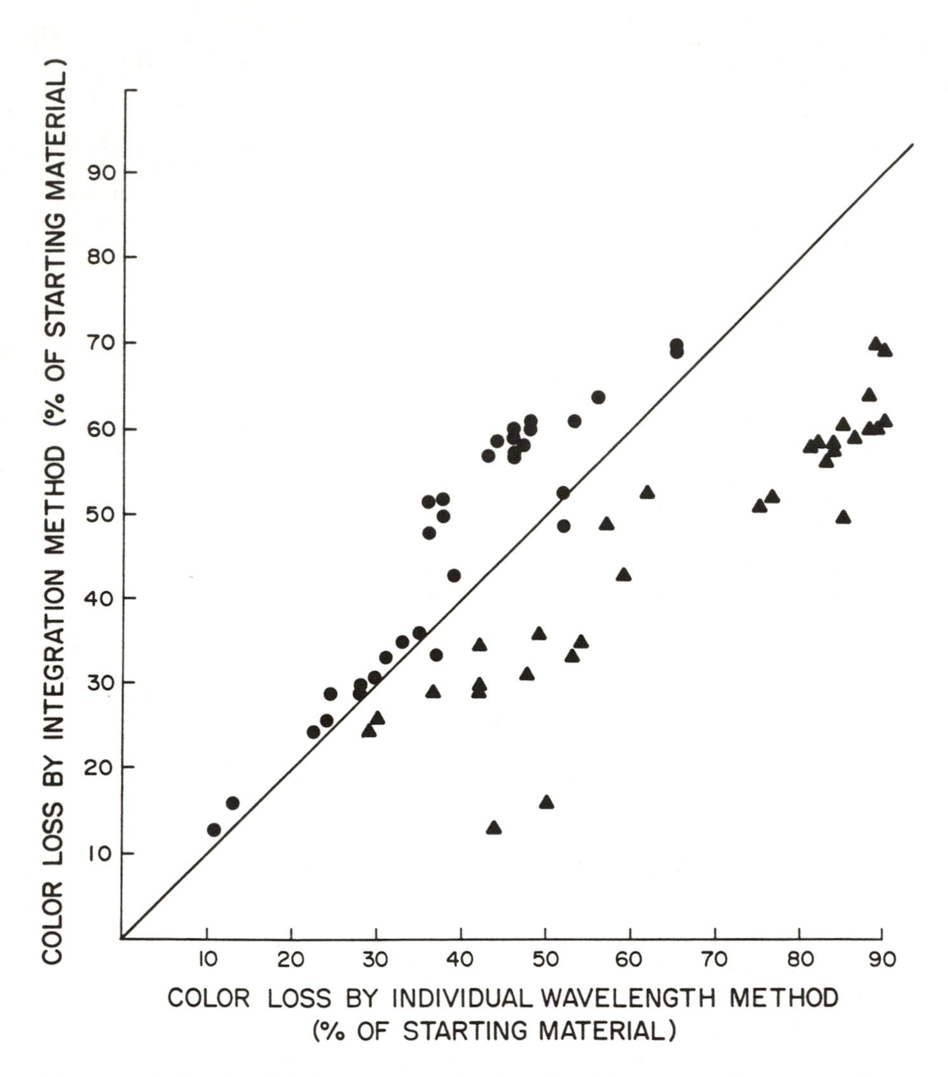

Figure 2. Relationship between color loss by the integration method as compared to that by the individual wavelength method; selected values of several lignin types; 350 nm (●) and 500 nm (▲).

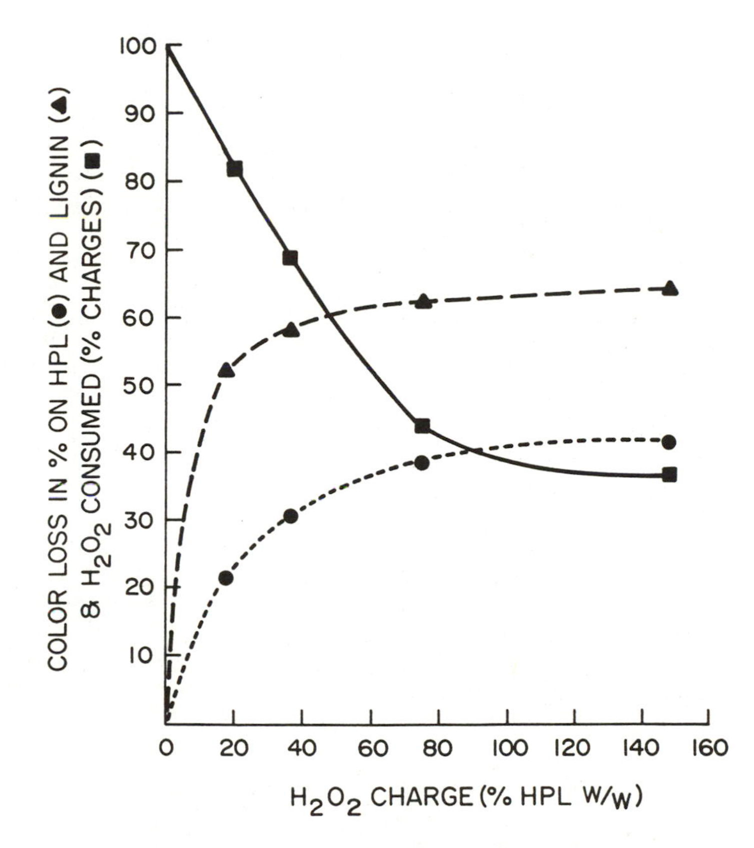

Figure 3. Relationship between color loss and H_2O_2 consumption and H_2O_2 charge.

(in 30 min.) under those conditions. Most of the color loss is achieved with 25-50% H_2O_2 on HPL, and this results in the consumption of 70-80% of the hydrogen peroxide charged. Larger amounts of hydrogen peroxide are not justified by the bleaching effect, and the bleaching agent remains largely unconsumed.

A 25-50% charge of hydrogen peroxide to lignin corresponds to ca. 2-4% H_2O_2 on pulp having a Kappa number of about 50. This is within normal bleaching practice.

Reaction time. The reaction of H_2O_2 with non-phenolic lignin derivative in homogeneous phase and at 80°C is rapid. More than 50% of the color is lost within the first 30 min. period, after which the reaction slows considerably (Fig. 4). The rate of color loss at constant H_2O_2 charge, pH, and temperature varies between >50% per half-hour initially to between 0.5 and 2% after 8 hours. Significant differences are detected for response rate to bleaching with H_2O_2 for lignins from different sources (Fig. 4). This may be attributed to differences in chemical functionality and other factors.

pH. Lignin-retaining bleaching in the pulp and paper industry commonly employs H_2O_2 at alkaline pH levels (3,4,12). This preference must be attributed to the suspectibility of the phenoxide anions to H_2O_2. These ions are absent in HPL due to the non-phenolic character of this derivative type. Thus all pH levels have been explored in the current study. The results indicate that effective color loss is achieved at both extremes of the pH scale, at pH 2, and >12-13 (Fig. 5). Under alkaline conditions, the initial pH must exceed that of the pK_a of H_2O_2 (i.e., 11.6). It is, however, recognized that the pH drops rapidly in the course of the reaction due apparently to the formation of acidic functionality which neutralizes the alkali. On the acid side, effective use of H_2O_2 requires strongly acidic conditions (pH 2). A pH of 1, however, was found to result in less than optimal bleaching, probably because of chromophore-creating side-reactions. The chemistry of H_2O_2 involves the formation of active reaction species from hydrogen peroxide under the extreme pH conditions:

$$H_2O + [HO]^{\oplus} \rightleftharpoons [H_2OOH]^{\oplus} \overset{\text{Acid}}{\rightleftharpoons} \mathbf{H_2O_2} \overset{\text{Base}}{\rightleftharpoons} [HO_2]^{\ominus}$$

Whereas the reactive species under acidic conditions are the $[H_2OOH]^{\oplus}$ and $[HO]^{\oplus}$ cations, the reactive species under alkaline conditions is the $[HO_2]^{\ominus}$ anion. Whereas the cationic oxidation species are stronger oxidants, with oxidation potential rising with increasing acidity favoring aromatic ring hydroxylation by electrophilic substitution (3,13), the anionic species mediate a nucleophilic attack on quinones according to the scheme illustrated in Fig. 6 (3,10,11). The formation of new acidic functionality results in a rapid decline in pH.

The effect of H_2O_2 on HPL chemistry and functionality can be assessed conveniently by FTIR spectroscopy. It has recently been demonstrated that quantitative structural differences between lignins can be revealed by

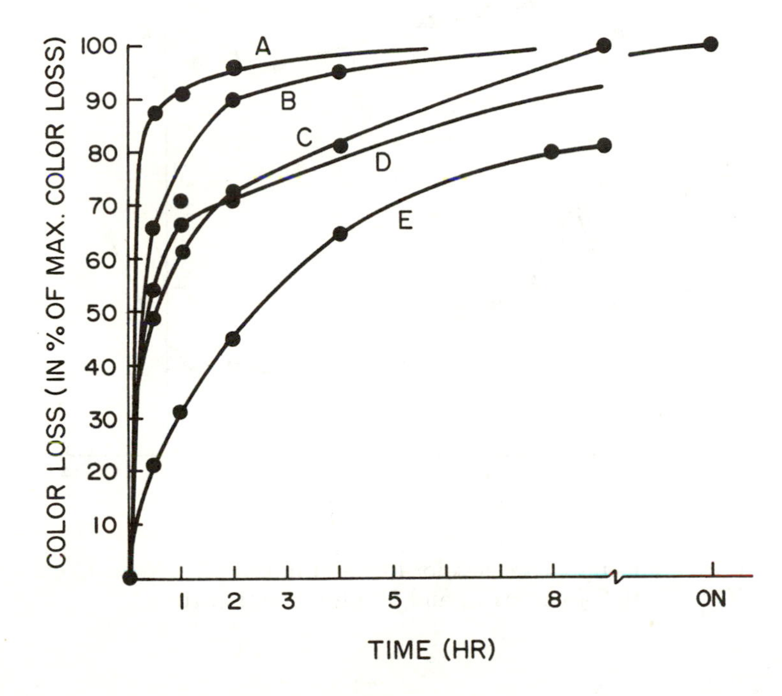

Figure 4. Rate of color loss for HPL derivatives from red oak organosolv lignin (A), aspen organosolv lignin (B), pine kraft lignin (C), hardwood kraft lignin (D), and spruce organosolv lignin (E). (ON stands for overnight.)

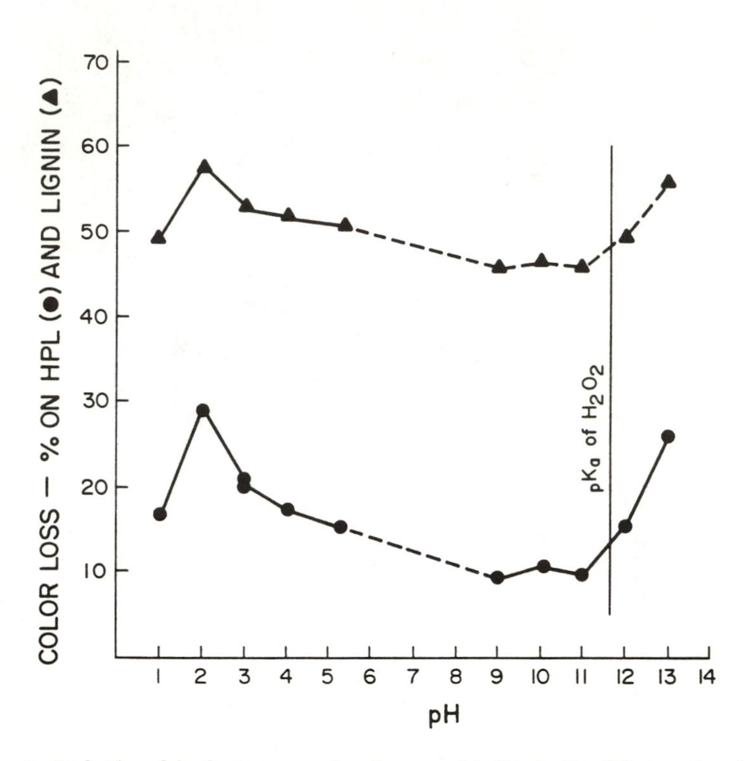

Figure 5. Relationship between color loss and initial pH. (Note: Conditions were 80°C, 95% H_2O_2 per HPL, and 30 min. reaction time.)

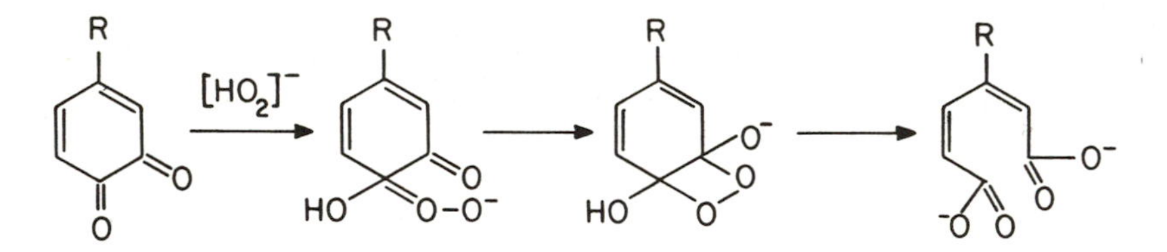

Figure 6. Suggested mechanism for the reaction of ortho-quinonoid structures in lignin derivatives with hydrogen peroxide under alkaline conditions (according to Gierer, ref. 3).

this method (14,15). Differences between bleached and unbleached hydroxypropyl lignins are illustrated in Figures 7 and 8. Using a baseline-corrected absorptivity spectrum which was autoscaled so that the ether peak at 1125 cm^{-1} was adjusted to 100% absorption, spectral differences between the bleached (pH 2) and the unbleached lignin derivative are revealed by the shaded areas of Figure 7. The most apparent differences are in the regions between 1600 and 1800 cm^{-1}, and between 3000 and 3600 cm^{-1}. Minor differences also appear between 1200 and 1300 cm^{-1}. Subtraction spectra for acid and alkaline-bleached lignin derivatives are shown in Figures 8A and 8B, respectively. The aromatic peak at 1505 cm^{-1} was taken as the absorption standard in both samples. This enhances spectral differences and emphasizes structural characteristics which differ in both samples. The preparation which was bleached with H_2O_2 under acidic conditions exhibits a broad peak centered at 1725 cm^{-1}, and this strongly suggests carbonyl and carboxyl group formation (Figure 8A). The sample treated with $[HO_2]^{\ominus}$ ions, by contrast, raises a strong signal at 1625 cm^{-1}, and this reflects changes in aromaticity. Both difference spectra also display weaker signals at lower wavenumbers, but they are less pronounced. They generally support, however, the numerous side reactions which have been attributed to hydrogen peroxide and its degradation products (3). Prominent among them are reactions caused by oxygen anion radical and hydroxy radical (3).

These results indicate that differences exist between the H_2O_2-bleaching chemistry depending on pH; and that difference-FTIR spectroscopy is a useful method for revealing the effect of chemical modification on chemical structure.

Thus, bleaching of non-phenolic lignin derivatives is possible at both pH extremes, and conditions may be chosen according to process convenience. The formation of new functionality during color removal is inevitable, and this can be assessed by FTIR spectroscopy. Acidic conditions appear to result in the formation of carbonyl and carboxyl groups, and alkaline H_2O_2 is suggested to alter aromatic character. Hydroxylation and hydrolysis are encountered also.

Various Parameters. Other factors affecting bleaching were explored briefly, and pragmatic decisions were made to determine standard reaction conditions. The effect of temperature was explored over the range of 20 to 80°C, and results favored elevated temperature for standard conditions. Using a 20% charge of H_2O_2 at pH 2 for 30 minutes, color loss increased from 3.5% at 40°C to 29.6% at 80°C. Ethanol was chosen as the solvent for reasons relating to other process factors (i.e., isolation). (Choice of solvent must not ignore possible reaction with H_2O_2.) The selection of a 20% concentration level reflects an effort to produce as concentrated solutions as possible and still permit homogeneous phase reactions at all pH levels.

Yield. The gravimetric determination of lignin yield following bleaching with hydrogen peroxide poses difficulties through the need to remove inorganic reagents. An alternative assay of lignin involves determination of UV absorbance at 280 nm. This method, however, does not produce infor-

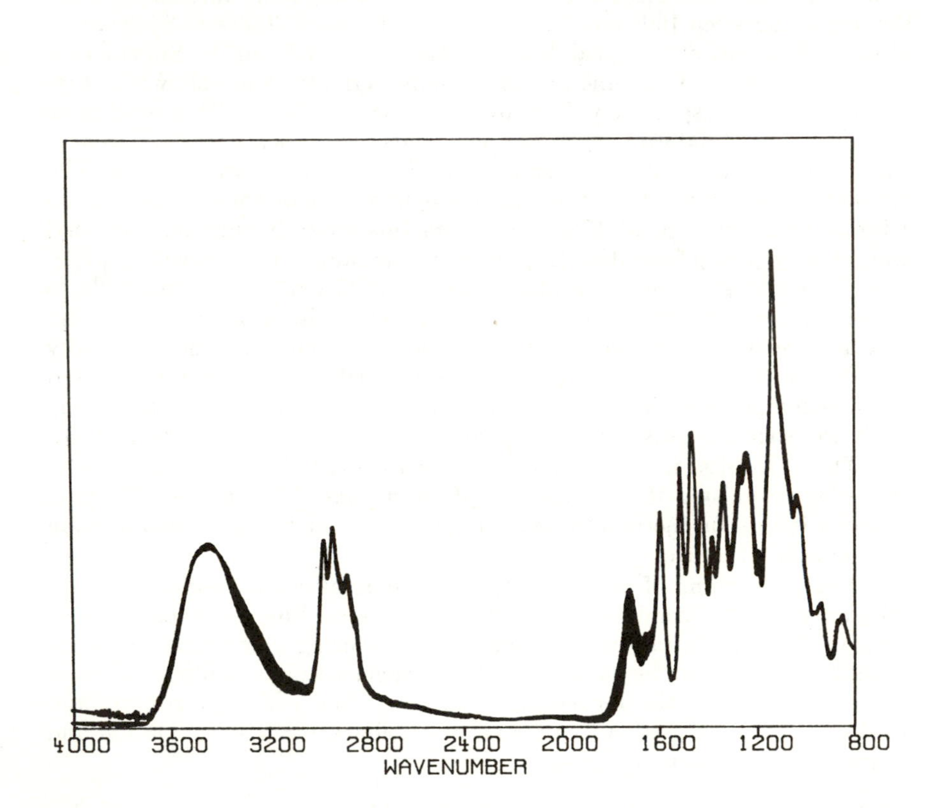

Figure 7. FTIR spectra of unbleached HPL and of HPL bleached with H_2O_2 under acidic conditions; overlaid and scaled so that the 1125 cm^{-1} peak reflects 100% absorption. Shaded areas represent quantitative structural differences.

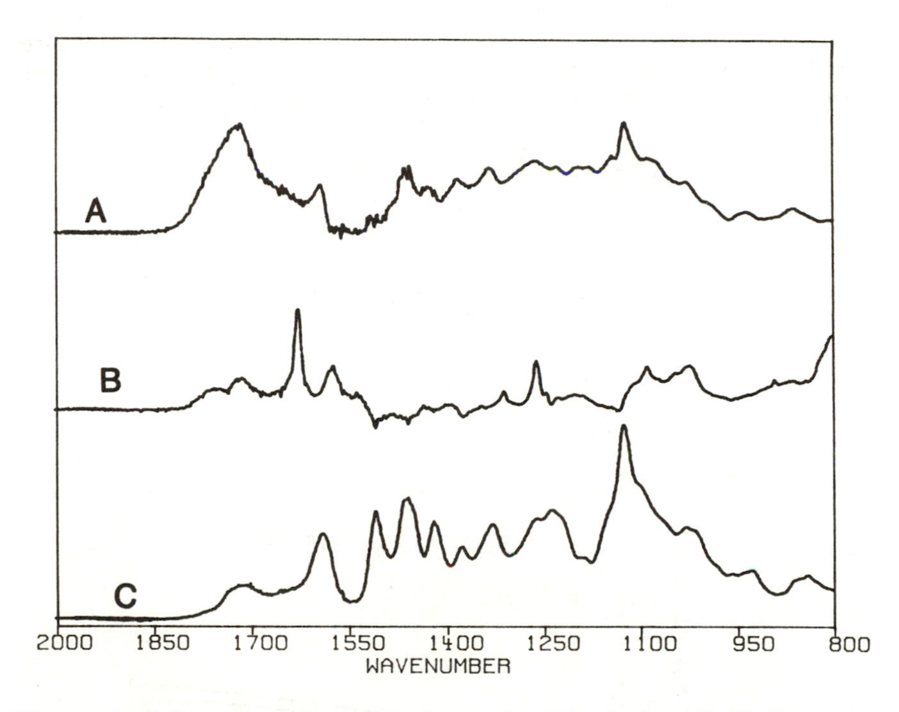

Figure 8. Subtraction spectra of samples bleached with H_2O_2 under acidic (A) and alkaline (B) conditions. Unbleached HPL control (normal spectrum) is shown in (C). (Subtraction spectra were scaled using the 1505 cm^{-1} peak as unity.)

mation on *gravimetric* yield but rather on the yield of *aromatic* (i.e., UV-absorbing) component. This ignores the expected conversion of aromatic features into aliphatic structures that escape detection by UV spectroscopy (at 280 nm). Nevertheless, the common UV spectroscopic assay produces meaningful results on the yield of aromatic components. The data of Fig. 9 reveal a loss of aromatic functionality in relation to time (Fig. 9A) and H_2O_2 charge (Fig. 9B) for an organosolv and a kraft lignin based HPL at constant pH (2), time (4 hours), and temperature (80°C). A consistent and expected loss with time and H_2O_2 charge is revealed. Judged by UV absorptivity, 15-20% of the samples' aromatic character is lost under bleaching conditions. It is doubtful that this implies that one of five aromatic structures in lignin has quinonoid nature, and this may more reasonably be explained with the removal of several types of UV absorbing constituents.

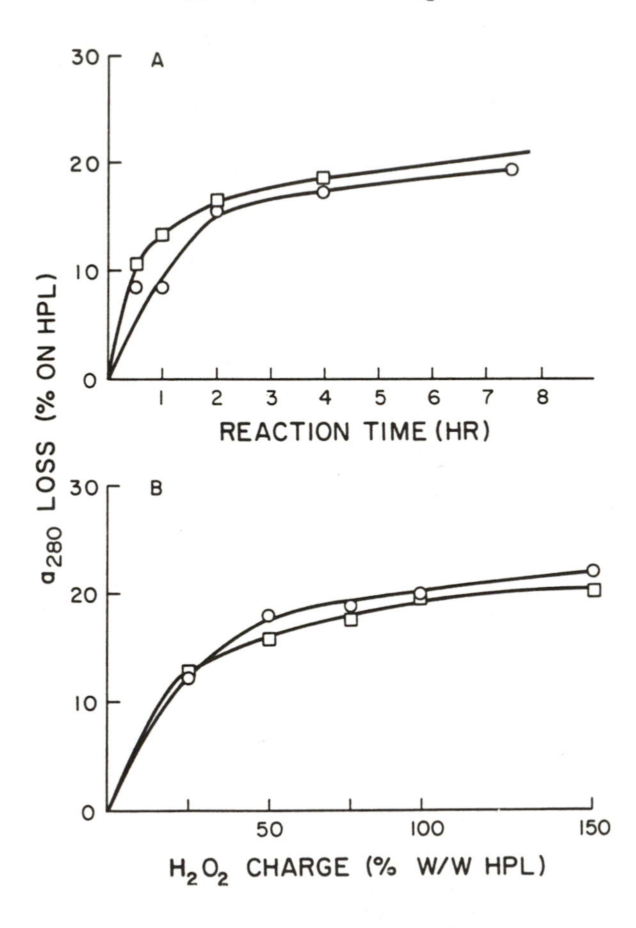

Figure 9. Relationship between loss of UV absorptivity at 280 nm and (A) reaction time and (B) H_2O_2 charge. Pine kraft HPL (○) and aspen organosolv HPL (□).

Other Blocking Agents. Although propylene oxide is a convenient and effective blocking agent for lignin, other alternatives are available. Among them, most notably, is acetylation. A comparison of two organosolv lignin derivatives, from aspen and from red oak, is given in Fig. 10. The data reveal that both chemical modification reactions result in derivatives with reduced light absorption characteristics, even without oxidative post treatment. Propylene oxide and acetic anhydride produce about identical color loss values. Bleaching of the acetylated red oak sample resulted in superior values as compared to the corresponding propoxylated derivatives. It appears that bleaching with hydrogen peroxide can be achieved with non-phenolic lignin derivatives regardless of the nature of the blocking group.

Conclusions

1. An amount of 25-50% hydrogen peroxide per lignin derivative constitutes a reasonable compromise between maximum color loss and efficient peroxide use. This achieves a color loss of about 60% on lignin, if color is defined as the weighted absorptivity average in the range of 350-500 nm.

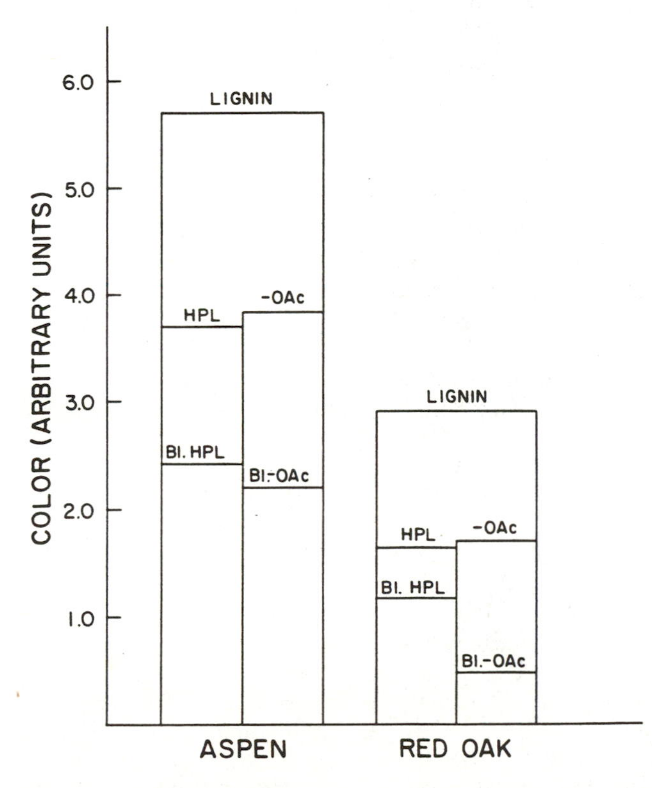

Figure 10. Color loss of lignin derivatives in relation to method of modification and blocking of phenolic OH groups.

2. While more than 90% of color loss is achieved after 30 min. in homogeneous phase at 80°C with some lignin preparations, others, especially softwood lignins, take longer.
3. Both low (pH 2) and high (pH 12-13) pH levels are suitable for bleaching. FTIR spectroscopy indicates, however, that the two different conditions produce differences in the chemistry of the bleached derivatives.
4. Although gravimetric yield may not be significantly affected by the bleaching treatment, aromaticity as indicated by UV absorption at 280 nm declines by as much as 20%.

Acknowledgment

This study was financially supported by a grant from an industry cooperative in which the following companies were involved: ARCO Chemical Corporation, Bio-Regional Energy Associates, Alberta Forest Service, and Borregaard Industries; and the Center of Innovative Technology of Virginia. This support is gratefully acknowledged.

This is part 18 of a publication series dealing with engineering plastics from lignin. Earlier parts have been published in *J. Appl. Polym. Sci.*, *J. Wood Chem. Technol.*, and *Holzforschung*.

Literature Cited

1. Niederdellmann, G., *Gewinnung von Lignin aus Holzabfallen und Verwertung fur Herstellung von Polyurethanhartschaumen*, Forschungsbericht 03 VM 619-C 077 (Bayer AG, Leverkusen) an das Bundesministerium fur Forschung und Technologie, 1981, 48 pg.
2. Hon, D. N.-S.; Glasser, W. G. *Polym.-Plast. Technol. Eng.* 1979, **12**(2), 159-179.
3. Gierer, J. *Holzforschung* 1982, **36**(2), 55-64.
4. Loras, V. Chapter 5 in *Pulp and Paper Chemistry and Chemical Technology*; J. P. Casey, Ed.; J. Wiley & Sons: New York, 1980; Vol. 1, pp. 633-764.
5. Gellerstedt, G.; Petterson, E.-L. *Svensk Papperstidn.* 1977, **80**(1), 15.
6. Lin, S. Y. U.S. Patent #4,184,845, 1980.
7. Dilling, P.; Sarjeant, P. T. U.S. Patent #4,454,066, 1984.
8. Wu, L. C.-F.; Glasser, W. G. *J. Appl. Polym. Sci.* 1984, **29**(4), 1111-23.
9. Glasser, W. G.; Wu, L. C.-F.; Selin, J.-F. Chapter in *Wood and Agricultural Residues: Research on Use for Feed, Fuels, and Chemicals*; Soltes, Ed J., Ed.; Academic Press, 1983, pp. 149-166.
10. Spittler, T. D.; Dence, C. W. *Svensk Papperstidn.* 1977, **80**(9), 275-284.
11. Bailey, C. W.; Dence, C. W. *Tappi* 1969, **52**(3), 491-500.
12. Dence, C. W. In *Reactions of Hydrogen Peroxide with Lignin: Current Status in Chemistry of Delignification with Oxygen, Ozone and Perox-*

ides; Gratzl, J. S.; Nakano, J.; Singh, R. P., Eds.; Uni: Tokyo, 1980; pp. 199-205.
13. Levitt, L. S. *J. Org. Chem.* 1955, **20**, 1297-1310.
14. Schultz, T. P.; Nichols, D. D. *International Analyst* 1987, **1**(9), 35-39.
15. Schultz, T. P.; Glasser, W. G. *Holzforschung* 1986, **40**, 37-44.

RECEIVED March 17, 1989

Chapter 35

Polymer Blends with Hydroxypropyl Lignin

Scott L. Ciemniecki[1-3] and Wolfgang G. Glasser[1,2]

[1]Department of Wood Science and Forest Products, Virginia Polytechnic Institute and State University, Blacksburg, VA 24061
[2]Polymeric Materials and Interfaces Laboratory, Virginia Polytechnic Institute and State University, Blacksburg, VA 24061

Polymer blends of hydroxypropyl lignin (HPL) with polyethylene (PE), with ethylene vinyl acetate copolymer (EVA), with poly(methyl methacrylate) (PMMA), and with poly(vinyl alcohol) (PVA) have been examined in relation to morphological, thermal, and mechanical characteristics. The results suggest that lignin gains the ability to contribute to the properties of thermoplasts, especially modulus and occasionally strength, when the co-component to lignin contains some type of polar functionality. Whereas polyethylene shows no signs of interaction with HPL, rising vinyl acetate content produces blends with superior strength properties. Similar observations are made with PMMA; and PVA fails to display any sign of phase separation. The greater than expected compatibility with thermoplasts containing polar functionality produces blends which are either plasticized (especially by low molecular weight lignin fractions) or antiplasticized.

Interest in polymer blends has been rising during the past decades because blending is seen as an inexpensive method for the modification of polymer properties. Blending is the cheaper alternative to the engineering of plastics from a multitude of components, in order to achieve target specifications. Given the ideal combination of polymer-polymer interactive forces, a polymer blend may be as useful as a block copolymer with microphase separation producing strength, toughness, and durability.

A blend's properties are largely determined by the compatibility of the component polymers since this influences morphology. When two polymers are blended polymer-polymer interactions will be exhibited if the thermodynamic equilibrium state permits. The basic thermodynamic equation

[3]Current address: Union Camp, P.O. Box 570, Savannah, GA 31402

0097-6156/89/0397-0452$06.00/0

which describes this state is the Gibbs free-energy equation:

$$\Delta G_{mix} = \Delta H_{mix} - T\Delta S_{mix} \quad (1)$$

where ΔG_{mix} is the Gibbs free energy of mixing, ΔH_{mix} is the enthalpy of mixing, ΔS_{mix} is the entropy of mixing, and T is the absolute temperature. In order for a polymer mixture to achieve intimate mixing or "miscibility," ΔG_{mix} must be negative. Thus, the miscibility of a two-component mixture depends on the sign and magnitude of ΔH_{mix} and, to a lesser extent, on ΔS_{mix}. When dealing with polymer blends, three important parameters affect the magnitude and sign of ΔH_{mix} and ΔS_{mix}: (1) solubility parameter; (2) specific interactions; and (3) molecular weight.

The solubility parameter δ is a measure of the cohesion of a material, or of the strength of molecular attractive forces between like molecules. The relationship between the solubility parameter and enthalpy of mixing is given by the Hildebrand equation (1):

$$H_{mix}/V = (\delta_1 - \delta_2)^2 \phi_1 \phi_2 \quad (2)$$

where V is the molar volume of the mixture, $\delta_{1,2}$ are the solubility parameters of components 1 and 2, and $\phi_{1,2}$ are the volume fractions of components 1 and 2. This equation expresses the concept that two polymers are miscible when their solubility parameters become perfectly matched. Equation (2) indicates that the predicted ΔH_{mix} is always positive or at best zero (in the absence of other attractive forces). Thus, the Hildebrand equation suggests that phase separation rather than miscibility is favored when ΔS is small. However, many polymers have been reported to be miscible (2,5). The mechanisms responsible for polymer-polymer miscibility are attributed to such specific interactions between two macromolecules as hydrogen bonding, π-π complex formation, charge transfer and ionic interactions. They have all been found to be responsible for miscibility in polymer blends (6). Considering that ΔH_{mix} values are the net result of breaking solvent-solvent (1-1) and polymer-polymer (2-2 and 3-3) secondary bonds and creating new polymer-polymer (2-3) interactions, the importance of the solubility parameter and specific interactions between two polymers is clearly understood (7,8).

The role that molecular weight plays in determining the miscibility of two polymers is most easily described by the ΔS_{mix} term in the Gibbs free energy equation. When mixing small molecules, the increase in randomness, and thus in entropy, is high and positive. This allows the $T\Delta S$ product to outweigh the endothermic heat of mixing, ΔH_{mix}. A negative free energy term, ΔG, is therefore favored, and miscibility is possible. However, when large polymer molecules are mixed, the thousands of atoms in each molecule must remain together, and mixing cannot be nearly as random. Thus, it is rarely possible for the $T\Delta S_{mix}$ term to exceed the endothermic heat of mixing, ΔH_{mix}. In fact, in those rare cases in which polymer blends exhibit complete miscibilty and homogeneity, it is usually due to specific interactions between the two polymers.

It was the objective of this work to examine the contribution hydroxypropyl lignin can make to the properties of polymer blends with several commercial thermoplastics. This has been the topic of three earlier publications (9-11).

Experimental

Materials. Polyethylene (PE): Low density polyethylene (PE) and ethylene-vinyl acetate (EVA) copolymers with vinyl acetate (VA) contents of 9, 18, 25, 28, 33, and 40% were obtained from Scientific Polymer Products, Inc.

Polymethyl methacrylate (PMMA): An atactic PMMA with a T_g of 110C and an intrinsic viscosity of 1.4 was obtained from Scientific Polymer Products, Inc.

Polyvinyl alcohol (PVA): Four commercially available (Scientific Polymer Products, Inc.) PVA's were used in this study. Important parameters for each were as shown in Table I:

Table I. Poly(vinyl alcohol) Characteristics

PVA Type	T_g(C)	δ ($cal^{1/2}cm^{3/2}$)	$\overline{M}_w$ ($g\ mol^1$)
96% hydrolyzed	77	12.6	96,000
88% hydrolyzed	70	11.8	96,000
75% hydrolyzed	60	10.6	3,500
0% hydrolyzed	32	9.6	unknown

Hydroxypropyl lignin (HPL): Two different HPL preparations were used in this study (12). Blends containing PE and EVA utilized an HPL from organosolv (methanol) lignin from red oak; and blends containing PVA and PMMA utilized an HPL from kraft lignin (Indulin-AT from Westvaco Corporation, Charleston, SC).

Methods. Polyblend samples were prepared by solution casting from tetrahydrofuran (THF), or chloroform; or by injection molding using a "Mini-Max Molder" by Custom Scientific Instruments. Specific experimental details are given elsewhere (9, 10, 11).

Physical properties were determined by *differential scanning calorimetry (dsc)* with a Perkin-Elmer DSC-4 instrument; and by *dynamic mechanical thermal analysis (dmta)* using a Polymer Laboratories dmta instrument. *Stress-strain tests* were performed on an Instron Table Model TM-M at a cross-head speed of 1mm min^1. The Young's modulus was obtained from the tangent of the initial slope of the force vs. elongation curve. *Scanning electron microscopy (s.e.m.)* employed an AMR 900 instrument. Specific experimental details are given elsewhere (9, 10, 11).

Results and Discussion

HPL/PE and EVA Blends. Polyethylene (PE) is by far the most important industrial polymer today. Its availability, its widely understood processability, and its low cost all contribute to the interest in mixing in the melt polyethylene with other polymeric components. Injection molded blends of HPL and PE (11) were found to be incompatible over the entire blend range, by SEM, DMTA, and optical inspection. The results of Figure 1 illustrate that blends of HPL with PE (coded as EVA(0))result in a rapid loss of tensile strength which far exceeds the rule of mixing (11). If, however, rising amounts of vinyl acetate are incorporated into the backbone of PE, thus producing ethylene-co-vinyl acetate polymer with vinyl acetate contents rising from 9 to 40%, a gradual increase in interaction is indicated by a rise in tensile strength of the polyblend. The data of Figure 1 reveal that EVA with 18-40% vinyl acetate content produces positive deviation from the rule of mixing if hydroxypropyl lignin content is high (i.e., between 20 and 30%) in the case of low vinyl acetate contents, and low (i.e., in the range of 10-20%) if vinyl acetate content is high. These results (11) illustrate that, although the lignin derivative produces immiscible polyblends with polyethylene and its copolymer with vinyl acetate, phase compatibility, and thus strength, increases with vinyl acetate content rising. The presence of polar carbonyl groups is suspected to be responsible for enhancing this observed polymer-polymer interaction.

HPL/PMMA Blends. Blend Structure. All blends of HPL and PMMA revealed a two-phase (immiscible) morphology regardless of solvent, weight fraction, molecular weight (of PMMA), or method of preparation (9). Extruded HPL/PMMA blends exhibited a less obvious two-phase morphology (9), and this was attributed to differences in the rate of vitrification (9).

Thermal Characteristics. Typical dsc thermograms for solution-cast and injection-molded HPL/PMMA blends (Fig. 2) reveal an initial scan (Fig. 2, scan A) with two endotherms, at 63 and at 104C, corresponding to HPL and PMMA, respectively. A second scan (Fig. 2, scan B) indicates only a single transition at 104C. The endotherm at 63C seen in scan A has previously been attributed to enthalpy relaxation (13, 14). This occurs when a polymer is cooled from the melt, and when the rapid rise in viscosity as the polymer approaches T_g causes the polymer chains to freeze into non-equilibrium conformations. By annealing at temperatures above the two T_g values, the respective endotherms are replaced with a single broad T_g. Considering the SEM results, which indicated immiscibility and showed no signs of interaction between the two phases, the presence of a single T_g cannot be taken as an indication of phase uniformity (15). Rather, this broad T_g, which in the case of low molecular weight HPL spans a range of 90C, is believed to be the result of "transitional smearing." This phenomenon has previously been observed with blend components that have T_g values within 20C of each other, and it simply represents the overlapping of two T_g's to produce a single broad T_g.

The tan δ peak of a dmta thermogram provides a second measure of

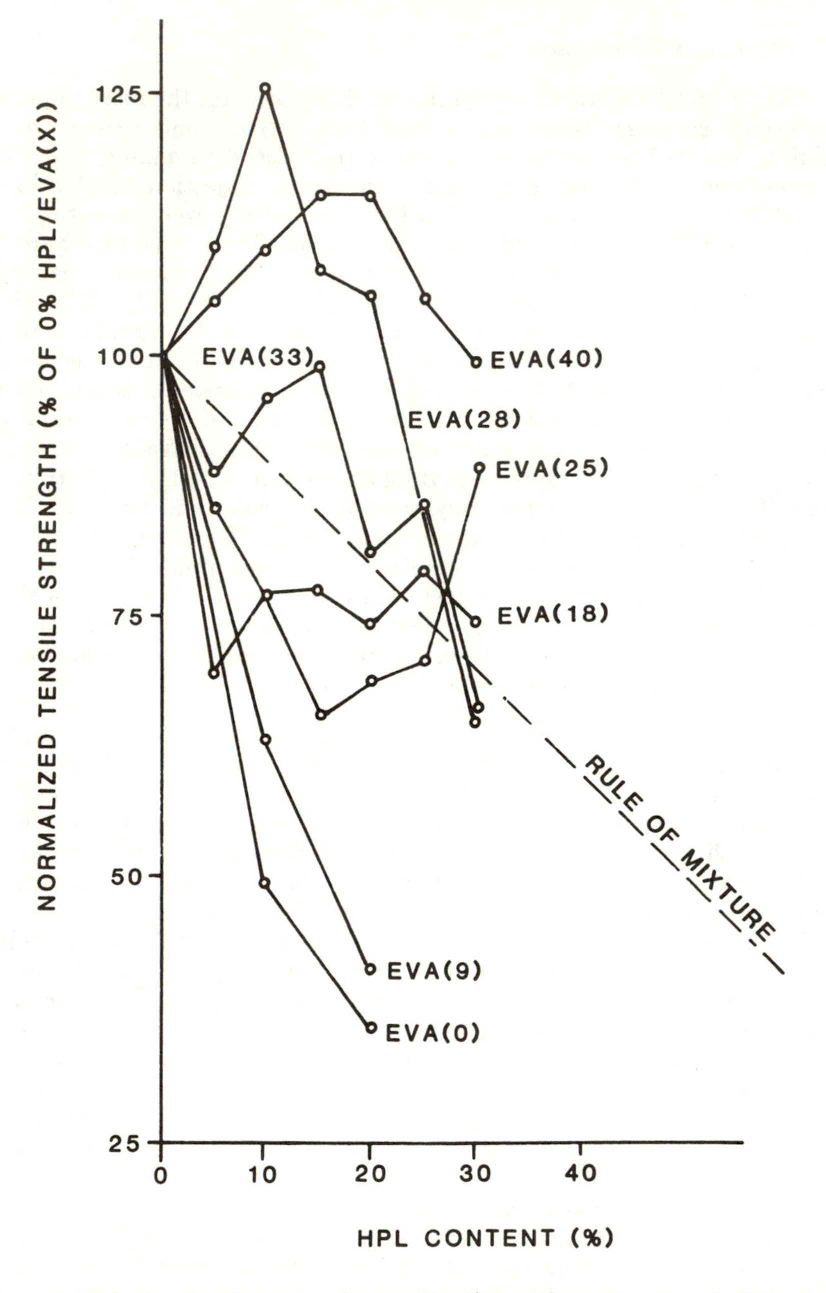

Figure 1. Relationship between (normalized) tensile strength and HPL content of injection molded blends with ethylene vinyl acetate copolymer. (Vinyl acetate content of the thermoplastic copolymer is given in parentheses.) (Strength values expressed in percent of unblended copolymer.) (From Ref. 11, with permission by Marcel Dekker, Inc.)

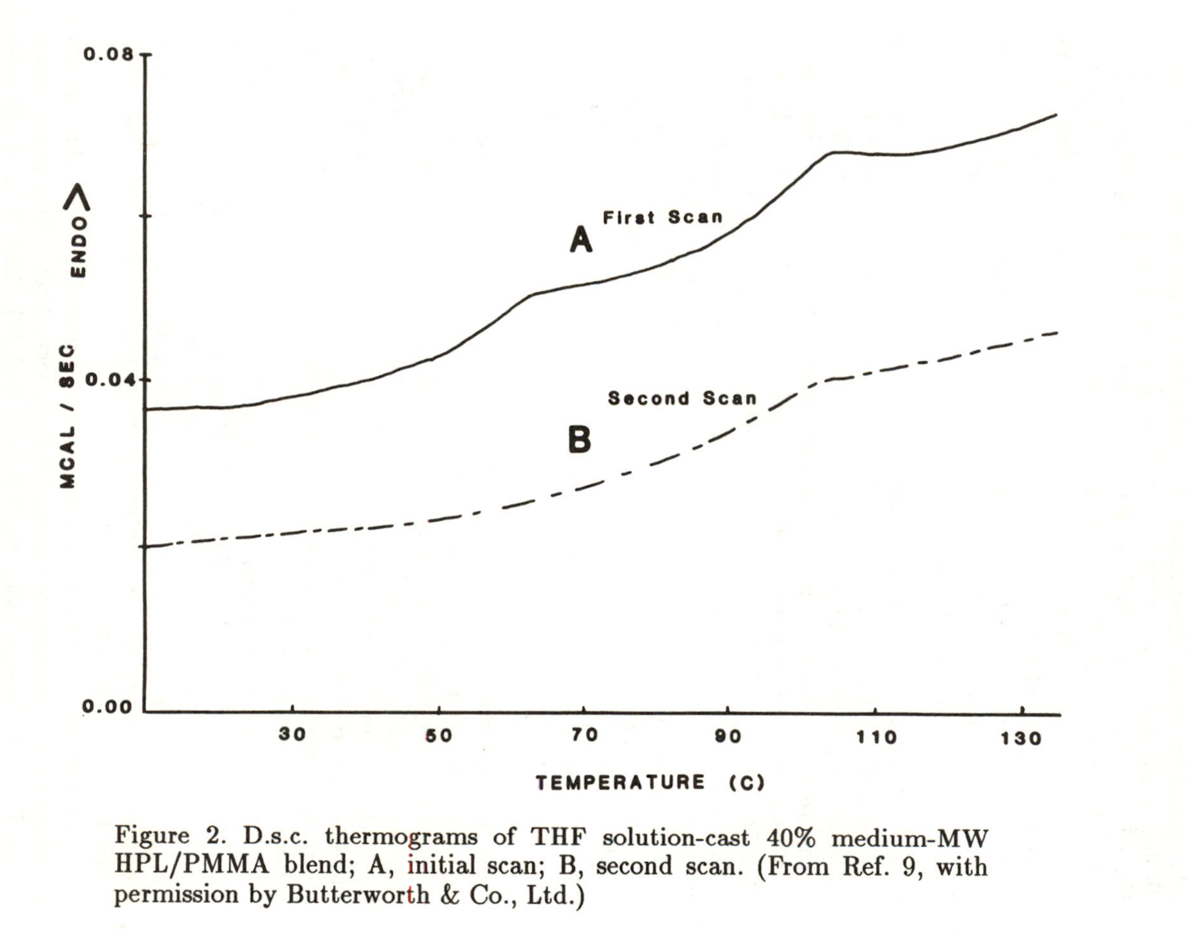

Figure 2. D.s.c. thermograms of THF solution-cast 40% medium-MW HPL/PMMA blend; A, initial scan; B, second scan. (From Ref. 9, with permission by Butterworth & Co., Ltd.)

blend T_g and morphology. For PMMA, the α or glass transition appears at 112C for molded samples, and at 108 and 104C for solution cast (THF and chloroform, respectively) samples. A β-transition which is centered around 20-50C is generally attributed to rotation of the methyl ester side group. The effects of increased HPL content, and of increasing molecular weight, on the shape of the tan δ curves is shown in Figure 3. Blend transitions are seen to increase in width with increasing HPL fraction, and this observation is independent of HPL molecular weight. Where low molecular weight HPL fractions lower the common T_g of the blend, i.e., plasticize the material, high molecular weight fractions have the opposite effect, i.e., anti-plasticize the molecular composite structure (9).

HPL/PVA Blends. Blend Structure. All blends of HPL with all three (hydrolyzed) PVA's had a characteristically brown color. They were fully transparent and clear. This is symptomatic of compatibility. SEM micrographs revealed distinct phase separation with PVA(0) blends, but they displayed smooth fracture surfaces with 25% HPL/PVA(96), PVA(88) and PVA(75) blends (10). This is characteristic of homogeneous materials. SEM results thus suggest that partially hydrolyzed PVA blends with HPL produce at least partially miscible systems. This can be explained with secondary associations between the hydroxy groups of HPL and PVA (> 0). (The number preceding the designation of blend components represents the weight fraction of the first-mentioned component and the figure in parentheses following "PVA" designates the extent of hydrolysis.)

Thermal Properties. A typical dsc thermogram of an HPL/PVA blend (Fig. 4) shows a single T_g and T_m (10). Differences in the shape of the melting endotherms of PVA(96), (88), and (75) can be attributed to different degrees of crystallinity in the three polymers. Changes in crystalline structure of polymer blends usually result from polymer-polymer interactions in the amorphous phase. Such interactions result in a reduction of crystallinity, thereby reducing the enthalphy of the phase change (16,17). The observed reductions in melt endotherm area of HPL blends with PVA (> 0) may therefore indicate the existence of polymer-polymer interactions between the two types of macromolecules.

T_g-analysis of the blends was hampered by the fact that HPL had a T_g of 63C, and PVA(96), PVA(88) and PVA(75) had T_g values of 77, 70, and 60C, respectively. This is too close for the resolution of two separate transitions. But analysis of the actual transition temperatures, rather than the shape of the transitions, can still provide information about the state of the HPL/PVA blends.

Theoretically miscible polymer blends will show T_g values that are intermediate between those of the parent polymers. They follow such models as the Fox or Gordon-Taylor relationships (18,19). However, in the case of HPL/PVA blends, the T_g data did not follow any of these well known models, and T_g values *above* those of the parent polymers were observed (10). The quotient of the experimental blend-T_g divided by the predicted (Fox) T_g consistently rose above 1.00 for blends exceeding 5% HPL content (10). This indicates molecular interactions between HPL and PVA. An

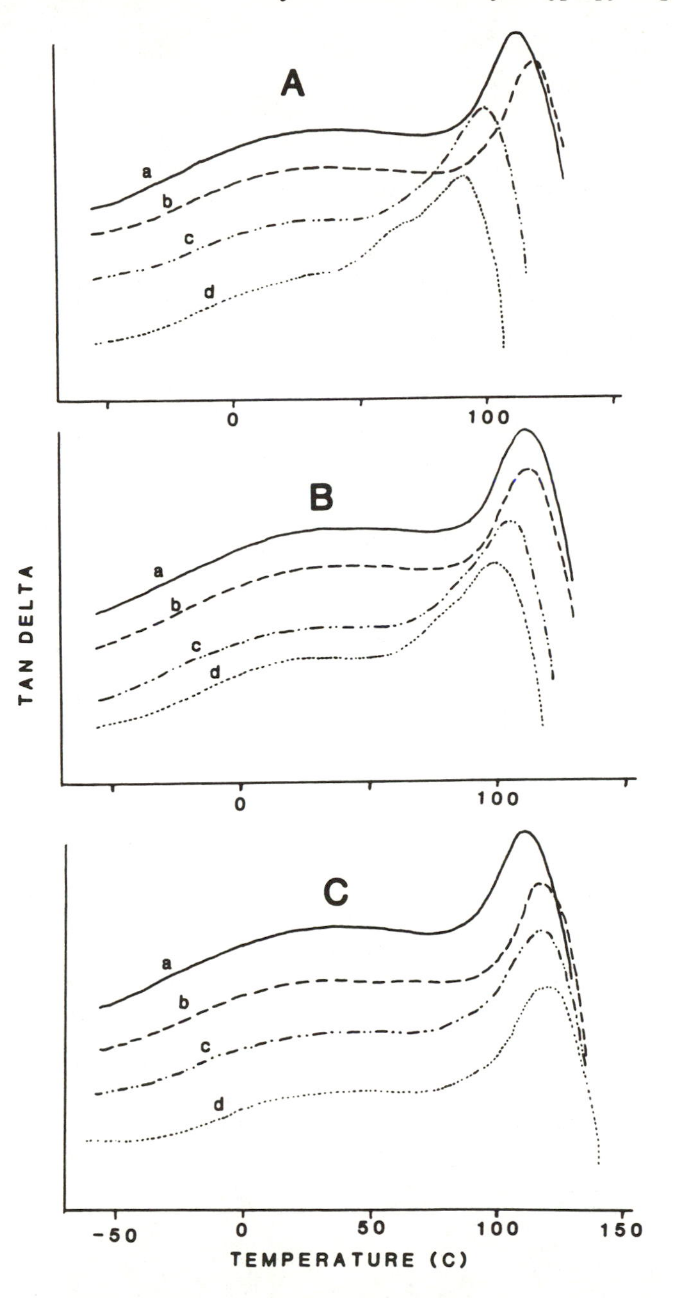

Figure 3. D.m.t.a. thermograms of HPL/PMMA blends; (A) low-MW HPL; (B) medium-MW HPL; and (C) high-MW HPL. Individual curves are for: a, PMMA (control); b, 5% HPL/PMMA blend; c, 25% HPL/PMMA blend; and d, 40% HPL/PMMA blend. (Reprinted with permission from ref. 9. Copyright 1988 Butterworth & Co.)

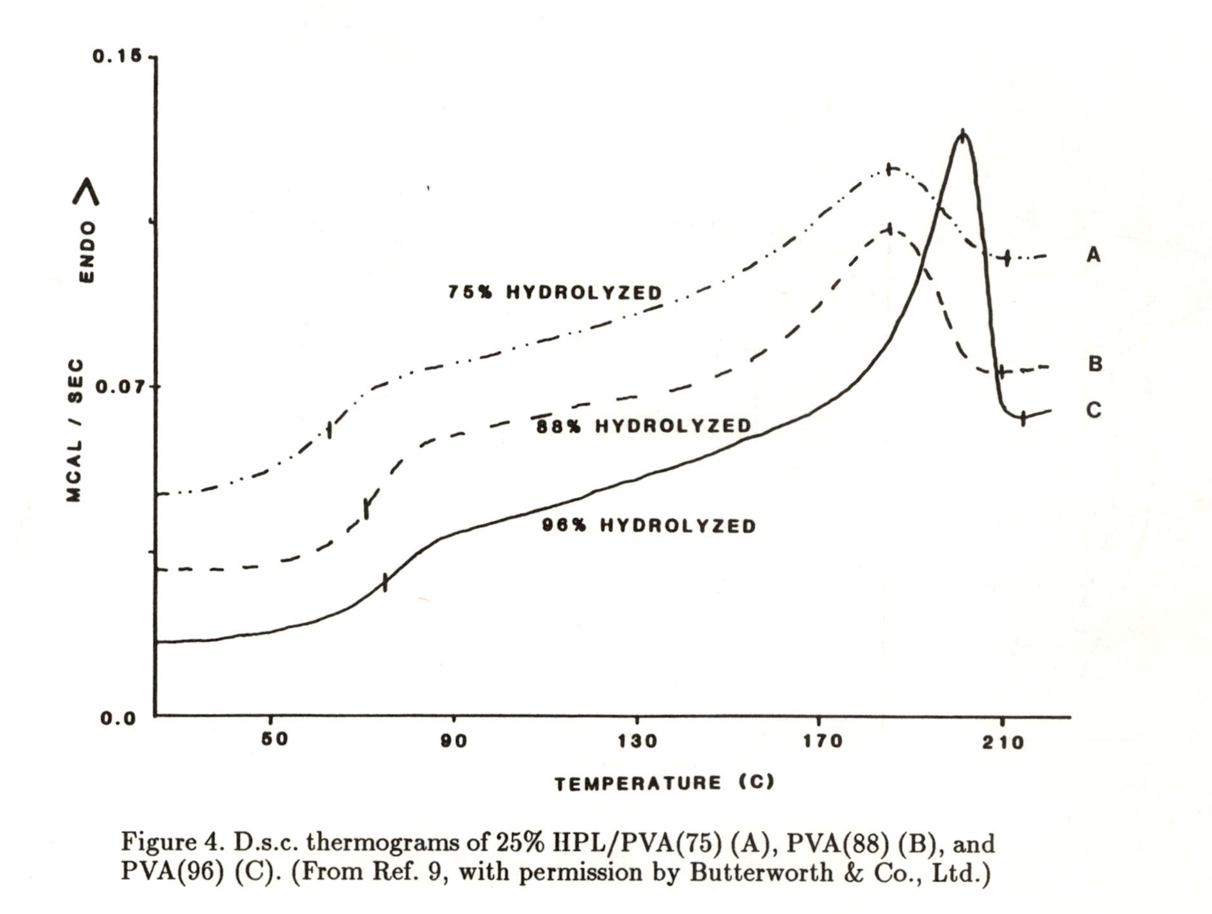

Figure 4. D.s.c. thermograms of 25% HPL/PVA(75) (A), PVA(88) (B), and PVA(96) (C). (From Ref. 9, with permission by Butterworth & Co., Ltd.)

amorphous component seems to be created in which the polymers coexist in a closely associated state, thereby reducing free volume and raising T_g. Another possible explanation for the observed increase in T_g is that the secondary bonds act as quasi-crosslinks which restrict Brownian motion of the long chain molecules, thereby raising T_g.

Tan δ curves from dmta thermograms of blends based on combinations of HPL/PVA(> 0) are shown in Figure 5 (10). The tan δ transition, which is another measure of T_g, is broadened by the presence of HPL component, and it increases above that of the parent polymers consistent with dsc data. This can again be explained with the presence of strong interactions between the amorphous components which coexist in a closely associated state.

Conclusions

1. Blends of PE, of PMMA, and of PVA(0) with HPL of varying molecular weight produce two-phase materials. Blends of HPL with PVA (> 0) exhibit at least partial miscibility.

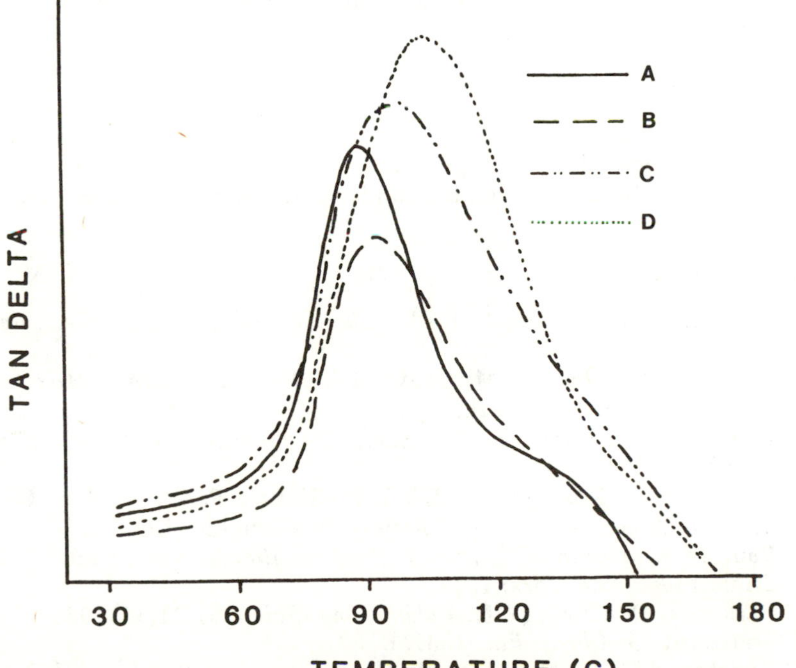

Figure 5. Relationship between tan δ of the HPL/PVA(96) blends and HPL content; A, 0% HPL; B, 5% HPL; C, 25% HPL; and D, 40% HPL. (From Ref. 9, with permission by Butterworth & Co., Ltd.)

2. Enthalpy relaxation is often responsible for two endothermic transitions in the initial dsc thermograms, and a single, broad transition is observed in most second scans and in dmta thermograms.
3. The interphase between the continuous and the discontinuous polymer phases differs with respect to blend preparations. Solution-cast blends produce inclusions that are pulled away from the matrix, whereas injection-molded blends show HPL striations that are closely associated with the matrix.
4. Ultimate properties, tensile strength, and ultimate strain typically decrease with the addition of HPL; however, combinations of molecules with strong interaction between amorphous components may also exhibit enhanced ultimate properties. Injection molding produces superior material properties.
5. Variations in the molecular weight of HPL have no significant effect on the material properties of HPL/PMMA blends.
6. Some commercial, linear (thermoplastic) polymers produce blends with lignin and lignin derivatives that fail to result in phase separation on macroscopic scale. Polyblends with lignin derivatives sometimes resemble plasticized or anti-plasticized materials. The greatest contribution lignin can make to thermoplastic systems is that of modulus; and this is the same as that which lignin makes to the amorphous component of wood.

Acknowledgment

Valuable assistance was provided by Dr. S. S. Kelley and Dr. T. G. Rials, and this is acknowledged with gratitude. Financial support was provided by the National Science Foundation under grant number CBT-8512636.

Literature Cited

1. Hildebrand, J. H.; Scott, R. L. *The Solubility of Non-Electrolytes*; Dover: New York, 1964.
2. Olabisi, O.; Robeson, L. M.; Shaw, M. T. *Polymer-Polymer Miscibility*; Academic Press: New York, 1979.
3. Klempner, D.; Frisch, K. C. (Eds.). *Polymer Alloys*; Plenum Press: New York, 1980.
4. Cooper, S. L.; Estes, G. M. (Eds.). *Multiphase Polymers*; Adv. Chem. Ser. 176, American Chemical Society: Washington, DC, 1979.
5. Paul, D. R.; Newman, S. (Eds.). *Polymer Blends*; Vols. 1 and 2; Academic Press: New York, 1978.
6. Barlow, J. W.; Paul, D. R. *Polym. Eng. Sci.* 1981,**21**(15), 985.
7. Scatchard, G. *Chem. Rev.* 1931, **8**, 321.
8. Hildebrand, J. H.; Pravsnitz, J. M.; Scott, R. L. *Regular and Related Solutions*; Van Nostrand Reinhold: New York, 1970.
9. Ciemniecki, S. L.; Glasser, W. G. *Polymer* 1988, **29**, 1021.
10. Ciemniecki, S. L.; Glasser, W. G. *Polymer* 1988, **29**, 1030.

11. Glasser, W. G.; Knudsen, J. S.; Chang, C.-S. *J. Wood Chem. Tech.* 1988, **8**(2), 221.
12. Wu, L. C.-F.; Glasser, W. G. *J. Appl. Polym. Sci.* 1984, **29**, 1111.
13. Hatakeyama, T.; Nakamura, K.; Hatakeyama, H. *Polymer* 1982, **23**, 1801.
14. Rials, T. G.; Glasser, W. G. *J. Wood Chem. Technol.* 1984, **4**(3), 331.
15. Walsh, D. J.; Higgins, J. S.; Maconnachie, A. (Eds.). *Polymer Blends and Mixtures*; NATO ASI Series E, Appl. Sci. No. 89, Nijhoff: The Hague, 1985.
16. Wenig, W.; Karasz, F. E.; MacKnight, W. J. *J. Appl. Phys.* 1975, **46**, 4166.
17. Hammel, R.; MacKnight, W. J.; Karasz, F. E. *J. Appl. Phys.* 1975, **46**, 4199.
18. Fox, T. G. *Bull. Am. Phys. Soc.* 1956, **1**, 123.
19. Gordon, M.; Taylor, J. S. *J. Appl. Chem.* 1952, **2**, 493.

RECEIVED May 29, 1989

Chapter 36

Phase Morphology of Hydroxypropylcellulose Blends with Lignin

Timothy G. Rials[1–3] and Wolfgang G. Glasser[1,2]

[1]Department of Wood Science and Forest Products, Virginia Polytechnic Institute and State University, Blacksburg, VA 24061
[2]Polymeric Materials and Interfaces Laboratory, Virginia Polytechnic Institute and State University, Blacksburg, VA 24061

The incremental elimination of hydroxy functionality in an organosolv lignin by ethylation and acetylation dramatically influenced the state of miscibility of blends prepared with hydroxypropyl cellulose (HPC). The various blends were characterized by three distinct morphologies. At the lowest levels of component compatibility, the morphology resulted from classical phase separation of the lignin component and HPC liquid crystal (LC) mesophase formation. A miscible, amorphous blend resulted when the polymer-polymer interaction was maximized. At intermediate levels, however, the blend morphology was dominated by entropic effects leading to the development of LC superstructure dispersed in a lignin-reinforced HPC matrix. This latter case demonstrates the potential for developing high-strength cellulosic composites.

In recent years, some of the more significant developments in materials technology have come in the area of multi-component polymer systems. One of the most exciting topics is that of rigid-rod composites which addresses the replacement of the macroscopic reinforcing fiber of conventional composites with a single polymer chain (1). This accomplishment would retain the high strength of the composite, while eliminating such detrimental aspects as property anisotropy and interfacial defects.

Significant advances to this end have been made using primarily liquid crystal copolyesters as the rigid-rod chain in various matrix polymers (2,3). Surprisingly, while cellulose and its derivatives also exhibit LC phenomena

[3]Current address: U.S. Department of Agriculture Forest Service, Southern Forest Experiment Station, 2500 Shreveport Highway, Pineville, LA 71360

0097–6156/89/0397–0464$06.00/0

(4), their utility in this novel material system has received relatively little attention. Recent reports have addressed this interest by polymerizing an acrylic monomer in which the cellulosic component is dissolved (5), and have provided some encouraging results. A more convenient approach involves blending the cellulosic component with a flexible coil polymer serving as the matrix. Although the phase behavior of binary cellulosic systems have been extensively studied (6-8), the relationships in ternary systems remain largely unanswered, particularly as they extend to the bulk morphology.

Recent studies in this laboratory on hydroxypropyl cellulose (HPC) blends with lignin (9,10) have indicated that the composite morphology and properties are extremely sensitive to the degree of miscibility between the two components. This paper summarizes the observations that have been made with particular regard to the development of high-strength composites. For experimental details, the reader is referred to the above referenced publications.

Experimental

Materials. The hydroxypropyl cellulose (Klucel 'L') used in this study was supplied by Hercules, Inc. The manufacturer reported a molar substitution of 4 propylene oxide units/anhydroglucose unit, and a nominal molecular weight of 10^5 g· mol^{-1}.

The lignin component was an organosolv lignin (OSL) (schematically represented in Figure 1) supplied by Repap Technologies of Valley Forge, PA, which had been isolated from aspen wood. Neither reaction conditions nor yield were made available. The polystyrene equivalent number ($<M_n>$) and weight [($<M_w>$)] average molecular weight were determined by gel permeation chromatography as 900 and 3,000 g· mol^{-1}, respectively.

The hydroxyl content of the lignin was incrementally eliminated by acetylation and ethylation according to previously described procedures (11). In addition, modification with propylene oxide was also used to alter the original lignin structure (12). Table I presents a summary of the hydroxy content, and pertinent physical properties, of the derivatized lignins.

Methods. Individual solutions of the blend components in dioxane (or tetrahydrofuran) were mixed, and stirred for about 12 hours before casting into a Teflon mold. Solvent evaporation proceeded under ambient conditions for 24 hours followed by transferral to a vacuum oven at 60°C for further removal of solvent. The dried blends were then stored in a vacuum desiccator over P_2O_5.

The glass transition (T_g) and melting (T_m) temperature of the pure component polymers and their blends were determined on a Perkin-Elmer (DSC-4) differential scanning calorimeter and Thermal Analysis Data Station (TADS). All materials were analyzed at a heating and cooling rate of 20°C · min^{-1} under a purge of dry nitrogen. Dynamic mechanical properties were determined with a Polymer Laboratories, Inc. dynamic mechanical thermal analyzer interfaced to a Hewlett-Packard microcomputer. The

Table I. Hydroxyl content and glass transition temperature of lignins derivatized with acetic anhydride (Ac), diethyl sulfate (Et), and propylene oxide (Pr)

Sample ID	Molar Ratio (Reagent/OH)	OH-Groups/C9		T_g(°C)
		Total OH	Phenolic OH	
OSL	—	1.47	0.59	115
Ac-1	29	1.14	0.41	110
Ac-2	61	0.52	0.26	106
Ac-3	80	0.35	0.21	103
Ac-4	100	0.19	0.14	101
Ac-5	192	0.0	0.03	96
Et-1	23	1.22	0.34	109
Et-2	50	1.08	0.20	96
Et-3	67	0.96	0.08	75
Et-4	172	0.88	0.00	57
Pr-1	—	1.47	0.00	69

spectra were collected at a heating rate of 4°C · min^{-1} from −50°C to 150°C utilizing a single cantilever beam geometry.

Results and Discussion

Blend Properties. Of the modification reactions utilized in the study, acetylation yielded the widest range of substituted lignins (Table I) and resulted in the greatest variety of blend morphologies. A series of thermograms for 20 wt.% blends prepared from this family of lignins is presented in Figure 2. Although data for the pure HPC are not shown in the figure, the T_g was found at about 25 °C and T_m at 212 °C (X=16%). An additional second-order transition for this material is located at about 90 °C which has been described (13,14) as originating from a liquid crystal (LC) mesophase. The unmodified lignin blend has a broad glass transition centered at 42°C, resulting from the convoluting effects of these two transition as revealed by dynamic mechanical analysis. Also, relative to the pure HPC component, a depression in the melting temperature (T_m) of about 20° results along with a reduction in the degree of crystallinity. Interestingly, the blend prepared from the lowest DS lignin exhibits a much more intense T_g with an associated loss of crystalline and LC superstructure. As the lignin hydroxy content is further reduced, DSC analysis of the 20 wt.% blend again resembles that of the HPC/OSL system in that a crystalline melt is observed at about 190 °C, and a second-order transition occurs at 40°C. Upon complete acetylation of lignin an additional feature can be identified in the thermogram. While the deviation at 90°C (common to the HPC) is still evident, an additional transition can be found at 115°C originating from from a pure lignin phase. Further evidence of the limited compatibility between these

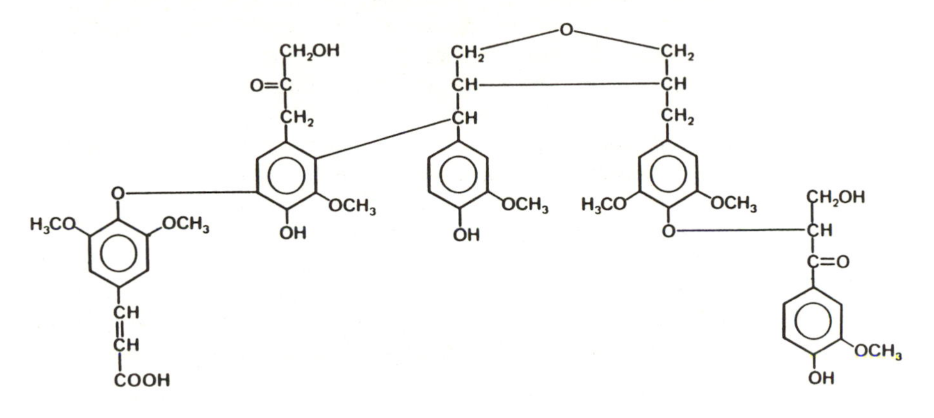

Figure 1. Schematic representation of organosolv lignin.

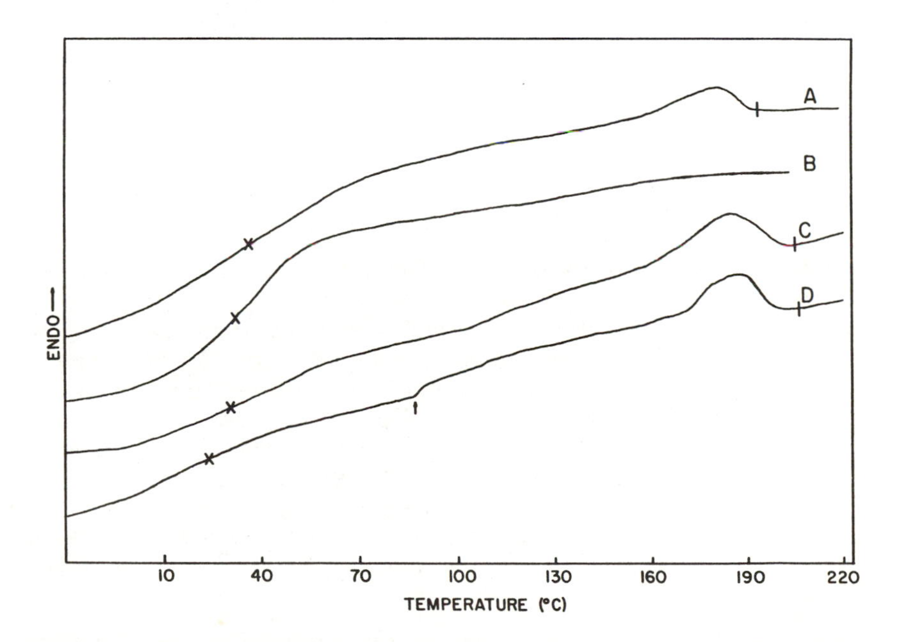

Figure 2. DSC thermograms of HPC blends containing 20 wt.% of acetylated lignins with a degree of substitution of: A) 0%, B) 23%, C) 76%, D) 100

two polymers can be cited by the similarity in both intensity and location of the melting peak.

The above observations indicate that as the degree of substitution of lignin hydroxy functionality increases, there is an initial compatibility increase with HPC followed by a rapid decrease until complete incompatibility results after complete acetylation. This is further illustrated by monitoring the degree of crystallinity of the blends as shown in Figure 3. Relative to the unmodified lignin blend, there occurs a much more rapid disruption of crystallinity in the materials prepared from low DS lignins. As the level of substitution increases, however, the level of crystallinity retained in the blend increases until at DS = 87% it remains above that of the unmodified lignin system. Recognizing that the decrease in crystallinity of the blend is attributable to the interaction between the individual components, this approach is consistent with the above scenario of an initial increase in compatibility followed by a rapid decrease in polymer-polymer miscibility*.

Melting Point Depression. A more quantitative evaluation of the relationships existing between lignin structure and blend miscibility is possible through the T_m depression observed in these materials. For semi-crystalline blend systems, such as these, the polymer-polymer interaction parameter, 'B', can be determined through the following simplified expression (15):

$$T^{o}_{M2} - T_{M2} = \frac{-B \cdot V_{2u}}{(\Delta H_{2u})} \cdot T^{o}_{m2}\phi_1^2 \quad (1)$$

where T^{o}_{M2} is the equilibrium melting point of HPC (= 213.1°C, from ref. 16), T_{M2} is the melting point of the blend, $\Delta H_{2u}/V_{2u}$ is the heat of melting per unit volume of 100% crystalline material (= 7.52 cal · cm^{-3}, from ref. 16), and ϕ_1 is the volume fraction of the lignin component. A plot of ΔT_{M2} vs. ϕ_1^2 should then yield a straight line with a slope proportional to 'B'. The results of this treatment for the blends are presented in Table II below.

Although all of the blend systems are adequately modeled through this linear assumption (R^2 = 0.91-0.99), it is of interest that the intercept fails to pass through the origin, as expected. Generally, this behavior has been attributed to a T_m depression resulting from extraneous concerns such as reduction in lamellar thickness of the crystallites (15). Since no attempt was made here to obtain equilibrium melting points, this is not out of the question. Unfortunately, this effect eliminates the purely quantitative aspect of this analysis, but the figures remain valid for direct comparison of these blends.

Discussion. The overall effect of lignin hydroxy content on the interaction with HPC is illustrated in Figure 4. As the hydroxy content is reduced, the interaction energy increases until a maximum is reached at a DS of ca.

---

*It should be noted that results on blends prepared with the ethylated lignins are consistent with this view, as will become evident in later discussion.

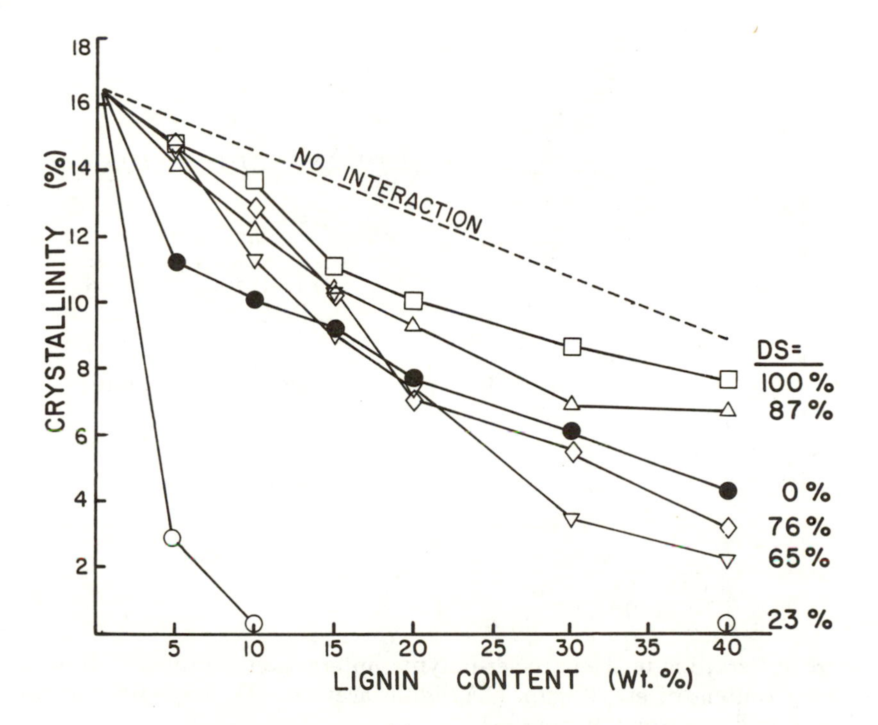

Figure 3. Relationship between % of crystallinity and lignin content for blends prepared with the acetylated lignin series. The degree of substitution of hydroxy functionality is given with each curve.

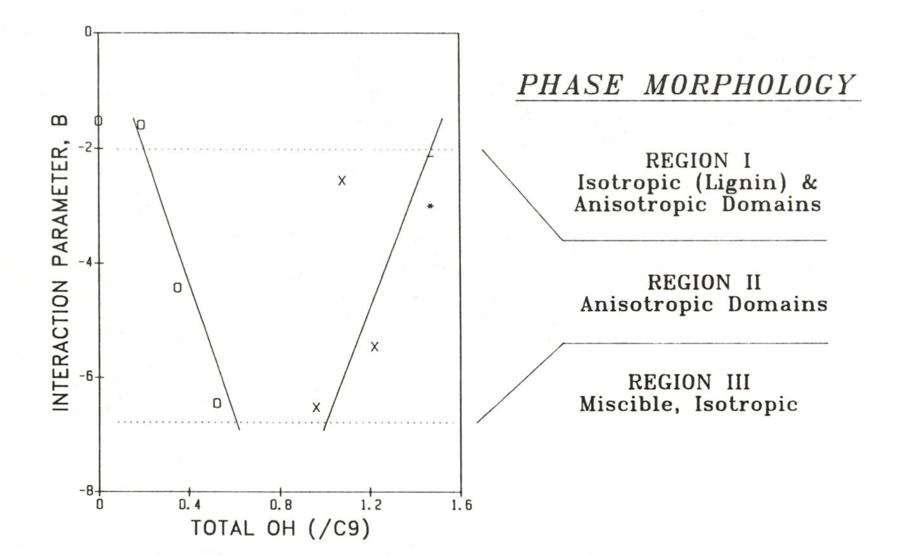

Figure 4. Variation in the polymer-polymer interaction parameter, 'B' with hydroxy content of ethyl lignin (X), lignin acetate (O), unmodified lignin (*), and hydroxypropyl lignin (—).

Table II. Summary of melting point depression analysis for blends of HPC and lignin modified by acetylation (Ac), ethylation (Et) and propoxylation (Pr)

Sample ID	Intercept	Slope	R^2	–'B' (cal/cm^3)
HPC/OSL	9.675	196.8	0.98	3.00
HPC/Pr	6.820	137.6	0.95	2.13
HPC/Ac-2	2.322	415.6	0.95	6.45
HPC/Ac-3	3.187	285.5	0.98	4.43
HPC/Ac-4	5.337	102.4	0.91	1.59
HPC/Ac-5	6.104	97.8	0.94	1.52
HPC/Et-1	5.281	351.9	0.99	5.46
HPC/Et-2	5.514	165.6	0.97	2.56
HPC/Et-3	5.212	420.7	0.98	6.52

40%. (This region is represented by the HPC/Ac-1 and HPC/Et-4 blends which were largely amorphous, making it impossible to determine the T_m depression). The extent of interaction then rapidly diminishes as hydroxy groups are further eliminated until a value of −1.5 is reached upon complete substitution. It is worth emphasizing that this curve is generated from data obtained on both the ethylated and acetylated lignin blends; consequently, the overlap that is encountered in this analysis suggests that the type of modification applied does not appreciably influence the behavior. Furthermore, the parabolic shape of the curve indicates that the hydroxy functionality of lignin does not favorably impact component interaction through the establishment of intermolecular hydrogen bonds. Rather, the observed maximum occurs at a degree of substitution which provides a very close match in the solubility parameter δ of the two polymers [for the lignin derivatives, 10.0-10.5; HPC = 10.1 (17)]. There does exist an interesting exception. The propoxylated lignin derivative blend actually reveals a slight decrease in interaction, relative to the parent lignin system, even though its solubility parameter ($\delta = 9.7$) (Table I) would suggest a much higher level of compatibility. The implications of this observation are not yet clear, unfortunately, and will not be expanded on at this time.

Further consideration of these materials as a whole provides information on those factors contributing to the blend morphology and, subsequently, impacting the development of high-strength composites. Within the interaction diagram of Figure 4, three distinct regions of morphological behavior can be identified. At low levels of interaction between the lignin component and HPC (Region I), phase separation occurs due to the unfavorable energy considerations as well as LC mesophase formation. The heterogeneity introduced by the lignin domains has a catastrophic effect on material properties. At the other extreme (Region III), a miscible blend is obtained presumably representing the ideal situation; however, it appears

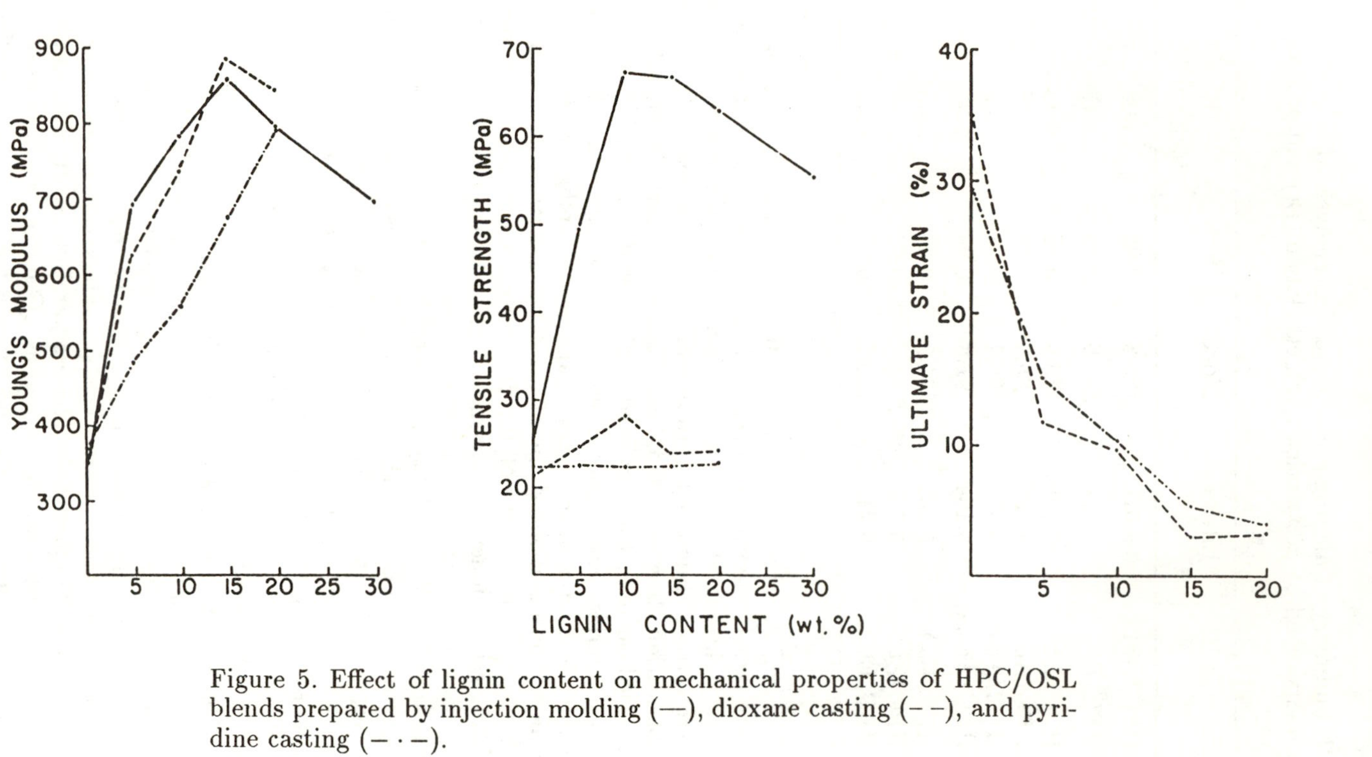

Figure 5. Effect of lignin content on mechanical properties of HPC/OSL blends prepared by injection molding (—), dioxane casting (– –), and pyridine casting (– · –).

Figure 6. Scanning electron micrographs of freeze-fracture surfaces for injection molded (A), dioxane cast (B), and pyridine cast (C) blends at a composition of 15% lignin. (2000X) (From ref. 10, with permission of Butterworth & Co., Ltd.)

as though the rigidity of the cellulosic chain is largely derived from intermolecular interactions. The disruption of these chain interactions yields a much more flexible chain, negating the advantages afforded by the rigid-rod. Morphology development in the region of intermediate interaction (Region II) originates primarily from LC mesophase formation in a miscible phase of HPC and lignin, and presents the most favorable conditions for enhanced properties.

Figure 5 presents the results of tensile tests for the HPC/OSL blends prepared by solvent-casting and extrusion. All of the fabrication methods result in a tremendous increase in modulus up to a lignin content of ca. 15 wt.%. This can be attributed to the T_g elevation of the amorphous HPC/OSL phase leading to increasingly glassy response. Of particular interest is the tensile strength of these materials. As is shown, there is essentially no improvement in this parameter for the solvent cast blends, but a tremendous increase is observed for the injection molded blend. Qualitatively, this behavior is best modeled by the presence of oriented chains, or mesophase superstructure, dispersed in an amorphous matrix comprised of the compatible HPC/OSL component. The presence of this fibrous structure in the injection molded samples is confirmed by SEM analysis of the freeze-fracture surface (Figure 6). This structure is not present in the solvent cast blends, although evidence of globular domains remain in both of these blends appearing somewhat more coalesced in the pyridine cast material.

Conclusions

Although specific intermolecular interactions do not appear to contribute in this polymer blend system, the elimination of lignin's hydroxy functionality does dramatically influence the level of interaction between these polymers. Consequently, it is possible to define three morphological regions for these materials which directly impact the potential development of high-strength composite materials. At intermediate levels of interaction, the blend morphology is characterized by a dispersion of liquid crystal superstructure in an amorphous HPC/OSL matrix. Upon orientation of this supermolecular structure, a dramatic enhancement of both modulus and tensile strength is observed indicating the validity of this approach for the development of high performance composites based on cellulosics.

Acknowledgments

The authors would like to acknowledge the assistance of Dr. S. S. Kelley (Tennessee-Eastman Co.), Dr. T. C. Ward, and Dr. G. L. Wilkes (PMIL, Virginia Tech), and Mr. C. Price (Dept. of Forest Products, Virginia Tech).

Literature Cited

1. Hwang, W. F.; Wiff, D. R.; Verschoore, C.; Price, G. E.; Helminiak, T. E.; Adams, W. W. *Polym. Eng. Sci.* 1983, **23**, 784.
2. Siegman, A.; Dagan, A.; Kenig, S. *Polymer* 1985, **26**, 1326.

3. Paci, M.; Barone, C.; Magagnini, P. *J. Polym. Sci.-Polym.* 1987, **B25**, 1559.
4. Gray, D. G. In *Polymeric Liquid Crystals*; Blumstein, A., Ed.; Plenum: New York, 1985; p 369.
5. Nishio, Y.; Susuki, S.; Takahashi, T. *Polymer J.* 1985, **17**, 753.
6. Navard, P.; Hardin, J.-M. In *Polymeric Liquid Crystals*; Blumstein, A., Ed.; Plenum: New York, 1985; p 389.
7. Tseng, S.-L.; Valente, A.; Gray, D. G. *Macromolecules* 1981, **14**, 715.
8. Werbowyj, R. S.; Gray, D. G. *Mol. Cryst. Liq. Cryst* 1976, **34**, 97.
9. Rials, T. G.; Glasser, W. G. *J. Appl. Polym. Sci.*, in press.
10. Rials, T. G.; Glasser, W. G. *Polymer*, in press.
11. Rials, T. G. Ph.D. Thesis, Virginia Polytechnic Institute and State University, Blacksburg, VA, 1986.
12. Wu, L. C.-F.; Glasser, W. G. *J. Appl. Polym. Sci.* 1984, **29**, 1111.
13. Kyu, T.; Mukheija, P.; Park, H.-S. *Macromolecules* 1985, **18**, 2331.
14. Rials, T. G.; Glasser, W. G. *J. Appl. Polym. Sci.* 1988, **36**, 749.
15. Harris, J. E.; Paul, D. R.; Barlow, J. W. In *Polymer Blends and Composites in Multiphase Systems*; Han, C.D., Ed.; American Chemical Society: Washington, DC, 1984; p 17.
16. Samuels, R. J. *J. Polym. Sci.: A2* 1969, **7**, 1197.
17. Aspler, J. S.; Gray, D. G. *Polymer* 1982, **23**, 43.

RECEIVED March 17, 1989

Chapter 37

Engineering Lignopolystyrene Materials of Controlled Structures

Ramani Narayan[1], Nathan Stacy[1], Matt Ratcliff[2], and Helena Li Chum[2]

[1]Laboratory of Renewable Resources Engineering, Purdue University, West Lafayette, IN 47907
[2]Solar Energy Research Institute, 1617 Cole Boulevard, Golden, CO 80401

Well characterized, low molecular weight lignins of narrow polydispersity were reacted with a predefined molecular weight polystyryl carbanion, which was prepared by anionic polymerization. The carbanion was shown to displace mesylate groups on the lignin macromolecule forming ligno-polystyrene graft copolymers of controlled structures. The toluene extract of the reaction mixture contained largely unreacted polystyrene and about 5% of lignin-g-PS. The residue (toluene insoluble) also contained lignin-polystyrene graft copolymer.

The grafting of synthetic polymers to lignin and other natural polymers offers the potential of preparing new class of engineering plastics because grafting frequently results in the superposition of properties relating to backbone and side chain (1,2). An example is the grafting of 8-10% polyacrylic acid onto high density polyethylene, which gave the copolymer increased modulus and softening point while leaving the melting point and crystallinity of the polyethylene unchanged. The properties of the graft copolymers will, primarily, be dependent on the molecular weight of the side chain graft and that of the lignin. It will also depend on the nature and number of side chain grafts and the type of backbone-graft linkage. These tailor-made lignin-synthetic polymer graft copolymers of controlled structures can function as compatibilizers/interfacial agents for blending lignin with synthetic thermoplastics. Polymer blends have been successfully used in an increasing number of applications in recent years as a means of combining the useful properties of different materials to meet new market applications with minimum development cost (3).

Current approaches to grafting synthetic polymers onto lignin (4-6) and other natural polymers like cellulose (7-11) involve radical polymerization methods (chemical or radiation). Using radical approaches, it is

0097-6156/89/0397-0476$06.00/0

usually more difficult to control or change the molecular weight of the grafts, the molecular weights are very high and different types of backbone-graft linkage exist. Furthermore, the graft contains considerable amounts of homopolymers. It is more difficult to reproduce the graft yields, properties, and other features of the graft copolymer (12). Because of these problems, such graft copolymers cannot function effectively as compatibilizers/interfacial agents.

Glasser and co-workers (see publications in these Proceedings for references) have been using approaches with good structural control through the production of chain-extended hydroxypropylated lignins reacted with other polymeric systems. We are developing new synthetic approaches that allow us to prepare renewable polymer-synthetic polymer graft copolymers of controlled structures with precise control over molecular weights of the grafts and backbone polymer, degree of graft substitution, and backbone-graft linkage (13-18). In this paper we report on the preliminary synthesis and characterization of lignin-polystyrene graft copolymers of controlled structures.

Experimental

Materials. Styrene (Aldrich) was purified by distillation from CaH_2 before use. n-Butyllithium (Aldrich) was used without further purification. Tetrahydrofuran (Fisher Scientific) was purified in a solvent still by distillation from a sodium/benzophenone mixture. Toluene (Fisher Scientific) was used without further purification. Reagent grade methylene chloride (Baker Chemical Co.) was dried on 5A° molecular sieves. Reagent grade triethylamine (Baker) was dried over KOH. Methanesulfonyl chloride (Aldrich, 98%) was used without further purification.

Instrumentation. UV/Visible spectra were collected on a Perkin-Elmer Lambda 4 spectrophotometer. IR spectra were collected on a Perkin-Elmer 1640 spectrophotometer. NMR spectra were taken on a Nicolet NT-200 spectrometer. Differential scanning calorimetry was run on a Perkin-Elmer DSC-4 unit, equipped with a system 4 microprocessor controller and a 3600 data station. Elemental analyses were run by the Purdue microanalysis laboratory in the Department of Chemistry at Purdue University and by Huffman Labs (Golden, CO). Lignin group analysis techniques are described in references 19-21.

Preparation of Mesylated Lignin. In a 500 ml flask, 3.0 g of the organosolv lignin (elemental: C, 64.5%; H, 6.3%; O, 29.2%) was dissolved in a solution of 60 ml CH_2Cl_2 containing 21.8 mmole (~3 ml) dry triethylamine. The flask was sealed and purged with nitrogen, while cooling to 0°C. 17.4 mmole (1.3 ml) of methanesulfonyl chloride (MSCl) was added dropwise over a 20 min period while stirring. The reaction mixture was then allowed to stir for an additional 60 min. The mesylated lignin was purified by successive extractions at room temperature in a separatory funnel, using one 30 ml H_2O wash followed by two washes with dilute (10%) HCl. The aqueous fractions were combined and back extracted with CH_2Cl_2. The combined

CH_2Cl_2 phase was then washed with saturated $NaHCO_3$. Back extraction of the combined $NaHCO_3$ washes recovered some of the emulsified products in CH_2Cl_2. The combined organic phase was then washed with NaCl, and the mesylated lignin was recovered from the CH_2Cl_2 by freeze drying with N_2 in 67% yield. The mesylated lignin had an approximate elemental analysis of C: 55.3%, H: 5.6%, O: 29.8%, S: 9.5% (after discounting for the presence of salt impurities). Group analysis for the lignin before mesylation gave 6.3% phenolic OH and 13.6% methoxyl groups, while after mesylation the results were 14.2% methoxyl and 23.5% SO_2CH_3 groups, respectively. ^{1}H NMR yielded peaks at 1.27 ppm (broad m, side chain methyls), 2.98 ppm (broad m, side chains), 3.12 ppm (s, C-O_3SCH_3), 3.28 (broad m, side chain), 3.88 (broad m, OCH_3), and 6.4-7.5 ppm (aromatics). In ^{13}C NMR, main peaks are seen in the range of 153.2-104.5 ppm (aromatic carbons), 56.3 ppm (methoxyl), 39.8 ppm (O_3SCH_3 in syringyl units), 38.2 ppm (O_3SCH_3 in guaiacyl units) and 37.3 ppm (O_3SCH_3 in side chain). The NMR assignments were confirmed through the synthesis of 2-methoxyphenol and 2.6-dimethoxyphenol, -(3-methoxy, 4-methane sulfonatephenyl) ethylene glycol-methanesulfonate-β-guaiacyl ether and α-(3-methoxy, 4-methanesulfonate phenyl) glycerol bis methanesulfonate-β-guaiacyl ether. NMR analysis indicated that the mesylated lignin is largely free of methanesulfonyl chloride and methanesulfonic acid.

Preparation of the "Living" Polystyrene. 18 g of the "living" polymer was prepared by standard anionic polymerization using n-butyl lithium. The reaction was carried out by the dropwise addition of 20 ml of styrene to 5 ml of the initiator solution in 150 ml of neat THF at -78°C. The styrene drip was adjusted to take approximately 30 min for completion and then the reaction was allowed to stir for two hours before the grafting reaction with mesylated lignin was carried out. The number average molecular weight of the polystyrene, as determined by HPSEC, was 9500 with polydispersity of 1.2.

Preparation of the Lignin-Polystyrene Graft Copolymer. 75 mg of the mesylated lignin was weighed and placed in a 50 ml round bottom flask. This flask was then purged by repeated vacuum/nitrogen cycles and left under a positive pressure of nitrogen. Approximately 50 ml of the "living" polystyrene solution, containing approximately 6 g of the polymer, was then transferred into the flask by a double-ended needle. The resulting solution turned dark brown in around 5-10 min and was allowed to warm to room temperature. After stirring for several hours, the solvent was evaporated by heating under a nitrogen flow. After a water/methanol wash to remove ionic residue, the solid residue was placed in soxhlet extraction apparatus and extracted with toluene for 48 hrs to remove excess polystyrene from the reaction mixture. After extraction, the toluene was removed by nitrogen flow and both the toluene soluble extract and the insoluble residue were dried in a vacuum oven for 3-4 hrs at 55°C.

Results and Discussion

The new synthetic approach being adopted for the preparation of the lignin-polystyrene graft copolymer involved three steps.

1. Introduction of reactive mesylate groups onto the lignin macromolecule by mesylation of the free –OH groups on the lignin (Fig. 1). The mesylate groups are good leaving groups in nucleophilic displacement reactions.
2. Preparation of desired molecular weight polystyryl carbanion ("Living Polystyrene") by anionic polymerization (Fig. 2). Anionic polymerization has been used extensively to provide control over molecular weight with narrow molecular weight distribution.
3. Reaction of the mesylated lignin prepared in step 1 (Fig. 1) with the polystyryl carbanion (living polystyrene) from step 2 (Fig. 2). The carbanion displaces the mesylate groups on the lignin in a nucleophilic displacement reaction with the formation of the polystyrene-lignin graft copolymer (Fig. 3).

The lignin selected for study was prepared by pulping aspen with methanol/water (70% methanol) with a liquor pH of 2.4 and isolated as described earlier (19-21). The acetone soluble fraction had a narrow polydispersity index (apparent $\overline{M}_n = 864$, apparent $\overline{M}_w = 1690$) and contained high aliphatic hydroxyl and free phenolic content. Based on detailed chemical and spectroscopic analysis, three types of structural units were identified as comprising the lignin macromolecule and are shown in Fig. 1 (22,23). The lignin was mesylated with methanesulfonyl chloride following the procedure of Crossland and Servis (24) with a modified work-up procedure, as described in the Experimental section. The lignin mesylate was characterized by elemental analysis, proton and carbon NMR. Based on NMR analysis, the lignin was 85% mesylated.

The narrow dispersity polystyryl carbanion of desired molecular weight was prepared by anionic polymerization using n-butyl lithium as the initiator and was reacted with the mesylated lignin as described in the Experimental section. The reaction product was extracted with toluene to remove unreacted polystyrene. Analysis of the extract showed small amounts of lignin present along with the polystyrene. Since lignin is insoluble in toluene, the presence of lignin in the toluene extract must be due to the formation of a toluene-soluble lignin-polystyrene graft copolymer fraction. Analysis of the residue from the toluene extraction revealed the presence of polystyrene. Since polystyrene is readily soluble in toluene, the residue must contain lignin-polystyrene graft copolymer. Lignin is a complex irregular macromolecule composed of different types of structural units with varying reactivities and molecular weights. This can easily affect the amount of polystyrene grafted onto the different structural units leading to a toluene-soluble and a toluene-insoluble graft copolymer product.

Proof of Grafting. The presence of the lignin-PS graft copolymer in the extract was shown by absorbance spectroscopy, as shown in Figure 4. In Figure 4(a) a dilute solution of the extract in toluene shows an absorbance

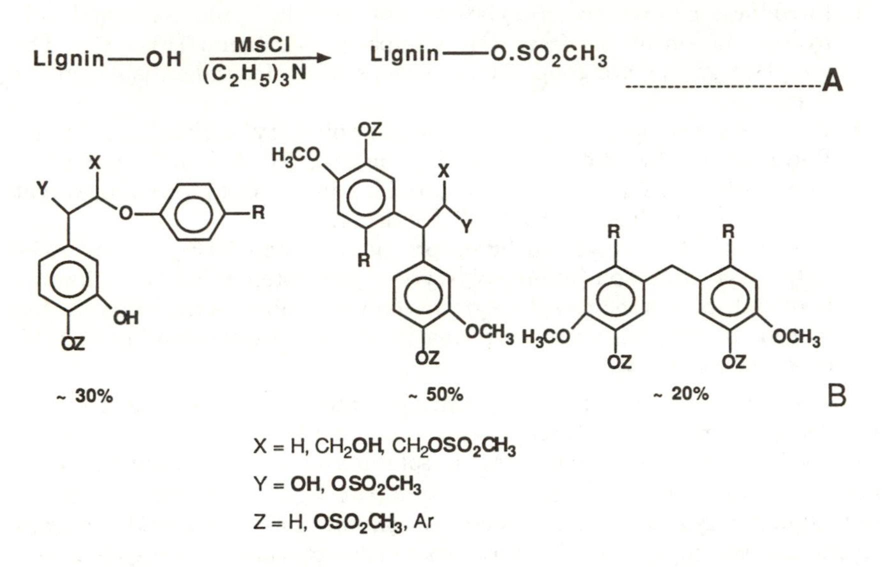

Figure 1. Functionalization of the lignin macromolecule. (A) Mesylation of –OH groups. (B) Structural units comprising the lignin macromolecule.

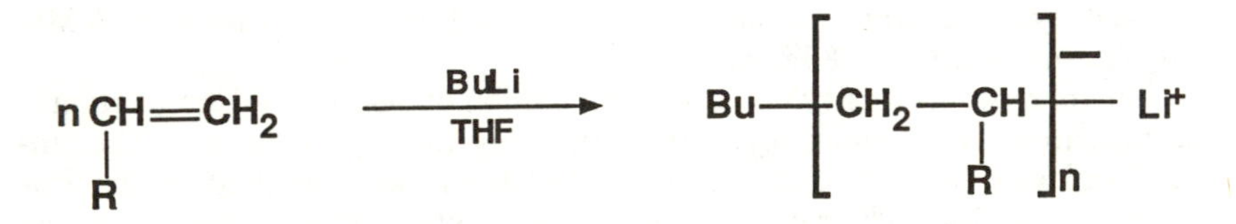

Figure 2. Preparation of the monodisperse polystyrylcarbanion of desired molecular weight.

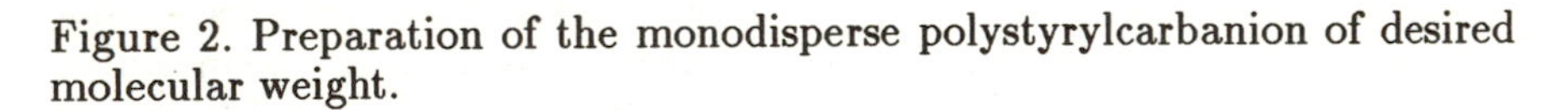

Figure 3. Coupling of the polystyrylcarbanion with the mesylated lignin.

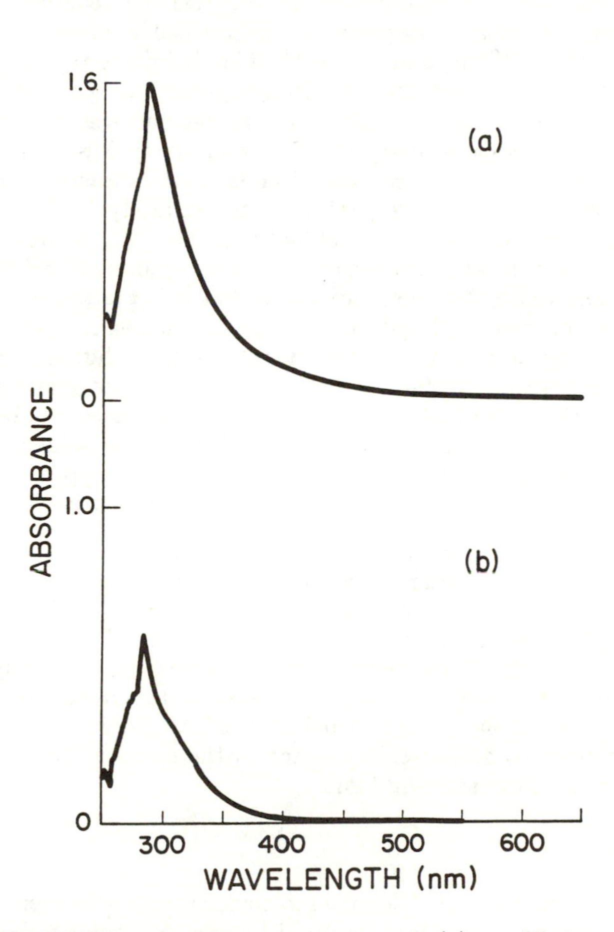

Figure 4. UV/Visible absorbance spectra of (a) a dilute solution of lignin/polystyrene extract in toluene and (b) a saturated solution of mesylated lignin in toluene.

of approximately 1.6. In comparison, Figure 4(b) shows a saturated solution of the mesylated lignin used in the grafting reaction. The absorbance of this solution is less than 0.6, indicating that the lignin is only slightly soluble in toluene. The observed higher absorbance for the extract demonstrates that the lignin has been solubilized by the grafting reaction, most likely due to formation of a graft product. Elemental analysis of the extract gave 90.28% C, 7.94% H, and 1.70% O. The definitive presence of oxygen confirmed lignin in the extract, since polystyrene has no oxygen.

The presence of a graft product in the residue was demonstrated by differential scanning calorimetry (DSC), as shown in Figure 5. Figure 5(a) shows the DSC run for the residue after toluene extraction had removed excess polystyrene. The large peak at approximately 105°C can be compared to the reference spectrum of polystyrene ($\overline{M}_n$ = 4000) shown in Figure 5(b). This shows the presence of a large amount of polystyrene that cannot be removed by toluene, indicating that it is grafted onto the lignin. The increase in carbon and hydrogen content of the residue (76.5% C, 6.75% H, 16.78% O+S) over that of the starting mesylated lignin, confirmed the presence of a grafting product. Further confirmation of grafting was given by IR and NMR, both of which showed peaks characteristic of both types of polymer in the extract and the residue. NMR of the residue showed sharp peaks at 41 ppm, 45 ppm, 126 ppm, 129 ppm and 147 ppm, characteristic of polystyrene in addition to characteristic broad lignin peaks.

Quantitation of Graft Copolymer. The approximate amounts of polystyrene and lignin in the residue and extract were determined by the deconvolution of the absorbance data at 270 nm and 300 nm. This yielded the ratios of 26% polystyrene to 74% lignin for the residue and 96% polystyrene to 4% lignin for the extract. The toluene extract is a mixture of largely unreacted polystyrene and lignin-g-PS. This is in reasonable agreement with the elemental analysis of the extract which showed an oxygen content of 1.7%. This corresponds to 5.8% lignin content in the extract, based on a 29.2% oxygen content of the starting lignin.

Conclusions

Monodisperse polystyrene of defined molecular weight has been grafted onto a well characterized mesylated lignin of known molecular weight and relatively narrow polydispersity. The chemistry involved the nucleophilic displacement of mesylate groups on lignin by the polystyryl carbonion. Preparation of polystyryl carbanion by anionic polymerization allows monodisperse polystyrene of any desired molecular weight to be grafted onto the lignin in a reproducible and consistent manner. By using well characterized, low molecular weight lignins of narrow polydispersity, tailor-made lignin-polystyrene graft copolymers can now be prepared. These engineered lignin graft copolymers of controlled structures can function as compatibilizers/interfacial agents in preparing blends of kraft lignins ($0.10-0.30/lb.) with polystyrene ($0.55-0.74/lb.) leading to new materials. Future work addresses synthesis of lignins with lower degrees of mesylation and other good

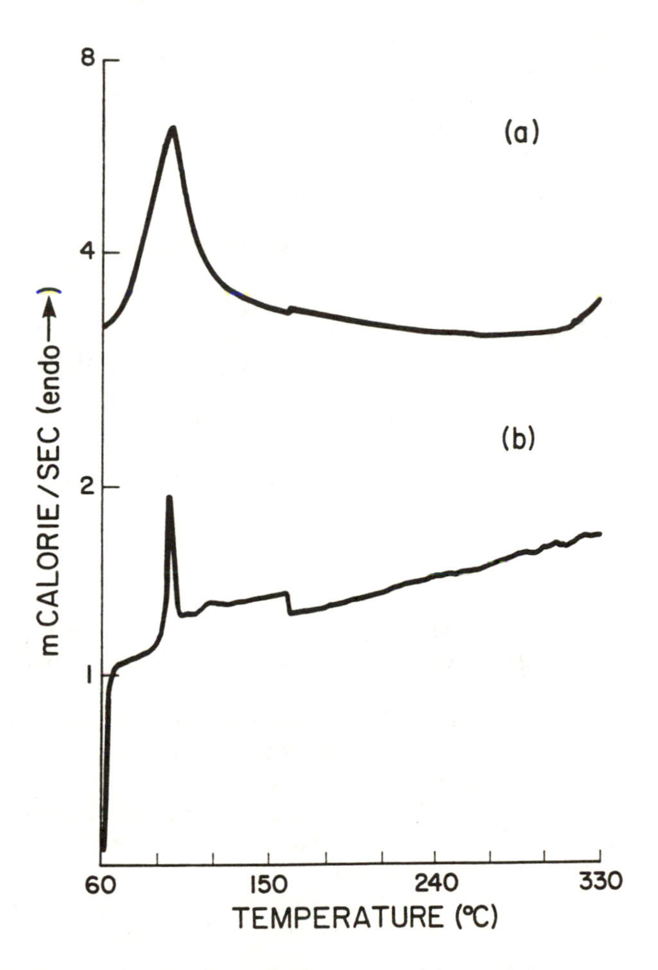

Figure 5. Differential scanning calorimetry of (a) lignin/polystyrene residue and (b) the polystyrene 4000 reference.

leaving groups, and the grafting of different molecular weight polystyrene chains. In addition to this, detailed characterization of the graft copolymer will be carried out and structure-property relationships will be established.

Acknowledgment

The work was funded by the Biobased Materials Project of the Department of Energy's (DOE) Energy Conversion and Utilization Technologies through Solar Energy Research Institute (FWP D071), Mr. Stan Wolf, DOE Manager, and T. Kollie, ORNL Manager.

Literature Cited

1. Hopfenberg, H. P.; Stannett, V.; Kumura-Yeh, F.; Rigney, P. T. *Appl. Polym. Symp.* 1970, **13**, 139.
2. Stannett, V. In *Block and Graft Copolymers*, Burke, J. J., and Weist, V., Eds.; Syracuse University Press: Syracuse, 1973, p. 281.
3. Paul, D. R.; Newman, S., Eds. *Polymer Blends*, Vol. 182; Academic Press: New York.
4. Simlonescu, Cr.; Cernatescu-Asandel, A.; Stoleru. *Cell. Chem. Technol.* 1975, **9**(4), 363-380.
5. Koshijima, T.; Muraki, E. *Zairyo* 1976, **16**(169), 834-838.
6. Meister, J. J.; Patil, D. R. *Ind. Eng. Chem. Prod. Res. Ser.* 1985, **24**(2), 313.
7. Arthur, J. C., Jr. *Appl. Polym. Symp.* 1981, **36**, 201.
8. Mansour, O. Y.; Nagaty, A. *Prog. Polym. Sci.* 1985, **11**, 91.
9. Bhattcharyya, S. N.; Maldas, D. *Prog. Polym. Sci.* 1984, **10**, 171.
10. Morin, B. R.; Breusova, I. P.; Rogovin, Z. A. *Adv. Polym. Sci.* 1982, **42**, 139.
11. Hebish, A.; Guthrie, J. T. *The Chemistry and Technology of Cellulosic Copolymers*; Springer-Verlag: New York, 1981.
12. Stannett, V. *ACS Symp. Ser.* 1982, **187**, 1.
13. Narayan, R.; Tsao, G. T. In *Advances in Polymer Synthesis*, Culbertson, B. M., and McGrath, J. E., Eds.; Plenum Press: New York, 1985; *Polym. Sci. Tech.* 1985, **31**, 405.
14. Narayan, R.; Tsao, G. T. In *Cellulose: Structure, Modification, and Hydrolysis*, Young, R. A., and Rowell, R. M., Eds.; Wiley: New York, 1986, pp. 117-185.
15. Narayan, R.; Shay, M. In *Renewable Resource Materials, New Polymer Sources*, Carraher, C. E., Jr., and Sperling, L. H., Eds.; Plenum Press: New York, 1986; *Polym. Sci. Tech.* 1986, **33**, 137.
16. Narayan, R.; Shay, M. In *Recent Advances in Anionic Polymerization*, Hogen-Esch, T. E., and Smid, J., Eds.; Elsevier: New York, 1987, pp. 441-450.
17. Biermann, C. J.; Chung, J. B.; Narayan, R. *Macromolecules* 1987, **20**, 954.
18. Biermann, C. J.; Narayan, R. *Polymer* 1987, **28**, 2176.

19. Chum, H. L.; Johnson, D. K.; Ratcliff, M.; Black, S.; Schroeder, H. A.; Wallace, K. International Symposium on Wood and Pulping Chemistry, Vancouver, August 26-30, 1985, pp. 223-225. Chum, H. L.; Ratcliff, M.; Schroeder, H. A.; Sopher, D. W. *J. Wood Chem. Tech.* 1984, **4**, 505.
20. Hawkes, G. E.; Smith, C. Z.; Utley, J. H. P.; Chum, H. L. *Holzforschung* 1986, **40**, 115-123.
21. Chum, H. L.; Johnson, D. K., Tucker, M. P.; Himmel, M. E. *Holzforschung* 1987, **41**, 97-107.
22. Chum, H. L.; Johnson, D. K.; Black, S.; Baker, J.; Sarkanen, K. V.; Schroeder, H. A. *Proc. Inter. Symp. Wood and Pulping Chem.* Paris, 1987, Vol. 1, 91-93.
23. Chum, H. L.; Black, S.; Johnson, D. K.; Sarkanen, K. V.; Robert, D. Organosolv Pretreatment for the Enzymatic Hydrolysis of Poplars. II. Isolation and Quantitative Structural Studies of Lignins. Submitted.
24. Crossland, R. K.; Servis, K. L. *J. Org. Chem.* 1970, **35**, 3195-3196.

RECEIVED February 27, 1989

EPOXIES AND ACRYLICS

Chapter 38

Recent Progress in Wood Dissolution and Adhesives from Kraft Lignin

Nabuo Shiraishi

Department of Wood Science and Technology, Kyoto University, Sakyo-ku, Kyoto 606, Japan

It is becoming apparent that wood components, especially lignin, are chemically modified by solvents during wood dissolution, and that the resulting wood tars or pastes become highly reactive. Attempts have therefore been made to prepare effective adhesives, moldable resins and other products from wood after dissolution in phenols or polyhydric alcohols. This review presents recent progress on wood dissolution, and on the preparation of epoxy and phenol resin adhesives from kraft lignin.

Various attempts have been made to prepare adhesives from lignin. The preparation of resol resin adhesives has been studied especially extensively. The introduction of phenols into the α or β-position of the sidechain of the phenylpropane unit (phenolation of lignin) has been considered a key reaction for the formulation of these types of adhesives with adequate gluability.

It is known that the phenolation of lignin takes place under acidic conditions at elevated temperatures. For example, about 0.36 moles of phenol were found to be combined with each phenylpropane unit of a kraft lignin when this was heated in a glass ampule containing 5 g of lignin, 5 g of phenol and 6 mmol hydrochloric acid at 130°C for 6 h (1). Another experiment (2) indicated that more than 0.43 moles of phenol were introduced into lignin by reaction with boron trifloride (BF_3) at 60°C for 4 h. A slight modification of this reaction (80°C for 3 h) was found to raise the incorporation of phenol into lignin to 0.62 moles (3).

On the other hand, it has recently been demonstrated that untreated wood and/or wood modified by, for example, esterification or etherification can be dissolved in several organic solvents including phenols (3-10). Characterization of the resulting wood tars has revealed a high reactivity and the products can be converted readily into adhesives, moldable resins, etc.

0097–6156/89/0397–0488$06.00/0

(5-7, 9-11). It is therefore very likely that the lignin itself is highly modified by solvents, especially by phenols. In consequence, effective modification of lignin can be considered to enhance its reactivity.

In this study, kraft lignin was chemically modified with phenol and with bisphenol-A in order to enhance its potential as adhesive.

Dissolution of Wood

The dissolution of chemically modified wood has been developed recently (3-7). The solubilization of untreated wood was found to be possible as well (7-10).

At least three methods have been found to be applicable to the solubilization of chemically modified wood. The first experiment (4) (Direct method) employed severe dissolution conditions. For example, in 20-150 min at 200-250°C, wood samples esterified by a series of aliphatic acids could be dissolved in benzyl ether, styrene oxide, phenol, resorcinol, benzaldehyde, aqueous phenol solutions, etc. For carboxymethylated, allylated and hydroxyethylated woods, the conditions provided for dissolution in phenol, resorcinol or their aqueous solutions, formalin, etc., by standing or stirring at 170°C for 30 to 60 min (5).

The second method of dissolution is based on solvolysis (Solvolysis method) (6, 7, 11, 12). Under milder conditions (80°C for 30 to 150 min) phenolation was accomplished with an appropriate catalyst, and the chemically modified wood was dissolved in phenol (11). Under similar conditions, woods derivatized by allylation, methylation, ethylation, hydroxylation and acetylation have also been found to dissolve in polyhydric alcohols, such as 1,6-hexanediol, 1,4-butanediol, 1,2-ethanediol, 1,2,3-propane triol (glycerol), and bisphenol-A (6).

The third method of dissolution is based on the mild chlorination of chemically modified wood according to Sakata and Morita (13) (Post-chlorination method). Chlorination, in fact, resulted in enhanced solubility of chemically modified wood in solvents, including phenol. For example, chlorinated cyanoethylated wood does not only dissolve in cresol even at room temperature, but it also dissolves (under heating) in resorcinol, phenol and a LiCl-dimethylacetamide solution.

Unmodified wood has been dissolved in various neutral organic solvents or solvent mixtures by adopting the "Direct method" described above. This observation was made during an investigation into the relationship between the degree of chemical modification of wood and its solubility in organic chemicals. So far, we have found that under the conditions of 200-250°C for 30-150 min, both untreated wood chips and ground wood can be dissolved in the following solvents: phenols, bisphenols, polyhydric alcohols such as 1,6-hexanediols, 1,4-butanediol, oxyethers such as methyl cellosolve, ethyl cellosolve, diethylene glycol, triethylene glycol, polyethylene glycol, and others.

The dissolution of modified or unmodified wood gives rise to tar or paste-like solutions with a high wood concentration (70%, for example). The dissolved wood components were found to be degraded to a certain extent, and this resulted in mixtures having high reactivity, with the exception of those obtained by the "Post-chlorination method." The solutions with high wood concentration can be used for the preparation of adhesives and moldings. This has opened a new field for the utilization of wood materials.

Attempts (6, 11, 12,14-17) have concentrated on the preparation of wood-based adhesives by curing degraded or chemically modified wood components with reactive solvents. Phenols, bisphenols and polyhydric alcohols have all been shown to be effective reactive solvents. With these solutions (tars or pastes), phenol-formaldehyde resins (such as resol resins), polyurethane resins, epoxy resins, etc., have successfully been prepared with high contents of chemically modified or unmodified wood. Cure reactions take place with ease. Gluability studies have demonstrated excellent performance as wood-based adhesives. Adhesives were found to perform satisfactorily even in exterior-grade applications. A resol-type wood-based adhesive has recently satisfied the requirements of the Japanese Agricultural Standard for waterproof binders (22). Solubilization conditions for its preparation were 120 min at 250°C with equal amounts of phenol and untreated Makamba (birch) wood chips. By adding small amounts (1.6-4 parts per 100 parts of wood-based adhesive) of alkylresorcinol as hardener, the same adhesive was found to bind plywoods even at 120°C. This is almost 20°C below the temperature used for commercial resol-type adhesives. The adhesive satisfied exterior-grade standards. This means that the resol-type wood-based adhesive is similar in its reactivity to urea-formaldehyde adhesives, and higher than commercial resol resin adhesives.

Wood solubilization has already been discussed in a recent review (18) with special emphasis on reactivity. The lignin component of wood gains reactivity by modification with solvent during solubilization. Nucleophilic substitution with phenol takes place at the α-position of lignin sidechains. This derivatization of lignin is associated with degradation that plays an important role in making the wood-phenol solution reactive. This point was studied in detail by reacting phenol with isolated lignin, and converting it into an adhesive. The following summarizes this study.

Epoxy Resin Adhesives from Kraft Lignin

Epoxy resin adhesives from lignin were reported by Tai *et al.* (19,20) in 1967. Satisfactory gluability was found. The solubility in organic solvents, however, was found to be poor. In essence, the phenolic hydroxyl groups of kraft lignin were glycidylated directly.

We have also made an attempt to prepare lignin-epoxy resin adhesives (21). However, in order to improve its reactivity, kraft lignin was first phenolated with bisphenol-A. For phenolation, a small amount of aqueous hydrochloric acid or BF_3-ethyl etherate was used as catalyst, and thus two kinds of glycidylation methods were adopted (21). The phenolation with bisphenol-A was found to enhance the solubility of the lignin derivative. In fact, the lignin-epoxy resins obtained were found to be completely soluble in certain organic solvents, including acetone.

The gluability of the lignin-epoxy resin was also found to be improved by phenolation, and waterproof adhesives resulted (21). The temperature dependence of the dynamic viscoelastic response was studied on cured lignin-epoxide films prepared with varying degrees of phenolation (21). The results revealed that well-defined differences were found in the glass transition temperature (T_g), the storage modulus (G') in the rubbery plateau region, and the peak height of the logarithmic decrement (α_T) curve of the cured films. T_g and G' in the rubbery plateau region were found to increase with an increase in degree of phenolation. Furthermore, the peak height of the α_T curve was found to be higher for samples from less phenolated lignin-epoxide films. This suggests that the more bisphenolated lignin forms, the more crosslinks during cure, thereby producing a better developed three-dimensional network structure (21).

The gluability of two types of epoxy-resins, prepared from lignins with different degrees of purity, was examined. Tokai Pulp kraft lignin F (< 4.5% sugars) and Oji kraft lignin (11.7% sugars) were used. The difference in lignin purity did not reveal any inferiority in either dry or wet bond strengths, but enhanced gluability was noted for the less purified lignin, suggesting active participation of the sugar components. Both epoxy resins gave satisfactory dry- and wet-bond strengths after 5 min of hot-pressing at 140°C. All requirements for "First-Class Plywood" according to the Japanese Agriculture Standard (JAS) were satisfied.

The gluability of the lignin-epoxy resin adhesives was found to be improved by the addition of calcium carbonate (50% by weight) to the liquid resin. This must be attributed to the nature of the weak alkali in calcium carbonate as a cure accelerator, and to the reinforcement effect of fillers. Since wood surfaces are acidic, the addition of alkaline fillers effectively alters the pH of the glue line.

These results were obtained by using hydrochloric acid as the catalyst for the phenolation reaction with lignin. Besides hydrochloric acid, BF_3 is known to be an effective catalyst for the introduction of phenol groups into lignin. BF_3 is also known as a catalyst which can promote glycidylation not only of phenolic but also of alcoholic hydroxyl groups. Thus, the preparation of lignin epoxy resins with BF_3 as catalyst was studied also. The epoxy value found for the standard lignin-epoxide prepared in the presence of BF_3 was 0.48, whereas that of the epoxide prepared with HCl as catalyst was 0.38.

The lignin-epoxide resins from the reaction with BF_3 were also tested as adhesives for plywood, using triethylenetetramine as curing agent with hot-pressing at 140°C. The results of adhesion tests showed that the waterproof adhesive strengths were improved by use of BF_3 as catalyst. The use of BF_3 as catalyst permitted the adoption of hot-pressing times as short as 3 min for preparing three-ply plywood panels with 6 mm thickness. This resulted in satisfactory waterproof adhesive performance. The addition of calcium carbonate (50% by weight) to the liquid adhesive was again found to enhance the waterproof gluability.

Resol-Type Phenol Resin Adhesives from Kraft Lignin

There have been many attempts to prepare resol resin adhesives from kraft lignin (23). Phenolation of lignin can, in these cases, also be considered a key reaction. That is, effective modification of lignin with phenol can enhance its reactivity, resulting in good gluability.

Lignin was chemically modified prior to resinification, and the effect of phenolation was examined on the gluability of the resol resin adhesives (23). The phenolation was performed either with HCl (80°C, 60 min) or without catalyst (200°C, 60 min). Prior to adhesive testing with three-ply plywood, 5 parts of coconut husk powder were mixed with 100 parts of the resin.

Results are shown in Figure 1. Wet-bond tensile-shear adhesion strengths are compared for two kinds of adhesives prepared using different phenolation conditions. The lignin-resol resin adhesive prepared without catalyst revealed sufficient waterproof adhesion strength to meet or exceed the Japanese Industrial Standard (JIS) requirement (0.98 MPa) with only 6 min hot-pressing at 120°C. By contrast, the adhesive prepared with HCl as catalyst required 9 min of hot-pressing at the same temperature before it reached the same level of strength. This result emphasizes that the phenolation without catalyst results in better modification of lignin compared to that with hydrochloric acid as catalyst. Furthermore, the lignin-resol resin adhesives prepared without catalyst can be cured under conditions under which amino resin adhesives are generally used (i.e., 120°C with a hot-pressing rate of 1 min per 1 mm plywood thickness).

In order to prepare better adhesives, the effect of phenolation time, phenolation temperature, reaction time for resol resinification, and degree of lignin purity were examined on the adhesive properties. The results indicated that optimum conditions for phenolation involve 200°C for 60 min and 60 min resol resinification at 90°C and at pH 9. The conditions for resol resinification correspond well to those of the conventional manufacturing method (24).

Tests for the effect of lignin purity on the gluability involved two kinds of resol resin adhesives prepared from Oji kraft lignin (87.8% purity) and Tokai Pulp kraft lignin F (> 95% purity). Both lignins produced satisfactory wet-bond adhesion strength that met JIS requirements by a hot-pressing rate of 1 min per 1 mm plywood thickness. The difference in lignin purity had no effect on gluability. The same conclusions were drawn for epoxy resin adhesives from kraft lignin. This suggests that polysaccharide components may be converted into reactive materials, such as 5-hydroxymethyl-2-furfural, furfural, etc., through hydrolysis and dehydration during phenolation (18).

For comparison, a conventional resol resin adhesive without lignin was prepared (24), and its gluability was examined. This resin was found to require a hot-pressing rate of at least 1.5 min per 1 mm plywood thickness before a satisfactory wet-bond adhesion strength was achieved at the low hot-pressing temperature of 120°C. This indicates that replacing a part of the phenol with lignin does not imply a mere extender addition, but that a positive role is achieved which enhances the reactivity of the adhesive.

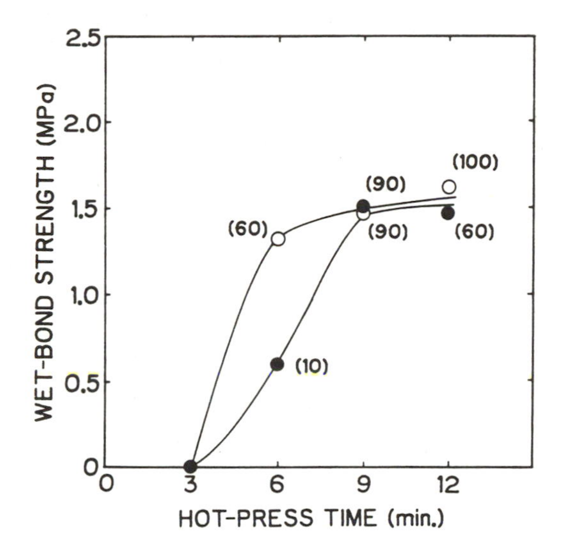

Figure 1. Comparison of wet-bond adhesion strengths of lignin-resol resin adhesives phenolated with and without acid catalyst. Legend: ● : phenolation with acid catalyst at 80°C for 60 min; ○ : phenolation without catalyst at 200°C for 60 min. *Note:* Numerical values in parentheses are percentages of wood failure; hot-press temperature: 120°C.

For the purpose of further enhancing the reactivity and gluability of kraft lignin-resol resin adhesives, the addition of alkylresorcinol was examined. The results of adding 2-10 parts of alkylresorcinol to 100 parts of lignin-resol resin adhesive prior to application indicate a significant improvement in gluability, especially in waterproof gluability. The waterproof gluability as described by wet-bond adhesion strength tesing of two types of adhesives prepared with and without alkylresorcinol (10 parts) are compared in Figure 2. The addition of alkylresorcinol resulted in an improvement of wet-bond adhesion strength. The adhesive met the requirements for "First-Class Plywood" according to JAS even after only 3 min of hot-pressing (a hot-pressing rate of 0.5 min per 1 mm plywood thickness) at 120°C.

It was also found that the lignin-resol resin adhesives satisfied the JIS requirements for non-volatile content, pH value, viscosity, gel time, and dry and wet adhesion strength. Furthermore, the low temperature curability typically found in amino resin adhesives could also be achieved. Thus, it can be concluded that an effective utilization of lignin is possible with simultaneous improvement of the properties of resol resin adhesives.

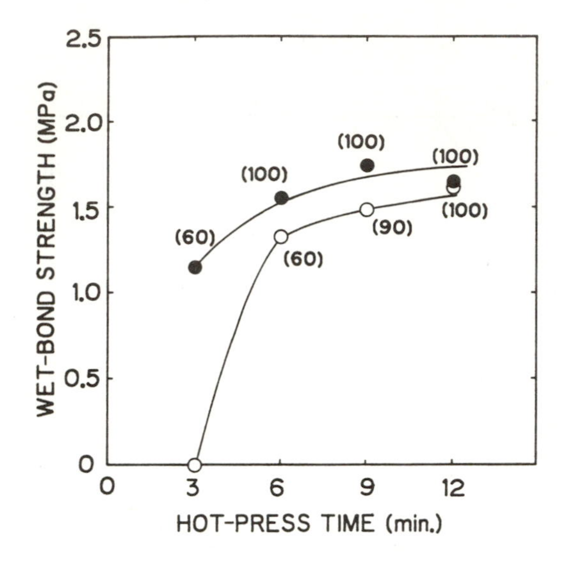

Figure 2. Relationship between hot-press time and wet-bond adhesion strength for lignin-resol resin adhesives with and without alkylresorcinol. Legend: ○ : without alkylresorcinol; ● : with 10 parts alkylresorcinol. *Note:* Numerical values in parentheses are percentages of wood failure; phenolation: 200°C, 60 min, without catalyst; hot-press: 120°C, 6 min.

Literature Cited

1. Kobayashi, A.; Haga, T.; Sato, K. *Mokuzai Gakkaishi* 1966, **12**, 305; *Ibid.* 1967, **13**, 60.
2. Abe, I. *Bull. Hokkaido For. Prod. Res. Inst.* 1970, **55**, 1.
3. Shiraishi, N. *Kobunshi Kako* 1982, **31**(11), 500.
4. Shiraishi, N. Japan Patent 1988-1992 (Appl. June 6, 1980).
5. Shiraishi, N. ; Goda, K. *Mokuzai Kogyo* 1984, **39**(7), 329.
6. Shiraishi, N.; Onodera, S.; Ohtani, M.; Masumoto, T. *Mokuzai Gakkaishi* 1985, **31**(5), 418.
7. Shiraishi, N. *Tappi Proc. 1987 Inter. Dissolving Pulps Conf.*, 95-102.
8. Shiraishi, N. ; Tsujimoto, N.; Pu, S. Japan Patent (Open) 1986-261358 (Appl. May 14, 1985).
9. Pu, S.; Shiraishi, N. ; Yokota, T. *Abst. Papers Presented at 36th Natl. Mtg., Japan Wood Res. Soc.* 1986, 179-180.
10. Shiraishi, N. *Mokuzai Kogyo* 1987, **42**(1), 42.
11. Shiraishi, N. ; Kishi, H. *J. Appl. Polym. Sci.* 1986, **32**, 3189.
12. Shiraishi, N.; Ito, H.; Lonikar, S. V.; Tsujimoto, N. *J. Wood Chem. Technol.* 1987, **7**(3), 405.
13. Morita, M.; Shigematsu, K.; Sakata, I. *Abst. Papers Presented at 35th Natl. Mtg., Japan Wood Res. Soc.* 1985, 215-216.
14. Kishi, H.; Shiraishi, N. *Mokuzai Gakkaishi* 1986, **32**(7), 520.

15. Shiraishi, N. *Mokuzai Gakkaishi* 1986, **32**(10), 755.
16. Ono, H.; Sudo, K.; Karasawa, J. *Proc. 31st Lignin Forum* 1986, 133.
17. Ono, H.; Sudo, K.; Karasawa, J. *Abst. Papers Presented at 37th Natl. Mtg., Japan Wood Res. Soc.* 1987, 288.
18. Shiraishi, N.; Tamura, Y.; Tsujimoto, N. *Mokuzai Kogyo* 1987, **42**(11), 492.
19. Tai, S.; Nagata, M.; Nakano, J.; Migita, N. *Mokuzai Gakkaishi* 1967, **13**(3), 102.
20. Tai, S.; Nakano, J.; Migita, N. *Mokuzai Gakkaishi* 1967, **13**(6), 257.
21. Ito, H.; Shiraishi, N. *Mokuzai Gakkaishi* 1987, **33**(5), 393.
22. Pu, S.; Shiraishi, N. *Abst. Papers Presented at 37th Natl. Mtg., Japan Wood Res. Soc.* 1987, 239.
23. Kato, K.; Shiraishi, N. *Holzforschung*, submitted.
24. Nakarai, Y.; Watanabe, T. *Mokuzai Gakkaishi* 1965, **11**, 137.

RECEIVED February 27, 1989

Chapter 39

Ozonized Lignin–Epoxy Resins

Synthesis and Use

B. Tomita[1], K. Kurozumi[1], A. Takemura[1], and S. Hosoya[2]

[1]Department of Forest Products, Faculty of Agriculture, University of Tokyo, 1–1–1 Yayoi, Bunkyo-ko, Tokyo 113, Japan
[2]Forestry and Forest Products Research Institute, Inashiki-gun, Ibaraki-ken 305, Japan

Ozonization of lignin forms derivatives of muconic acid that have the unique chemical structure of conjugated double bonds with two carboxyl groups. These derivatives have great potential for chemical modification. The ozonized lignin of white birch was soluble in epoxy resin at 120°C, and the free carboxyl groups were found to react with epoxide. This paper discusses developmental work on the preparation of pre-reacted ozonized lignin/epoxy resins; the dynamic mechanical properties of cured resins; and preliminary results of the application of these resins as wood adhesives.

Ozone has recently become industrially available at low cost. From the point of view of the development of complete wood utilization, the pretreatment of wood with ozone enhances the yield of glucose by enzymatic saccharification. Ozonized lignin is thereby obtained as a by-product.

The ozonization of lignin produces derivatives of muconic acid that have two conjugated double bonds terminated by two carboxyl groups. It has been shown that the cleavage of aromatic nuclei of phenylpropane units in lignin occurs between C-3 and C-4 as shown in Figure 1 (1,2). When the substituents at C-3 and C-4 are hydroxy groups, free carboxyl groups are obtained after ozonization. On the other hand, the methyl ester results when these substituents are methoxy groups. Since many chemical modifications can be considered on these structures, new utilization developments are expected.

The lignin isolated from white birch after steaming was ozonized. The ozonized lignin was soluble in epoxy resins at 120°C. It was also found that free carboxyl groups (introduced by ozonization) react with epoxide only by heating. Previous developmental work on the blending of lignin with epoxy resin has been reported by Ball and his co-workers (3). They synthesized

0097–6156/89/0397–0496$06.00/0

kraft lignin/epoxy resin blends in melt state at 190°C and cured them with phthalic anhydride at 160°C. Here the preparation of pre-reacted, ozonized lignin/epoxy resins; the dynamic mechanical properties of the polyamine-cured networks; and some applications of these resins as wood adhesives are discussed.

Experimental

Isolation and Ozonization of Lignin. The wood meal of white birch (*Betula papyrifera* Marsh.) was steamed for 15 min at 180°C, and extracted with water at 60°C, followed by extraction with methanol as shown in Figure 2. The methanol-soluble part was dried under vacuum after removal of methanol. The yield of methanol extract was 75% based on the amount of lignin present in the original wood.

The lignin (50 g) was dissolved in a mixture of dioxane (500 ml) and methanol (1,000 ml), and ozonized at 0°C with an oxygen flow rate of 0.5 ml/min and ozone concentration of 3% as shown in Figure 3. After treatment with ozone, the solution was treated with an excess amount of ether, and the insoluble fraction was filtered off, followed by drying under vacuum. Three samples (No. 1, No. 2, and No. 3) differed in the extent of ozone treatment as shown in Table I. The molar equivalents were based on the ratio of ozone to each phenylpropane (C_9) unit. The yield of each sample is also shown in Table I.

Table I. Ozonization Condition and Yield

Sample No.	Molar Equivalent (O_3/C_9-unit)	Yield (%)
No. 1	0.3	108.2
No. 2	0.6	84.8
No. 3	1.0	77.0

Gel Permeation Chromatography. Each ozonized lignin was analyzed by liquid chromatograph (ALC/GPC 201 with R-401 Differential Refractometer Water-Associates). Columns and solvent conditions were as follows: μ-Styragel 1,000, 500, 500, 100Å in series; and tetrahydrofuran at 2.0 ml/min. Molecular weight was estimated by use of polystyrene standards.

Measurement of Gel Time of Ozonized Lignin and Epoxy Resin Mixture. The epoxy resin used was Epikote 828 (Shell Chemical Co.), diglycidyl ether of bisphenol-A (DGEBA). Ozonized lignins No. 1 and No. 3 were mixed with DGEBA at three levels of 20, 40 and 80 PHR (parts per hundred parts of resin), respectively, and maintained at 120°C with stirring. The ozonized lignin dissolved into DGEBA by heating at 120°C for several minutes. The gel time was determined as the time when the mixture solidified at 120°C. Every solidified sample was completely insoluble in acetone.

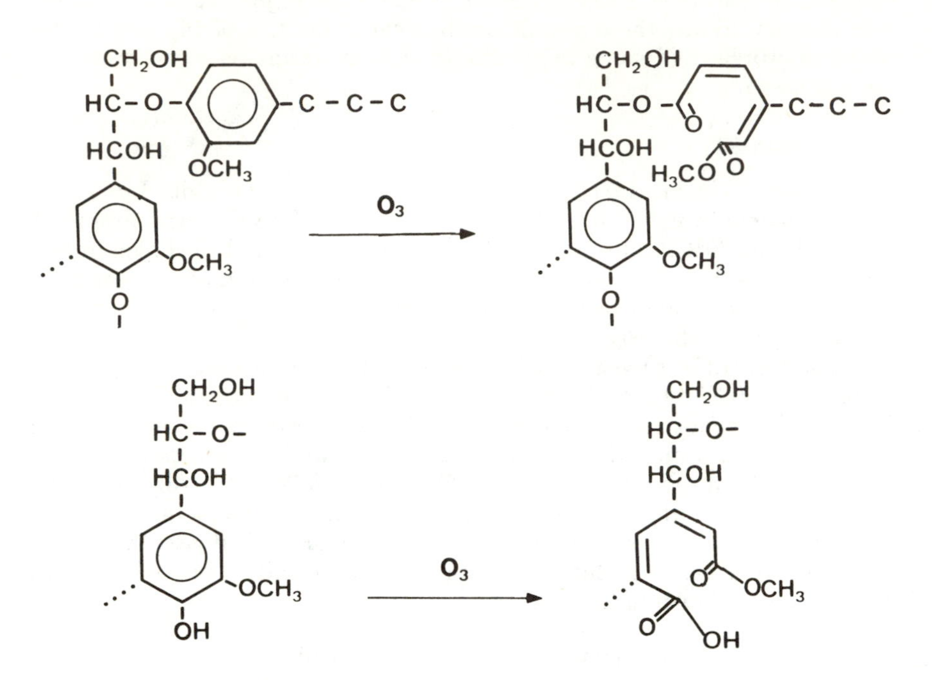

Figure 1. Reaction of lignin with ozone.

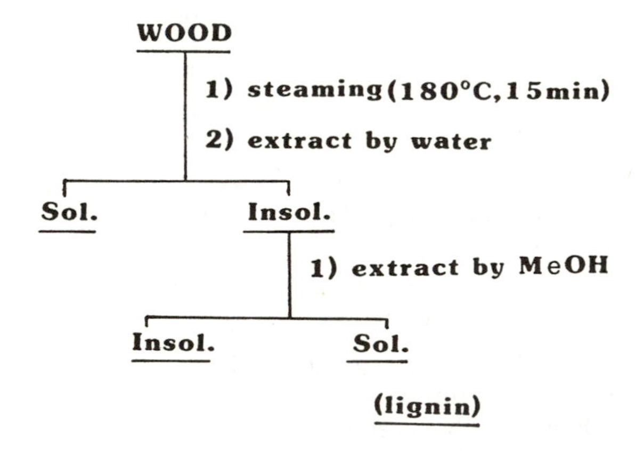

Figure 2. Lignin isolation scheme.

Preparation of Ozonized Lignin/Epoxy Resins. Each ozonized lignin (1.0 g) was mixed with DGEBA and heated at 120°C with stirring as described in the previous section. After heating for 30 min, the mixture was cooled to room temperature. The solidified reactants were dissolved in acetone (2 ml) and the curing reagents, diethylenetriamine (DETA) or hexamethylenediamine (HMDA), were added at 90% of the stoichiometric amount to epoxy equivalent.

Curing was generally done by heating at 130°C for 2 hours and allowing the product to stand at room temperature for one day. All cured resins had no solubility in acetone after extraction for one week.

Dynamic Mechanical Measurements. Films were prepared by casting the acetone solution of sample No. 2 onto a Teflon sheet after adding curing agents. The sample was allowed to stand at room temperature for one day, and then cured at 130°C for 2 hours. The dynamic mechanical spectroscopic data were measured in tension with a Rheovibron DDV-II (Toyo Baldwin Co. Ltd.) at a frequency of 110 Hz with a heating rate of about 1°C/min.

Adhesive Strength Tests. The adherends were birch (*Betula maximowicziana* Regel). The sizes of tensile shear specimens and bonding areas were as follows: 25mm (width) × 65mm (length) × 5mm (thickness) and 3.75 cm^2. Acetone solutions of the pre-reacted ozonized lignin/epoxy resins were spread just after adding curing agents. The hot press conditions were as follows: bonding pressure, 0.96 N/mm^2; press time, 1 hour; temperature 130°C. The specimens were conditioned at 20°C and 65 RH for one week before being tested. Two types of tests were applied. One is a normal test and the other a cyclic boil test (soaking in boiling water for 4 hours; drying at 60°C for 20 hrs; soaking in boiling water for 4 hrs; cooling and testing in the wet state). The tensile shear adhesive strengths were measured with a Tensilon testing machine (Toyo Baldwin Co. Ltd.) under a crosshead speed of 10 mm/min.

Results and Discussion

Isolation and Ozonization of Lignin. The yield of lignin obtained with steaming and successive water and methanol extraction was about 75% of the amount of lignin present in wood. The lower molecular-weight lignin produced by steaming can be removed by water extraction. The isolated lignin contained a small amount of low molecular-weight hemicellulose and polysaccharides, though it was treated with water. However, it was used without further purification in these experiments. The gel permeation chromatogram of the lignin is shown in Figure 4, and the number average molecular weight was estimated to be about 2,000 using polystyrene standard calibration.

The ozonization was performed at the three levels of molar equivalents to phenylpropane unit shown in Table I. The highest yield of over 100% in sample No. 1 was caused by the addition of ozone to lignin. On the other hand, the yield decrease in samples No. 2 and No. 3 was considered to be the result of removal of a considerable amount of ozonized lignin by

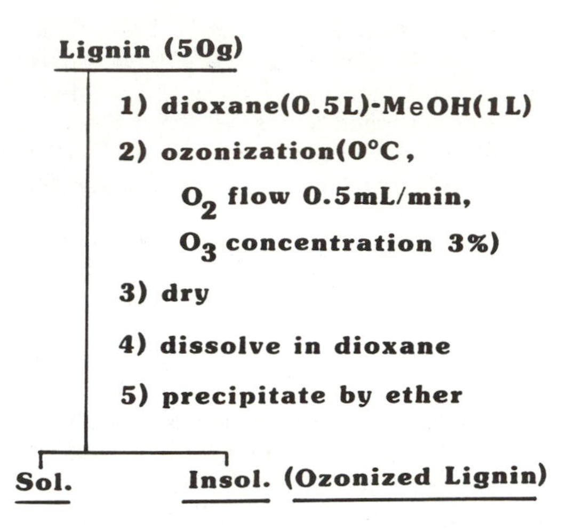

Figure 3. Process scheme for the preparation of ozonized lignin.

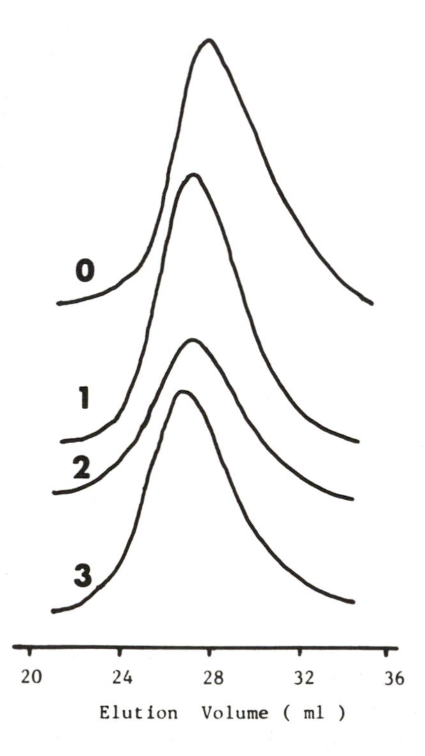

Figure 4. Gel permeation chromatograms of lignin isolated from birch and its ozonized lignins. Sample 0: untreated lignin; 1: ozonized lignin No. 1; 2: ozonized lignin No. 2; 3: ozonized lignin No. 3.

ether extraction. The molecular weight distributions of ozonized lignin are also compared with the original lignin in Figure 4. No obvious difference could be observed between ozonized lignins and untreated lignin, partly because the lower molecular weight portion of ozonized lignin was removed by ether extraction. The structural characteristics of ozonized lignins, such as distribution of functional groups, have not been determined on these samples, but results with similar samples have been given previously (1,2).

Reaction of Ozonized Lignin and Epoxy Resin. The ozonized lignin dissolved in DGEBA with heating. It began to dissolve at about 100°C, and gelation took place at 170°C. Gelation was also observed at 120°C under longer reaction times. After gelation, the reactants were completely insoluble in acetone and other solvents. Table II shows the relation between gel time and the amount of ozonized lignin added to DGEBA at 120°C. As shown in Table II, either mixing of highly ozonized lignin (No. 3) or increasing the amount of ozonized lignin resulted in a decrease in gel time. On the other hand, in the case of the untreated lignin, gelation did not occur even at 200°C under the same conditions. These results suggest that carboxyl groups introduced into lignin by ozonization react with the epoxide to form three-dimensionally crosslinked networks. It has been reported by Ball and his co-workers that kraft lignin is soluble in DGEBA at 190°C (3). However, no reference was made to the occurrence of gelation. This fact also supports the conclusion that the gelation takes place by the reaction of carboxyl groups that have been introduced into lignin by ozonolysis.

Table II. Gel Time of Ozonized Lignin and DGEBA at 120°C

Ozonized lignin	Gel time			
	Lignin content (PHR)			
	20	40	60	80
No. 1	> 300 min	290 min	140 min	65 min
No. 3	> 300 min	120 min	60 min	15 min

The limit of the amount of ozonized lignin that can be added to DGEBA was about 100 PHR in sample No. 1, and 80 PHR in No. 2 and No. 3. Further addition of ozonized lignin caused incompatibility between the two polymers. After pre-reacting at 120°C for 30 min, the products solidified by cooling to room temperature, but they dissolved in acetone. The pre-reactant was cured at room temperature with common curing agents, such as DETA (diethylenetriamine) and HMDA (hexamethylenediamine). The amount of curing agent was set to 90% of the stoichiometric amount of epoxy equivalent, because the free carboxyl groups of the ozonized lignin were considered to consume at most 10% epoxide of DGEBA.

No acetone soluble extractives were detected in completely cured resins. The films cured at 130°C for 1 hour had good rigidities and toughness as well as high transparencies. On the other hand, the resins obtained from the original lignin and DGEBA under similar curing conditions were not transparent, and they were too brittle for film preparation. This suggests that the crosslinking reaction is only between DGEBA and the amine used as crosslinking agent. These results also suggest that the unmodified lignin was only mechanically blended with the epoxy networks, without chemical bonding at the molecular level. Therefore, it was concluded that the carboxyl groups introduced by ozonization react chemically with epoxide resulting in single phase network materials.

Dynamic Mechanical Properties of Ozonized Lignin/Epoxy Resins. The temperature dependence of the viscoeleastic properties are shown in Figures 5 and 6 at various amounts of ozonized lignin (No. 2) content for HMDA and DETA curing reagents, respectively. The dispersion absorption peaks observed at 120°C in HMDA and at 150°C in DETA were attributed to the glass transition. Rubbery plateaus were recognized in the higher temperature region. Generally the absorption peak due to the glass transition shifted to a higher temperature and broadened as the content of ozonized lignin increased. Only one dispersion peak is observed in every sample, indicating that the ozonized lignin/epoxy resin system has complete compatibility after curing with amines. However, glass transition peaks tend to spread broadly as the ozonized lignin content increases (Figs. 5 and 6). These results could be due to a broad distribution of crosslinking density derived from the addition of ozonized lignin. A spreading of peaks to a lower temperature region is thought to be caused by the steric hindrance of lignin. The main chain of the epoxy resin adjacent to the epoxide linked to an ozonized lignin fragment may become hard to move. Therefore, a considerable amount of epoxide remained unreacted resulting in lower crosslinking density. On the other hand, the ozonized lignin behaves as a more highly crosslinked material, and this causes the absorption peak to spread to a higher temperature.

These observations suggest that the new resin system forms an interpenetrating polymer network; an AB crosslinked copolymer according to the definition of Sperling (4). It should also be noted that the glass transition temperature can be modified at any levels of lignin addition by changing crosslinking agent as shown in Figure 7.

Bonding Tests. The results of wood bonding tests are shown in Tables III and IV. The bond strength remained almost at the same level as obtained with epoxy resin with lignin additions up to 40 of PHR, though it decreased gradually with increasing blending ratios of ozonized lignin. The comparatively low values shown in the normal test of sample No. 3 were not considered significant because the cyclic boil test showed the same strength level as sample No. 2. The difference in the degree of ozonization was not found to affect the bond strength dramatically. Further work, such as investigations on bonding conditions, will be necessary in order to develop practical uses for this adhesive system.

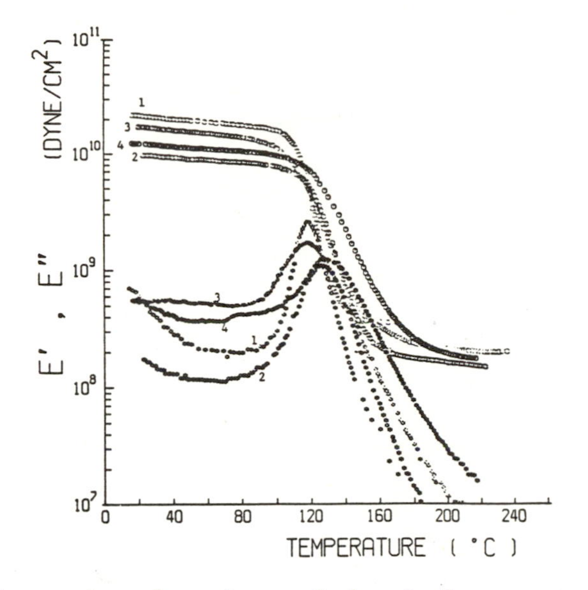

Figure 5. Temperature dependence of viscoelastic property of ozonized lignin/epoxy resins cured with hexamethylenediamine. Ozonized lignin content in DGEBA: (1) 0 PHR; (2) 20 PHR; (3) 40 PHR; (4) 80 PHR.

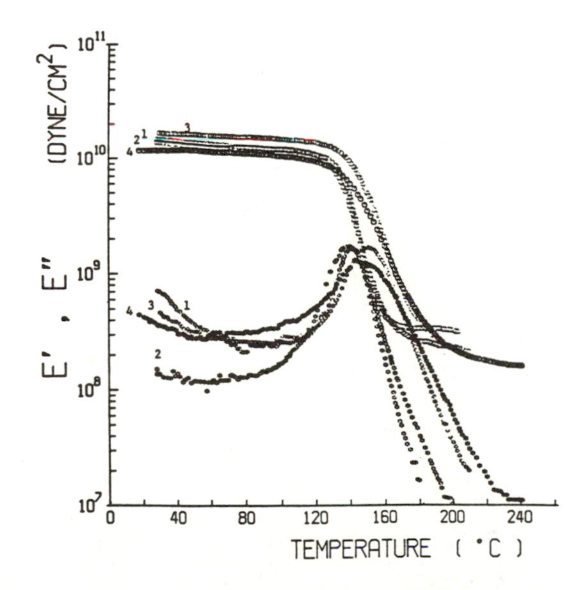

Figure 6. Temperature dependence of viscoelastic property of ozonized lignin/epoxy resins cured with diethylenetriamine. Ozonized lignin content in DGEBA: (1) 0 PHR; (2) 20 (PHR); (3) 40 PHR; (4) 80 PHR.

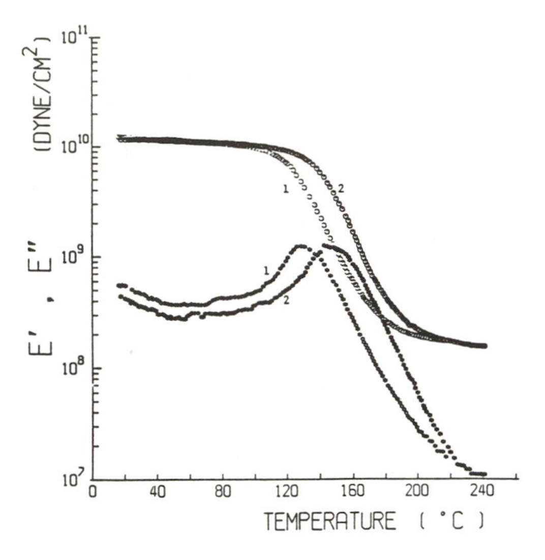

Figure 7. Temperature dependence of viscoelastic property of ozonized lignin/epoxy resins cured with hexamethylenediamine (1) and diethylenetriamine (2) at ozonized lignin content of 80 PHR.

Table III. Tensile Shear Adhesive Strength of Ozonized Lignin/Epoxy Resins. (Normal Test)

		Shear Strength (wood failure) (N/mm^2)		
Sample	0 PHR	20 PHR	40 PHR	80 PHR
Epikote 828	7.79 (65%)	—	—	—
No. 1	—	8.32 (93%)	7.63 (45%)	5.62 (8%)
No. 2	—	8.25 (100%)	7.73 (63%)	6.05 (56%)
No. 3	—	6.66 (68%)	6.49 (33%)	5.89 (22%)

Table IV. Tensile Shear Adhesive Strength of Ozonized Lignin/Epoxy Resins. (Cyclic Boil Test)

		Shear Strength (wood failure) (N/mm^2)		
Sample	0 PHR	20 PHR	40 PHR	80 PHR
Epikote 828	5.72 (3%)	—	—	—
No. 1	—	5.11 (22%)	3.89 (8%)	3.06 (4%)
No. 2	—	4.39 (12%)	3.91 (19%)	3.65 (2%)
No. 3	—	5.08 (28%)	4.51 (8%)	3.56 (2%)

Conclusions

Ozonized lignins are soluble in epoxy resins when heated at 120°C, where the carboxyl groups react with epoxide. Therefore, pre-reacted ozonized lignin/epoxy resins were developed. The dynamic mechanical measurements showed complete compatibility between ozonized lignin and epoxy resins after curing with amines with the formation of interpenetrating polymer networks (IPN's). It was also found that the glass transition temperatures of cured resins could be modified over wide ranges by selecting different curing agents. Preliminary wood bonding tests on this new adhesive system indicate good potential. Further investigations on bonding conditions to enhance adhesion of wood are warranted.

Acknowledgments

This work was performed under the special research plan on "Development of Effective Utilization on Biomass Resources," 1986, by Ministry of Agriculture, Japan.

Literature Cited

1. Kaneko, H.; Hosoya, S.; Nakano, J. *Mokuzai Gakkaishi* 1980, **26**, 752.
2. Kaneko, H.; Hosoya, S.; Nakano, J. *Mokuzai Gakkaishi* 1981, **27**, 678.
3. Ball, F. T.; Doughty, W. K.; Moorer, H. H. Canadian Patent 654 728, 1962.
4. Sperling, L. H. *J. Polym. Sci., Macromol. Rev.* 1977, **12**, 141.

RECEIVED February 27, 1989

Chapter 40

Lignin Epoxide

Synthesis and Characterization

World Li-Shih Nieh[1-3] and Wolfgang G. Glasser[1,2]

[1]Department of Wood Science and Forest Products, Virginia Polytechnic Institute and State University, Blacksburg, VA 24061
[2]Polymeric Materials and Interfaces Laboratory, Virginia Polytechnic Institute and State University, Blacksburg, VA 24061

An epoxide resin was synthesized on the basis of lignin by reaction of epichlorohydrin (ECH) with hydroxypropyl lignin. A mixture of a quarternary ammonium salt (QAS) and potassium hydroxide (KOH) was used as catalyst. Additional KOH was added stepwise at a rate which compensated for KOH consumption during dechlorohydrogenation. The epoxidation reaction was first studied using hydroxypropyl guaiacol (HPG) as a lignin-like model compound. At room temperature, and when ECH was in excess, the reaction was completed in five days. The reaction was found to be highly dependent on the stepwise addition of KOH, and it was independent of ECH concentration. The maximum conversion of hydroxy to epoxy functionality was found to be 100 and 50% for model compound and lignin derivative, respectively. The lignin epoxide resin was crosslinked with diethylenetriamine (DETA), amine terminated poly (butadiene-co-arylonitrile) (ATBN) and phthalic anhydride (PA). Sol fraction and swelling behavior, and dynamic mechanical characteristics of the cured lignin epoxides were studied in relation to cure conditions.

The major functional groups in lignin are methoxy, phenolic hydroxy, aliphatic hydroxy, and carboxy groups. During reaction with alkylene oxide, most of the functional groups of lignin, besides methoxy groups, become

---

[3]Current address: Mississippi State Forest Products Utilization Laboratory, Mississippi State University, Mississippi State, MS 39762

0097-6156/89/0397-0506$06.00/0

substituted by hydroxyalkyl groups (1). Reactions of aliphatic hydroxy groups with compounds having α-epoxy groups, such as epichlorohydrin (ECH), have been investigated. With acids as catalyst, the reaction product consists of a mixture of the 1,2- and the 1,3-chlorohydrin derivative of the parent compound. When base is used as catalyst, the 1,2-chlorohydrin is the only product and upon dechlorohydrogenation, the α-epoxy derivative of the parent compound is formed. Bases such as lithium hydroxide, sodium hydroxide, sodium alkoxide, and Lewis bases such as tributylamine and triisopropylolamine have all been used in this reaction. This synthetic route to epoxy-functionalized derivatives of OH-functional polymers has been successfully applied to cellulose, among others (2,3).

Epoxy resins from lignin and lignin derivatives have been explored as well (4-10). In some cases, kraft lignin epoxidized with ECH in the presence of sodium hydroxide as catalyst underwent irreversible hardening (4,5) at 100°C since epoxides are capable of undergoing homopolymerization. Alternatively, phenolated kraft lignin, mixtures of phenol with kraft lignin, and mixtures of phenol with phenolated kraft lignin were all epoxidized with ECH in 25-40% NaOH which formed a solid epoxy polymer that softened at 72-95°C (4-7). Both the degree of epoxidation and the extent of side reactions increased with increasing amount of NaOH. When tested as adhesive, the lignin/phenol-based epoxy resin gave "good" adhesion to wood. However, when tested as coating material to steel, the best results came from lignin/phenol-based epoxy resins modified with urea-formaldehyde and melamine-formaldehyde varnishes. In a study by Tai *et al.* (8,9), lignin-based epoxy resins were synthesized from kraft lignin, bisguaiacylated kraft lignin and phenolated kraft lignin. The epoxidation was conducted at 97 to 117°C with the addition of 7-20 moles of ECH per mole of lignin repeat unit (i.e., C_9-unit) using NaOH as catalyst. The epoxy content of the resulting lignin derivatives was found to be insensitive to ECH concentration, but dependent on the amount of NaOH added. The epoxy content was found to increase to 0.16 eq/100g with NaOH content increasing to 40%. At higher NaOH contents, the epoxy content declined due to homopolymerization. Maximum degree of functionalization with epoxide groups was never achieved. The solubility of the lignin epoxides in organic solvents was reported to be in the order of phenolated lignin epoxide > lignin epoxide > bisguaiacylated lignin epoxide. The lignin epoxides were tested as adhesives on aluminum and beech wood using anhydride and diamine as curing agents. The lignin-based epoxy resins showed an adhesive strength equivalent to resol resin when wood was used as the adherent. D'Alelio synthesized a series of lignin epoxides from lignins that had their functional groups partially blocked by esterification with carboxylic acids (10). The epoxidizing agent was ECH using NaOH as catalyst and dimethyl sulfoxide (DMSO) as solvent at 90-100°C for 14-24 hours. The degree of epoxidation was controlled by the amount of NaOH added.

It is obvious that several options exist for the synthesis of lignin-based epoxide resins. The use of ECH under alkaline conditions seems to be favored. The degree of epoxidation is controlled by the amount of catalyst added since the dechlorohydrogenation step consumes the catalyst. The degree of epoxidation is independent of the amount of ECH as long as this is present in excess. Crosslinking through homopolymerization is a potential problem which limits the degree of epoxidation achievable.

Experimental

Synthesis and Characterization of Lignin-like Model Epoxide. A series of epoxidation reactions were performed with hydroxypropyl guaiacol (HPG) as model compound and ECH. Molar ratios of ECH:HPG varied from 1 to 10. Pelletized KOH and a quarternary ammonium salt (QAS) served as catalyst and reagent. Toluene was the solvent. Catalyst concentration and method of addition were varied. Experimental details are given elsewhere (11).

The epoxidation reaction was monitored by high performance liquid chromatograph (HPCL) on a reverse phase column. The reaction product was identified by IR, proton NMR and carbon-13 NMR spectroscopy. In the IR spectrum, absorption bands at 904 cm^{-1} and 842 cm^{-1} were attributed to the epoxy groups (12); peaks at 2.60, 2.78, and 3.15 ppm on the ^{1}H-NMR spectrum were assigned to the three protons of the epoxide group (13); and in the ^{13}C-NMR spectrum, the three glycidyl carbons were identified at 44.5, 50.6, and 70.6 ppm (14).

Synthesis and Characterization of Lignin-Based Epoxide. Lignin-based epoxides were prepared by reacting HPL from a mixed hardwood organosolv lignin with ECH following the procedure developed with the model compound. Details of the reaction have been disclosed elsewhere (15). The reaction was monitored by titrating the epoxide groups with HBr according to Durbetaki (16). Samples were taken from the reaction mixture at three-day intervals, and they were titrated for their epoxy content. The fraction of the total hydroxy groups that were converted to epoxide groups was defined as *degree of conversion*. Protons on the epoxy ring were identified by ^{1}H-NMR, and glycidyl carbons were detected by ^{13}C-NMR spectroscopy.

Epoxy Resin Characterization. Epoxy films were prepared by crosslinking 2 g of lignin-based epoxide having an epoxy content of 0.11 eq/100g with stoichiometric amounts of diethylenetriamine (DETA) and amine-terminated poly(butadiene-co-acrylonitrile) (ATBN) as crosslinking agents. The films were solvent cast (methylene chloride) and cured at 105°C for 24 hours following solvent evaporation. To determine sol fraction and degree of swelling, films were oven dried and swollen to equilibrium in dimethylformamide (DMF). Weight loss was taken as *sol fraction*, and weight increase (due to swelling) determined the *degree of swelling*. Dynamic mechanical analysis was performed on a Polymer Laboratories Dynamic Mechanical Thermal Analyzer (DMTA). Solvent-cast lignin epoxide films were scanned at a heating rate of 4°C min^{-1}. The frequency was 1 Hz and the strain level was 4%. Uniaxial stress-strain testing was performed on a standard Instron testing machine employing a cross-head speed of 1 mm min^{-1}. Samples were cut with a die in a dog bone shape, from solvent cast film. Tensile characteristics were calculated on the basis of initial dimensions.

Results and Discussion

Epoxidation of HPG. The reaction scheme for the epoxidation of HPG is shown in Figure 1. The generation of oxyanion II in toluene is assisted by (solid) KOH and QAS. II reacts rapidly with ECH to produce the 1,2-chlorohydrin of HPG, III. Dechlorohydrogenation of III proceeds immediately under the prevailing reaction conditions with formation of epoxide IV and KCl. The process can conveniently be monitored by HPLC. Completion is indicated when I is depleted.

The degree of conversion of I to IV was followed in relation to time and catalyst addition in a series of experiments which employed toluene as solvent at room temperature. The results are summarized in Figure 2. The reaction of I with excess ECH in the presence of KOH (1 eq mol^{-1} of I) and an equivalent amount (by weight) of QAS produced IV in 65% yield in about 5 days (experiment A). Failure to add additional KOH is obviously responsible for the low degree of conversion. By raising the KOH content to 1.25 eq mol^{-1} of I and including QAS in an amount equivalent (by weight) to the (initial) KOH content, the reaction becomes 75% complete in about 24 hours (experiment C). Subsequent and repeated additions of KOH (1.25 eq mol^{-1} of I, each) help the degree of conversion to reach 100% in another two days. Presence of QAS in the secondary KOH charges was found to be unnecessary. However, total absence of QAS from the reaction medium (experiment B) prevented the reaction from ever reaching completion by reducing the degree of conversion at each step. Single additions of large excesses of KOH to the reagent mixture (experiment B) failed likewise to produce high degrees of conversion. Stepwise addition of KOH (with or without QAS) is clearly seen as necessary for compensating for KOH consumption during dechlorohydrogenation. The KOH/QAS mixture was found to be more efficient than KOH without QAS in generating the nucleophilic hydroxyanion of HPG. QAS did not interfere with the dechlorohydrogenation step as long as KOH was compensated for. In experiment C (Fig. 2), the epoxidation reaction reached 100% conversion on the fifth day, two days after the addition of the first additional amount of KOH. All room temperature reactions had higher degrees of conversion as compared to the elevated temperature epoxidation reactions of HPG. The dechlorohydrogenation step (i.e., III to IV in Figure 1) was still found to be rapid, even at the milder reaction temperature.

Epoxidation of HPL. Results from the epoxidation of HPL with ECH in the presence of KOH and QAS using methylene chloride (at room temperature) as solvent are shown in Figure 3. The degree of conversion of (aliphatic) hydroxy groups of HPL to epoxide functionality was monitored by titration. Parameters important to the success of this reaction included (a) stepwise addition of KOH, approximately paralleling the formation of KCl by dechlorohydrogenation; (b) presence of QAS in the reaction mixture; (c) an at least five-fold stoichiometric excess of ECH over available

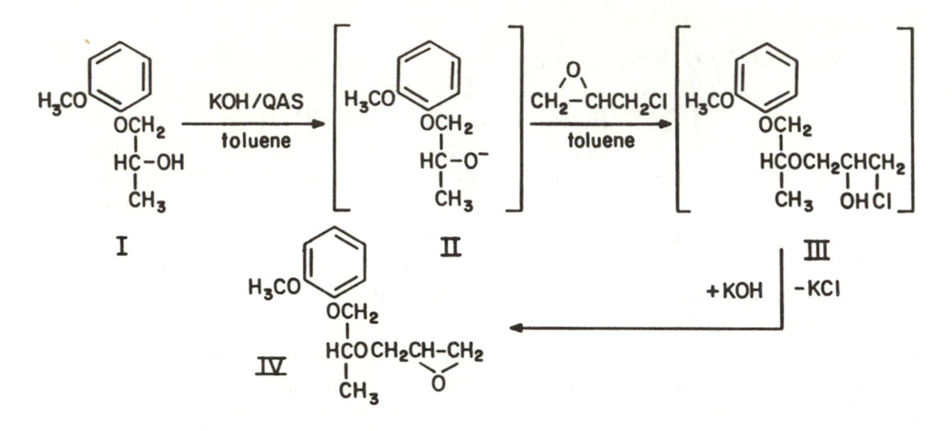

Figure 1. Reaction scheme of the epoxidation of HPG.

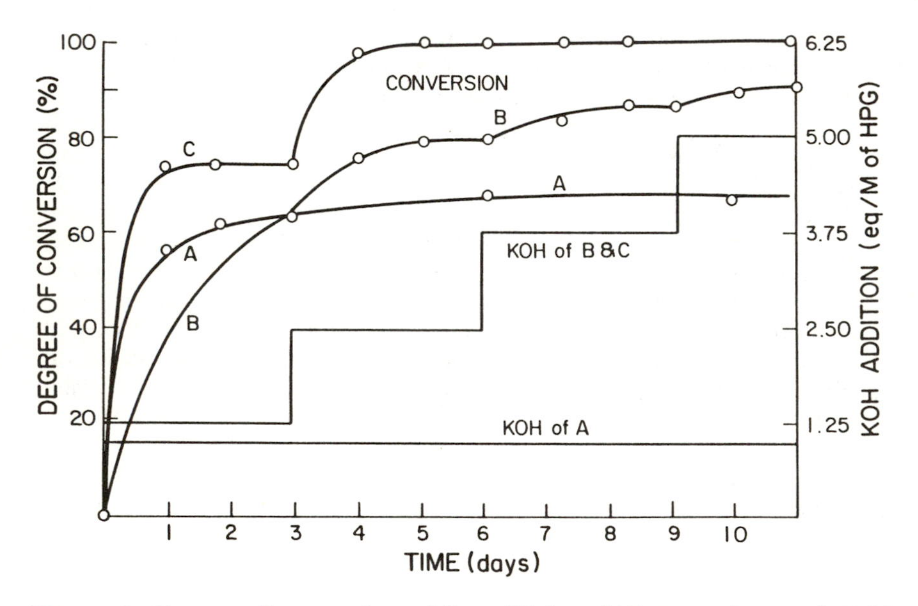

Figure 2. Degree of conversion of I to IV in relation to time and KOH addition. Experiment B employed KOH alone; and experiments A and C employed KOH and QAS.

oxyanion functional groups; (d) mild reaction temperature; and (e) absence of moisture from the reaction mixture. Figure 3 indicates that epoxidation of HPL may reach a degree of conversion of ca. 50% after 12 days. Assuming that HPL has a hydroxy content of about 6-8% and a number average molecular weight of 1500 g mol^{-1}, a 50% conversion of hydroxy groups to epoxide functionality can be expected to produce a lignin-based epoxy resin with an epoxide functionality of about 0.2 eq/100g; with an epoxy equivalent weight of 400 to 600 g; and with an average of three functional groups per (number average) molecular fragment. High hydroxy functionality and degree of conversion, and absence of low molecular weight components (below tri- or tetrameric lignin structures) are critical parameters for lignin selection for successful epoxidation.

Network Characterization of Lignin-Based Epoxides. The formation of a polymer network requires that multifunctional lignin epoxide molecules are reacted with di- or multifunctional crosslinking agents, such as amines or anhydrides. Monofunctional components become chain ends that contribute to brittleness and reduce strength; and fragments without any reactive functional site become fillers, plasticizers or anti-plasticizers. When a hydroxybutyl lignin (HBL)-based epoxide with an epoxy content of 0.11 eq/100g was crosslinked with (a) amine terminated poly(butadiene-co-acrylonitrile) (ATBN) and (b) phthalic anhydride (PA), insoluble materials were formed which had average sol fractions (in DMF) of 10 and 14%, respectively. This indicates that crosslinking was quite complete, and that little lignin remained unincorporated. Weight gained by swelling increased linearly with ATBN content, and this was not surprising since the ATBN used was of relatively high molecular weight (6400 g mol^{-1}) which introduced larger free spaces into the film.

Preliminary results with epoxy networks cured with combinations of DETA and ATBN are given in Table I and Figure 4.

Table I. Composition of Lignin Epoxides

	Composition of Crosslinking Agent				
	by Functionality[1]		by Weight		
Sample No.	ATBN (%)	DETA (%)	ATBN (%)	DETA (%)	Lignin-Epoxide Content (wt. %)
A	0	100	0.00	5.4	94.6
B	10	90	14.3	4.2	81.5
C	25	75	38.3	3.7	58.0
D	50	50	46.1	1.5	52.4
E	100	0	67.9	0.0	32.1

[1] Distribution of amine functional groups equivalent to epoxide functionality.

Using a low molecular weight crosslinking agent such as DETA, a ma-

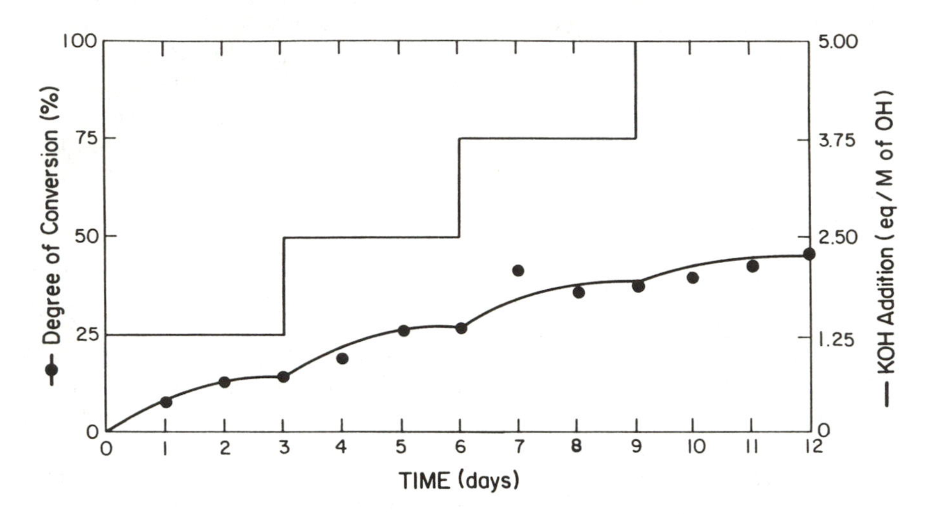

Figure 3. Degree of conversion of hydroxy groups of HPL to epoxide functionality in relation to reaction time and KOH addition.

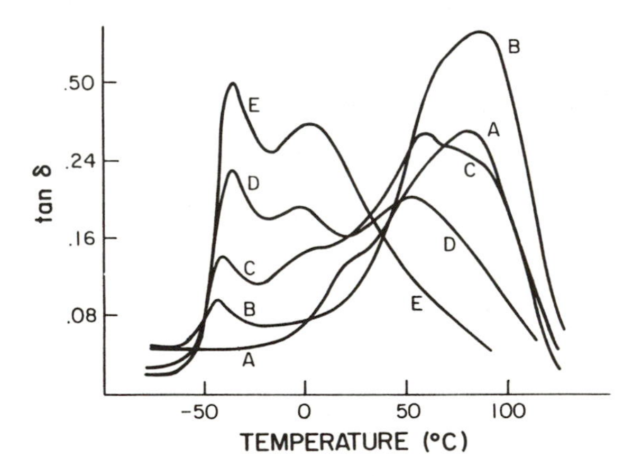

Figure 4. DMTA curves of lignin epoxides cured with combinations of DETA and ATBN. (Samples identified in Table I.)

terial with nearly 95% lignin derivative content is produced. Crosslinking epoxide functionality with amine-terminated rubber segments (i.e., ATBN) reduces the lignin derivative content to about 30% (Table I). Whereas the ATBN-free cured epoxy network forms a glassy, high-modulus material at room temperature (Fig. 4), rising rubber component introduces a secondary thermal transition at about −45°C which serves to reduce the glassy modulus at ambient temperatures. This amounts to a rubber-toughening effect with two-phase morphology. Whether a third transition at around 0°C, between the rubber transition at −45°C and that signifying lignin at 90°C, can be attributed to a mixed interphase between lignin and ATBN remains the subject of future studies. The mechanical properties of the cured lignin epoxides will be discussed elsewhere.

Conclusions

A lignin epoxide resin was synthesized from hydroxyalkyl lignin and ECH using KOH and a QAS mixture as catalyst at room temperature. Since rapid dechlorohydrogenation under reaction conditions consumes KOH, stepwise addition of KOH during epoxidation is necessary. A five-fold excess of ECH over stoichiometric requirements was required in order to prevent intramolecular crosslinking of multifunctional hydroxyalkyl lignin. The degree of conversion of hydroxy to epoxide groups was controlled by the stepwise addition of KOH rather than by ECH concentration.

Swelling experiments showed that a lignin epoxide resin of 0.11 epoxy equivalents per 100g formed a network polymer when cured with DETA, PA, or ATBN. Phase separation was observed in the rubber-toughened lignin epoxide network. Cured epoxides had lignin derivative contents of up to 95%.

This is part 19 of a publication series dealing with engineering plastics from lignin. Earlier parts have been published in *J. Appl. Polym. Sci., J. Wood Chem. Technol.,* and *Holzforschung.*

Literature Cited

1. Wu, L. C.-F.; Glasser, W. G. *J. Appl. Polym. Sci.* 1984, **29**, 1111.
2. Wing, R. E.; Doane, W. M.; Rist, C. E. *Carbohyd. Res.* 1970, **12**, 347.
3. Lee, D.-S.; Perlin, A. S. *Carbohyd. Res.* 1982, **106**, 1.
4. Mihailo, M.; Budevska, Ch. *Compt. Rend. Bulgare Sci.* 1962, **15**, 155; through *Chem. Abstr.* 1963, **58**, 4452d.
5. Mikhailov, M.; Budevska, Khr. *Izv. Inst. po Obshcha Neorg. Khim., Org. Khim. Bulgar. Akad. Nauk.* 1962, **9**, 187; through *Chem. Abstr.* 1963, **59**, 4159f.
6. Mikhailov, M.; Gerdzhikova, S. *Compt. Rand. Acad. Bulgare Sci.* 1965, **18**, 829; through *Chem. Abstr.* 1966, **64**, 14416h.
7. Mikhailov, M.; Gerdzhikova, S. *Compt. Rand. Acad. Bulgare Sci.* 1965, **18**, 43; through *Chem. Abstr.* 1965, **63**, 794c.
8. Tai, S.; Nagata, M.; Nakano, J.; Migita, N. *J. Jap. Wood Res. Soc.* 1967, **13**, 102.
9. Tai, S.; Nakano, J.; Migita, N. *J. Jap. Wood Res. Soc.* 1967, **13**, 257.
10. D'Alelio, G. F. U.S. Patent 3,984,353, Oct, 5, 1976.

11. Nieh, W. L.-S. M.S. Thesis, Virginia Tech, 1986.
12. Dobinson, B.; Hofmann, W.; Stark, B. P. *The Determination of Epoxide Groups*; Pergamon: New York, 1969.
13. Anteunis, M.; Borremans, F.; Van Den Bossche, R.; Verhegge, G. *Org. Magn. Resonance* 1972, **4**, 481.
14. Everatt, B.; Haines, A. H.; Stark, B. P. *Org. Mang. Resonance* 1976, **8**, 275.
15. Glasser, W. G.; Nieh, W. L.-S.; Kelley, S. S.; de Oliveira, W. U.S. Patent Appl.: Ser. No. 183,213 (April 19, 1988).
16. Durbetaki, A. J. *Anal. Chem.* 1956, **28**, 2000.

RECEIVED March 17, 1989

Chapter 41

Derivatives of Lignin and Ligninlike Models with Acrylate Functionality

Wolfgang G. Glasser and Hong-Xue Wang

Department of Wood Science and Forest Products
and
Polymeric Materials and Interfaces Laboratory,
Virginia Polytechnic Institute and State University, Blacksburg, VA 24061

Hydroxybutyl lignin, as well as lignin and lignin derivative-like model compounds, were chemically modified by reaction with isocyanatoethyl methacrylate (IEM). The acrylated derivatives were studied with regard to copolymerization characteristics with styrene (S) and methyl methacrylate (MMA). The reactivity ratios obtained suggest that lignin (model compound) acrylates form preferentially alternating copolymers with MMA and S, and that the degree of deviation from azeotropic composition varies with chemistry. Azeotropic points were between 17 and 30 mole % lignin (model compound) acrylate. Reaction of acrylated lignin derivatives with vinyl monomers produces network polymers with gel structure that shows the expected rise in sol fraction as vinyl equivalent weight increases.

Lignin, especially that derived from organosolv pulping and steam explosion, is beginning to prove itself as a useful polymeric component of structural materials (1-3). Its usefulness increases dramatically following chemical modification by reaction with propylene or other alkylene oxides, since this treatment enhances lignin's solubility in organic solvents and, at the same time, reduces its glass transition temperature (4). Lignin's molecular weight characteristics and functionality features have been the basis for viewing it as a "macromeric" segment for block copolymers and segmented thermosets (5-7).

Although most previous research on the use of lignin in structural polymers has dealt with its contribution to polyphenolics and polyurethanes (8), alternatives for crosslinking it with other polymer systems exist as well. These have been summarized recently (9). Among these alternatives are lignin derivatives with acrylate functionality.

Acrylation of lignin has been reported by Naveau (10), who employed

0097–6156/89/0397–0515$06.00/0

both methacrylic anhydride and methacrylyl chloride with "kraft" lignin to achieve substitution levels of between 17 and 72% of the monomeric units in lignin. Another avenue to acrylation presents itself through the availability of difunctional monomers having both acrylate (or vinyl) and isocyanate functionality. Isocyanatoethyl methacrylate (IEM) is one such monomer that may be used for converting terminal hydroxyl functionality into acrylate functionality (11, 12).

This study examines the synthesis of acrylated lignin derivatives using IEM, and the copolymerization characteristics of these derivatives with methylmethacrylate (MMA) and styrene (S).

Experimental

Materials. Difunctional Reagents: Isocyanatoethyl methacrylate (IEM) was supplied by Dow Chemical Co. as an experimental reagent. Meta TMI is available from Cyanamid, Inc.

Model compounds: Guaiacol, MMA, and styrene were obtained from commercial sources. Hydroxypropyl guaiacol was prepared according to earlier studies (13).

Hydroxybutyl (organosolv) lignin (HBL): The organosolv lignin derivative with butylene oxide was prepared in accordance with earlier work (14). The derivative had a hydroxyl content of 6.0% and a number average molecular weight ($\overline{M}_n$) of 1500 g mol^{-1}.

Methods. Reaction with IEM: Reactions with IEM were performed in methylene chloride using dibutyl tin dilaurate (ca. 1 wt.%) as catalyst. The reaction mixture with model compound or HBL was reacted at 40°C under dry nitrogen. The reaction product was obtained by precipitation with n-hexane. Detailed experimental and analytical results are given elsewhere (15).

Copolymerization reactions: Copolymerization experiments with styrene and MMA employed molar fractions of 20, 40, 60, and 80% comonomers, which were reacted in ethanol: 1,2-dichlorethane 60:40 (by volume) mixtures and benzoyl peroxide as catalyst. Polymerizations were carried out at 70°C. The reactions were quenched by the addition of methanol as non-solvent, and the copolymer was isolated by centrifugation. Copolymer analysis employed UV spectroscopy for copolymers with MMA, and methoxyl content determination according to a procedure by Hodges *et al.* (16) in the case of styrene copolymers. Reactivity ratios were determined in accordance with the method by Kelen-Tüdos (17) and that by Yezrielev-Brokhina-Roskin (YBR) (18). Experimental details and results are presented elsewhere (15).

Acrylated lignin derivatives were copolymerized with MMA and S in dry methylene chloride with benzoyl peroxide and N,N-dimethyl anyline as catalyst. Reaction mixtures were poured onto Teflon molds, the solvent was evaporated at room temperature in the fume hood, and films were cured in an oven at 105°C for 6 hrs. Experimental details and results are given elsewhere (15).

Results and Discussion

The synthesis of acrylated lignin derivatives has been achieved earlier using acrylic acid derivatives (10). A more convenient method for the synthesis of organic solvent soluble, powderous or tarry (i.e., after chain extension with propylene oxide) (19) acrylated lignin derivatives employs the reaction of hydroxyalkyl lignin derivatives with isocyanatoethyl methacrylate (IEM). IEM represents one among several commercially available difunctional monomers combining isocyanate and acrylate (or vinyl) functionality (Fig. 1). The reaction of the isocyanate group of the acrylating agent with the OH groups of the lignin derivative can be completed under mild reaction conditions catalytically. The use of hydroxyalkyl lignin as lignin component, and an isocyanate-functional acrylating agent, both contribute to superior solubility (in organic solvents) and prepolymer characteristics of the modified lignin derivative. For the purpose of determining (a) reaction conditions; (b) analytical methodology; and (c) reactivity ratios with common comonomer species, model compound studies were performed using acrylated guaiacol and hydroxypropyl guaiacol as functionalized derivatives representative of acrylated lignin (Figure 2). Reaction products of the model compounds were isolated in crystalline form, and their FTIR, H-NMR, and ^{13}C NMR spectra revealed that the expected model acrylate had been formed (15).

The usefulness of an acrylated lignin derivative as structural segment in polyacrylic networks depends to a large extent on its copolymerization characteristics. These are conveniently defined in terms of *reactivity ratios* vis-a-vis common vinylic comonomers (20,21). Reactivity ratios are defined in Figure 3. These ratios, which are available in the literature for a wide range of monomer species (20,21), are important for predicting the dependence of copolymer composition on monomer feed composition. Wall (22,23) has pioneered this relationship, and his suggested graphical presentation is adopted in Figures 4 and 5.

Several boundary combinations of reactivity ratios are illustrated in Figure 4. For the case in which neither active center shows any preference for any monomer species, i.e., $r_1 = r_2 = 1$, copolymer composition is determined at all times solely by the concentration of the respective monomer species in the feed composition. This defines an *azeotropic composition* over the entire range of monomer feed compositions. For the other extreme case, the one in which each active center adds exclusively the monomer other than that which represents the terminus on the active chain, i.e., $r_1 = r_2 = 0$, an alternating copolymer is formed in which half of the polymer chain consists of homopolymer of M_1 and the second half of homopolymer of M_2. In the case in which either active center shows preference for one, the more reactive, monomer, i.e., $r_1 > 1$ and $r_2 < 1$, the copolymer is always richer in that monomer and the feed composition is conversely richer in the less reactive monomer. In the final case (Fig. 4), in which each active center prefers cross-propagation to homo-propagation, i.e., $r_1 < 1$ and $r_2 < 1$, a tendency exists toward alternation. Copolymerizations with preferred alternation

H_3C
$OCN-CH_2-CH_2-OOC-C(CH_3)=CH_2$

ISOCYANATO-ETHYL METHACRYLATE (IEM)

$OCN-C(CH_3)_2-C_6H_4-C(CH_3)=CH_2$

meta-TMI

Figure 1. Two commercially available difunctional compounds featuring isocyanate and acrylate (vinyl) functionality.

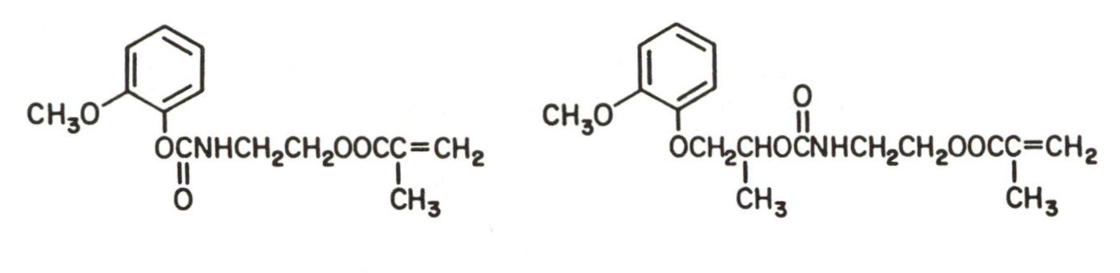

Figure 2. Acrylated lignin-like model compounds, GA and HPGA.

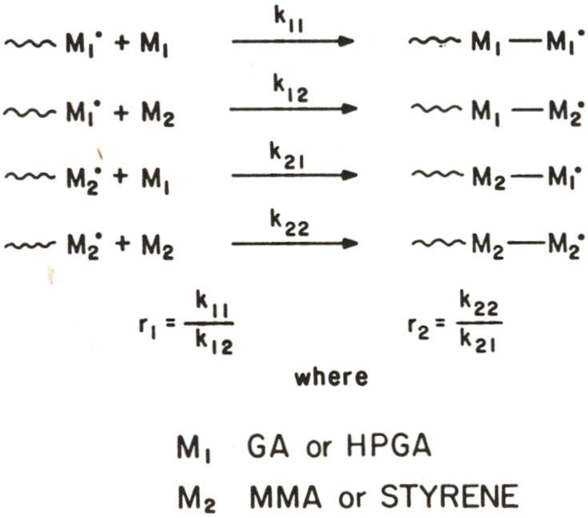

$$\sim M_1^{\cdot} + M_1 \xrightarrow{k_{11}} \sim M_1-M_1^{\cdot}$$

$$\sim M_1^{\cdot} + M_2 \xrightarrow{k_{12}} \sim M_1-M_2^{\cdot}$$

$$\sim M_2^{\cdot} + M_1 \xrightarrow{k_{21}} \sim M_2-M_1^{\cdot}$$

$$\sim M_2^{\cdot} + M_2 \xrightarrow{k_{22}} \sim M_2-M_2^{\cdot}$$

$$r_1 = \frac{k_{11}}{k_{12}} \qquad r_2 = \frac{k_{22}}{k_{21}}$$

where

M_1 GA or HPGA

M_2 MMA or STYRENE

Figure 3. Definition of reactivity ratio.

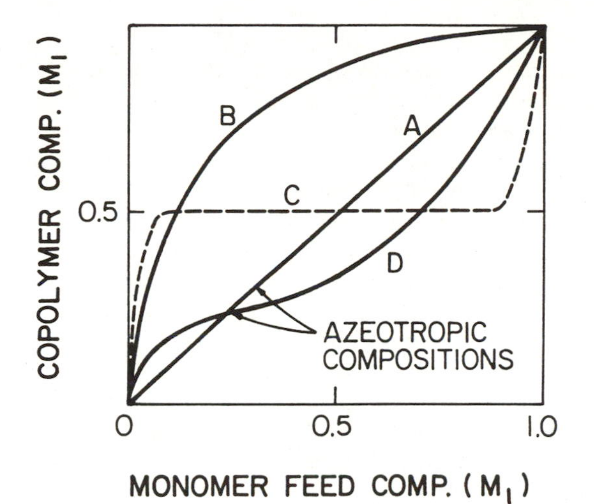

Figure 4. Relationship between copolymer composition (mole fraction M_1 in copolymer) and monomer feed composition (mole fraction M_1 in feed) for various reactivity ratio combinations. A, $r_1 = r_2 = 1$; B, $r_1 > 1$, $r_2 < 1$; C, $r_1 \sim r_2 \sim 0$; D, $r_1 < 1$, $r_2 < 1$. (Adopted from ref. 21.)

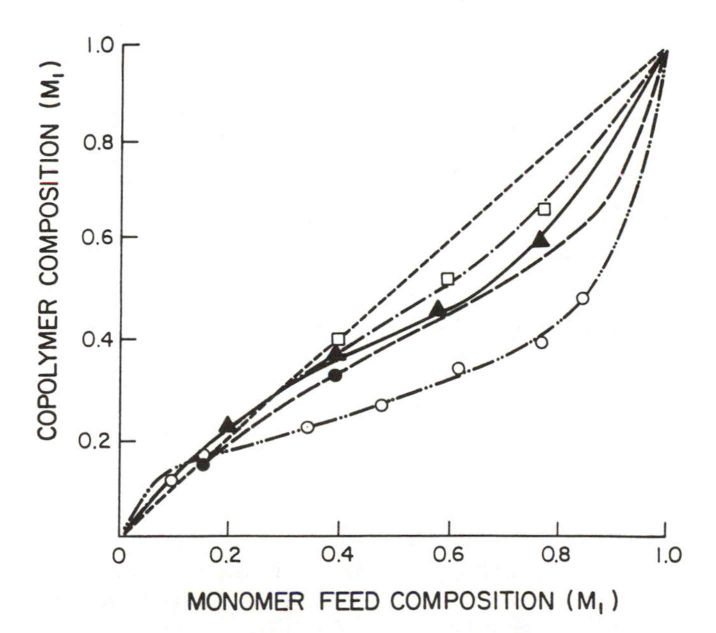

Figure 5. Relationship between copolymer composition and monomer feed composition (molar fraction of M_1 in both) for the combinations HPGA-co-MMA (—□—); GA-co-MMA (—▲—); GA-co-S (—●—); and HPGA-co-S (—○—).

have an azeotropic composition point at which copolymer composition corresponds to feed composition; and below and above which the copolymer is richer and poorer, respectively, in the monomer which is present in the lower (molar) concentration.

Reactivity ratios between acrylated lignin model compound (Fig. 2), defined as M_1, with either MMA or S, defined as M_2, were determined experimentally in accordance with standard procedures (15). These involve mixing two different vinyl monomers in various molar ratios with catalyst (i.e., benzoyl peroxide) and solvent, heating the mixture to achieve polymerization, and recovering the polymer by the addition of non-solvent, and centrifugation. The respective molar monomer fractions of the copolymer were determined by UV-spectroscopy in the cases where MMA served as M_2, and by methoxyl content analysis in those cases in which S was the M_2-species. The results were subjected to numerical treatments according to the established relationships of Kelen-Tüdos (17) and Yezrielev-Brokhina-Roskin (YBR) (18), and this is described elsewhere (15).

The reactivity ratios listed in Table I indicate that both r_1 and r_2 are consistently < 1.0, and that the products of r_1 and r_2 are also always < 1.0. This suggests (20, 21) that preference is given to the formation of alternating copolymers, and that azeotropic points exist at which copolymer composition equals monomer feed composition. The relationship between copolymer composition and monomer feed composition is illustrated in Figure 5 for the various combinations of M_1 and M_2. All combinations reveal a deviation from the azeotropic composition at higher (i.e., $> 40\%$) contents of lignin-like acrylate (M_1) in monomer feed, thereby indicating preference for the formation of alternating copolymers. This deviation is greatest for the combination HPGA-co-S, and it is least for HPGA-co-MMA. Azeotropic points lie between 17 and 30 mole % of lignin-like model acrylate in the monomer feed with the balance representing either MMA or S.

Important structural parameters that influence material properties of polyacrylics based on lignin derivative segments with terminal acrylate functionality are determined by degree of acrylation, and thus vinyl equivalent weight, and monomer ratio in the copolymer. Figure 6 illustrates the relationship between maximum gel fraction and vinyl equivalent weight of acrylated lignin derivative. This relationship reveals the expected increase in sol fraction as degree of substitution with acrylate groups on lignin declines (i.e., vinyl equivalent weight increases).

Conclusions

Solvent soluble and liquid (i.e., tars with T_g below ambient) acrylated lignin derivatives have been formulated by reaction of hydroxyalkyl lignin derivatives with isocyanotoethyl methacrylate (IEM).

Copolymerization behavior of acrylated lignin-like model compounds with MMA and styrene indicated a preference for the formation of alternating copolymers.

Depending on monomer combination, azeotropic points were found

Table I. Experimental Reactivity Ratios between Acrylated Lignin Model Compounds and MMA and S

Copolymer[1]	r_1 [3]	r_2 [3]	$r_1 r_2$
GA(1)-co-MMA (2)	0.17	0.68	0.12
GA(1)-co-S(2)[2]	0.07	0.79	0.06
HPGA(1)-co-MMA(2)	0.38	0.67	0.25
HPGA(1)-co-S(2)	0.01	0.73	0.01

[1] Acrylated guaiacol (GA) and acrylated hydroxypropyl guaiacol (HPGA).
[2] Wide error margin.
[3] Determined according to the Kelen-Tüdos (17) and the YBR (18) relationships.

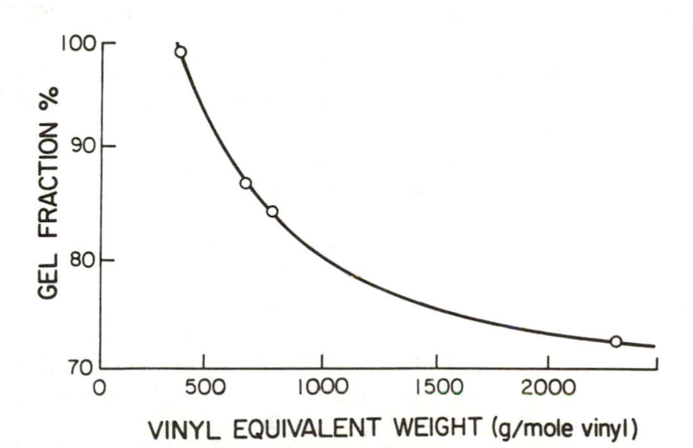

Figure 6. Relationship between gel fraction and vinyl equivalent weight of several acrylated lignin derivatives. (40 wt. % lignin acrylate and 60 wt. % MMA).

between 17 and 30 mole % acrylated lignin model compound; only modest differences were noted regarding deviations from azeotropic compositions.

Acrylated lignin derivatives were copolymerized with vinyl monomers to yield network materials (gels) with gel fraction declining as vinyl equivalent weight increased.

Acrylated lignin derivatives are viewed as a new and promising class of functional macromers from lignin useful for the formulation of block and segmented copolymers.

Acknowledgment

This study was financially supported by a grant from the National Science Foundation.

This is part 20 of a publication series dealing with engineering plastics from lignin. Earlier parts have been published in *J. Appl. Polym. Sci.*, *J. Wood Chem. Technol.*, and *Holzforschung*, among others.

Literature Cited

1. Glasser, W. G.; Barnett, C. A.; Muller, P. C.; Sarkanen, K. V. *J. Agric. Food Chem.* 1983 **31**(5), 921-30.
2. Glasser, W. G.; Barnett, C. A.; Sano, Y. *Appl. Polym. Symp.* 1983, **37**, 441-60.
3. Rials, T. G.; Glasser, W. G. *Holzforschung* 1986, **40**, 353-60.
4. Wu, L. C.-F.; Glasser, W. G. *J. Appl. Polym. Sci.* 1984, **29**(4), 1111-23.
5. Saraf, V. P.; Glasser, W. G.; Wilkes, G. L.; McGrath, J. E. *J. Appl. Polym. Sci.* 1985, **30**, 2207-24.
6. Saraf, V. P.; Glasser, W. G.; Wilkes, G. L. *J. Appl. Polym. Sci.* 1985, **30**, 3809-23.
7. Oliveira, W. de; Glasser, W. G. *J. Appl. Polym. Sci.*, in press.
8. Glasser, W. G.; Sarkanen, S., Eds. *Lignin: Properties and Materials*; ACS Symp. Ser., in press.
9. Glasser, W. G. In *Adhesives from Renewable Resources*; Hemingway, R. A.; Conner, A. H., Eds.; ACS Symp. Ser. No. 385, 1989, Ch. 4, p. 43.
10. Naveau, H. P. *Cell. Chem. Technol.* 1975, **9**, 71.
11. Hoffman, D. K. U.S. Patent 4 320 221, 1982.
12. Frisch, K. C. U.S. Patent 4 374 969, 1983.
13. Glasser, W. G.; Robert, D.; Hyatt, J. A. Unpublished manuscript.
14. Glasser, W. G.; Wu, L. C.-F.; Selin, J.-F. In *Wood and Agricultural Residues: Research on Use for Feed, Fuels, and Chemicals*; Soltes, J., Ed.; Academic Press: New York, 1983; p. 149.
15. Wang, H. X. M.S. Thesis, Virginia Tech, Blacksburg, VA, 1987.
16. Hodges, K. L.; Kester, W. E.; Wiederrich, D. L.; Grover, T. A. *Anal. Chem.* 1979, **51**, 2179.
17. Kelen, T.; Tüdos, F. *J. Macromol. Chem.* 1975, **A9**, 1.
18. Yezrielev, A. I.; Brokhina, E. L.; Roskin, Ye. S. *Polymer Sci. USSR* 1969, **11**, 1894.
19. Kelley, S. S.; Glasser, W. G.; Ward, T. C. *J. Wood Chem. Technol.* 1988, **8**(3), 341.
20. Doak, K. W. "Ethylene Polymers," in *Encyclopedia of Polymer Science and Engineering*; Kroschwitz, J. I., Ed.; John Wiley & Sons: New York, 1986; Vol. 6, p. 383.
21. Tirrell, D. A. "Copolymerization," in *Encyclopedia of Polymer Science and Engineering*; Kroschwitz, J. I., Ed.; John Wiley & Sons: New York, 1986; Vol. 4, p. 192.
22. Wall, F. T. *J. Am. Chem. Soc.* 1941, **63**, 1862.
23. Marvel, C. S.; Schertz, G. L. *J. Am. Chem. Soc.* 1943, **65**, 2054.

RECEIVED March 17, 1989

APPENDIX AND INDEXES

Appendix

Industrial lignin and lignin derivative products are made by these corporations.

EMPRESA NACIONAL DE CELULOSAS SA (ENCE)

Empresa Nacional de Celulosas SA is the largest producer of pulp in the European Economic Community. Jointly with Litchem Limited, it produces *Eucalin Lignin*, derived from the production of kraft pulp from Eucalyptus trees. This is a unique product which is currently not available from any other source.

Early research and development work has indicated that the structure of *Eucalin Lignin* should make it ideally suited for use as a chemical intermediate. Research in South Africa has indicated that *Eucalin Lignin* ranks between softwood and bagasse lignin in terms of the number of reactive sites with formaldehyde (under acidic conditions), and that it allows a high percentage of substitution of phenol in phenol-formaldehyde resins.

Eucalin Lignin has the same repeating unit structure of aryl propane as is typical of other lignins. However, the ratio of guaiacyl to syringyl groups is close to one which gives, on oxidation, a high percentage of syringaldehyde in addition to vanillin.

For further information contact Empresa Nacional de Celulosas SA, Juan Bravo, 49-Dpdo, Madrid 6, Spain; or Litchem Limited, Lamberhead Industrial Estate, Wigan, Lancs WN5 8EG, Great Britain.

0097–6156/89/0397–0524$06.00/0

GEORGIA-PACIFIC

Lignosite Lignosulfonate products are derived from the lignin in softwood trees by the calcium bisulfite pulping process at Georgia-Pacific's mill in Bellingham, Washington.

Georgia-Pacific's lignin sulfonate products are known under the tradename *Lignosite*. During the production of most *Lignosite* products, the hexose sugars (glucose, mannose and galactose) are converted to ethanol by continuous fermentation. Distillation of the ethanol increases the calcium lignin sulfonate content to approximately 80% of the liquor solids.

Lignosite Calcium Lignosulfonate is an excellent starting material for many modified products because of its high purity and because calcium may easily be replaced by any number of other cations using the appropriate soluble sulfate salts. Georgia-Pacific produces sodium, ammonium, and potassium lignosulfonates. Products with higher levels of purity can be provided by chemical modifications or by ultrafiltration. The molecular weight, sugar content and ionic charge can all be adjusted to meet specific needs.

For further information contact Technical Center, 1754 Thorne Road, Tacoma, Washington 98421.

REED LIGNIN/DAISHOWA CHEMICALS

Reed Lignin was formed in 1982 when Reed, Inc., acquired the Lignin Chemicals Division of American Can and merged it with their Lignosol Chemicals operation. Earlier this year Reed became part of Daishowa.

Daishowa is the largest, most diversified producer of lignin products in the world, with two plants located in Wisconsin and a third located in Quebec. Research and Development effort is carried out in major laboratories located in Rothschild, Wisconsin, and Quebec City, Quebec.

Daishowa has developed more than 50 distinctly different lignin sulphonates using a broad range of unit operations. Among the processing techniques are:

- base exchange
- ozonation
- polymerization
- fractionation
- depolymerization by alkaline hydrolysis or high pressure treatments
- reduction, and
- other chemical treatments.

The products are used in numerous industrial applications but can generally be classified as either:

- dispersants
- complexing agents
- protein reactives
- binders, and
- copolymers, or
- extrusion aids.

Daishowa's products are grouped under the following trade names:

- Lignosol
- Marasperse
- Norlig
- Maracarb
- Peltex
- Ameribond
- Marabond
- Additive-A
- Kelig
- Goulac
- Glutrin
- Maracell
- Petrolig.

For further information contact Daishowa Chemicals, P. O. Box 2025, Quebec, P.Q. G1K 7N1, Canada, or Daishowa Chemicals, Inc., Highway 51, Rothschild, WI 54474.

WESTVACO POLYCHEMICALS

The Westvaco Polychemicals Group produces and markets a wide variety of lignochemicals derived from a sugar-free kraft lignin source. Both non-sulfonated and sulfonated derivatives are manufactured in Charleston, South Carolina, USA.

Non-sulfonated lignins find utility as emulsifiers and stabilizers in water-based asphalt emulsions, as coreactants in phenolic binder applications, as negative plate expanders in lead acid storage batteries, as protein coagulants in fat rendering, and as flocculants in waste water systems.

▪ **Emulsifiers and Stabilizers:** Both anionic (INDULIN C) and cationic (INDULIN W-1) derivatives of non-sulfonated kraft lignin are commonly used as stabilizers for making asphalt, wax, or oil-in-water emulsions.

▪ **Resin Coreactants:** Functional groups in products such as REAX 27 or REAX 38 have led to their use as extenders for and coreactants with phenolic resins. Common end uses include binders for thermal insulation, ceiling tile, hardboard, and other thermal-pressure phenolic applications.

The *sulfonated lignin* products function primarily as dispersants in aqueous systems and help to form stable dispersions of a number of insoluble materials. For example, lignin dispersants find use in pigments, carbon black, gypsum, ceramics, coal slurry and water treatment systems to mention some of the more prominent applications.

▪ **Dye and Pigment Dispersants:** The REAX sulfonated lignin derivatives offer a wide range in degree of sulfonation and molecular weight. These attributes impart dispersion stability at high temperatures, improve grinding efficiency, and reduce staining in both liquid dispersed and vat dye systems.

▪ **Agrichemical Dispersants:** Chemical crosslinking to increase molecular weight and a high degree of sulfonation give POLYFON H, O, T, and F the surface active properties necessary to disperse specific pesticides. The REAX Series of products also finds application in flowable and water dispersible granule formulations where high active ingredient levels are sought.

In addition to these defined functions, lignin products are also known for their ability to reduce the size of particles, reduce slurry viscosities, inhibit crystal growth, sequester and complex metals, and function as interactive carriers.

▪ **Sequestrants:** Specially modified sulfonated lignins can sequester alkaline earth and heavy metals and still remain soluble, while with non-sulfonated lignins, the complexes become insoluble and can be removed. The metal sequestering properties of REAX and POLYFON products find application in dispersing iron in cooling water, prevention of scale deposits from hard water in boilers, and in forming metal complexes for agricultural micronutrients.

For further information contact Westvaco Corporation, Polychemicals Department, P.O. Box 70848, Charleston Heights, South Carolina 29415-0848.

RECEIVED May 29, 1989

Author Index

Affiliation Index

Subject Index

T

U

Production: Donna Lucas
Indexing: Deborah H. Steiner
Acquisition: Cheryl Shanks

Elements typeset by Hot Type Ltd., Washington, DC
Printed and bound by Maple Press, York, PA